Lecture Notes in Computer Science 6372

Commenced Publication in 1973
Founding and Former Series Editors:
Gerhard Goos, Juris Hartmanis, and Jan van Leeuwen

Hartmut Ehrig   Arend Rensink
Grzegorz Rozenberg   Andy Schürr (Eds.)

Graph Transformations

5th International Conference, ICGT 2010
Enschede, The Netherlands, September 27–October 2, 2010
Proceedings

 Springer

Volume Editors

Hartmut Ehrig
Technische Universität Berlin
Institut für Softwaretechnik und Theoretische Informatik
Franklinstr. 28/29, 10587 Berlin, Germany
E-mail: ehrig@cs.tu-berlin.de

Arend Rensink
Universiteit Twente
Afdeling Informatica
Drienerlolaan 5, 7522 NB Enschede, The Netherlands
E-mail: rensink@cs.utwente.nl

Grzegorz Rozenberg
Universiteit Leiden
Leiden Institute of Advanced Computer Science (LIACS)
Niels Bohrweg 1, 2333 CA Leiden, The Netherlands
E-mail: rozenber@liacs.nl

Andy Schürr
Technische Universität Darmstadt
Fachgebiet Echtzeitsysteme
Merckstraße 25, 64283 Darmstadt, Germany
E-mail: andy.schuerr@es.tu-darmstadt.de

Library of Congress Control Number: 2010934298

CR Subject Classification (1998): G.2, D.2, E.1, F.3, F.1, F.4

LNCS Sublibrary: SL 1 – Theoretical Computer Science and General Issues

ISSN   0302-9743
ISBN-10   3-642-15927-3 Springer Berlin Heidelberg New York
ISBN-13   978-3-642-15927-5 Springer Berlin Heidelberg New York

springer.com

© Springer-Verlag Berlin Heidelberg 2010

Typesetting: Camera-ready by author, data conversion by Scientific Publishing Services, Chennai, India
Printed on acid-free paper   06/3180

Preface

Graphs are among the simplest and most universal models for a variety of systems, not just in computer science, but throughout engineering and the life sciences. When systems evolve we are interested in the way they change, to predict, support, or react to their evolution. Graph transformation combines the idea of graphs as a universal modelling paradigm with a rule-based approach to specify their evolution. The area is concerned with both the theory of graph transformation and their application to a variety of domains.

The biannual International Conferences on Graph Transformation aim at bringing together researchers and practitioners interested in the foundations and applications of graph transformation. The fifth conference, ICGT 2010, was held at the University of Twente (The Netherlands) in September/October 2010, along with several satellite events. It continued the line of conferences previously held in Barcelona (Spain) in 2002, Rome (Italy) 2004, Natal (Brazil) in 2006 and Leicester (UK) in 2008, as well as a series of six International Workshops on Graph Transformation with Applications in Computer Science from 1978 to 1998. Also, ICGT alternates with the workshop series on Application of Graph Transformation with Industrial Relevance (AGTIVE). The conference was held under the auspices of EATCS and EASST.

In response to the call for papers, 48 papers were submitted. The papers were all reviewed by at least four, and in the majority of cases five, PC members or co-reviewers. After the reviewing phase and the ensuing discussion, the committee selected 22 papers for presentation at the conference and publication in the proceedings. These papers mirror the wide-ranged ongoing research activities in the theory and application of graph transformation. They are concerned with different kinds of graph transformation approaches, their algebraic foundations, composition and analysis, the relation to logic, as well as various applications, mainly to model transformation and distributed systems. The paper submission and reviewing, as well as part of the preparation of the printed volume, were very well supported by the free conference management system EasyChair.

In addition to the presentation of technical papers the conference featured three invited speakers, a doctoral symposium and four workshops. Moreover, this year ICGT was organised as a joint event with the 17th International SPIN Workshop on Software Model Checking (SPIN), and an associated workshop on Parallel and Distributed Methods in verifiCation and High-performance computational systems Biology (PDMC/HiBi).

Invited Speakers. Javier Esparza gave a joint ICGT/SPIN keynote speech, dedicated to the memory of Carl Adam Petri, in which he reviewed the path of ideas that led from the theory of true concurrency, a semantic theory about the nature of concurrent computation, to the unfolding approach to model-checking, a pragmatic technique for reducing the state-explosion problem in automatic verification. In his invited presentation, Krzysztof Czarnecki discussed the problems involved in model synchronisation, with illustrations from industrial practice, and reported an ongoing effort to create both a theoretical framework addressing these challenges and tools based on the framework. Finally, Christoph Brandt in his presentation proposed graph transformation techniques as a formal framework to address security, risk and compliance issues in banking.

Satellite Events. The successful Doctoral Symposium of ICGT 2008 was repeated at this year's event, organised by Andrea Corradini and Maarten de Mol. A total of 12 young researchers had the opportunity to present their work and interact with established researchers of the graph transformation community. In addition four workshops were organized where participants of the ICGT could exchange ideas and views on some subareas of graph transformation:

- 3rd Workshop on Graph Computation Models (GCM 2010), organised by Annegret Habel, Mohamed Mosbah and Rachid Echahed;
- 4th International Workshop on Graph-Based Tools (GraBaTs 2010), organised by Juan de Lara and Dániel Varró;
- 4th Workshop on Petri Nets and Graph Transformations (PNGT 2010), organised by Claudia Ermel and Katrin Hoffmann;
- Workshop and Tutorial on Natural Computing (WTNC 2010), organised by Ion Petre, Bogdan Iancu and Andrzej Mizera.

We would like to thank AmirHossein Ghamarian, Maarten de Mol and Eduardo Zambon for their valuable help throughout the preparation and organization of the conference and the proceedings.

July 2010

Hartmut Ehrig
Arend Rensink
Grzegorz Rozenberg
Andy Schürr

Organization

Programme Committee

Paolo Baldan	Padova, Italy
Luciano Baresi	Milan, Italy
Michel Bauderon	Bordeaux, France
Artur Boronat	Leicester, UK
Paolo Bottoni	Rome, Italy
Andrea Corradini	Pisa, Italy
Juan de Lara	Madrid, Spain
Hartmut Ehrig	Berlin, Germany
Gregor Engels	Paderborn, Germany
Claudia Ermel	Berlin, Germany
Holger Giese	Potsdam, Germany
Annegret Habel	Oldenburg, Germany
Reiko Heckel	Leicester, UK
Dirk Janssens	Antwerp, Belgium
Garbor Karsai	Vanderbilt, USA
Ekkart Kindler	Hyngby, Denmark
Barbara König	Duisburg-Essen, Germany
Hans-Jörg Kreowski	Bremen, Germany
Ralf Lämmel	Koblenz, Germany
Mark Minas	München, Germany
Ugo Montanari	Pisa, Italy
Mohamed Mosbah	Bordeaux, France
Manfred Nagl	Aachen, Germany
Fernando Orejas	Barcelona, Spain
Francesco Parisi-Presicce	Rome, Italy
Rinus Plasmeijer	Radboud, The Netherlands
Detlef Plump	York, UK
Arend Rensink (Co-chair)	Twente, The Netherlands
Leila Ribeiro	Rio Grande, Brazil
Andy Schürr (Co-chair)	Darmstadt, Germany
Gabriele Taentzer	Marburg, Germany
Pieter Van Gorp	Eindhoven, The Netherlands
Dániel Varró	Budapest, Hungary
Gergely Varró	Darmstadt, Germany
Jens-Holger Weber-Jahnke	Victoria, USA
Albert Zündorf	Kassel, Germany

Subreviewers

Peter Achten	Ábel Hegedüs	Henry Muccini
Zoltán Balogh	Tobias Heindel	Muhammad Naeem
Mayur Bapodra	Frank Hermann	Manfred Nagl
Basil Becker	Kathrin Hoffmann	Stefan Neumann
Gábor Bergmann	Mathias Hülsbusch	Christopher Poskitt
Clara Bertolissi	Paola Inverardi	István Ráth
Enrico Biermann	Ruben Jubeh	Luigi Santocanale
Dénes Bisztray	Lucasz Kaiser	Andreas Scharf
Christoph Blume	Joost Pieter Katoen	Ildikó Schlotter
Paul Brauner	Tamim Khan	Andreas Seibel
Sander Bruggink	Pieter Koopman	Pawel Sobocinski
Roberto Bruni	Joerg Kreiker	Christian Soltenborn
Antonio Bucchiarone	Sabine Kuske	Michael Spijkerman
Mike Dodds	Steve Lack	Wolfgang Thomas
Adwoa Donyina	Leen Lambers	Paolo Torrini
Frank Drewes	Pascale Le Gall	Angelo Troina
Jörn Dreyer	Bas Lijnse	Caroline von Totth
Gregor Gabrysiak	Michael Löwe	András Vörös
Fabio Gadducci	Melanie Luderer	Bob Walters
Ulrike Golas	Alberto Lluch Lafuente	Manuel Wimmer
Davide Grohmann	Olivier Ly	Zhilin Wu
Stefan Gruner	Sonja Maier	Eduardo Zambon
Regina Hebig	Tony Modica	

Table of Contents

Session 3. Models and Model Transformation

Session 4. Algebraic Foundations

Session 5. Applications

Session 6. Rule Composition

Doctoral Symposium

A False History of True Concurrency: From Petri to Tools

Javier Esparza

Institut für Informatik, Technische Universität München
Boltzmannstr. 3, 85748 Garching, Germany

Carl Adam Petri passed away on July 2, 2010. I learnt about his death three days later, a few hours after finishing this text. He was a very profound and highly original thinker, and will be sadly missed. This note is dedicated to his memory.

This is an abstract of [1], a brief note describing the path of ideas that lead from the *theory of true concurrency*, a semantic theory about the nature of concurrent computation, to the *unfolding approach to model-checking*, a pragmatic technique for palliating the state-explosion problem in automatic verification. While the note provides a very incomplete and hence "false" view of true concurrency, it also includes several pages of references.

The theory of true concurrency started with two fundamental contributions by Carl Adam Petri in the 60s and 70s: *Petri nets*, the first mathematical formalism for asynchronous computation, and *nonsequential processes*, a semantics that proposes to order events not by means of global timestamps, but according to the *causality relation*. Then, in the early 80s Nielsen, Plotkin, and Winskel showed that, in the same way that the executions of a nondeterministic system can be bundled together into a *computation tree*, its nonsequential processes can be bundled together into the *unfolding* of the system, an object providing information about the causality and choice relations between events. The theory of unfoldings was further developed by Engelfriet, Nielsen, Rozenberg, Thiagarajan, Winskel, and others.

The goal of all this research was purely semantic: to mathematically define the behaviour of a concurrent system. During the 80s, model checking introduced a new (and very successful) view of the semantics of a system as an object that can be constructed and stored in a computer for verification purposes. However, model-checking concurrent systems faced the problem that the number of global states of a system may grow exponentially in the number of its sequential components. This *state-explosion* problem was attacked by McMillan in the early 90s in his PhD thesis. The thesis famously proposed the use of Binary Decision Diagrams, but in a different chapter it presents a second idea: instead of computing an initial part of the computation tree containing all global states (a *complete prefix*), McMillan suggested to construct a complete prefix *of the unfolding* containing the same information as the complete prefix of the computation tree, but encoded far more succinctly. This prefix becomes a data structure for compactly representing the set of global states of a concurrent system.

H. Ehrig et al. (Eds.): ICGT 2010, LNCS 6372, pp. 1–2, 2010.

McMillan's approach still faced two problems. First, while the complete prefix of the unfolding was usually much smaller than the complete prefix of the computation tree, it could also be *exponentially bigger* in the worst case. Second, the approach could at first only check specific properties, like deadlock freedom or conformance. Both problems were overcome during the 90s: improved algorithms for constructing complete prefixes were found, and extensions to (almost) arbitrary properties expressible in LTL were proposed.

Since 2000 the algorithms for constructing complete prefixes have been extended to many concurrency formalisms: a large variety of Petri net classes, communicating automata, process algebras, graph transformation systems, and others. They have also been parallelized and distributed. Unfolding-based verification techniques have been applied to conformance checking, analysis and synthesis of asynchronous circuits, monitoring and diagnose of discrete event systems, analysis of asynchronous communication protocols, and other problems. Today, the unfolding approach is a good example of how abstract, speculative considerations about the nature of computation can evolve into pragmatic techniques for automatic verification.

Reference

1. Esparza, J.: A False History of True Concurrency: from Petri to Tools. In: Weber, M. (ed.) SPIN 2010. LNCS, vol. 6349, pp. 180–186. Springer, Heidelberg (2010)

How Far Can Enterprise Modeling for Banking Be Supported by Graph Transformation?

Christoph Brandt[1] and Frank Hermann[2]

[1] Université du Luxembourg, SECAN-Lab, Campus Kirchberg,
6, rue Richard Coudenhove-Kalergi, L-1359 Luxembourg-Kirchberg, EU
christoph.brandt@uni.lu
http://wiki.uni.lu/secan-lab
[2] Technische Universität Berlin, Fakultät IV, Theoretische Informatik/Formale
Spezifikation, Sekr. FR 6-1, Franklinstr. 28/29, 10587 Berlin, EU
frank@cs.tu-berlin.de
http://www.tfs.tu-berlin.de

Abstract. This keynote paper presents results coming out of an on-going research project between Credit Suisse Luxembourg and the University of Luxembourg. It presents an approach that shows good potential to address security, risk and compliance issues the bank has in its daily business by the use of integrated organizational models build up by enterprise modeling activities. Such organizational models are intended to serve to describe, evaluate, automate, monitor and control as well as to develop an organization respecting given organizational security, risk and compliance side-constraints. Based on the empirical scenario at Credit Suisse, real-world requirements towards a modeling framework as well as the modeling process are developed. Graph Transformation techniques are proposed as formal framework for this purpose and they are evaluated in the sense of how far they can support the identified enterprise modeling activities in the context of the new enterprise modeling framework.

Keywords: enterprise modeling, graph transformation, triple graph grammars, model transformation and integration, graph constraints, correctness, completeness, termination, functional behavior.

1 Introduction and Historical Background

This paper presents and summarizes key results that came out of an ongoing research project between the Credit Suisse Luxembourg and the University of Luxembourg about the question of how far enterprise modeling [20,36,27] can be supported by graph transformation [8,7,4,6,12,31,13,17,5]. However, the original questions were about security, risk and compliance of IT systems in banks. But after a deep analysis, interdependencies between the organizational business universe and the IT universe of the bank came up that made it necessary to reformulate the research questions. The central insight here was that organizational security, risk and compliance can only be discussed given a holistic and

H. Ehrig et al. (Eds.): ICGT 2010, LNCS 6372, pp. 3–26, 2010.

integrated organizational model encompassing the business and IT universe of a bank. However, such a model was not available. In contrast to that, the bank implemented lots of different kinds of "best practices" that used fragmented organizational knowledge at various degrees of abstraction and quality. The empirical observation was that the legally enforced organizational "best practices" regarding security, risk and compliance used by banks today are not automatable, do not scale well, are not integrated, are not consistent and are not complete. This leads to unnecessary costs and insufficient quality of operational results regarding organizational security, risk and compliance. Therefore, the approach in this paper has been developed as a counter-proposal to the given situation at Credit Suisse. It aims to establish a modeling framework for organizational services, processes and rules encompassing the business and IT universe using human-centric and machine-centric models that serve to collect the distributed organizational knowledge in order to derive one holistic organizational model which can serve to answer the above questions.

From a historical point of view this Credit Suisse research project dates back to 2005 when first meetings between Credit Suisse Luxembourg and the Luxembourg Institute for Advanced Studies in Information Technology (LIASIT) toke place. During the process of the ongoing development of the University of Luxembourg, which was founded in 2003, the LIASIT structure got completely absorbed by the newly founded Interdisciplinary Center for Security, Reliability and Trust (SnT) that is now integral part of the University of Luxembourg. It carries out interdisciplinary research and graduate education in secure, reliable, and trustworthy Information and Communication Technology (ICT) systems and services.

Given this organizational context, this Credit Suisse research project aims to generate competetive advantages for banks operating at Luxembourg, to support Luxembourg to become an international leading place in finance and banking, and to generate highly relevant research results.

The paper is organized as follows. In a first step, the differences between enterprise modeling today and tomorrow are discussed. Afterwards, potential domain modeling techniques are introduced and explained. Subsequently, model transformation and model integration techniques are presented and discussed that can help to build up a holistic organizational model. Thereafter, Sec. 5 presents a short evaluation of the graph transformation techniques and future extensions of the modeling framework based on the obtained results are addressed in Sec. 6. Finally, open issues and conclusions are presented that aim to identify relevant research questions for the near future.

2   Enterprise Modeling Today and Tomorrow

Enterprise modeling from its perspective of today and tomorrow is discussed here from the point of view of Credit Suisse. In this context, enterprise modeling happens today in an ad hoc way. It is either used to check for security, risk and compliance issues the bank has to respect because of external legal requirements

	Today	Tomorrow
Approach	"Best Practices"	Formal Methods
Focus	Stakeholders	Organization
Control	End-of-pipe	Begin-of-pipe
Judgment	Checklists	Prove, Simulation, Test
Coverage	Partial	Complete

Fig. 1. Security, Risk and Compliance – Today and Tomorrow

and internal organizational policies or because business processes are going to be automated. Given that "best practices" are usually driven by the specific interest of certain stakeholders they only cover parts of an organization. And because they are usually applied when the organizational structures are already set up and working they can only provide an end-of-pipe control. Finally, because lots of "best practices" that are used today to check for security, risk and compliance issues are build on top of simple checklists their judgment will never surpass the informal level.

Tomorrow, the situation can expected to be different. By collecting lots of small decentralized fragments of an organizational model that are soundly integrated security, risk and compliance can be discussed by using one single organizational model. At the same time this model can be used to automate business processes. Because it will serve as an instance representing the whole organization in an integrated way, it is not longer limited towards certain stakeholders' interest. By using fully implemented formal methods for specifying certain organizational aspects sound methods in the area of proving, simulating and testing can be used to check for organizational security, risk and compliance properties. In addition to that, there is a good chance that such methods are able to cover all possible model states. By doing this these properties could be checked ex-ante, meaning, in a begin-of-pipe control mode which is more aligned with real-world requirements.

Based on the assumption that a banking organization is represented by one integrated instance model certain changes in the enterprise modeling process can be expected. Because of the informal and semi-formal nature of enterprise models today that belong to different contexts such models can only be synchronized by

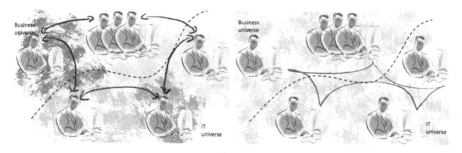

Fig. 2. Context and Synchronization – Today and Tomorrow

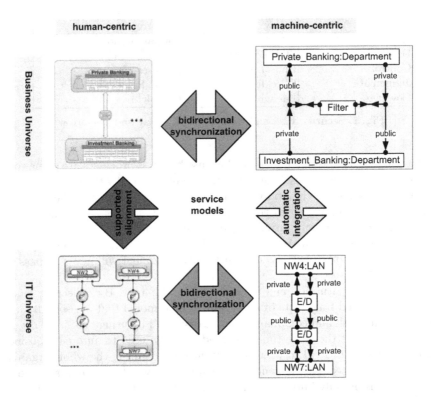

Fig. 3. Human- and machine-centric business and IT service models of tomorrow

the help of direct dialogs between people. This implies that there is very little automation available. Tomorrow, however, there is a good chance that model fragments of an enterprise model will be formally grounded and share the same context. This will enable direct model synchronization between models without or with limited human interaction. At the end, this is likely to result in more efficient solutions for model synchronization and enterprise models of higher quality as of today.

In the following, a possible future situation is briefly sketched by looking at service landscapes. Such service landscapes are required for the business and IT universe of the bank in order to keep the different life cycles of the models as well as their domains separated and to reflect that they are organizational independent. Because people like to codify their knowledge using open diagram languages, a type of model is used that we like to call human-centric. However, because organizational security, risk and compliance predicates should be evaluated using formal methods, models that we like to name machine-centric are generated based on the entered human-centric models. As it can be seen in Fig. 3 a mutual alignment between the human-centric business and IT service model should be supported. In order to reflect changes in the human-centric models, they should be synchronized with already available machine-centric models. Because machine-centric models can communicate certain evaluation results into human-centric models this synchronization is bidirectional. Once, machine-centric models are fully synchronized with their human-centric counterparts, they can be mutually integrated to check for inconsistencies. Therefore, human-centric models are allowed to lack well-formedness, they can be incomplete or inconsistent, whereas, machine-centric models are aimed to be well-defined, complete and consistent. As a consequence, this approach allows to combine the best of the human-centric and the machine-centric world in a flexible way.

3   Domain Specific Modeling Techniques

In order to allow for the specification of practically relevant models, we need a domain modeling technique that is powerful enough to capture all relevant properties of models, and simple enough to be practically useful. Possible techniques [37,10,19,26] can be distinguished into techniques of meta-modeling driven by type graphs and constraints and techniques driven by graph rules.

Meta-modeling has gained a wide acceptance, especially by using it for the definition of UML [29,30]. It is easy to use, and provides procedures to check automatically whether a model is valid or not. However, it is less suited for proving properties of all valid models, or for generating large sets of example models. It is therefore non-constructive.

Graph grammars, in contrast, offer a natural procedure for generating example models, and they support proofs because they define a graph language inductively. However, not all graph grammars that allow to specify practically relevant models are easily parseable. Hence, the main idea of graph transformation is to generalize well-known rewriting techniques from strings and trees to graphs, leading to graph transformations and graph grammars and to rule-based modifications of graphs, where each application of a graph rules leads to a graph transformation step.

An example for the use of the meta modeling approach is UML. Here, the meta model describes the abstract syntax of UML models. In this context, a meta model can be considered as a class diagram on the meta level. It may contain meta classes, meta associations and cardinality constraints. Besides that, further

	type graphs and constraints	graph grammars
Approach	descriptive	constructive
Use	easy to use	requires formal knowledge
Editing	free-hand	rule-driven
Validity Checks	automatic	automatic
Generative	no	yes
Parsing	(visual alphabet only)	partly
Proving Support	no	yes

Fig. 4. Domain specific modeling techniques

features include associations like aggregations, compositions and inheritance as well as abstract meta classes which cannot be instantiated. An UML instance model must conform to the cardinality constraints. Instances of meta models may be further restricted by the use of additional constraints specified by the Object Constraint Language (OCL) [28].

Examples for the use of the graph transformation approach are control flow diagrams, entity relationship and UML diagrams, Petri Nets, visualizations of software and hardware architectures, evolution diagrams of nondeterministic processes, SADT diagrams, and many more [37,10,19,26]. This shows a wide applicability of graph transformations as a very natural way explaining complex systems on an intinuitve level. Therefore, graph transformation has become attractive as a modelling and programming paradigm for complex-structured software systems. In particular, graph rewriting is promising as a comprehensive framework in which the transformation of all these very different structures can be modeled and studied in a uniform way.

Two aspects of the formal technique of algebraic graph transformation are presented in the following. The first aspect is about the rule based modification of graphs. The second one is about the definition of a graph constraint. The graph rules define how to construct an IT service landscape using the ABT/Reo language [3,2,7], which is based on Abstract Behaviour Types and Reo connectors. The graph constraint defines a security policy for those landscapes that must be respected.

In Fig. 5 a fragment of an IT service landscape is given before and after the application of three graph rules as presented in the figure. As it can be seen the rules address the abstract syntax of models. In detail, an encoding/decoding node ("E/D") is added to a public network connection. The lower graph is the result of a transformation starting at the upper graph via the rule sequence ⟨addABT, addInPort, glue⟩ using the algebraic approach in [12], also called double pushout (DPO) approach. In Fig. 6 a graph constraint $P \rightarrow C$ is given that

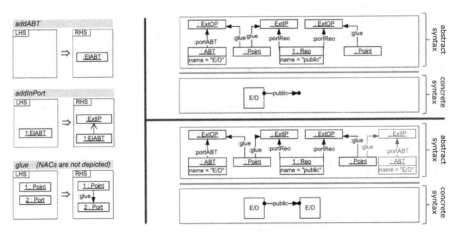

Fig. 5. Application of graph rules

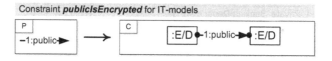

Fig. 6. Example for a graph constraint: P → C

checks whether each public network connection (premise P) is enclosed by an encoding/decoding device (conclusion C) on both sides. This graph constraint is satisfied for graphs in Fig. 3 and the lower graph in Fig. 5.

4   Interrelationship between Different Model Classes

Enterprise modeling is discussed in the following by focusing on the integration relation in the SEM-Square (Service Enterprise Modeling-Square) presented in Fig. 3. The integration relation serves to check for inconsistencies between the machine-centric business service model and the machine-centric IT service model. It also helps to create the second model out of the first one by model transformation. This section shows how these inter-model techniques are applied in a formal and consistent way in order to ensure important properties, such as correctness and completeness and to be able to analyze whether a model transformation has functional behaviour.

The techniques are based on the abstract syntax graphs of the models. The abstract syntax is usually defined by meta modelling. The meta model can be formalized as a type graph with inheritance and attribution [12,26] and the language constraints are specified as graph constraints. Alternatively to this descriptive language definition the constraints can be ensured by providing a set of graph rules that are used for syntax directed editing [19] as sketched in Sec. 3.

4.1   Triple Graph Grammars

The inter-model techniques have to ensure correctness and completeness in order to produce reliable results. For this reason, we use triple graph grammars presented by Schürr et. al. in [33,32,25] with various applications in [21,22,24,25,35]. The formal approach to TGGs has been developed in [11,14,11,15,18,9,16] and was shown to provide the correctness, completeness and termination results. Furthermore, analysis techniques for checking functional behaviour were developed in [18,23]. Furthermore, triple graph grammars convince by its intuitive specification of compact patterns that show how typical model fragments shall be related. Based on these patterns operational rules for model transformation and integration are derived automatically.

Example 1 (Triple graph). The triple graph in Fig. 7 shows an integrated model consisting of a machine-centric business service model in the source component (left) and a machine-centric IT service model in the target component (right).

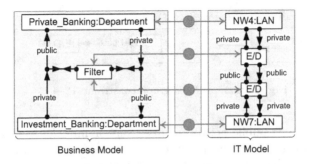

Fig. 7. Triple Graph with the business and IT model

Both models are ABT-Reo diagrams, i.e. they contain abstract behaviour type nodes that are connected by Reo connectors [3,2]. The source model specifies that the data in the communication channels between the private banking and investment banking departments is filtered. The filter has to ensure that private data do not go public. The target model on the other side specifies the IT service structure as a machine-centric service model. The local area networks "NW4" and "NW7", which are used in the private banking and investment banking departments, are connected via an encrypted communication channel shown by the ABT nodes of type "E/D" for encryption and decryption. The corresponding elements of both models are related by graph morphisms (indicated in grey) from the correspondence graph (light blue) to source and target, respectively. This is done in order to check the consistency between the security policy in the machine-centric business model (filtering out private data) and the security policy in the machine-centric IT service model (encrypting and decrypting of private data) for sound correspondence.

Model transformation as well as model integration do not require deletion during the transformation. The result of a model integration is the triple graph of the triple transformation sequence. In the case of model transformations the result is obtained by restricting the resulting triple graph to its target component. Thus it is sufficient to consider triple rules that are non-deleting.

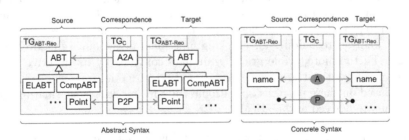

Fig. 8. Triple Type Graph TG_{B2IT}

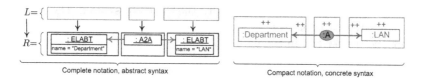

Fig. 9. Triple Rule "DepartmentToLAN"

The triple graph grammar $TGG_{B2IT} = (TG_{B2IT}, S_{B2IT}, TR_{B2IT})$ specifies how business service models and IT service models given as ABT-Reo diagrams are related and its type graph is shown in Fig. 8 in abstract and concrete syntax. The language of ABT-Reo diagrams is used for both, the source and the target language and the type graph shows a correspondence between the relevant types, which are "ABT" for abstract behaviour type elements and "Point" for the gluing points between input and output ports of ABT elements and Reo connectors. The start graph S_{B2IT} is empty and Figures 9 and 10 show some of the triple rules of TR_{B2IT}. Each rule specifies a pattern that describes how particular fragments of business and IT models shall be related.

The first rule "DepartmentToLAN" synchronously creates two ABT elements and a correspondence node that relates them. This reflects the general correspondence between departments in the business view and the installed local area networks in the IT view. The rule is presented in complete notation as well

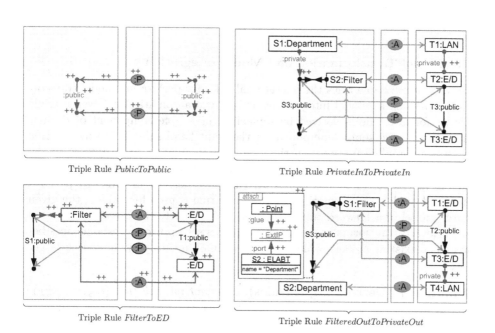

Fig. 10. Further Triple Rules of TGG_{B2IT}

as in compact notation using additionally the concrete syntax. The complete notion shows that a triple rule is given by a triple graph morphism from the left hand side (upper row in the figure) to the right hand side (lower row one in the figure). Thus, triple rules are non-deleting. The compact notation for triple rules combines the left and the right hand side of a rule, i.e. the rule is shown by a single triple graph with special annotations. All elements that are created by the rule, i.e. which appear in the right hand side only, are marked by green line colour and double plus signs. The concrete syntax of the rule shows the ABT diagrams in visual notation, where the attribute "name" is used as label of the visual elements.

Figure 10 shows further rules in compact notation and in concrete syntax. The rule "PublicToPublic" synchronously creates two Reo connectors on both sides and they are related by the points at their input and output ports. The next rule "FilterToED" is slightly more complex and shows how filters in business models correspond to encrypted connections in the related IT model. This reflects the abstract business requirement of hiding confidential information and its possible implementation by encryption in the IT domain.

Private connections leading to related and secured public connections are related by the rule "PrivateInToPrivateIn". The last rule "FilteredOutToPrivate-Out" in Fig. 10 specifies how outgoing communication from a secured connection is handled. The additional box "attach" explicitly specifies the creation of some elements using the abstract syntax, because these changes are not visible in the concrete syntax. The triple rule "PublicToPublicExtend" for symmetric communication is omitted, because it is similar to "FilterToED" and it is not needed for the example transformations.

4.2 Model Transformation and Model Integration

Based on these triple rules the operational rules for performing model transformations as well as model integrations are derived automatically. Given a triple rule its forward rule is derived by replacing the source component of the left hand side by the source component of the right hand side. This way the rule requires a complete fragment in the source component of an integrated model and completes the missing parts for the correspondence and target components. Similar to the forward case backward rules are derived in order to perform backward model transformations. The source rule is used for parsing the given source model of a forward transformation and similarly the target rule is used for backward transformations. Integration rules set up the correspondences between two existing models, which are parsed simultaneously by the source-target rules.

Definition 1 (Derived Triple Rules). *Given a triple rule* $tr = (tr_S, tr_C, tr_T)$: $L \to R$ *the source, target, forward, backward as well as source-target and integration rules are derived according to the diagrams below.*

$$L = (L_S \xleftarrow{s_L} L_C \xrightarrow{t_L} L_T) \qquad (L_S \leftarrow \varnothing \rightarrow \varnothing) \qquad (\varnothing \leftarrow \varnothing \rightarrow L_T)$$
$$tr\downarrow \quad tr_S\downarrow \quad tr_C\downarrow \qquad \downarrow tr_T \qquad tr_S\downarrow \quad\downarrow \quad\downarrow \qquad \downarrow \quad\downarrow \; tr_T\downarrow$$
$$R = (R_S \xleftarrow{s_R} R_C \xrightarrow{t_R} R_T) \qquad (R_S \leftarrow \varnothing \rightarrow \varnothing) \qquad (\varnothing \leftarrow \varnothing \rightarrow R_T)$$

triple rule tr   source rule tr_S   target rule tr_T

$$L_F = (R_S \xleftarrow{tr_S \, \circ \, s_L} L_C \xrightarrow{t_L} L_T) \qquad\qquad L_B = (L_S \xleftarrow{s_L} L_C \xrightarrow{tr_T \circ t_L} R_T)$$
$$tr_F\downarrow \quad id\downarrow \qquad tr_C\downarrow \qquad \downarrow tr_T \qquad\qquad tr_B\downarrow \quad tr_S\downarrow \quad tr_C\downarrow \qquad\qquad \downarrow id$$
$$R_F = (R_S \xleftarrow{} R_C \xrightarrow{t_R} R_T) \qquad\qquad R_B = (R_S \xleftarrow{s_R} R_C \xrightarrow{t_R} R_T)$$

forward rule tr_F   backward rule tr_B

$$L_{ST} = (L_S \longleftarrow \varnothing \longrightarrow L_T) \qquad\qquad L_I = (R_S \xleftarrow{tr_S \, \circ \, s_L} L_C \xrightarrow{tr_T \, \circ \, t_L} R_T)$$
$$tr_{ST}\downarrow \quad tr_S\downarrow \qquad \downarrow \qquad \downarrow tr_T \qquad\qquad tr_I\downarrow \quad id\downarrow \qquad tr_C\downarrow \qquad\qquad \downarrow id$$
$$R_{ST} = (R_S \longleftarrow \varnothing \longrightarrow R_T) \qquad\qquad R_I = (R_S \xleftarrow{s_R} R_C \xrightarrow{t_R} R_T)$$

source-target rule tr_ST   integration rule tr_I

Fig. 11 shows the derived forward and integration rules of the triple rule
"*FilterToED*" in Fig. 10. The source component of the forward rule is identical
in the left and right hand side and in the case of the integration rule, only the
correspondence component is changed by the rule. A model transformation based
on forward rules is performed using the on-the-fly construction in [15] leading to
a source consistent forward transformation sequence.

Example 2 (Model Transformation). Figures 12 and 13 show a model transformation sequence from a machine-centric business service model into its corresponding machine-centric IT service model, which is obtained by restricting
the final triple graph to its target component. At each transformation step a
part of the source model is completed by the missing elements in the correspondence and target component. The matches of the forward rules do not overlap
on their effective elements, which are given by $R_S \setminus L_S$ of the specified triple
rule and visualized by green colour and plus signs in the source components
of the triple rules. Furthermore, each element in the machine-centric business
service model is matched by an effective element of a forward rule. Both properties are ensured by a formal condition, called source consistency, which controls

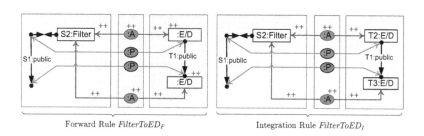

Forward Rule FilterToED_F   Integration Rule FilterToED_I

Fig. 11. Derived forward and integration rules

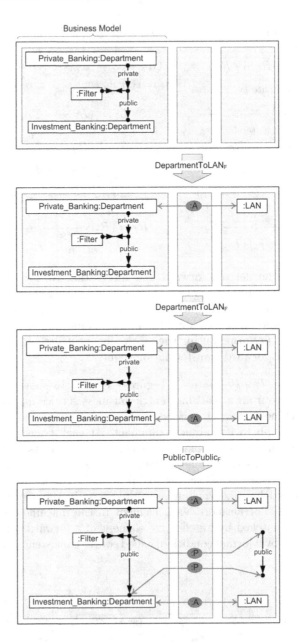

Fig. 12. Model Transformation Sequence Part 1

the forward transformation. This way each source element is translated exactly once and the resulting triple graph is a well formed integrated model. Figure 14 vizualizes the effect of the model transformation by showing the given business service model and the resulting IT service model.

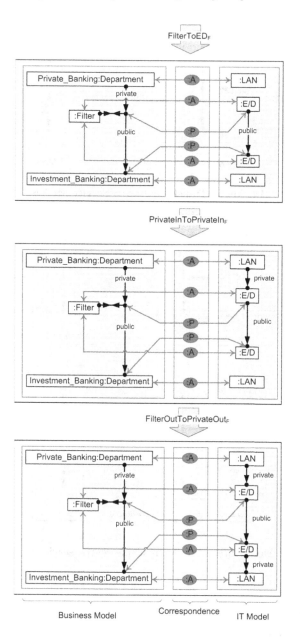

Fig. 13. Model Transformation Sequence Part 2

Model integrations based on TGGs are defined by integration sequences, in which the derived integration rules are applied. Given a source and a target model their integration is performed by completing the correspondence structure that relates both models. The consistency of a model integration is ensured by

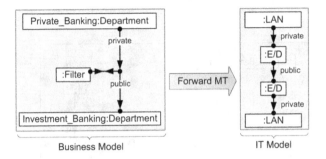

Fig. 14. Model Transformation of Service Models

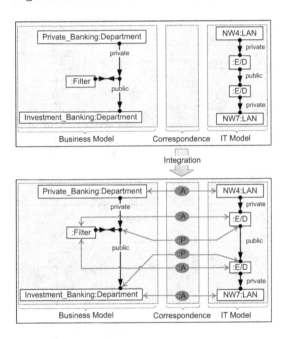

Fig. 15. Model integration for machine-centric service models

a formal condition, called source-target consistency [11]. This condition ensures that each fragment of the models is processed and integrated exactly once.

Example 3 (Model Integration). The model integration of the machine-centric business service model and the machine-centric IT service model is shown in Fig. 15 and is based on the following source-target consistent integration sequence using the derived integration rules.

$$G_0 \xrightarrow{DepartmentToLAN_I} G_1 \xrightarrow{DepartmentToLAN_I} G_2 \xrightarrow{PublicToPublic_I} G_3$$
$$\xrightarrow{FilterToED_I} G_4 \xrightarrow{PrivateInToPrivateIn_I} G_5 \xrightarrow{FilteredOutToPrivateOut_I} G_6 \text{ with}$$
$$G_0 = (M_{BusinessModel} \leftarrow \emptyset \rightarrow M_{ITModel}).$$

In the first four steps some fragments of the source and target components are completed by the missing elements in the correspondence component. In the two last steps no correspondence nodes are created. The reason is the following. The triple rules "*PrivateInToPrivateIn*" and "*FilteredOutToPrivateOut*" extend integrated fragments in the source and target model by connecting Reo elements and they do not create any correspondence node. Therefore, the derived integration rules do not create correspondences either. But these integration rules are necessary to ensure correct integrations in the way that the positions of the corresponding private Reo connectors are checked.

The presented forward transformation and integration approach are shown to be correct and complete with respect to the triple patterns [15,11], i.e. with respect to the language $VL = \{G \mid \emptyset \Rightarrow^* G \text{ in } TGG\}$ containing the integrated models generated by the triple rules. More precisely, each model transformation translates a source model into a target model, such that the integrated model that contains both models can be created by applications of the triple rules to the empty start graph. Vice versa, each source model that is part of an integrated model in the generated triple language VL is transformed into a corresponding valid target model.

The languages of translatable source models VL_S and of reachable target models VL_T are given by $VL_S = \{G_S \mid (G_S \leftarrow G_C \rightarrow G_T) \in VL\}$ and $VL_T = \{G_T \mid (G_S \leftarrow G_C \rightarrow G_T) \in VL\}$. Based on these definitions there is the correctness and completeness result below according to Theorems 2 and 3 in [16].

Theorem 1 (Correctness and Completeness)

- Correctness: *Each model transformation sequence given by* $(G_S, G_0 \xrightarrow{tr_F^*} G_n, G_T)$, *which is based on a source consistent forward transformation sequence* $G_0 \xrightarrow{tr_F^*} G_n$ *with* $G_0 = (G_S \leftarrow \emptyset \rightarrow \emptyset)$ *and* $G_n = (G_S \leftarrow G_C \rightarrow G_T)$ *is correct, i.e.* $G_n \in VL$.
- Completeness: *For each* $G_S \in VL_S$ *there exists* $G_n \in VL$ *with a model transformation sequence* $(G_S, G_0 \xrightarrow{tr_F^*} G_n, G_T)$ *where* $G_0 \xrightarrow{tr_F^*} G_n$ *is source consistent with* $G_0 = (G_S \leftarrow \emptyset \rightarrow \emptyset)$ *and* $G_n = (G_S \leftarrow G_C \rightarrow G_T)$.

These fundamental results are based on the Composition and Decomposition Thm. for TGGs [9] and visualized below. Each correct forward sequence $G_{n,0} \xrightarrow{tr_F^*} G_{n,n} = G_n$ requires the existence of a corresponding triple sequence $G_0 = G_{0,0} \xrightarrow{tr^*} G_{n,n} = G_n$, which is obtained by composing the forward sequence with a source sequence $G_{0,0} \xrightarrow{tr_S^*} G_{n,0}$ via source rules. Furthermore, each triple sequence can be decomposed into corresponding source and a forward sequences and as shown in [9] this result holds exactly if the forward sequence is source consistent meaning that the corresponding source sequence exists and its comatches coincide with the matches of the forward sequence in the source component.

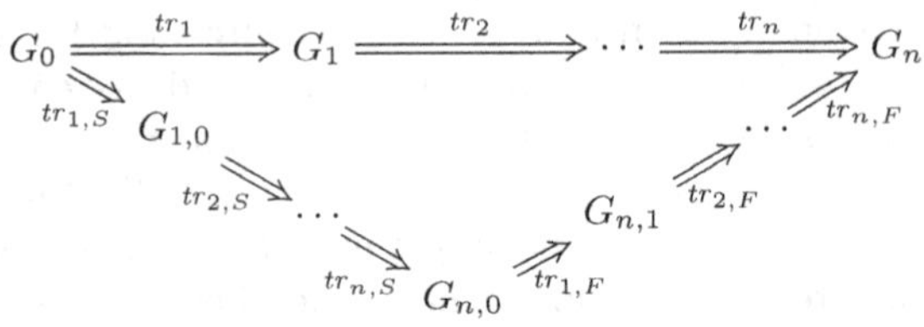

$$G_0 \overset{tr_1}{\Longrightarrow} G_1 \overset{tr_2}{\Longrightarrow} \cdots \overset{tr_n}{\Longrightarrow} G_n$$

Fig. 16. Source, forward and triple rule sequence

Analogously, model integrations based on source-target consistent integration sequences are correct and complete as shown in [11], i.e. a pair of source and target models is integrated if and only if this pair is contained in the language VL of integrated models generated by the triple rules.

4.3 Summary of Achievements for Inter-model Techniques

The described and illustrated techniques for model transformation and integration in this section improve the interoperability between the machine-centric business and the machine-centric IT service models given by ABT-Reo diagrams. The techniques are general, such that they should be applicable to the other visual languages and coordinates in the full EM-Cube that is presented in the section about the extensions later in this paper. The benefits of the techniques can be described as follows. Model transformation enables the consistent construction of models in related domains that can be further refined by domain experts. Model integration establishes the correspondences between the existing models e.g. for data interchange and furthermore, detection of inconsistencies between models. Both techniques are based on formal constructions and shown to be correct and complete.

5   Evaluation of Graph Transformation Techniques

An evaluation of the selected graph transformation techniques from the point of view of enterprise modeling is presented in the following. We list typical problems in enterprise modeling and show which kind of potential solutions in the area of graph transformation are available to address them.

In detail, typical problems in the area of enterprise modeling are to properly define a domain specific modeling language for enterprise models. Graph transformation supports this by the help of graph grammars or type graphs with constraints. Another typical issue is the need for syntax-supported editing operations. Graph transformation is able to help here offering graph grammars plus corresponding deletion rules. Complex modifications that can encompass the specification of refactorings and the encoding of construction knowledge showing up during enterprise modeling activities can be supported using sequences of rules or concurrent productions. The problem to support language evolution and model migration can be addressed by graph rules, too. Inter-model operations like integration, synchronization and alignment can be implemented by triple graph rules encompassing their derived forward, backward and integration rules. Finally, the problem to handle modeling policies can be supported by the use of graph constraints. A summary is given in the table in Fig. 17.

Problem	Solution
language definition	graph grammar or type graph with constraints
syntax-supported editing	grammar plus corresponding deletion rules
complex modifications: 1. specification of refactory operations 2. encoding of construction knowledge	rule sequences or concurrent productions
language evolution and model migration	graph rules for language evolution and model migration
integration, synchronization, alignment	TGGs and derived forward, backward, integration and synchronization rules
modeling policies	graph constraints

Fig. 17. Problems in enterprise modeling and possible solutions

6   Extensions

Possible future extensions encompass the vision of a full Enterprise Modelling Cube, short EM-Cube, for enterprise modelling consisting of the SEM square (Service Enterprise Modeling-Square), the PEM square (Process Enterprise Modeling-Square) and the REM square (Rule Enterprise Modeling-Square) as well as further relationships between the models of theses squares. Finally, extensions like a *Life* EM-Cube which is an EM-Cube that is able to represent real-world states can be imagined, or further modeling aspects and completely different modeling domains.

6.1   The Full EM-Cube

The full EM-Cube contains human-centric and machine-centric models of services, processes and rules of the Business and IT universe of a banking organization. Service, process and rule models are organized in their corresponding squares, as just mentioned.

The main philosophy behind the full EM-Cube is to provide a modeling framework that is able to collect the distributed knowledge about an organization from different people in their own domain language and to support the integration of all parts towards one holistic enterprise model. By completing human-centric

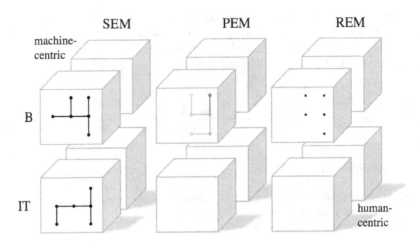

Fig. 18. The full EM-Cube

models with machine-centric models with the help of synchronization and integration techniques the best out of both worlds can be combined enabling semiformal ad hoc domain specific modeling languages used to collect organizational knowledge and well-defined formal methods used to evaluate organizational predicates like security, risk and compliance.

As sketched in Fig. 18 service models can be imagined as street maps describing possible connections between service nodes, whereas processes may look like bus lines running on such streets maps and rules appearing as specific traffic regulations for the intersections. Models of the business and the IT universe in Fig. 18 do not follow a clear top-down relationship as often assumed by model driven software development techniques. Instead of a top-down strategy we have a relationship between equals which is ideally supported by graph based alignment techniques which support mutual synchronizations rather than model driven generation approaches.

6.2   Relationships between the SEM and PEM-Square

Given the full EM-Cube it is possible to focus on the different relationships between models of the SEM-Square and the PEM-Square which can be characterized as supported integration and supported alignment as well as optional and automatic integration. In detail, we give a short characterization of these relationship types from the point of view of the application domain of enterprise modeling for banks.

Supported alignment between the human-centric business and IT service models as well as between the human-centric business and IT process models helps to allocate IT services and processes to run their business counterparts. Because this activity is inherently based on human choices, there can be no full

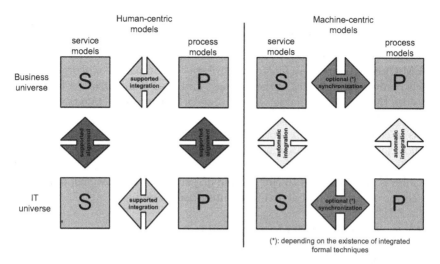

Fig. 19. Relationships between the SEM-Square and PEM-Square

automation but a support like it is available in today's recommender systems to facilitate human decision making in complex environments.

Supported integration helps to ensure the consistency between human-centric service and process models in the business as well as the IT universe. Because human-centric models are using open diagram languages that enable language evolution and that support model migration any integration approach is unlikely to be fully automated. Therefore, incomplete information will require human intervention in the integration process on a case-by-case basis.

Automatic integration applies when machine-centric service models of the business and IT universe are put in relationship. The same applies to machine-centric process models. The rationale here is that machine-centric models are build using well-defined formal languages that are fully implemented. Their language definition and semantics is therefore known and stable. Secondly, machine-centric process models are integrated with other machine-centric process models, the same is true for service models, which facilitates the integration. Hence, once the human-centric counterparts are aligned the machine-centric models are assumed to be checked for consistency issues in an automatic fashion.

Optional integration comes up when machine-centric models for services and those for processes are put into relationship. Here, it may happen quite often, that there is no intersection. Hence, no integration issues will show up. However, in case that the formal methods, e.g. used for machine-centric process and service models are formally integrated, their models can be integrated, too, and can be checked for consistency issues. Therefore, this integration is optional.

6.3   Further Extensions

Further extensions encompass the concept of a *Life* EM-Cube, new modeling aspects and completely different domains.

A *Life* EM-Cube reflects real-world states in instance models of an organization represented in the EM-Cube. In addition to that, it incorporates rules that will enable the simulation of real-world state changes in a given model. Further on, a *Life* EM-Cube will serve as a master control station for instances of organizational services, processes and rules of the business and the IT universe of a bank.

Besides the already available service, process and rule aspects new aspects can extend the EM-Cube, e.g. the document and the resource aspects. Document models in combination with process models could properly reflect the real-world data flow in a bank. Resource models can serve to codify available business and IT functions to make them re-usable in service and process models. Finally, the EM-Cube is not restricted to the domain of banking organizations, but can also be used for other suitable domains in enterprise modelling.

7   Open Issues and Conclusions

The open issues and possible conclusions concern likewise the scientific community as well as the banking industry. They are presented here according to the specific interests of both groups.

7.1   Scientific Community

For the scientific community the development of highly relevant theory from an application point of view is one driving interest. To further develop enterprise modeling using the formal techniques of algebraic graph transformation in combination with the full EM-Cube additional support is needed. In detail, models that are distributed over organizational boundaries enriched by security aspects, the semantics of models, model integration crossing multiple dimensions and information and behavior preservation as well as semantically enriched model transformation, integration and synchronization need to be further developed. Beyond that, there is a need for formal theories in the area of language evolution and model migration, model synchronization and the mutual dependencies between languages of different degrees of freedom based on model synchronization. Furthermore, analysis capabilities need to be extended and an integrated family of domain-specific modeling languages must be developed for the EM-Cube. Finally, there is a need for tool support assuring the formal results of correctness and completeness for the inter-modelling challenges, like model transformation and integration. The tools AGG [1] based on Java and AGT [8,7] based on Mathematica are based on the formal categorical constructions for algebraic graph transformation and will be extended in future work to cover the triple graph transformation constructions presented in this paper. Furthermore, future work will also encompass the analysis of how the formal results can be ensured within the existing tools for TGGs, like e.g. FUJABA [34].

7.2   Banking Industry

For the banking industry long term costs reductions, quality enhancements and flexibility gains are important. The presented formal modeling techniques based

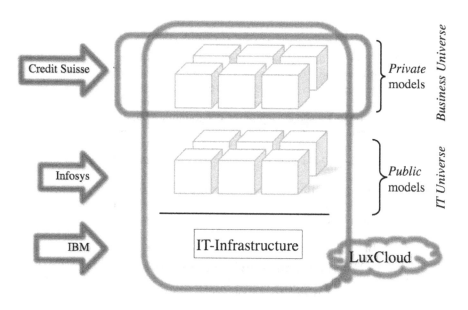

Fig. 20. Possible conclusions for the banking industry

on algebraic graph transformation in combination with the full EM-Cube could enable banks to focus on codifying their business universe only – freeing them from the need to model IT services, processes and rules. By keeping human-centric models and machine-centric model apart, the bankers themselves will be – in principle – able to perform the modeling operations without the help of experts for formal methods.

Assuming that there is a kind of public-private cloud infrastructure, similar to LuxCloud[1], organizational IT models could be provided as a service by companies like Infosys[2] and hosted in such a cloud, whereas the underlying IT infrastructure could be managed by companies like IBM[3] as visualized in Fig. 20. Assuming further, that such a public-private infrastructure is fully embedded in a legal framework, the use of provided IT services would have legal consequences, not only automation and scalability advantages. This would result into a complete new quality of IT services, not available today. As a consequence, the in-house production depth of today's banks could be dramatically reduced, leading to economics of scale regarding the cost of ownership of the IT infrastructure and IT services, better quality of results of IT services and much more adaptable cost models that will help changing today's fix costs into flexible costs of tomorrow.

---

[1] http://www.luxcloud.com
[2] http://www.infosys.com
[3] http://www.ibm.com

References

1. AGG Homepage, http://tfs.cs.tu-berlin.de/agg
2. Arbab, F.: Reo: A Channel-based Coordination Model for Component Composition. Mathematical Structures in Computer Science 14(3), 329–366 (2004), http://homepages.cwi.nl/~farhad/MSCS03Reo.pdf
3. Arbab, F.: Abstract Behavior Types: A Foundation Model for Components and Their Composition. Science of Computer Programming 55, 3–52 (2005)
4. Braatz, B., Brandt, C.: Graph Transformations for the Resource Description Framework. In: Ermel, J.d.L.C., Heckel, R. (eds.) Proc. Workshop on Graph Transformation and Visual Modeling Techniques (GT-VMT 2008), EC-EASST, vol. 10 (2008)
5. Braatz, B., Brandt, C.: How to modify on the semantic web? - a web application architecture for algebraic graph transformations on rdf. In: 2nd International Workshop on Semantic Web Information Management, Vienna. Springer, Heidelberg (to appear 2010)
6. Braatz, B., Brandt, C., Engel, T., Hermann, F., Ehrig, H.: An approach using formally well-founded domain languages for secure coarse-grained IT system modelling in a real-world banking scenario. In: Proc. 18th Australasian Conference on Information Systems (ACIS 2007), Toowoomba, Queensland, Australia (December 2007)
7. Brandt, C., Hermann, F., Ehrig, H., Engel, T.: Enterprise Modelling using Algebraic Graph Transformation - Extended Version. Technical Report 2010-6, Technische Universität Berlin,Fakultät IV (2010), http://www.eecs.tu-berlin.de/menue/forschung/forschungsberichte/2010
8. Brandt, C., Hermann, F., Engel, T.: Security and Consistency of IT and Business Models at Credit Suisse realized by Graph Constraints, Transformation and Integration using Algebraic Graph Theory. In: Proc. Int. Conference on Exploring Modeling Methods in Systems Analysis and Design 2009 (EMMSAD 2009). LNBIP, vol. 29, pp. 339–352. Springer, Heidelberg (2009)
9. Ehrig, H., Ehrig, K., Ermel, C., Hermann, F., Taentzer, G.: Information Preserving Bidirectional Model Transformations. In: Dwyer, M.B., Lopes, A. (eds.) FASE 2007. LNCS, vol. 4422, pp. 72–86. Springer, Heidelberg (2007)
10. Ehrig, H., Ehrig, K., Ermel, C., Prange, U.: Consistent Integration of Models Based on Views of Visual Languages. In: Fiadeiro, J.L., Inverardi, P. (eds.) FASE 2008. LNCS, vol. 4961, pp. 62–76. Springer, Heidelberg (2008)
11. Ehrig, H., Ehrig, K., Hermann, F.: From Model Transformation to Model Integration based on the Algebraic Approach to Triple Graph Grammars. In: Ermel, C., de Lara, J., Heckel, R. (eds.) Proc. Workshop on Graph Transformation and Visual Modeling Techniques (GT-VMT 2008), EC-EASST, vol. 10 (2008)
12. Ehrig, H., Ehrig, K., Prange, U., Taentzer, G.: Fundamentals of Algebraic Graph Transformation. In: EATCS Monographs in Theoretical Computer Science. Springer, Heidelberg (2006)
13. Ehrig, H., Engels, G., Kreowski, H.-J., Rozenberg, G. (eds.): Handbook of Graph Grammars and Computing by Graph Transformation. Applications, Languages and Tools, vol. 2. World Scientific, Singapore (1999)
14. Ehrig, H., Ermel, C., Hermann, F.: On the Relationship of Model Transformations Based on Triple and Plain Graph Grammars. In: Karsai, G., Taentzer, G. (eds.) Proc. Third International Workshop on Graph and Model Transformation (GraMoT 2008). ACM, New York (2008)

15. Ehrig, H., Ermel, C., Hermann, F., Prange, U.: On-the-Fly Construction, Correctness and Completeness of Model Transformationsbased on Triple Graph Grammars. In: Schürr, A., Selic, B. (eds.) MODELS 2009. LNCS, vol. 5795, pp. 241–255. Springer, Heidelberg (2009)
16. Ehrig, H., Hermann, F., Sartorius, C.: Completeness and Correctness of Model Transformations based on Triple Graph Grammars with Negative Application Conditions. In: Heckel, R., Boronat, A. (eds.) Proc. Workshop on Graph Transformation and Visual Modeling Techniques (GT-VMT 2009), EC-EASST (2009)
17. Ehrig, H., Kreowski, H.-J., Montanari, U., Rozenberg, G. (eds.): Handbook of Graph Grammars and Computing by Graph Transformation. Concurrency, Parallelism and Distribution, vol. 3. World Scientific, Singapore (1999)
18. Ehrig, H., Prange, U.: Formal Analysis of Model Transformations Based on Triple Graph Rules with Kernels. In: Ehrig, H., Heckel, R., Rozenberg, G., Taentzer, G. (eds.) ICGT 2008. LNCS, vol. 5214, pp. 178–193. Springer, Heidelberg (2008)
19. Ermel, C., Biermann, E., Ehrig, K., Taentzer, G.: Generating Eclipse Editor Plug-Ins using Tiger. In: Schürr, A., Nagl, M., Zündorf, A. (eds.) AGTIVE 2007. LNCS, vol. 5088, pp. 583–584. Springer, Heidelberg (2008)
20. Giaglis, G.M.: A taxonomy of business process modeling and information systems modeling techniques. International Journal of Flexible Manufacturing Systems 13(2), 209–228 (2001)
21. Guerra, E., de Lara, J.: Attributed typed triple graph transformation with inheritance in the double pushout approach. Technical Report UC3M-TR-CS-2006-00, Universidad Carlos III, Madrid, Spain (2006)
22. Guerra, E., de Lara, J.: Model view management with triple graph grammars. In: Corradini, A., Ehrig, H., Montanari, U., Ribeiro, L., Rozenberg, G. (eds.) ICGT 2006. LNCS, vol. 4178, pp. 351–366. Springer, Heidelberg (2006)
23. Hermann, F., Ehrig, H., Orejas, F., Golas, U.: Formal Analysis of Functional Behaviour for Model Transformations Based on Triple Graph Grammars. In: Ehrig, H., et al. (eds.) ICGT 2010. LNCS, vol. 6372, pp. 154–169. Springer, Heidelberg (2010)
24. Kindler, E., Wagner, R.: Triple Graph Grammars: Concepts, Extensions, Implementations, and Application Scenarios. Technical Report TR-ri-07-284, Universität Paderborn (2007)
25. Königs, A., Schürr, A.: Tool Integration with Triple Graph Grammars - A Survey. In: Proc. SegraVis School on Foundations of Visual Modelling Techniques. ENTCS, vol. 148, pp. 113–150. Elsevier Science, Amsterdam (2006)
26. Lara, J., Bardohl, R., Ehrig, H., Ehrig, K., Prange, U., Taentzer, G.: Attributed Graph Transformation with Node Type Inheritance. Theoretical Computer Science 376(3), 139–163 (2007)
27. Lillehagen, F., Krogstie, J.: State of the art of enterprise modeling. In: Active Knowledge Modeling of Enterprises, pp. 91–127. Springer, Heidelberg (2008)
28. Object Management Group. UML 2.0 OCL Specification (2003), http://www.omg.org/docs/ptc/03-10-14.pdf
29. OMG. Meta-Object Facility 2.0 (2006), http://www.omg.org/mof/
30. OMG. Unified Modeling Language 2.1.2: Superstructure (2007), http://www.uml.org/
31. Rozenberg, G.: Handbook of Graph Grammars and Computing by Graph Transformations. Foundations, vol. 1. World Scientific, Singapore (1997)
32. Schürr, A.: Specification of Graph Translators with Triple Graph Grammars. In: Mayr, E.W., Schmidt, G., Tinhofer, G. (eds.) WG 1994. LNCS, vol. 903, pp. 151–163. Springer, Heidelberg (1995)

33. Schürr, A., Klar, F.: 15 years of triple graph grammars. In: Ehrig, H., Heckel, R., Rozenberg, G., Taentzer, G. (eds.) ICGT 2008. LNCS, vol. 5214, pp. 411–425. Springer, Heidelberg (2008)
34. Software Engineering Group, University of Paderborn. Fujaba Tool Suite (2010), http://wwwcs.uni-paderborn.de/cs/ag-schaefer/Lehre/PG/Fujaba/projects/tgg/index.html
35. Taentzer, G., Ehrig, K., Guerra, E., de Lara, J., Lengyel, L., Levendovsky, T., Prange, U., Varro, D., Varro-Gyapay, S.: Model Transformation by Graph Transformation: A Comparative Study. In: Proc. Workshop Model Transformation in Practice, Montego Bay, Jamaica (October 2005)
36. Whitman, L., Ramachandran, K., Ketkar, V.: A taxonomy of a living model of the enterprise. In: Winter Simulation Conference, pp. 848–855 (2001)
37. Winkelmann, J., Taentzer, G., Ehrig, K., Küster, J.: Translation of Restricted OCL Constraints into Graph Constraints for Generating Meta Model Instances by Graph Grammars. In: Varro, D., Bruni, R. (eds.) Proc. International Workshop on Graph Transformation and Visual Modeling Techniques (GT-VMT 2006), Vienna, Austria. ENTCS. Elsevier Science, Amsterdam (2006)

Graph Transformation Units
Guided by a SAT Solver⋆

Hans-Jörg Kreowski, Sabine Kuske, and Robert Wille

University of Bremen, Department of Computer Science
P.O. Box 33 04 40, 28334 Bremen, Germany
{kreo,kuske,rwille}@informatik.uni-bremen.de

Abstract. Graph transformation units are rule-based devices to model graph algorithms, graph processes, and the dynamics of systems the states of which are represented by graphs. Given a graph, various rules are applicable at various matches in general, but not any choice leads to a proper result so that one faces the problem of nondeterminism. As countermeasure, graph transformation units provide the generic concept of control conditions which allow one to cut down the nondeterminism and to choose the proper rule applications out of all possible ones. In this paper, we propose an alternative approach. For a special type of graph transformation units including the solution of many *NP*-complete and *NP*-hard problems, the successful derivations from initial to terminal graphs are described by propositional formulas. In this way, it becomes possible to use a SAT solver to find out whether there is a successful derivation for some initial graph or not and how it is built up in the positive case.

1   Introduction

Graph transformation units are rule-based devices to model algorithms and processes on graphs and the dynamic behavior of systems the states of which are represented by graphs [11]. Such a unit provides descriptions of initial and terminal graphs, a set of rules, and a control condition. The initial graphs are the inputs of the modeled computational processes, the terminal graphs are the potential outputs, the rules can be applied to graphs yielding derived graphs so that the iteration of rule applications establishes the running processes. In general, rule applications are quite nondeterministic because various rules may be applicable at various matches. The control condition is a feature to get rid of undesired nondeterminism. A typical example of control conditions are regular expressions which specify that the rules are applied in a certain order. Another kind are priorities that make sure that a rule with highest priority is applied in each step.

⋆ The first two authors would like to acknowledge that their research is partially supported by the Collaborative Research Centre 637 (Autonomous Cooperating Logistic Processes: A Paradigm Shift and Its Limitations) funded by the German Research Foundation (DFG).

H. Ehrig et al. (Eds.): ICGT 2010, LNCS 6372, pp. 27–42, 2010.

In this paper, we propose an alternative approach to deal with nondeterminism. Graph transformation units of a special type (using only rules that keep the set of nodes invariant) are transformed into propositional formulas. If the formula corresponding to some graph transformation unit is satisfied by some assignment of truth values to the Boolean variables, then the assignment tells which rule must be applied at which match in which step such that a successful derivation is efficiently constructed. If the formula is unsatisfiable, then there is no successful derivation, and, thus, no need to start the derivation process at all. If the formula is extended properly, then the same correspondence between successful derivations and satisfying truth assignments holds for a single initial graph. As a consequence, one can employ a SAT solver to find successful derivations.

The proposed approach has a theoretical and a practical inspiration. In [4], Cook proved his seminal theorem that the satisfiability problem of the propositional calculus is NP-complete by describing the runs of a polynomial Turing machine in terms of propositional formulas. Despite this proven complexity, in the last decade efficient SAT solvers have been developed which can handle instances including hundreds of thousands variables and clauses in very short run-time. As a consequence, SAT solvers are successfully applied in the domain of verification, planning, and many more (see, e.g., [2]). In, e.g., [1,8] it is shown that chip designs can be verified by translating them into propositional formulas such that the unsatisfiability of the latter corresponds to the correctness of the respective chip. In contrast, if the formula turns out to be satisfiable, a counterexample showing the error can be derived. Recently, SAT solvers are applied in order to verify system descriptions given in the *Unified Modeling Language* (UML) [15]. Thus, SAT solvers have been proved to be practical for many relevant problems. Using this as motivation, the key idea of this paper is to replace Turing machines on one hand and chip designs or UML specifications on the other hand by a special type of graph transformation units. But the use of a SAT solver is only feasible if the size of the input formulas is polynomially bounded. Hence, we restrict the consideration to derivations the lengths of which are bounded by a polynomial. Therefore, the present approach applies particularly to polynomial graph transformation units including many solutions of NP-complete and NP-hard graph problems.

This paper is organized as follows. In Section 2, we recall the concepts of graph transformation and graph transformation units used in this paper. Section 3 shows how polynomial derivations can be modeled by propositional formulas in such a way that each satisfying instantiation of the formula represents a derivation. Section 4 generalizes Section 3 to graph transformation units, i.e, instead of polynomial derivations we model polynomial graph transformation units by propositional formulas. Moreover, an example of a polynomial graph transformation unit is given that can be represented as a propositional formula. This unit checks for any input graph whether it contains a Hamiltonian path. In Section 5 a very first implementation of the Hamiltonian path example is presented and tested for a series of concrete input graphs. The paper ends with the conclusion in Section 6.

2   Preliminaries

In this section, we recall basic notions and notations of graphs, graph transformation, and graph transformation units.

Directed edge-labeled simple graphs. For a set Σ of labels, a (*directed edge-labeled simple*) *graph* over Σ is a pair $G = (V, E)$ where V is a finite set of *nodes* and $E \subseteq V \times \Sigma \times V$ is a finite set of labeled *edges*. An edge $e = (v, x, v')$ is called a *loop* if $v = v'$. The components of G are also denoted by V_G and E_G. The set of all graphs over Σ is denoted by $\mathcal{G}_\Sigma$. We reserve a specific label $*$ which is omitted in drawings of graphs. In this way, graphs where all edges are labeled with $*$ may be seen as *unlabeled graphs*. The number of nodes is the *size* of G, denoted by $size(G)$. If Σ is finite with $|\Sigma|$ labels, the number of edges of a graph is bounded by $|\Sigma| \cdot size(G)^2$ because of the simplicity.

Graph morphisms. For graphs $G, H \in \mathcal{G}_\Sigma$, a *graph morphism* $g \colon G \to H$ is a mapping $g_V \colon V_G \to V_H$ that is structure- and label-preserving, i.e., for all $(v, x, v') \in E_G$, $(g_V(v), x, g_V(v')) \in E_H$. If the mapping g_V is an inclusion, then G is called a *subgraph* of H, denoted by $G \subseteq H$. For a graph morphism $g \colon G \to H$, the image $g(G) = (g_V(V_G), g_E(E_G)) \subseteq H$ of G in H with $g_E(E_G) = \{(g_V(v), x, g_V(v')) \mid (v, x, v') \in E_G\}$ is called a *match* of G in H.

Rules. A *rule* $r = (L \supseteq K \subseteq R)$ consists of three graphs $L, K, R \in \mathcal{G}_\Sigma$ such that K is a discrete subgraph of L and R, i.e., $E_K = \emptyset$. The components L, K, and R of r are called *left-hand side*, *gluing graph*, and *right-hand side*, respectively.

Rule application. Let $r = (L \supseteq K \subseteq R)$ be a rule and let $G \in \mathcal{G}_\Sigma$. Moreover, let $g \colon L \to G$ be a graph morphism satisfying the following conditions:

- *dangling condition*: If a node v in $g(L)$ is the source or the target of an edge in $E_G - g_E(E_L)$, then $v \in g_V(V_K)$.
- *identification condition*: $g_V(v) = g_V(v')$ for $v, v' \in V_L$ implies $v = v'$ or $v, v' \in V_K$.

Then the application of r to G with respect to g consists of the following three steps.

1. Remove the nodes of $g_V(V_L - V_K)$ and the edges of $g_E(E_L - E_K)$ yielding the *intermediate graph* $Z \subseteq G$.
2. Let $d \colon K \to Z$ be the restriction of g to K and Z, then the pushout of d and the inclusion of K into R yields the resulting graph H and graph morphisms $h \colon R \to H$ and $i \colon Z \to H$. Without loss of generality, one can assume that i is the inclusion of Z into H and that h is the identity on $R - K$. This provides an explicit construction of H because $Z \cup h(R) = H$ and $Z \cap h(R) = d(K) = h(K)$.

The application of a rule r to a graph G with respect to g is denoted by $G \underset{r,g}{\Rightarrow} H$, where H is the graph resulting from this application. As usual, a rule application is called a *direct derivation*. The subscript r, g may be omitted if it is clear from the context. The notion of a direct derivation coincides with a rule application in the double-pushout approach (cf., e.g., [5,7]), but we need the explicit set-theoretic construction in the next section.

The sequential composition $d = G_0 \underset{r_1,g_1}{\Longrightarrow} G_1 \underset{r_2,g_2}{\Longrightarrow} \cdots \underset{r_n,g_n}{\Longrightarrow} G_n$ $(n \in \mathbb{N})$ is called a *derivation* from G_0 to G_n. The derivation from G_0 to G_n can also be denoted by $G_0 \underset{P}{\overset{n}{\Longrightarrow}} G_n$ where $\{r_1, \ldots, r_n\} \subseteq P$, or just by $G_0 \underset{P}{\overset{*}{\Longrightarrow}} G_n$. The subscript P may be omitted if it is clear from the context. The string $r_1 \cdots r_n$ is the *application sequence* of the derivation d.

Given a finite set of rules and a graph G, the number of graph morphisms from left-hand sides to G is bounded by a polynomial in the size of G because the sizes of left-hand sides of rules are bounded by a constant. Given a match, the check, whether the dangling and the identification condition hold, and the construction of the directly derived graph is linear in the size of G. Therefore, polynomial time is needed to find a match and to construct a direct derivation, and there is a polynomial number of matches. Moreover, the size of the resulting graph differs from the size of the host graph by a constant.

Control conditions. The nondeterminism of rule application can be restricted via control conditions. A typical example is a regular expression over a set of rules (or any other string-language-defining device). Let C be a regular expression specifying the language $L(C)$. Then a derivation with application sequence s is *permitted* by C if $s \in L(C)$.

Graph class expressions. The initial and terminal graphs of derivations can be specified by graph class expressions. For every graph class expression e the set of graphs specified by e is denoted by $SEM(e)$. A typical example is a subset $\Delta \subseteq \Sigma$ with $SEM(\Delta) = \mathcal{G}_\Delta \subseteq \mathcal{G}_\Sigma$. Requested or forbidden subgraphs and graphs that are reduced with respect to some rule set are also frequently used.

Graph transformation unit. Every graph transformation unit transforms initial graphs to terminal graphs via the successive application of rules according to a control condition. A (*simple*) *transformation unit* is a system $tu = (I, P, C, T)$, where I and T are graph class expressions, P is a finite set of rules, and C is a control condition. tu is *polynomial* if:

- there is a polynomial p such that for each initial graph $G \in SEM(I)$ and each derivation $G \underset{P}{\overset{n}{\Longrightarrow}} G'$, $n \leq p(size(G))$,
- the membership problems of $SEM(initial)$ and $SEM(T)$ are polynomial, and
- it can be checked in polynomial time whether the $G \underset{P}{\overset{n}{\Longrightarrow}} G'$ is permitted by C.

Each transformation unit tu specifies a binary relation $SEM(tu) \subseteq SEM(I) \times SEM(T)$ that contains a pair (G, H) of graphs if and only if there is a derivation $G \overset{*}{\underset{P}{\Longrightarrow}} H$ permitted by C.

Transformation units are presented in [10,11]. In this first approach towards translating transformation units into propositional formulas, we consider simple transformation units where the structuring component is omitted. More about control conditions for transformation units can be found in, e.g., [12,9].

Further notions and notations. For each language $L \subseteq P^*$ and each number $m \in \mathbb{N}$, the finite sublanguage $L(m) = \{w \in L \mid 0 \le |w| \le m\}$ denotes the set of all words in L of length at most m. For each $n \in \mathbb{N}$, $[n]$ denotes the set $\{1, \ldots, n\}$. For each propositional formula Q, the set of variables of Q is denoted by $\mathcal{V}_Q$ and the number of literals by $L^\#(Q)$.

3   From Polynomial Derivations to Propositional Logic

In this section, we show how derivations can be modeled by propositional formulas. More precisely, every finite rule set P together with a given initial graph G_0 and a polynomial p can be automatically transformed into a propositional formula that describes all derivations from G_0 that are not longer than $p(size(G_0))$.

In this first approach, we restrict the consideration to graphs with nodes that are numbered from 1 to n for some $n \in \mathbb{N}$, to rules which keep the set of nodes invariant, and to injective graph morphisms. Moreover, we assume that the label alphabet Σ is finite. A well-known instance of graph transformation systems with invariant node sets are graph relabelling systems (see, e.g., [13]).

It is worth noting that all constructions in this section work for arbitrary lengths bounds of derivations and not only for polynomial bounds, but then the resulting propositional formulas can become intractable.

3.1   Representing Simple Graphs as Propositional Formulas

Every propositional formula Q with $\mathcal{V}_Q = [n] \times \Sigma \times [n]$ specifies a set of graphs because each instantiation $f \colon \mathcal{V}_Q \to \text{BOOL}$ represents the graph $graph(f) = ([n], E)$ where $E = \{(v, a, v') \in [n] \times \Sigma \times [n] \mid f(v, a, v') = \text{TRUE}\}$. Hence, the set $Graphs(Q)$ of graphs specified by Q is equal to $\{graph(f) \mid f \colon \mathcal{V}_Q \to \text{BOOL}, f(Q) = \text{TRUE}\}$. In the following, for each $n \in \mathbb{N}$, a propositional formula Q with $\mathcal{V}_Q = [n] \times \Sigma \times [n]$ is called a *propositional graph formula of n*. The set of all propositional graph formulas of n is denoted by $FG(n)$. A particular form of propositional graph formulas of n is $\bigwedge_{e \in E} e \wedge \bigwedge_{e \in ([n] \times \Sigma \times [n]) - E} \neg e$ where E is a subset of $[n] \times \Sigma \times [n]$. For each such subset E, this formula represents the single graph $G = ([n], E)$ and will be denoted by $fg(G)$.

3.2   Representing Derivations as Propositional Formulas

As said before, for every rule $r = (L \supseteq K \subseteq R)$, we require that $V_L = V_K = V_R$. This implies that every morphism from L into some graph G satisfies the dangling condition. Additionally, since we use only injective graph morphisms in rule applications the identification condition is also fulfilled. In the following, the set V_L will be also denoted by V_r. The number of nodes in V_r will be denoted by $size(r)$. For an underlying rule set P its maximal size $\max\{size(r) \mid r \in P\}$ will be denoted by max.

Given a polynomial p, a polynomial derivation with respect to p has the form $G_0 \underset{r_1,g_1}{\Longrightarrow} G_1 \underset{r_2,g_2}{\Longrightarrow} \ldots \underset{r_m,g_m}{\Longrightarrow} G_m$ with $m \leq p(size(G_0))$. Due to the assumption for rules, the set of nodes $[n]$ is invariant through the whole derivation. Therefore, the occurring graphs are fully described by fixing their edges where each edge is a triple $(i,a,j) \in [n] \times \Sigma \times [n]$. This is possible by means of Boolean variables of the form $edge(i,a,j,k)$ which should be true if and only if (i,a,j) is an edge of G_k.

The actual edges of all the derived graphs depend on the initial graph G_0. Let $G_0 = ([n], E_0)$. Then G_0 can be directly described via the formula

$$\bigwedge_{(v,a,v')\in E_0} edge(v,a,v',0) \wedge \bigwedge_{(v,a,v')\in([n]\times\Sigma\times[n])-E_0} \neg edge(v,a,v',0).$$

Consider for example, the derivation in Figure 2 where the rule of Figure 1 is applied twice. The graph G_0 of this derivation is described by the preceding formula by choosing $E_0 = \{(1,ok,1),(1,*,2),(2,*,2),(2,*,3),(3,*,3)\}$, $n = 3$ and $\Sigma = \{ok,*\}$.

In the following, for each propositional graph formula $Q \in FG(n)$ and each $k \in \mathbb{N}$ the term $Q(k)$ denotes the formula that is obtained from Q by replacing every variable (v,a,v') by $edge(v,a,v',k)$. Hence, the above formula for G_0 is equal to $fg(G_0)(0)$. Moreover, for $k \geq 1$, let $\mathcal{V}(n,k) = \{edge(v,a,v',k) \mid v,v' \in [n], a \in \Sigma\}$. Then for every mapping $f\colon \mathcal{V}(n,k) \to \text{BOOL}$, the graph induced by f is $graph(f) = ([n], E)$ with $E = \{(v,v') \mid f(edge(v,a,v',k)) = \text{TRUE}\}$.

The edges of the derived graphs depend not only on the initial graph but also on the rules and matches in the derivation steps. Let $r = (L \supseteq K \subseteq R)$ be a rule and let $g\colon V_L \to [n]$ be an injective function that maps the nodes of L to the nodes of G_{k-1}. The fact that g is a graph morphism from L to G_{k-1} is expressed by the formula

Fig. 1. A rule

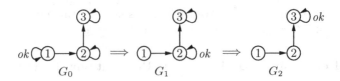

Fig. 2. Example of a derivation describable by a propositional formula

$$morph(r, g, k) = \bigwedge_{(v,a,v')\in E_L} edge(g(v), a, g(v'), k - 1),$$

which means that every edge in E_L must have an image in G_{k-1} through g. Concerning the first derivation step in Figure 2, there exist six injective mappings from V_L to the set $[3] = \{1, 2, 3\}$ but the formula $morph(r, g, 1)$ only holds if $g(i) = i$ for $i = 1, 2$. In this case we get $morph(r, g, 1) = edge(1, ok, 1, 0) \wedge edge(1, *, 2, 0) \wedge edge(2, *, 2, 0)$.

The application of r to G_{k-1} removes the image of every edge of the left-hand side L from G_{k-1} provided that it is not re-inserted via the right-hand side R. This is expressed by the propositional formula

$$rem(r, g, k) = \bigwedge_{(v,a,v')\in E_L - E_R} \neg edge(g(v), a, g(v'), k).$$

With respect to our example derivation we get

$$rem(r, g, 1) = \neg edge(1, ok, 1, 1) \wedge \neg edge(2, *, 2, 1)$$

which evaluates to TRUE.

Afterwards, the edges of the right-hand side R are added, which corresponds to the formula

$$add(r, g, k) = \bigwedge_{(v,a,v')\in E_R} edge(g(v), a, g(v'), k).$$

For our example derivation we get $add(r, g, 1) = edge(1, *, 2, 1) \wedge edge(2, ok, 2, 1)$.

The number of literals in each of the formulas $morph(r, g, k)$, $rem(r, g, k)$, and $add(r, g, k)$ is bounded by $\max^2 \cdot |\Sigma|$, i.e.,

$$L^\#(Q) \in O(1), \text{ for } Q \in \{morph(r, g, k), rem(r, g, k), add(r, g, k)\}.$$

All edges of G_{k-1} that are neither deleted nor inserted must be kept in the derivation step k. This leads to the formula

$$keep(r, g, k) =$$
$$\bigwedge_{(v,a,v')\in([n]\times\Sigma\times[n]) - g(E_L\cup E_R)} \left(edge(v, a, v', k) \leftrightarrow edge(v, a, v', k - 1) \right),$$

where $g(E_L \cup E_R) = \{(g(v), a, g(v')) \mid (v, a, v') \in E_L \cup E_R\}$. The number of literals in $keep(r, g, k)$ is bounded by $2n^2 \cdot |\Sigma|$, i.e., $L^\# \in O(n^2)$. With respect to the example we get

$$\begin{aligned}
keep(r, g, 1) = &\ (edge(1, *, 1, 1) \leftrightarrow edge(1, *, 1, 0)) \wedge \\
&\ (edge(1, ok, 2, 1) \leftrightarrow edge(1, ok, 2, 0)) \wedge \\
&\ \bigwedge_{x\in\{*,ok\}} ((edge(1, x, 3, 1) \leftrightarrow edge(1, x, 3, 0)) \wedge \\
&\ (edge(2, x, 1, 1) \leftrightarrow edge(2, x, 1, 0)) \wedge \\
&\ (edge(2, x, 3, 1) \leftrightarrow edge(2, x, 3, 0)) \wedge \\
&\ (edge(3, x, 1, 1) \leftrightarrow edge(3, x, 1, 0)) \wedge \\
&\ (edge(3, x, 2, 1) \leftrightarrow edge(3, x, 2, 0)) \wedge \\
&\ (edge(3, x, 3, 1) \leftrightarrow edge(3, x, 3, 0))).
\end{aligned}$$

Summarizing, the fact that g is a graph morphism from L to G_{k-1} such that G_k is obtained from G_{k-1} via the application of r with respect to g is expressed by the conjunction of the formulas *morph*, *rem*, *add*, and *keep*:

$$apply(r, g, k) = morph(r, g, k) \land rem(r, g, k) \land add(r, g, k) \land keep(r, g, k).$$

The following lemma concerning the semantics and the number of literals of the formula *apply* can be proved in a straightforward way by using the introduced definitions.

Lemma 1.   1. Let $f \colon \mathcal{V}(n, k-1) \cup \mathcal{V}(n, k) \to$ BOOL. Then

$$graph(f|\mathcal{V}(n, k-1)) \underset{r,g}{\Longrightarrow} graph(f|\mathcal{V}(n, k))$$

if and only if $f(apply(r, g, k)) = $ TRUE.[1]
2. $L^\#(apply(r, g, k)) \in O(n^2)$.

In each step k of a derivation, the formula $apply(r, g, k)$ must hold for at least one rule r of the underlying rule set P and for at least one injective mapping $g \colon V_r \to [n]$. Let $M(r, n)$ be the set of injective mappings from V_r to $[n]$. Then a kth derivation step is described by the formula

$$step(k) = \bigvee_{r \in P, g \in M(r,n)} apply(r, g, k),$$

which has at most $|P| \cdot n^{\max} \cdot (3\max^2 \cdot |\Sigma| + 2n^2 \cdot |\Sigma|)$ literals, i.e., $L^\#(step(k)) \in O(n^{2+\max})$.

The following formula $fder(G_0, m)$ describes all derivations of length m that start in G_0, for each graph G_0 and each natural number m.

$$fder(G_0, m) = fg(G_0)(0) \land \bigwedge_{k=1}^{m} step(k),$$

where for $m = 0$ the formula $\bigwedge_{k=1,\dots,0} step(k)$ is the empty formula which evaluates to TRUE. The next theorem follows from Lemma 1 and the definitions of *step* and *fder*. It states that each instantiation of the formula $fder(G_0, m)$ represents a derivation from G_0 of length m. In particular, all graphs of this derivation can be explicitly constructed from the corresponding variable instantiation. The number of literals in $fder(G_0, m)$ are bounded by a polynomial.

Theorem 1. Let $G_0 = ([n], E)$ be a graph, let $m \in \mathbb{N}$, and let $f \colon \mathcal{V}_{fder(G_0,m)} \to$ BOOL. Then

1. $G_0 \underset{P}{\overset{m}{\Longrightarrow}} G$ if and only if $f(fder(G_0, m)) = $ TRUE;
2. $L^\#(fder(G_0, m)) \in O(n^{2+\max} \cdot m)$.

---

[1] For $A \subseteq B$ and a function $f \colon B \to C$, $f|A \colon A \to C$ is defined by $f|A(x) = f(x)$, for each $x \in A$.

Remarks

1. The formula $fder(G_0, m)$ can be used in the following more general formulas:
   (a) Let p be a polynomial. Then the set of all derivations of length at most $p(n)$ that start in G_0 is specified via the formula $fder(G_0)$ as follows: [2]

   $$fder(G_0) = \bigvee_{m=0}^{p(n)} fder(G_0, m).$$

   For the number of literals of $fder(G_0)$ we get

   $$L^{\#}(fder(G_0)) \in O(n^{\max+2} \cdot p(n)^2).$$

   The quadratic factor of the polynomial $p(n)$ in the number of literals can be reduced to a linear one. One may add the empty rule $mt = (\emptyset \supseteq \emptyset \subseteq \emptyset)$ to the set of rules which is always applicable with the empty match, but does never cause any change. Then the set of all polynomial derivations starting from G_0 is described by $fder(G_0, p(n))$ because each derivation of length $m < p(n)$ may be prolonged by empty steps to the length $p(n)$. The other way round, a successful derivation of length $p(n)$ with empty steps is also successful if the empty steps are removed.
   (b) Even more generally, the set of all polynomial derivations over the rule set P starting from an arbitrary graph with n nodes is described by the formula

   $$fder(n) = \left(\bigwedge_{e \in [n] \times \Sigma \times [n]} (e \vee \neg e) \right)(0) \wedge \left(\bigvee_{m=0}^{p(n)} \bigwedge_{k=1}^{m} step(k) \right).$$

   The number of literals of $fder(n)$ is also in $O(n^{2+\max} \cdot p(n)^2)$. Analogously to point (a), $p(n)^2$ can be replaced by $p(n)$ by adding the empty rule.
2. As mentioned above, the assumption that the length bound of derivations is a polynomial implies that the number of literals of the associated formulas is also polynomial and hence tractable. But the presented constructions work for every bounding function.

4   From Polynomial Transformation Units to Propositional Formulas

In this section, polynomial transformation units are translated into propositional formulas so that every valid instantiation of such a formula represents a successful derivation of the unit. More explicitly, we do not consider stand-alone sets of rules but also initial and terminal graph class expressions as well as control

---

[2] For not having to introduce too many different names for very similar functions, we overload the function name *fder* by making its semantics dependent of its parameter types.

conditions. Given a set of rules, polynomial graph class expressions for the initial and terminal graphs, as well as a polynomial control condition, the resulting propositional formula describes all polynomial derivations that start with an initial graph, end with a terminal graph, and are allowed by the control condition.

4.1   Propositional Graph Class Expressions

As graph class expressions we use the propositional graph formulas introduced in Subsection 3.1. More precisely, every graph class expression is a mapping e that associates with each $n \in \mathbb{N}$ a propositional graph formula $e(n)$ of n. The graph class expression e is called polynomial if for each $n \in \mathbb{N}$, the number of literals in $e(n)$ is bounded by $p(n)$ for some given polynomial p. The semantics of each such graph class expression e is the set $SEM(e) = \bigcup_{n \in \mathbb{N}} Graphs(e(n))$.

Examples of propositional graph class expressions

1. The set of all graphs in $\mathcal{G}_\Sigma$ is specified by the propositional graph class expression *all*, where for each $n \in \mathbb{N}$,

$$all(n) = \bigwedge_{e \in [n] \times \Sigma \times [n]} (e \vee \neg e).$$

2. The class of unlabeled graphs in $\mathcal{G}_\Sigma$ can be specified by the graph class expression *unlabeled*, where for each $n \in \mathbb{N}$,

$$unlabeled(n) = \bigwedge_{v,v' \in [n]} \Big((v, *, v') \vee \neg(v, *, v') \Big) \wedge \bigwedge_{e \in [n] \times (\Sigma - \{*\}) \times [n]} \neg e.$$

3. More generally, the set of all graphs in $\mathcal{G}_\Sigma$ that are labeled over a set $\Delta \subseteq \Sigma$ is specified by Δ with

$$\Delta(n) = \bigwedge_{e \in [n] \times \Delta \times [n]} (e \vee \neg e) \wedge \bigwedge_{e \in [n] \times (\Sigma - \{\Delta\}) \times [n]} \neg e.$$

4. For $\Sigma = \{*\}$, the set of graphs in which every node has a loop is specified by *loop* with

$$loop(n) = \bigwedge_{v \in [n]} (v, *, v) \wedge \bigwedge_{v,v' \in [n]} \Big((v, *, v') \vee \neg(v, *, v') \Big).$$

All these examples of graph class expressions are polynomial because their numbers of literals are bounded by $2n^2 \cdot |\Sigma|$.

4.2   Control Conditions of the Language Type

As control conditions we use language type conditions as introduced in Section 2. Every such control condition C specifies a language $L(C) \subseteq P^*$ where P is the rule set of the transformation unit in which C occurs. A particular case of a language type condition is *free* that does not restrict anything because it specifies the language P^*. A control condition of the language type is called polynomial if there is a polynomial p such that for every $m \in \mathbb{N}$, $|L(C)(m)| \leq p(m)$, i.e., the number of words specified by C that are shorter than or as long as m does not exceed $p(m)$.

Example of a class of polynomial control conditions. Consider the following subset $REG(P)$ of regular expressions over P.

  - $\emptyset, \lambda \in REG(P)$ with $L(\emptyset) = \emptyset$ and $L(\lambda) = \{\lambda\}$;
  - for each $r \in P$, $r, r^* \in REG(P)$ with $L(r) = \{r\}$ and $L(r^*) = \{r\}^*$;
  - for each $C_1, C_2 \in REG(P)$, $C_1; C_2 \in REG(P)$ with $L(C_1; C_2) = L(C_1) \cdot L(C_2)$.

For example, the regular expression $r_1^*; r_2$ specifies the language $\{r_1^m r_2 \mid m \in \mathbb{N}\}$. It is polynomial because $|L(r_1^*; r_2)(m)| = m$ for each $m \in \mathbb{N}$.

The following proposition states that all control conditions of the presented subclass of regular expressions are polynomial. More precisely, for every regular expression C in $REG(P)$ the number of words in $L(C)$ of length at most m is in $O(m^{size(C)})$ where $size(C)$ is the number of occurrences of r and r^* in C (for all $r \in P$).

Proposition 1. Let $C \in REG(P)$. Then $|L(C)(m)| \in O(m^{size(C)})$ for each $m \in \mathbb{N}$, where $size(\emptyset) = size(\lambda) = 0$, $size(r) = size(r^*) = 1$, and $size(C_1; C_2) = size(C_1) + size(C_2)$.

Proof. Obviously, for the conditions $\emptyset$, λ, and r, the number of specified words is bounded by 1, i.e., for each $m \in \mathbb{N}$, $|L(C)(m)| \in O(m^0)$ with $C \in \{\emptyset, \lambda, r\}$. Since $O(m^0) \subseteq O(m^1)$ we get that $|L(C)(m)| \in O(m^{size(C)})$. Moreover, by induction on m we get that $|L(r^*)(m)| = m + 1$, i.e., $|L(r^*)(m)| \in O(m)$. Finally, consider the condition $C_1; C_2$. Then by definition $L(C_1; C_2) = \{w_1 w_2 \mid w_i \in L(C_i), i = 1, 2\}$ which implies that the number of words in $L(C_1; C_2)(m)$ is bounded by $|L(C_1)(m)| \cdot |L(C_2)(m)|$. By induction hypothesis $|L(C_i)(m)| \in O(m^{size(C_i)})$ which implies that $|L(C_1; C_s)(m)| \in O(m^{size(C_1)+size(C_2)})$. Since $size(C_1) + size(C_2) = size(C_1; C_2)$, the condition $C_1; C_2$ satisfies the required property.

4.3 Representing the Semantics of Polynomial Transformation Units by Propositional Formulas

Let $tu = (I, P, C, T)$ be a transformation unit such that I and T are polynomial propositional graph class expressions, and $C = free$. Then for all $m, n \in \mathbb{N}$, the set of successful derivations of length m starting from a graph of size n can be described by the mapping *ftufree* defined as follows.

$$ftufree(n, m) = I(n)(0) \wedge \bigwedge_{k=1}^{m} step(k) \wedge T(n)(m).$$

This can be shown by using Theorem 1.

If the control condition C of tu is polynomial, then all successful derivations of length m are described by the formula

$$ftu(n, m) = I(n)(0) \wedge \left(\bigvee_{r_1 \ldots r_m \in L(C)} \left(\bigwedge_{k=1}^{m} step(k, r_k) \wedge T(n)(m) \right) \right),$$

where for each $r \in P$, $step(k, r) = \bigvee_{g \in M(r,n)} apply(r, g, k)$, i.e., $step(k, r)$ requires the application of rule r in step k..

The next theorem states that for every valid instantiation of the formula $ftu(n, m)$, there exists a successful derivation of length m in tu. Moreover, the number of literals in $ftu(n, m)$ is polynomial.

Theorem 2. Let $tu = (I, P, C, T)$ be a transformation unit with $I(n), T(n) \in O(n^j)$ and $|L(C)(m)| \in O(m^l)$ for $j, l \in \mathbb{N}$.

1. There is a derivation $G_0 \underset{r_1, g_1}{\Longrightarrow} G_1 \underset{r_2, g_2}{\Longrightarrow} \ldots \underset{r_m, g_m}{\Longrightarrow} G_m$ with $(G_0, G_m) \in SEM(I) \times SEM(T)$, $r_1 \cdots r_m \in L(C)$ if and only if there is an instantiation $\mathcal{V}_{ftu(n,m)} \to$ BOOL with $f(ftu(n, m)) = $ TRUE where $n = size(G_0)$.
2. The number of literals of $ftu(n, m)$ is in $O(m^{l+1} \cdot n^i)$ where $i = \max\{2 + \max, j\}$.

Remarks. Let p be a polynom.

1. If the control condition is *free*, the set of all successful polynomial derivations of tu can be described by $ftufree(n) = \bigvee_{m=0}^{p(n)} ftufree(n, m)$.
2. If the control condition is polynomial, all polynomial successful derivations can be described by generalizing the formula $ftu(n, m)$ analogously to the generalization of *ftufree* in point 1.

4.4   Example

An example for a polynomial (nondeterministic) unit that finds Hamiltonian paths is depicted in Figure 3. The underlying alphabet Σ is equal to $\{*, ok\}$. The class of initial graphs consists of all unlabeled simple graphs in which every node has a loop. These graphs are specified via the polynomial graph class expression *unlabeled&loops* with

$$unlabeled\&loops(n) = \bigwedge_{v \in [n]} (v, *, v) \wedge unlabeled(n),$$

where *unlabeled(n)* is the graph class expression introduced in the examples of Subsection 4.1. The first rule *start* labels an unlabeled loop with *ok*. The second rule *run* converts an unlabeled neighbor of an *ok*-node v into an *ok*-node while removing the *ok*-loop of v. (Please note that we call a node with a *-loop an unlabeled node and a node with an *ok*-loop an *ok*-node.) The third rule *stop* removes an *ok*-loop.

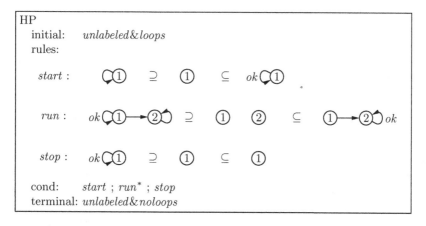

Fig. 3. A transformation unit for finding Hamiltonian paths

The control condition is the regular expression $start$; run^* ; $stop$ which guarantees that at first the rule $start$ is applied exactly once; afterwards, run is applied arbitrarily often; and finally, $stop$ is applied exactly once. Hence, the rule $start$ selects an arbitrary node v of the input graph G_0, the rule run traverses a path in G_0 starting from v, and the rule $stop$ terminates this process. The terminal graphs must not contain any loop which means that the rules must visit all nodes, i.e., the traversed path is Hamiltonian. The terminal graph class expression $unlabeled\&noloops$ is defined by

$$unlabeled\&noloops(n) = \bigwedge_{v \in [n]} \neg(v, *, v) \wedge unlabeled(n).$$

Successful derivations of HP can be found by feeding a SAT solver with the corresponding propositional formula.

5   Prototypical Implementation

The proposed methodology has been prototypically implemented for the Hamilton path problem. To this end, concrete unlabeled graphs (taken from [3]) have been considered. Having concrete instances available, the formulation as described in the former sections can be significantly simplified. In particular, the formulas for initial graphs as introduced in Subsection 3.2 can be removed and instead implicitly handled. Analogously, the formulas $morph$, rem, add, $keep$, and $step$ needed to formulate the respective rule applications can be significantly simplified. Thus, in the following a simpler formulation derived from these constraints is used. Nevertheless, the same principles as described above are thereby applied.

More precisely, the SAT instance encoding the Hamilton path problem for a given graph instance $G_0 = ([n], E)$ only includes

- for each node $v \in [n]$ and for each $k \in \{0, \ldots, n\}$ the variables $edge(v, ok, v, k)$ stating whether the node v has an ok-loop in G_k and

– for each edge $(v, *, v') \in E$ of the given graph instance the variables $run_{vv'}$ stating whether the *run*-rule (see Subsection 4.4) was applied for this edge.

Then, the application of the *run*-rule between two nodes $v, v' \in V$ with $(v, *, v') \in E$ is constrained by

$$run_{vv'} \leftrightarrow \bigvee_{k=1}^{n} (edge(v, ok, v, k - 1) \wedge edge(v', ok, v', k)).$$

That is, the *run*-rule can be applied between two nodes v, v' if and only if there is a node v with an *ok*-loop in G_{k-1} and a node v' with an *ok*-loop in G_k. By adding these constraints only for the edges from the given graph instance, all remaining $edge(v, a, v', k)$-variables as well as the respective *keep*-constraints are implicitly given and, thus, do not have to be added. Additionally, it is constrained that each node v is allowed to have an *ok*-loop in only one step, i.e., for each node $v \in [n]$ of the given graph instance $\sum_{k=1}^{n} edge(v, ok, v, k) = 1$ must hold which yields TRUE if and only if there is exactly one $k \in [n]$ with $edge(v, ok, v, k) = $ TRUE.[3] Finally, it must be ensured that the *run*-rule is applied at most once to a node v which can be expressed by $\sum_{(v,*,v')\in E} run_{vv'} \leq 1$. This implicitly represents the *start*-rule, the *stop*-rule, as well as the initial and terminal conditions introduced in Subsection 4.4. If the SAT solver determines a satisfying assignment to all these variables, a Hamilton path can be derived from the assignments to $run_{vv'}$. Otherwise, it has been proven that no such path exists for the given graph instance. Since current state-of-the-art solvers usually work on Boolean formulas in Conjunctive Normal Form (i.e. conjunctions of clauses), the constraints introduced above have to be converted, respectively. However, this can be done in linear time and space [16].

Table 1 gives the results obtained by the proposed formulation. The first columns list applied graph instances with their numbers of nodes as well as their numbers of edges and indicate whether the considered graphs have a Hamilton

Table 1. Results obtained by the prototypical implementation

Samples (cf. [3])	Nodes	Edges	HP?	SAT-Vars	SAT-Clauses	Time (s)
anna	138	1972	no	335185	1235420	22,1
david	87	1624	no	168868	629735	5,4
games120	120	2552	no	357097	1344256	12,6
miles250	128	1548	no	251947	917624	13,4
myciel3	11	40	yes	924	2751	0,1
myciel4	23	142	yes	5280	17535	134,5
queens5_5	25	640	no	19796	73095	0,2
queens6_6	36	1160	no	49129	184816	0,8
queens10_10	100	5880	no	635641	2467440	16,3

---

[3] Such kind of constraints are known as *cardinality constraints* for which a couple of efficient SAT formulations exist [14].

path or not. The number of Boolean variables and the number of clauses of the resulting SAT formulation are given in the next two columns. Finally, the needed run-time (in CPU seconds) is given in the last column. MiniSAT [6] on an Intel 2.4 GHz with 1.4 GB has been utilized to obtain the results.

The results confirm that for established benchmark functions, practical relevant problems (in this case determination of Hamilton paths) can be efficiently solved using the proposed approach. In fact, for all considered graphs the considered problem can be solved in very short run-time.

6   Conclusion

In this paper, we have described derivations of a special type of graph transformation units by propositional formulas in such a way that a SAT solver can be employed to find semantically significant derivations or to prove that none of them exists. This result provides the chance to overcome the inefficiency caused by the nondeterminism of rule applications in many cases.

Clearly, the existing SAT solvers are exponential in the worst case, but they often run very fast and yield good results in the verification of chip designs, etc. Our very first tests on a sample graph transformation unit that solves the Hamiltonian-path problem look quite promising. This encourages us to undertake further steps in this direction including the following:

1. As mentioned at the end of Section 3, the description of polynomial derivations by propositional formulas can be improved by the introduction of empty steps reducing the sizes of the formulas. An interesting question is whether there are further optimizations or improvements that yield higher efficiency or more insight into the derivation process.
2. There are some first ideas how to get rid of the restriction that the used rules keep the set of nodes invariant. This would enlarge the class of polynomial graph transformation units to which SAT solvers can be applied tremendously.
3. SAT solvers are mainly used to prove the correctness of some given specification. We would like to adopt the proof principle to the situation of graph transformation units to come up with a new verification technique for the framework of graph transformation.

References

1. Biere, A., Cimatti, A., Clarke, E., Fujita, M., Zhu, Y.: Symbolic model checking using SAT procedures instead of BDDs. In: Design Automation Conf., pp. 317–320 (1999)
2. Biere, A., Heule, M., van Maaren, H., Walsh, T. (eds.): Handbook of Satisfiability, Frontiers in Artificial Intelligence and Applications, vol. 185. IOS Press, Amsterdam (2009)
3. Carnegie Mellon University, Graph Coloring Instances, http://mat.gsia.cmu.edu/COLOR/instances.html

4. Cook, S.A.: The complexity of theorem-proving procedures. In: Proc. Third ACM Symposium on Theory of Computing, pp. 151–158 (1971)
5. Corradini, A., Ehrig, H., Heckel, R., Löwe, M., Montanari, U., Rossi, F.: Algebraic approaches to graph transformation part I: Basic concepts and double pushout approach. In: Rozenberg, G. (ed.) Handbook of Graph Grammars and Computing by Graph Transformation. Foundations, vol. 1, pp. 163–245. World Scientific, Singapore (1997)
6. Eén, N., Sörensson, N.: An extensible SAT solver. In: Giunchiglia, E., Tacchella, A. (eds.) SAT 2003. LNCS, vol. 2919, pp. 502–518. Springer, Heidelberg (2004)
7. Ehrig, H., Ehrig, K., Prange, U., Taentzer, G. (eds.): Fundamentals of Algebraic Graph Transformation. Springer, Heidelberg (2006)
8. Ganai, M., Gupta, A.: SAT-Based Scalable Formal Verification Solutions. Series on Integrated Circuits and Systems. Springer, Heidelberg (2007)
9. Hölscher, K., Klempien-Hinrichs, R., Knirsch, P.: Undecidable control conditions in graph transformation units. Electronic Notes in Theoretical Computer Science 195, 95–111 (2008)
10. Kreowski, H.-J., Kuske, S.: Graph transformation units with interleaving semantics. Formal Aspects of Computing 11(6), 690–723 (1999)
11. Kreowski, H.-J., Kuske, S., Rozenberg, G.: Graph transformation units – an overview. In: Degano, P., De Nicola, R., Meseguer, J. (eds.) Concurrency, Graphs and Models. LNCS, vol. 5065, pp. 57–75. Springer, Heidelberg (2008)
12. Kuske, S.: More about control conditions for transformation units. In: Ehrig, H., Engels, G., Kreowski, H.-J., Rozenberg, G. (eds.) TAGT 1998. LNCS, vol. 1764, pp. 323–337. Springer, Heidelberg (2000)
13. Litovski, I., Métivier, Y., Sopena, É.: Graph relabelling systems and distributed algorithms. In: Ehrig, H., Kreowski, H.-J., Montanari, U., Rozenberg, G. (eds.) Handbook of Graph Grammars and Computing by Graph Transformation. Concurrency, Parallelism, and Distribution, vol. 3, pp. 1–56. World Scientific, Singapore (1999)
14. Marques-Silva, J., Lynce, I.: Towards robust CNF encodings of cardinality constraints. In: Bessière, C. (ed.) CP 2007. LNCS, vol. 4741, pp. 483–497. Springer, Heidelberg (2007)
15. Soeken, M., Wille, R., Kuhlmann, M., Gogolla, M., Drechsler, R.: Verifying UML/OCL models using Boolean satisfiability. In: Design, Automation and Test in Europe, pp. 1341–1344 (2010)
16. Tseitin, G.: On the complexity of derivation in propositional calculus. In: Studies in Constructive Mathematics and Mathematical Logic, Part 2, pp. 115–125 (1968); Reprinted in: Siekmann, J., Wrightson, G. (eds.): Automation of Reasoning, vol. 2, pp. 466–483. Springer, Berlin (1983)

Delaying Constraint Solving in Symbolic Graph Transformation

Fernando Orejas[1],[*] and Leen Lambers[2]

[1] Universitat Politècnica de Catalunya, Spain
orejas@lsi.upc.edu
[2] Hasso Plattner Institut, Universität Potsdam, Germany
Leen.Lambers@hpi.uni-potsdam.de

Abstract. Applying an attributed graph transformation rule to a given object graph always implies some kind of constraint solving. In many cases, the given constraints are almost trivial to solve. For instance, this is the case when a rule describes a transformation $G \Rightarrow H$, where the attributes of H are obtained by some simple computation from the attributes of G. However there are many other cases where the constraints to solve may be not so trivial and, moreover, may have several answers. This is the case, for instance, when the transformation process includes some kind of searching. In the current approaches to attributed graph transformation these constraints must be completely solved when defining the matching of the given transformation rule. This kind of *early binding* is well-known from other areas of Computer Science to be inadequate. For instance, the solution chosen for the constraints associated to a given transformation step may be not fully adequate, meaning that later, in the search for a better solution, we may need to backtrack this transformation step.

In this paper, based on our previous work on the use of *symbolic graphs* to deal with different aspects related with attributed graphs, including attributed graph transformation, we present a new approach that allows us to delay constraint solving when doing attributed graph transformation. In particular we show that the approach is sound and complete with respect to *standard* attributed graph transformation. A running example, where a graph transformation system describes some basic operations of a travel agency, shows the practical interest of the approach.

Keywords: Attributed graph transformation, symbolic graph transformation, lazy transformation.

1 Introduction

Attributed graphs and attributed graph transformation play a significant role in most applications of graph transformation. In practice, an attributed graph

* This work has been partially supported by the CICYT project (ref. TIN2007-66523) and by the AGAUR grant to the research group ALBCOM (ref. 00516).

H. Ehrig et al. (Eds.): ICGT 2010, LNCS 6372, pp. 43–58, 2010.

transformation rule is like a normal rule, but some nodes or edges are labelled by expressions over some given variables. Then, defining a match m of a rule to a given object graph, whose attributes are some concrete values, means finding the values that must be assigned to the variables occurring in the rule, so that the value of each expression associated to each node or edge e on the left-hand side of the rule coincides with the value of the corresponding attribute associated to $m(e)$ in the object graph. That is, defining a match of a rule means solving a set of constraints. In many cases, these constraints are trivial. For instance, when a rule describes a transformation $G \Rightarrow H$, where the attributes in H are obtained by some simple computation from the attributes in G, e.g. when the expressions used as attributes in the left-hand side of the rules are just variables, and more general expressions, defined over these variables, only occur in the right-hand side. However there are many other cases where the constraints to solve may be not so trivial and, moreover, may have several answers. For instance, when the transformation process includes some kind of searching (e.g. when the right-hand side of a rule involves a variable which does not occur explicitly on the left-hand side). In existing approaches to attributed graph transformation these constraints must be completely solved when defining the matching of the given transformation rule. Then, finding a match means choosing one specific solution. This kind of *early binding* is well-known from other areas of Computer Science to be inadequate. One problem is that the solution chosen for the constraints associated to a given transformation step may not be fully adequate, meaning that we may need to backtrack this transformation step. The approach taken in areas like Constraint Logic Programming [8], in which our approach is inspired, is to postpone solving the constraints as much as possible, checking meanwhile their satisfiability. Then, not only may we avoid some useless backtracking, but we have other advantages. On the one hand, checking satisfiability may be computationally simpler than solving a set of constraints, meaning that it may also be simpler to apply a transformation step. On the other hand, some constraints which may be difficult to solve at a given moment, may become simpler, even trivial, because of the interaction with constraints defined by later steps.

In [12], when studying the problem of defining graph constraints over attributed graphs, we saw that the existing approaches [11,1,7,3,14] were not fully adequate for our purposes. These approaches presented different kinds of technical difficulties together with a limited expressive power for defining conditions on the attributes. To avoid these problems we presented a new formal approach which (we believe) is conceptually simpler and more powerful than existing approaches. It is simple because, in our approach, attributed graphs are not defined as some kind of combination of a graph and an algebra, as in [11,1,7,3], nor do we have to establish a difference between transformation rules and rule schemata, as in [14]. At the same time, our approach is expressively more powerful, not only because we can define graph constraints with arbitrary conditions on the attributes as we aimed, but also because we can define transformation rules that cannot be defined in other approaches, as shown in [13]. Graphs in our approach are called *symbolic graphs*, because the attributes in the graph are represented symbolically by variables whose possible values are specified by a set of formulas.

In this paper, we continue the work in [13], where we studied symbolic graph transformation and we compared it with attributed graph transformation as presented in [4]. In particular, now we present a new approach to symbolic graph transformation that allows us to delay constraint solving when doing attributed graph transformation. In particular we show that the approach is sound and complete with respect to *standard* attributed graph transformation. Moreover, a running example, where a graph transformation system describes some basic operations of a travel agency, shows the practical interest of the approach.

The paper is organized as follows. In Section 2 we provide a reminder of some notions that are used in the rest of the paper. In Section 3 we present the category of symbolic graphs. Section 4 is dedicated to describe how (standard) symbolic graph transformation works, and Section 5 to present the new notion of lazy transformation and to prove its soundness and completeness. In Section 6, we compare our approach with other related work and we draw some conclusions.

2 Preliminaries

We assume that the reader has a basic knowledge on algebraic specification and on graph transformation. For instance, we advise to look at [5] for more detail on algebraic specification or at [4] for more detail on graph transformation.

2.1 Basic Algebraic Concepts and Notation

A signature $\Sigma = (S, \Omega)$ consists of a set of sorts S, and Ω is a family of operation and predicate symbols typed over these sorts. A Σ-algebra A consists of an S-indexed family of sets $\{A_s\}_{s \in S}$ and a function op_A (resp. a relation pr_A) for each operation op (resp. each predicate pr) in the signature. A Σ-homomorphism $h : A \to A'$ consists of an S-indexed family of functions $\{h_s : A_s \to A'_s\}_{s \in S}$ commuting with the operations.

Given a signature Σ, we denote by T_Σ the term algebra, consisting of all the possible Σ-(ground) terms. Given any Σ-algebra A there is a unique homomorphism $h_A : T_\Sigma \to A$. In particular, h_A yields the value of each term in A. Similarly, $T_\Sigma(X)$ denotes the algebra of all Σ-terms with variables in X, and given a variable assignment $\sigma : X \to A$, this assignment extends to a unique homomorphism $\sigma^\# : T_\Sigma(X) \to A$ yielding the value of each term after the replacement of each variable x by its value $\sigma(x)$. In particular, when an assignment is defined over the term algebra, i.e. $\sigma : X \to T_\Sigma$, then $\sigma^\#(t)$ denotes the term obtained by substituting each variable x in t by the term $\sigma(x)$. However, for simplicity, even if it is an abuse of notation, we will write $\sigma(t)$ instead of $\sigma^\#(t)$.

2.2   E-Graphs

E-graphs are introduced in [4] as a first step to define attributed graphs. An E-graph is a kind of labelled graph, where nodes and edges may be decorated with labels from a given set E. The difference with labelled graphs, as commonly understood, is that in labelled graphs it is usually assumed that each node or edge is labelled with a given number of labels, which is fixed a priori. In the case of E-graphs, each node or edge may have any arbitrary (finite) number of labels, which is not fixed a priori. Actually, in the context of graph transformation, the application of a rule may change the number of labels of a node or of an edge.

Formally, in E-graphs labels are considered as a special class of nodes and the labeling relation between a node or an edge and a given label is represented by a special kind of edge. For instance, this means that the labeling of an edge is represented by an edge whose source is an edge and whose target is a label.

Definition 1 (E-Graphs and morphisms). *An* E-graph *over the set of labels* X_G *is a tuple* $G = (V_G, X_G, E_G, E_{NL}, E_{EL}, \{s_j, t_j\}_{j \in \{G,NL,EL\}})$ *consisting of:*

- V_G *and* X_G, *the sets of* graph nodes *and of* label nodes, *respectively.*
- E_G, E_{NL}, *and* E_{EL}, *the sets of* graph edges, node label edges, *and* edge label edges, *respectively.*

and the source and target functions:

- $s_G : E_G \to V_G$ *and* $t_G : E_G \to V_G$
- $s_{NL} : E_{NL} \to V_G$ *and* $t_{NL} : E_{NL} \to X_G$
- $s_{EL} : E_{EL} \to E_G$ *and* $t_{EL} : E_{EL} \to X_G$

Given the E-graphs G and G', an E-graph morphism $f : G \to G'$ *is a tuple,* $\langle f_V : V_G \to V'_G, f_X : X_G \to X_{G'}, f_E : E_G \to E'_G, f_{E_{NL}} : E_{NL} \to E'_{NL}, f_{E_{EL}} : E_{EL} \to E'_{EL} \rangle$ *such that f commutes with all the source and target functions.*

E-graphs and E-graph morphisms form the category **E – Graphs.**

The following constructions on E-graphs are needed in the sections below. The first one tells us how to replace the labels of an E-graph:

Definition 2 (Label substitution). *Given a set of labels X', an E-graph $G = (V_G, X_G, E_G, E_{NL}, E_{EL}, \{s_j, t_j\}_{j \in \{G,NL,EL\}})$, and a function $h : X_G \to X'$ we define the graph resulting from the substitution of X_G along h, denoted $h(G)$, the E-graph $h(G) = (V'_G, X', E'_G, E'_{NL}, E'_{EL}, \{s'_j, t'_j\}_{j \in \{G,NL,EL\}})$ defined:*

- $V'_G = V_G, E'_G = E_G, E'_{NL} = E_{NL}, E'_{EL} = E_{EL}, \{s'_j = s_j\}_{j \in \{G,NL,EL\}}$, *and* $t'_G = t_G$
- *For every $e \in E'_{NL} : t'_{NL}(e) = h(t_{NL}(e))$*
- *For every $e \in E'_{EL} : t'_{EL}(e) = h(t_{EL}(e))$*

Moreover h induces the definition of the morphism $h^* : G \to G'$, with $h^* = \langle id_V, h, id_{E_G}, id_{E_{NL}}, id_{E_{EL}} \rangle$.

Reach(G) eliminates all the labels in an E-graph which are not bound to a node or an edge and *Restr(G, Y)* eliminates all the labels which are not in Y (we assume that they are not reachable):

Definition 3 (Reachable subgraph and restriction subgraph). *Given an E-graph $G = (V_G, X_G, E_G, E_{NL}, E_{EL}, \{s_j, t_j\}_{j \in \{G, NL, EL\}})$, we define the reachable subgraph of G, Reach(G) as the largest subgraphs of G where $X_{Reach(G)}$ consists of all labels x such that:*

- *There is a node label edge $nl \in E_{NL}$ such that $x = t_{NL}(nl)$, or*
- *there is an edge label edge $el \in E_{EL}$ such that $x = t_{EL}(el)$.*

Given a morphism $f : G \to G'$ we denote by $Reach(f) : Reach(G) \to G'$ the restriction of f to the subgraph $Reach(G)$.

Similarly, we define the restriction of G to a set of labels Y, Restr(G, Y), where $X_{Reach(G)} \subseteq Y$ as the largest subgraph of G where $X_{Restr(G,Y)} = Y$.

3   Symbolic Graphs

If we consider that an attributed graph is like an E-graph whose labels are values over the given data algebra, a symbolic graph can be seen as the specification of a class of attributed graphs. In particular, a symbolic graph consists of an E-graph G whose labels are variables, together with a set of formulas Φ that constrain the possible values of these variables. We consider that a symbolic graph denotes the class of all attributed graphs where the variables in the E-graph have been replaced for values that make Φ true in the given data domain. For instance, the symbolic graph below specifies a class of attributed graphs, including distances in the edges, that satisfy the well-known triangle inequality.

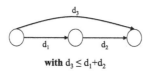

with $d_3 \leq d_1 + d_2$

Definition 4 (Symbolic graphs and morphisms). *A symbolic graph over the data Σ-algebra D, with $\Sigma = (S, \Omega)$, is a pair $\langle G, \Phi \rangle$, where G is an E-graph over an S-sorted set of variables $X = \{X_s\}_{s \in S}$, i.e. $X_G = \cup_{s \in S} X_s$, and Φ is a set of first-order Σ-formulas with free variables in X and including elements in D as constants.*

Given symbolic graphs $\langle G_1, \Phi_1 \rangle$ and $\langle G_2, \Phi_2 \rangle$ over D, a symbolic graph morphism $h : \langle G_1, \Phi_1 \rangle \to \langle G_2, \Phi_2 \rangle$ is an E-graph morphism $h : G_1 \to G_2$ such that $D \models \Phi_2 \Rightarrow h(\Phi_1)$, where $h(\Phi_1)$ is the set of formulas obtained when replacing in Φ_1 every variable x_1 in the set of labels of G_1 by $h_X(x_1)$.

Symbolic graphs over D together with their morphisms form the category **SymbGraphs$_D$**.

As said in the definition, we consider that Φ is a set of arbitrary first-order Σ-formulas. However, for practical purposes, we may want to restrict the class of formulas that can occur in this set, since first-order formula satisfiability is an undecidable problem. In principle, the only condition that we need to ensure that the results presented in this paper hold is that the given class of formulas is closed under conjunction, disjunction and existential quantification, because these are the connectives needed to ensure the existence of pushouts and pullbacks in the category of symbolic graphs.

Remark 1 (Symbolic typed graphs). Even if along the paper, for simplicity, we have only considered untyped graphs, in our running example we consider that our graphs are typed. We can easily extend all our theory to the case of typed graphs using the standard technique. In particular, we just have to see typed graphs as morphisms $SG \rightarrow TG$, where SG is a symbolic graph and TG is the given type graph being a specific symbolic graph (i.e. the category of typed symbolic graphs over the type graph TG is the slice category over TG). Moreover, in this case, type graphs TG must be symbolic graphs $TG = \langle ETG, \mathbf{False} \rangle$, where ETG is an E-graph over a set of labels X_{ETG} consisting of just one variable for each sort in the given data signature, and $\mathbf{False}$ is the set consisting just of the *false* formula. In this way, given a symbolic graph $SG = \langle G, \Phi \rangle$, any (typing) morphism $t : G \rightarrow ETG$ for G is also a (typing) morphism $t : SG \rightarrow TG$.

Definition 5 (Semantics of symbolic graphs). *The semantics of a symbolic graph $\langle G, \Phi \rangle$ over a data algebra D is a class of attributed graphs defined:*

$$Sem(\langle G, \Phi \rangle) = \{\langle \sigma(G), D \rangle \mid \sigma : X_G \rightarrow D \text{ and } D \models \sigma(\Phi)\}$$

where $\sigma(G)$ denotes the graph obtained according to Def. 2.

It may be noted that the class of attributed graphs denoted by a symbolic graph may be empty if the associated condition is unsatisfiable.

Every attributed graph may be seen as a symbolic graph by just replacing all its values by variables and by including an equation $x_v = v$, for each value v in the data algebra, into the corresponding set of formulas, where x_v is the variable that has replaced the value v. We call these kind of symbolic graphs *grounded symbolic graphs*. For instance, below on the right, we can see the symbolic representation of the attributed graph on the left.

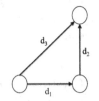

with $d_1 = 12$, $d_2 = 15$, $d_3 = 18$

In [13] we showed that the subcategories of attributed graphs and grounded symbolic graphs are equivalent, provided that the given data algebra is finitely generated. As a consequence, we may identify them. For instance, we may consider that, if $h : SG \to SG'$ is a symbolic graph morphism and SG' is grounded then we may say that, for a given variable x, $h(x) = v$, with $v \in D$, instead of saying that $h(x) = x_v$. In addition, in this case, the condition $D \models \Phi' \Rightarrow h(\Phi)$, that the symbolic morphism h must satisfy, can be simplified:

Fact 1. *Given a symbolic graph $SG = \langle G, \Phi \rangle$ and a grounded symbolic graph $SG' = \langle G', \Phi' \rangle$, an E-graph morphism $h : G \to G'$ is a symbolic graph morphism if and only if $D \models h'(\Phi)$, where $h' : X_G \to D$ is the mapping defined $\forall x \in X_G :$ $h'(x) = v$ if $h(x) = x_v$.*

The reason is quite obvious. If $X_G = \{x_1, \ldots x_n\}$ and $h(x_1) = x_{v_1}, \ldots, h(x_n) = x_{v_n}$, then $D \models \{x_{v_1} = v_1, \ldots, x_{v_n} = v_n\} \Rightarrow \Phi[x_1/x_{v_1}, \ldots, x_n/x_{v_n}]$ if and only if $D \models \Phi[x_1/v_1, \ldots, x_n/v_n]$, where $\Phi[x/a]$ denotes the set of formulas obtained replacing in Φ all occurrences of x by a. This means that defining a morphism from $SG = \langle G, \Phi \rangle$ into a grounded symbolic graph is (partially) a constraint satisfaction problem. In particular we need to find the values that must be assigned to the variables in Φ so that it is satisfied.

Sometimes, as we will see in Section 5, we want to work only with graphs where no variable is bound to two different elements of the graph (i.e. every variable occurs at most once in the graph). We call these graphs in *normal form*. It should be clear that every symbolic graph G is equivalent to some symbolic graph G' in normal form, in the sense that $Sem(G) = Sem(G')$. It is enough to replace each repeated occurrence of a variable x by a fresh variable y, and to include the equality $x = y$ in $\Phi_{G'}$.

In [13], we also showed that symbolic graphs are an adhesive HLR category [10,6] taking as M-morphisms all injective graph morphisms where the formulas constraining the source and target graphs are equivalent (in most cases they will just be the same formula).

Definition 6 (M-morphisms). *An M-morphism $h : \langle G, \Phi \rangle \to \langle G', \Phi' \rangle$ is a monomorphism such that $X_G \cong X_{G'}$, i.e. h_X is a bijection, and $D \models h(\Phi) \Leftrightarrow \Phi'$*

The intuition of this definition is based on the use of our category of symbolic graphs to define graph transformation. More precisely, we think that the most reasonable formulation of graph transformation rules in our context is based on defining a graph transformation rule as an E-graph transformation rule, together with a set of formulas that globally constrain and relate all the variables in the rule. This is equivalent to consider that the left and right-hand sides (and also the interface) of a rule are constrained by the same set of formulas.

We will not study in detail all the constructions and results that are needed to prove that **SymbGraphs$_D$** is an adhesive HLR category (the interested reader is addressed to [13]). However, it is important to see how pushouts work in order to understand graph transformation:

Proposition 1. *[13] Diagram (1) below is a pushout if and only if diagram (2) is also a pushout and $D \models \Phi_3 \Leftrightarrow (g_1(\Phi_1) \cup g_2(\Phi_2))$.*

$$
\begin{array}{ccc}
\langle G_0, \Phi_0 \rangle & \xrightarrow{\ h_1\ } & \langle G_1, \Phi_1 \rangle \\
h_2 \downarrow & (1) & \downarrow g_1 \\
\langle G_2, \Phi_2 \rangle & \xrightarrow[\ g_2\ ]{} & \langle G_3, \Phi_3 \rangle
\end{array}
\qquad
\begin{array}{ccc}
G_0 & \xrightarrow{\ h_1\ } & G_1 \\
h_2 \downarrow & (2) & \downarrow g_1 \\
G_2 & \xrightarrow[\ g_2\ ]{} & G_3
\end{array}
$$

Therefore, as said above, we have:

Theorem 1. *[13]* **SymbGraphs$_D$** *is adhesive HLR.*

Example 1 (Symbolic graphs). In our running example we specify a travel agency in terms of typed symbolic graphs and symbolic graph transformation rules. In particular, the data signature of our example is:

> **Sorts** $int, bool, city, hotel, flight$
> **Opns** $price : hotel \to int$
> $location : hotel \to city$
> $price : flight \to int$
> $departure : flight \to city$
> $destination : flight \to city$

and the data algebra, depicted by means of two tables, is:

hotel	price	city
h1	100	München
h2	90	Berlin
h3	85	Berlin
h4	70	Köln
h5	200	Berlin

flight	price	departure	destination
f1	175	Berlin	Barcelona
f2	115	Berlin	Barcelona
f3	150	Barcelona	Berlin
f4	85	Barcelona	Berlin

The type graph consists of a *Customer* that can request a hotel reservation *Hotel-Request* or a flight reservation *FlightRequest*. These requests can be responded by a *FlightReservation* or a *HotelReservation*, respectively. A *Customer* has the labels *name* and *bud*, describing the total budget that the customer is willing to spend for his reservations. A *HotelRequest* has as labels the number of *nights* and *loc*, describing the city where the hotel is located. A *FlightRequest* has as labels *dep* and *dest*, the departure and destination city of the flight, respectively. The *HotelReservation* and *FlightReservation* have only one label *h* and *f* for the hotel and flight that are reserved, respectively.

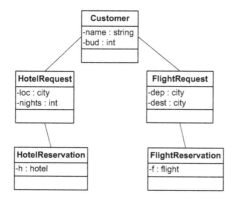

Finally, a possible start graph describing the needs of a customer may be a grounded graph (represented as an attributed graph, including directly the values in the nodes) describing a customer with a budget of 450, requesting a hotel for two nights in Berlin, a flight from Barcelona to Berlin and a flight from Berlin to Barcelona. We have not assigned a specific value to the name of the customer, because it is not relevant for the example:

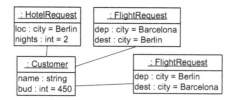

4   Symbolic Graph Transformation

As usual, symbolic graph transformation rules are spans of M-morphisms. This means, as explained in the previous section, that we may denote *symbolic graph transformation rules* as pairs $\langle L \hookleftarrow K \hookrightarrow R, \Phi \rangle$, where L, K and R are E-graphs over the same set of labels X and Φ is a set of formulas over X and over the values in D. Intuitively, Φ relates the attributes in the left and right-hand side of the rule and it may also impose some constraints on the matchings.

Definition 7 (Symbolic graph transformation rule). *A symbolic graph transformation rule is a pair $\langle L \hookleftarrow K \hookrightarrow R, \Phi \rangle$, where L, K, R are E-graphs over the same set of labels X, $L \hookleftarrow K \hookrightarrow R$ is a span of E-graph inclusions[1], and Φ is a set of formulas over X and over the values in the given data algebra D.*

Example 2 (Symbolic graph transformation rules). Below we can see the rules that describe some operations of our travel agency. The rule *reserveFlight* connects a *FlightReservation* with a *FlightRequest* for some *Customer*. Identical

---

[1] Or, in general, M-morphisms.

names in rules specify that the corresponding nodes are preserved. Edges between nodes with identical names are preserved as well. Its formula Φ, expresses
that the departure city *dep* and destination city *dest* of the *FlightRequest* should
be equal to the departure city *departure(f)* and destination city *destination(f)*
of the reserved flight *f*, respectively. Moreover the total budget *bud* of the customer is diminished by the price of the reserved flight *price(f)* and as an extra
constraint it is required that the new budget *bud'* is still bigger or equal to zero.
The rule *reserveHotel* connects a *HotelReservation* with a *HotelRequest* for some
Customer. It holds a formula Φ, expressing that the location *loc* of the *HotelRequest* should be identical to the location of the reserved hotel *location(h)*. The
total budget *bud* of the customer is diminished by the price of the reserved hotel
multiplied with the number of nights *price(h)*nights* and as an extra constraint
it is required again that the new budget *bud'* is still greater or equal to zero.

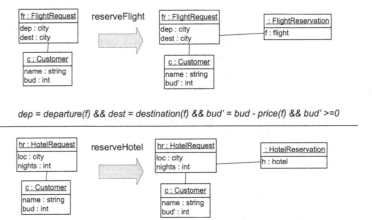

dep = departure(f) && dest = destination(f) && bud' = bud - price(f) && bud' >=0

loc = location(h) && bud' = bud − (price(h) * nights) && bud'>=0

As usual, the application of a graph transformation rule $\langle L \hookleftarrow K \hookrightarrow R, \Phi \rangle$ can
be defined by a double pushout in the category of symbolic graphs. However,
since M-morphisms are preserved by pushouts, the application of a symbolic
transformation rule can be defined in terms of a transformation of E-graphs.

Fact 2

1. *Given a symbolic graph transformation rule $\langle L \hookleftarrow K \hookrightarrow R, \Phi \rangle$ over a given
   data algebra D and an E-graph morphism $m : L \to G$, $\langle G, \Phi' \rangle \Longrightarrow_{p,m} \langle H, \Phi' \rangle$
   if and only if the diagram below is a double pushout in $\mathbf{E} - \mathbf{Graphs}$ and
   $D \models \Phi' \Rightarrow m(\Phi)$.*

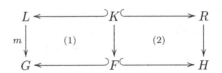

2. If $\langle G, \Phi' \rangle \Longrightarrow_{p,m} \langle H, \Phi' \rangle$ and $\langle G, \Phi' \rangle$ is grounded then $\langle H, \Phi' \rangle$ is grounded too, since $X_H = X_G$ and Φ' defines the values of all the variables in X_G.

According to Fact 1, if we apply the rule $\langle L \hookleftarrow K \hookrightarrow R, \Phi \rangle$ to a grounded symbolic graph $SG = \langle G, \Phi' \rangle$, the condition $D \models \Phi' \Rightarrow m(\Phi)$ can be seen as the constraint satisfaction problem where we find the values to assign to the variables in Φ, so that this set of formulas is satisfied in D.

Remark 2. The fact that the formula constraining the variables of the E-graph H after transformation is the same formula for the graph G before transformation does not mean that the values in the graphs do not change after transformation. The reason is that the transformation may change the bindings of variables to nodes and edges, as the example below shows.

Example 3. Below we can see how the rule reserveHotel presented in Example 2 is applied to a given start graph as presented in Example 1, describing a customer that would like to reserve a hotel for 2 days in Berlin and, in addition, would also like to reserve a flight from Barcelona to Berlin and return.

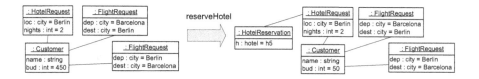

In this transformation, a hotel in Berlin is chosen non-deterministically. This choice is made when defining the matching between the left-hand side of the rule and the given graph. Even if it is not explicitly depicted, the left-hand side of ReserveHotel includes the variable h. Then, when defining the match morphism we must bind that variable to some hotel hi in D_{hotel} such that the formula

```
Berlin = location(hi) && bud'=450-(price(hi)*2) && bud'>=0
```

is satisfied. Here, the chosen hotel, $h5$, satisfies the above formula for $h = h5$ and $bud' = 50$. For this reason, the value of the attribute bud in the resulting graph is set to 50. This kind of early binding may not be very efficient. For instance, after the transformation defined above no flight reservation is possible, because the budget of the user would not be sufficient for reserving any flight connecting Barcelona and Berlin. Hence, we would need to backtrack this transformation and to try a different hotel. This situation also occurs when working with attributed graphs and attributed graph transformation rules as in [4].

5   Lazy Symbolic Graph Transformation

The problem shown in Example 3 is that, when defining a match for a rule $p = \langle L \hookleftarrow K \hookrightarrow R, \Phi_p \rangle$, this match must also be defined on variables that are bound to a node or edge in R but not bound to any node or edge in L (i.e. intuitively, these variables are not part of L). The basic idea of our approach is that, when applying a rule, we do not need to match these variables if we do not want to have early binding. This means that the match morphism for p is not defined on L, but on $Reach(L)$ (cf. Definition 3) and, then, when we apply the rule to a graph SG, we add to the result of the transformation the formulas in Φ_p, since these formulas define the possible values of the non-matched variables with respect to the values of the rest of the variables.

Definition 8 (Lazy graph transformation). *Given a symbolic graph $SG = \langle G, \Phi_G \rangle$, a transformation rule $p = \langle L \hookleftarrow K \hookrightarrow R, \Phi_p \rangle$, where L, K and R are in normal form and where the sets of labels in SG and in p are disjoint[2], and given a morphism $m : Reach(L) \to G$, we define the lazy transformation of SG by p via m, denoted $SG \Rightarrow_{p,m} SH$, as the symbolic graph $SH = \langle H, \Phi_H \rangle$, where H is obtained by the double pushout below in the category of E-graphs:*

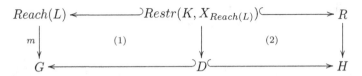

and where $\Phi_H = \Phi_G \cup m(\Phi_p)$.
    As usual, we write $SG \Rightarrow SH$ if the given rule and match can be left implicit, and we write $\overset{*}{\Rightarrow}$ to denote the reflexive and transitive closure of $\Rightarrow$.

Intuitively, H is similar to the E-graph obtained by a standard symbolic transformation, except that it would include all the variables in $X_R \setminus X_{Reach(L)}$ instead of their corresponding matchings.

In general, m is not defined on all the variables in Φ_p, but only on the variables in $Reach(L)$ and, therefore, we could consider that the definition of Φ_H is incorrect. However, we will consider that $m(\Phi_p)$ denotes the set of formulas obtained replacing in every formula $\alpha \in \Phi_p$ every variable x in $X_{Reach(L)}$ by $m(x)$, i.e. the variables which are not in $X_{Reach(L)}$ are not replaced.

It may also be noted that the resulting condition Φ_H may be unsatisfiable for the given data algebra D. This would mean that $Sem(SH)$ is empty. In general, we probably want to avoid this kind of situation, which means that *we assume that the lazy application of a rule preserves satisfiability.*

Fact 3. *If $SG \Rightarrow_{p,m} SH$, we can obtain SH from SG by applying a double pushout transformation in the category of symbolic graphs using the span $lazy(p) = \langle \langle Reach(L), Restr(\Phi_p, X_{Reach(L)}) \rangle, \hookleftarrow \langle K, Restr(\Phi_p, X_{Reach(L)}) \rangle \hookrightarrow \langle R, \Phi_p \rangle \rangle$ and the match m, i.e. $SG \Rightarrow_{lazy(p),m} SH$, where $Restr(\Phi, X)$ is the formula obtained quantifying existentially all free variables in Φ which are not in X, i.e. $\exists (freevar(\Phi) \setminus X) \Phi$. However, we may note that $lazy(p)$ is not a symbolic graph*

---

[2] If needed we can rename the labels apart and we can transform L, K and R into normal form, as described in Section 3.

transformation rule, since the embeddings involved in the span in general are not M-morphisms.

Example 4. For example, if we apply the rule ReserveHotel as in Example 3 without binding the variable h to some concrete value hi in D_{hotel}. Then we obtain a lazy graph transformation step as depicted in the first step of the lazy graph transformation in the figure in Example 5. After this first step the formula

```
Berlin = location(h) && bud=450-(price(h) * 2) && bud>=0
```

should be satisfied. Given the data algebra D as presented in Example 1, it is obvious that this formula is satisfiable.

Let us now study the relation between symbolic graph transformation and lazy transformation. The first result tells us that lazy transformation may be seen as a sound and complete implementation of symbolic graph transformation over grounded symbolic graphs (i.e. attributed graphs). More precisely, in the case of grounded symbolic graphs, for every transformation $SG \overset{*}{\Longrightarrow} SH$, if AH is the attributed graph denoted by SH then there is a lazy transformation $SG \overset{*}{\Rightarrow} SH'$ such that AH is in the semantics of SH', and conversely if there is a lazy transformation $SG \overset{*}{\Rightarrow} SH'$ then every attributed graph AH in $Sem(SH')$ can be obtained by a symbolic graph transformation from AG. This means that, semantically, it does not matter if, each time we apply a transformation rule, we solve the constraints associated to that application, or if we apply lazy transformation and, at the end, we solve the constraints of the resulting graph.

Theorem 2 (Soundness and completeness of lazy graph transformation). *Given a grounded symbolic graph $SG = \langle G, \Phi_G \rangle$, then:*

1. **Completeness.** *For every symbolic transformation $SG \overset{*}{\Longrightarrow} SH$ there is a lazy transformation $SG \overset{*}{\Rightarrow} SH'$ such that $Sem(SH) \subseteq Sem(SH')$.*

2. **Soundness.** *For every lazy transformation $SG \overset{*}{\Rightarrow} SH'$ and every attributed graph $AH \in Sem(SH')$ there is a symbolic transformation $SG \overset{*}{\Longrightarrow} SH$ such that $AH \in Sem(SH)$.*

If we consider the general case, where the transformations are applied to arbitrary symbolic graphs, lazy transformation is still complete in the above sense, but it may be considered, technically, not sound. In particular, it may be possible to apply a rule $p = \langle L \hookleftarrow K \hookrightarrow R, \Phi_p \rangle$ to a graph $SG = \langle G, \Phi_G \rangle$ to apply lazy transformation, but it may be impossible to apply the same rule to SG to do symbolic transformation. The reason is that for symbolic transformation we require that $D \models \Phi_G \Rightarrow m(\Phi_p)$, for a given match m, while in the case of lazy transformation it may be enough that $\Phi_G \cup m(\Phi_p)$ is satisfiable in D. This may happen, for instance, if the condition on G is $0 < X < 5$ and the condition on the rule is $3 < X < 7$. In that case, we would be unable to apply symbolic graph transformation, since $0 < X < 5$ does not imply $3 < X < 7$. However, it could be possible to apply lazy transformation and, then, the resulting condition would be $0 < X < 5 \wedge 3 < X < 7$, or equivalently $3 < X < 5$. Actually, we may consider that what happens in the case of lazy transformation is that, at each transformation step, we narrow the conditions of the given graphs, while in the case of symbolic transformations the conditions remain invariant. In this sense, we can state an extended soundness result for lazy transformation:

Theorem 3 (Extended soundness and completeness of lazy transformation). *Given a symbolic graph* $SG = \langle G, \Phi_G \rangle$, *then:*

1. **Completeness.** *For every symbolic transformation* $SG \overset{*}{\Longrightarrow} SH$ *there is a lazy transformation* $SG \overset{*}{\Rightarrow} SH'$ *such that* $Sem(SH) \subseteq Sem(SH')$.

2. **Extended soundness.** *For every lazy transformation* $SG \overset{*}{\Rightarrow} SH'$, *with* $SG = \langle G, \Phi_G \rangle$ *and* $SH' = \langle H', \Phi_{H'} \rangle$ *there is a symbolic transformation* $\langle G, \Phi_{H'} \rangle \overset{*}{\Longrightarrow} SH$ *such that* $SH \cong SH'$.

Example 5. Consider the lazy graph transformation as presented below, where a hotel is reserved for two nights in Berlin and a flight from Barcelona to Berlin and return is reserved. The labels h for the hotel, $fr1.f$ and $fr2.f$ for the flights are not bound to a concrete hotel or concrete flights yet. Note that we use the $*.*$ notation to distinguish labels having the same name in different nodes. At the end of this transformation a formula equivalent to:

```
Berlin = location(h) && Barcelona = departure(fr1.f) &&
Berlin = destination(fr1.f) && Berlin = departure(fr2.f) &&
Barcelona = destination(fr2.f) && bud = 450 - (price(h) * 2)
- price(fr1.f) - price(fr2.f) && bud >=0
```

should be satisfied[3]. Solving this formula for the data algebra D given in Example 1, we obtain the following solutions for the labels $(h, fr1.f, fr2.f, bud)$: $(h2, f4, f1, 10)$, $(h3, f4, f1, 20)$, $(h2, f3, f2, 5)$, $(h3, f3, f2, 15)$, $(h2, f4, f2, 70)$ and $(h3, f4, f2, 80)$. In particular, for the solution $(h2, f4, f1, 10)$ a corresponding symbolic transformation exists, where h, $fr1.f$ and $fr2.f$ are bound to $h2$, $f4$ and $f1$, respectively. Moreover, in the end the total budget bud of the customer is 10.

To solve the resulting constraints after a sequence of transformations we may need to use a specialized constraint solver. If that solver gives the possibility of defining some kind of optimality criteria (e.g. the overall price of the reservations) then it would return us the best solution according to these criteria. However, if the data algebra D can be considered to be implemented by a database, as it may happen in this example, then the resulting constraints after a sequence of transformations could be compiled into a query to that database, meaning that all the possible solutions would just be obtained as a result of that query.

---

[3] Actually,the formula that we would obtain is a bit more complex, including several (renamed) copies of the variable bud, since each application of a rule would add a new variable bud and a constraint defining it. For readability, we have preferred to use this simpler version. Anyhow, if we would allow to simplify the given formula after each transformation step, as it is sometimes done in Constraint Logic Programming, we could indeed obtain the above formula.

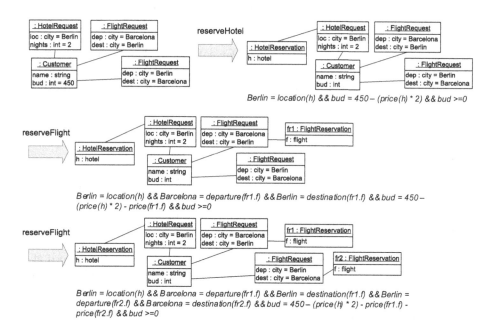

Berlin = location(h) && bud = 450 − (price(h) * 2) && bud >=0

Berlin = location(h) && Barcelona = departure(fr1.f) && Berlin = destination(fr1.f) && bud = 450 − (price(h) * 2) - price(fr1.f) && bud >=0

Berlin = location(h) && Barcelona = departure(fr1.f) && Berlin = destination(fr1.f) && Berlin = departure(fr2.f) && Barcelona = destination(fr2.f) && bud = 450 − (price(h) * 2) - price(fr1.f) - price(fr2.f) && bud >=0

6   Conclusion and Related Work

In this paper we have seen how we can delay constraint solving in the framework of attributed graph transformation by means of symbolic graphs. The idea of the technique comes from constraint logic programming [8]. Actually, the idea underlying symbolic graphs also comes from constraint logic programming, where the domain of values, and the conditions on them are neatly separated from symbolic clauses, in a similar way we separate in symbolic graphs, the graph part from the conditions defining the value part. Symbolic graph transformation can be seen as a generalization of attribute grammars [9]. In particular, attribute grammars can be considered to provide a method for computing attributes in trees. Moreover, these attributes must be computed in a fixed way: either top-down (inherited attributes) or bottom-up (synthesized attributes).

There are three kinds of approaches to define attributed graphs and attributed graph transformation. In [11,1] attributed graphs are seen as algebras. In particular, the graph part of an attributed graph is encoded as an algebra extending the given data algebra. In [7,3] an attributed graph is a pair (G, D) consisting of a graph G and a data algebra D whose values are nodes in G. In [2] these two approaches are compared showing that they are, up to a certain point, equivalent. Finally, [14] is based on the use of labelled graphs to represent attributed graphs, and of rule schemata to define graph transformations involving computations on the labels. That approach has some similarities with our approach, including the simplicity provided by the separation of the algebra and the graph part of attributed graphs. We do not think that there is a direct way of defining lazy graph transformations for those approaches, since there is no specific way of handling constraints explicitly.

There are further aspects related to symbolic graph transformations that deserve some future work. Addressing the implementation of these techniques could be an obvious goal. However, there are other fundamental aspects which may be interesting to study, like developing specific techniques for checking confluence for symbolic transformation or the definition of symbolic graph transformation with borrowing contexts, avoiding some existing technical difficulties.

References

1. Berthold, M., Fischer, I., Koch, M.: Attributed graph transformation with partial attribution. In: GRATRA 2000. Joint APPLIGRAPH and GETGRATS Workshop on Graph Transformation Systems, pp. 171–178 (2000)
2. Ehrig, H.: Attributed graphs and typing: Relationship between different representations. Bulletin of the EATCS 82, 175–190 (2004)
3. Ehrig, H., Ehrig, K., Prange, U., Taentzer, G.: Fundamental theory of typed attributed graph transformation based on adhesive HLR-categories. Fundamenta Informaticae 74(1), 31–61 (2006)
4. Ehrig, H., Ehrig, K., Prange, U., Taentzer, G.: Fundamentals of Algebraic Graph Transformation. In: EATCS Monographs of Theoretical Computer Science. Springer, Heidelberg (2006)
5. Ehrig, H., Mahr, B.: Fundamentals of Algebraic Specifications 1: Equations and Initial Semantics. In: EATCS Monographs of Theoretical Computer Science. Springer, Heidelberg (1985)
6. Ehrig, H., Padberg, J., Prange, U., Habel, A.: Adhesive high-level replacement systems: A new categorical framework for graph transformation. Fundamenta Informaticae 74(1), 1–29 (2006)
7. Heckel, R., Küster, J., Taentzer, G.: Towards automatic translation of uml models into semantic domains. In: APPLIGRAPH Workshop on Applied Graph Transformation, pp. 11–22 (2002)
8. Jaffar, J., Maher, M., Marriot, K., Stuckey, P.: The semantics of constraint logic programs. The Journal of Logic Programming 37, 1–46 (1998)
9. Knuth, D.: Semantics of context-free languages. Mathematical Systems Theory 2, 127–145 (1968)
10. Lack, S., Sobocinski, P.: Adhesive and quasiadhesive categories. Theoretical Informatics and Applications 39, 511–545 (2005)
11. Löwe, M., Korff, M., Wagner, A.: An algebraic framework for the transformation of attributed graphs. In: Term Graph Rewriting: Theory and Practice, pp. 185–199. John Wiley, New York (1993)
12. Orejas, F.: Attributed graph constraints. In: Ehrig, H., Heckel, R., Rozenberg, G., Taentzer, G. (eds.) ICGT 2008. LNCS, vol. 5214. Springer, Heidelberg (2008)
13. Orejas, F., Lambers, L.: Symbolic attributed graphs for attributed graph transformation. In: Int. Coll. on Graph and Model Transformation on the Occasion of the 65th Birthday of Hartmut Ehrig (2010)
14. Plump, D., Steinert, S.: Towards graph programs for graph algorithms. In: Ehrig, H., Engels, G., Parisi-Presicce, F., Rozenberg, G. (eds.) ICGT 2004. LNCS, vol. 3256, pp. 128–143. Springer, Heidelberg (2004)

A Dynamic Logic for Termgraph Rewriting[*]

Philippe Balbiani[1], Rachid Echahed[2], and Andreas Herzig[1]

[1] Université de Toulouse, CNRS
Institut de Recherche en Informatique de Toulouse (IRIT)
118 route de Narbonne, 31062 Toulouse Cedex 9, France
{balbiani,herzig}@irit.fr
[2] Laboratoire LIG
Bât IMAG C, BP 53
38041 Grenoble Cedex, France
Rachid.Echahed@imag.fr

Abstract. We propose a dynamic logic tailored to describe graph transformations and discuss some of its properties. We focus on a particular class of graphs called termgraphs. They are first-order terms augmented with sharing and cycles. Termgraphs allow one to describe classical data-structures (possibly with pointers) such as doubly-linked lists, circular lists etc. We show how the proposed logic can faithfully describe (i) termgraphs as well as (ii) the application of a termgraph rewrite rule (i.e. matching and replacement) and (iii) the computation of normal forms with respect to a given rewrite system. We also show how the proposed logic, which is more expressive than propositional dynamic logic, can be used to specify shapes of classical data-structures (e.g. binary trees, circular lists etc.).

1 Introduction

Graphs are common structures widely used in several areas in computer science and discrete mathematics. Their transformation constitute a domain of research per se with a large number of potential applications [13,9,10]. There are many different ways to define graphs and graph transformation. We consider in this paper structures known as *termgraphs* and their transformation via rewrite rules [5,12]. Roughly speaking, a termgraph is a first-order term with possible sharing (of sub-terms) and cycles. Below we depict three examples of termgraphs : G_0 is a classical first-order term. G_1 represents the same expression as G_0 but argument x is shared. G_1 is often used to define the function double $double(x) = G_1$. The second termgraph G_2 represents a circular list of two "records" (represented here by operator $cons$) sharing the same content G_1.

---

[*] This work has been partly funded by the project ARROWS of the French *Agence Nationale de la Recherche*.

H. Ehrig et al. (Eds.): ICGT 2010, LNCS 6372, pp. 59–74, 2010.

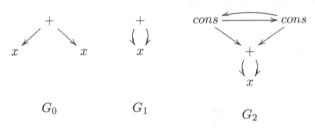

$$G_0 \qquad\qquad G_1 \qquad\qquad G_2$$

Termgraphs allow one to represent real-world data structures (with pointers) such as circular lists, doubly-linked lists etc. [8], and rewriting allows to efficiently process such graphs. They are thus a suitable framework for declarative languages dealing with such complex data structures. However, while there exist rewriting-based proof methods for first-order terms, there is a lack of appropriate termgraph rewriting proof methods, diminishing thus their operational benefits. Indeed, equational logic provides a logical setting for first-order term rewriting [4], and many theorem provers use rewrite techniques in order to efficiently achieve equational reasoning. In [6] an extension of first-order (clausal) logic dealing with termgraphs has been proposed to give a logic counterpart of termgraph rewriting. In such a logic operations are interpreted as continuous functions [14,15] and bisimilar graphs cannot be distinguished (when the edges are uniquely fixed by labels, two termgraphs are bisimilar if and only if they represent the same rational term). Due to that, reasoning on termgraphs is unfortunately much trickier than in first-order classical logic. For example, equational theories on termgraphs are not recursively enumerable whereas equational theories on terms are recursively enumerable [6].

In this paper, we investigate a dynamic logic [11] with a possible worlds semantics that better fits the operational features of termgraph rewriting systems. Termgraphs can easily be interpreted within the framework of possible worlds semantics, where nodes are considered as worlds and edges as modalities. Based on this observation, we investigate a new dynamic logic which has been tailored to fit termgraph rewriting. We show how termgraphs as well as rewrite rules can be specified by means of modal formulae. In particular we show how a rewrite step can be defined by means of a modal formula which encodes termgraph matching (graph homomorphism) and termgraph replacement (graph construction and modification). We show also how to define properties on such structures, such as being a list, a circular list, a tree, a binary tree. The computation of termgraph normal form is formulated in this new logic. In addition, we formulate invariant preservation by rewriting rules and discuss subclasses for which validity is decidable.

The next two sections respectively introduce the considered class of termgraph rewrite systems and our dynamic logic. In section 4 we discuss briefly the expressive power of our dynamic logic and show particularly how graph homomorphisms can be encoded. In section 5 we show how elementary graph transformations can be expressed as dynamic logic formulae whereas section 6 shows how termgraph rewriting can be specified as modal formulae. Section 7 gives some concluding remarks.

2   Termgraph Rewriting

This section defines the framework of graph rewrite systems that we consider in the paper. There are different approaches in the literature to define graph transformations. We follow here an algorithmic approach to termgraph rewriting [5]. Our definitions are consistent with [8].

Definition 1 (Graph). *Let Ω be a set of node labels, and let $\mathcal{F}$ be a set of edge labels. A* termgraph, *or simply a* graph *is a tuple $G = (\mathcal{N}, \mathcal{E}, \mathcal{L}^n, \mathcal{L}^e, \mathcal{S}, \mathcal{T})$ which consists of a finite set of nodes $\mathcal{N}$, a finite set of edges $\mathcal{E}$, a (partial) node labelling function $\mathcal{L}^n : \mathcal{N} \to \Omega$ which associates labels in Ω to nodes in $\mathcal{N}$, a (total) edge labelling function $\mathcal{L}^e : \mathcal{E} \to \mathcal{F}$ which associates, to every edge in $\mathcal{E}$, a label (or feature) in $\mathcal{F}$, a source function $\mathcal{S} : \mathcal{E} \to \mathcal{N}$ and a target function $\mathcal{T} : \mathcal{E} \to \mathcal{N}$ which specify respectively, for every edge e, its source $\mathcal{S}(e)$ and its target $\mathcal{T}(e)$.*

We assume that the labelling of edges $\mathcal{L}^e$ fulfills the following additional de-terminism condition:
$\forall\, e_1, e_2 \in \mathcal{E}, (\mathcal{S}(e_1) = \mathcal{S}(e_2)$ *and* $\mathcal{L}^e(e_1) = \mathcal{L}^e(e_2))$ *implies* $e_1 = e_2$. *This last condition expresses the fact that for every node n there exists at most one edge e of label a such that the source of e is n. We denote such an edge by the tuple (n, a, m) where m is the target of edge e.*

Notation: For each labelled node n the fact that $\omega = \mathcal{L}^n(n)$ is written $n : \omega$, and each unlabelled node n is written as $n : \bullet$. This 'unlabelled' symbol $\bullet$ is used in termgraphs to represent anonymous variables. $n : \omega(a_1 \Rightarrow n_1, \ldots, a_k \Rightarrow n_k)$ describes a node n labelled by symbol ω with k outgoing edges, $e_1, \ldots, e_k$, such that for every edge e_i, $\mathcal{L}^e(e_i) = a_i$, $\mathcal{S}(e_i) = n$ and $\mathcal{T}(e_i) = n_i$. In the sequel we will use the linear notation of termgraphs [5] defined by the following grammar. The variable A (resp. F and n) ranges over the set Ω (resp. $\mathcal{F}$ and $\mathcal{N}$) :

TERMGRAPH ::= NODE | NODE + TERMGRAPH
NODE   ::= n:A(F $\Rightarrow$ NODE,...,F $\Rightarrow$ NODE) | n:$\bullet$ | n

the operator + stands for the disjoint union of termgraph definitions. We assume that every node is labelled at most once. The expression $n : \omega(n_1, \ldots, n_k)$ stands for $n : \omega(1 \Rightarrow n_1, \ldots, k \Rightarrow n_k)$.

A *graph homomorphism*, $h : G \to G_1$, where $G = (\mathcal{N}, \mathcal{E}, \mathcal{L}^n, \mathcal{L}^e, \mathcal{S}, \mathcal{T})$ and $G_1 = (\mathcal{N}_1, \mathcal{E}_1, \mathcal{L}_1^n, \mathcal{L}_1^e, \mathcal{S}_1, \mathcal{T}_1)$ is a pair of functions $h = (h^n, h^e)$ with $h^n : \mathcal{N} \to \mathcal{N}_1$ and $h^e : \mathcal{E} \to \mathcal{E}_1$ which preserves the labelling of nodes and edges as well as the source and target functions. This means that for each labelled node m in G, $\mathcal{L}_1^n(h^n(m)) = \mathcal{L}^n(m)$ and for each edge u in G, $\mathcal{L}_1^e(h^e(u)) = \mathcal{L}^e(u)$, $\mathcal{S}_1(h^e(u)) = h^n(\mathcal{S}(u))$ and $\mathcal{T}_1(h^e(u)) = h^n(\mathcal{T}(u))$. Notice that the image by h^n of an unlabelled node may be any node.

Remark: Because of the determinism condition, a homomorphism $h : G \to G_1$ is completely defined by the function $h^n : \mathcal{N} \to \mathcal{N}_1$ which should satisfy the following conditions : for each labelled node m in G, $\mathcal{L}_1^n(h^n(m)) = \mathcal{L}^n(m)$ and for every outgoing edge from m, say (m, a, w), for some feature a and node w, the edge $(h^n(m), a, h^n(w))$ belongs to $\mathcal{E}_1$.

Example 1. *Let B_1, B_2 and B_3 be the following termgraphs.*

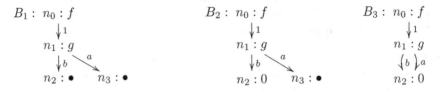

Let h_1^n and h_2^n be two functions on nodes defined as follows: $h_1^n(n_i) = n_i$ for i in $\{0, 1, 2, 3\}$, $h_2^n(n_i) = n_i$ for i in $\{0, 1, 2\}$ and $h_2^n(n_3) = n_2$. Let $e_0 = (n_0, 1, n_1)$, $e_1 = (n_1, a, n_3)$, $e_2 = (n_1, b, n_2)$ and $e_3 = (n_1, a, n_2)$ be the edges occurring in the termgraphs above and h_1^e and h_2^e be two functions on edges such that $h_1^e(e_i) = e_i$ for i in $\{0, 1, 2\}$, $h_2^e(e_0) = e_0$, $h_2^e(e_1) = e_3$ and $h_2^e(e_2) = e_2$.

$h_1 = (h_1^n, h_1^e)$ defines a homomorphism from B_1 to B_2. $h_2 = (h_2^n, h_2^e)$ defines a homomorphism from B_1 to B_3 and from B_2 to B_3. Notice that there is no homomorphism from B_3 to B_2 or to B_1, nor from B_2 to B_1.

The following definition introduces a notion of actions. Each action specifies an elementary transformation of graphs. These elementary actions are used later on to define graph transformations by means of rewrite rules.

Definition 2 (Actions). *An action has one of the following forms.*

- *a **node definition** or **node labelling** $n : f(a_1 \Rightarrow n_1, \ldots, a_k \Rightarrow n_k)$ where $n, n_1, \ldots, n_k$ are nodes and f is a label of node n. For $i \in \{1, \ldots, k\}$, a_i is the label of an edge, e_i, such that $(\mathcal{L}^e(e_i) = a_i)$ and whose source is n $(\mathcal{S}(e_i) = n)$ and target is node n_i $(\mathcal{T}(e_i) = n_i)$. This action first creates a new node n if n does not already exist in the context of application of the action. Then node n is defined by its label and its outgoing edges.*
- *an **edge redirection** or **local redirection** $n \gg_a m$
  where n, m are nodes and a is the feature of an edge e outgoing node n $(\mathcal{S}(e) = n$ and $\mathcal{L}^e(e) = a)$. This action is an edge redirection and means that the target of edge e is redirected to point to the node m (i.e., $\mathcal{T}(e) = m$ after performing the action $n \gg_a m$).*
- *a **global redirection** $n \gg m$
  where n and m are nodes. This means that all edges e pointing to n (i.e., $\mathcal{T}(e) = n$) are redirected to point to the node m (i.e., $\mathcal{T}(e) = m$).*

The result of applying an action α to a termgraph $G = (\mathcal{N}, \mathcal{E}, \mathcal{L}^n, \mathcal{L}^e, \mathcal{S}, \mathcal{T})$ is a termgraph, say $G_1 = (\mathcal{N}_1, \mathcal{E}_1, \mathcal{L}_1^n, \mathcal{L}_1^e, \mathcal{S}_1, \mathcal{T}_1)$, denoted by $\alpha[G]$ and is defined as follows :

- If $\alpha = n{:}f(a_1 \Rightarrow n_1, \ldots, a_k \Rightarrow n_k)$ then
  - $\mathcal{N}_1 = \mathcal{N} \cup \{n, n_1, \ldots, n_k\}$,
  - $\mathcal{L}_1^n(n) = f$ and $\mathcal{L}_1^n(m) = \mathcal{L}^n(m)$ if $m \neq n$,
  - Let $E = \{e_i \mid 1 \leq i \leq k, e_i$ is an edge such that $\mathcal{S}(e_i) = n, \mathcal{T}(e_i) = n_i$ and $\mathcal{L}^e(e_i) = a_i\}$. $\mathcal{E}_1 = \mathcal{E} \cup E$,
  - $\mathcal{L}_1^e(e) = \begin{cases} a_i & \text{if } e = e_i \in E \\ \mathcal{L}^e(e) & \text{if } e \notin E \end{cases}$
  - $\mathcal{S}_1(e) = \begin{cases} n & \text{if } e = e_i \in E \\ \mathcal{S}(e) & \text{if } e \notin E \end{cases}$
  - $\mathcal{T}_1(e) = \begin{cases} n_i & \text{if } e = e_i \in E \\ \mathcal{T}(e) & \text{if } e \notin E \end{cases}$

  $\cup$ denotes classical union. This means that the nodes in $\{n, n_1, \ldots, n_k\}$ which already belong to G are reused whereas the others are new.
- If $\alpha = n \gg_a m$ then
  - $\mathcal{N}_1 = \mathcal{N}$, $\mathcal{E}_1 = \mathcal{E}$, $\mathcal{L}_1^n = \mathcal{L}^n$, $\mathcal{L}_1^e = \mathcal{L}^e$, $\mathcal{S}_1 = \mathcal{S}$ and
  - Let e be the edge of label a outgoing n.
    $\mathcal{T}_1(e) = m$ and $\mathcal{T}_1(e') = \mathcal{T}(e')$ if $e' \neq e$.
- If $\alpha = n \gg m$ then $\mathcal{N}_1 = \mathcal{N}$, $\mathcal{E}_1 = \mathcal{E}$, $\mathcal{L}_1^n = \mathcal{L}^n$, $\mathcal{L}_1^e = \mathcal{L}^e$, $\mathcal{S}_1 = \mathcal{S}$ and

$$\mathcal{T}_1(e) = \begin{cases} m & \text{if } \mathcal{T}(e) = n \\ \mathcal{T}(e) & \text{otherwise} \end{cases}$$

A rooted *termgraph* is a termgraph G with a distinguished node n called its root. We write $G = (\mathcal{N}, \mathcal{E}, \mathcal{L}^n, \mathcal{L}^e, \mathcal{S}, \mathcal{T}, n)$. The application of an action α to a rooted termgraph $G = (\mathcal{N}, \mathcal{E}, \mathcal{L}^n, \mathcal{L}^e, \mathcal{S}, \mathcal{T}, n)$ is a rooted termgraph $G_1 = (\mathcal{N}_1, \mathcal{E}_1, \mathcal{L}_1^n, \mathcal{L}_1^e, \mathcal{S}_1, \mathcal{T}_1, n_1)$ such that $G_1 = \alpha[G]$ and root n_1 is defined as follows :

- $n_1 = n$ if α is not of the form $n \gg p$.
- $n_1 = p$ if α is of the form $n \gg p$.

The application of a sequence of actions Δ to a (rooted) termgraph G is defined inductively as follows : $\Delta[G] = G$ if Δ is the empty sequence and $\Delta[G] = \Delta'[\alpha[G]]$ if $\Delta = \alpha; \Delta'$ where ";" is the concatenation (or sequential) operation. Let $h = (h^n, h^e)$ be a homomorphism. We denote by $h(\Delta)$ the sequence of actions obtained from Δ by substituting every node m occurring in Δ by $h^n(m)$.

Example 2. This example illustrates the application of actions. Let H_1, H_2, H_3, H_4 and H_5 be the following termgraphs.

$H_1 :$ $n_1 : f$
$\downarrow b$
$n_2 : 0$

$H_2 :$ $n_1 : g$
$\downarrow b \quad \searrow a$
$n_2 : 0 \quad n_3 : \bullet$

$H_3 :$ $n_0 : f$
$\downarrow 1$
$n_1 : g$
$\downarrow b \quad \searrow a$
$n_2 : 0 \quad n_3 : \bullet$

$$H_4 : \quad n_0 : f$$
$$\downarrow 1$$
$$n_1 : g$$
$$\langle b \, \rangle a$$
$$n_2 : 0 \qquad n_3 : \bullet$$

$$H_5 : \quad n_0 : f$$
$$b \langle \, \downarrow 1 \, \rangle a$$
$$n_1 : g$$
$$n_2 : 0 \qquad n_3 : \bullet$$

H_2 is obtained from H_1 by applying the action $n_1 : g(b \Rightarrow n_2, a \Rightarrow n_3)$. n_1 is relabelled whereas n_3 is a new unlabelled node. H_3 is obtained from H_2 by applying the action $\alpha = n_0 : f(n_1)$. n_0 is a new node labelled by f. f has one argument n_1. H_4 is obtained from H_3 by applying the action $n_1 \gg_a n_2$. The effect of this action is to change the target n_3 of the edge (n_1, a, n_3) by n_2. H_5 is obtained from H_4 by applying the action $n_2 \gg n_0$. This action redirects the incoming edges of node n_2 to target node n_0.

Definition 3 (Rule, system, rewrite step). *A rewrite rule is an expression of the form $l \to r$ where l is a termgraph and r is a sequence of actions. A rule is written $l \to (a_1, \ldots, a_n)$ or $l \to a_1; \ldots; a_n$ where the $a_i's$ are elementary actions. A termgraph rewrite system is a set of rewrite rules. We say that the termgraph G rewrites to G_1 using the rule $l \to r$ iff there exists a homomorphism $h : l \to G$ and $G_1 = h(r)[G]$. We write $G \to_{l \to r} G_1$, or simply $G \to G_1$.*

Example 3. *We give here an example of a rewrite step. Consider the following rewrite rule:*

$$i_1 : g(a \Rightarrow i_2 : \bullet, b \Rightarrow i_3 : \bullet) \to n_0 : f(1 \Rightarrow i_1); i_1 \gg_a i_3; i_3 \gg n_0$$

The reader may easily verify that the graph H_2 of Example 2 can be rewritten by the considered rule into the graph H_5 of Example 2. The left-hand side of the rule matches the graph H_2 via the homomorphism $h = (h^n, h^e)$ such that $h^n(i_1) = n_1$, $h^n(i_2) = n_3$ and $h^n(i_3) = n_2$ (h^e being obvious from the context). $H_5 = h(n_0 : f(1 \Rightarrow i_1); i_1 \gg_a i_3; i_3 \gg n_0)[H_2]$. Thus $H_5 = n_0 : f(1 \Rightarrow h^n(i_1)); h^n(i_1) \gg_a h^n(i_3); h^n(i_3) \gg n_0[H_2]$.

Example 4. *We give here somme illustrating examples of the considered class of rewrite systems. We first define an operation, insert, which inserts an element in a circular list.*

$r : insert(m : \bullet, p_1 : cons(m_1 : \bullet, p_1)) \to p_2 : cons(m, p_1); p_1 \gg_2 p_2; r \gg p_2$
$r : insert(m : \bullet, p_1 : cons(m_1 : \bullet, p_2)) + p_3 : cons(m_2, p_1) \to p_4 : cons(m, p_1);$
$$p_3 \gg_2 p_4; r \gg p_4$$

As a second example, we define below the operation length which computes the number of elements of any, possibly circular, list.

$r : length(p : \bullet) \to r' : length'(p, p); r \gg r'$
$r : length'(p_1 : nil, p_2 : \bullet) \to r' : 0; r \gg r'$

$r : length'(p_1 : cons(n : \bullet, p_2 : \bullet), p_2) \to r' : succ(0); r \gg r'$
$r : length'(p_1 : cons(n : \bullet, p_2 : \bullet), p_3 : \bullet) \to r' : s(q : \bullet); q : length'(p_2, p_3); r \gg r'$

Pointers help very often to enhance the efficiency of algorithms. In the following, we define the operation reverse which performs the so-called "in-situ list reversal".

$o : reverse(p : \bullet) \to o' : reverse'(p, q : nil); o \gg o'$
$o : reverse'(p_1 : cons(n : \bullet, q : nil), p_2 : \bullet) \to p_1 \gg_2 p_2; o \gg p_1$
$o : reverse'(p_1 : cons(n : \bullet, p_2 : cons(m : \bullet, p_3 : \bullet), p_4 : \bullet) \to p_1 \gg_2 p_4;$
$$o \gg_1 p_2; o \gg_2 p_1$$

The last example illustrates the encoding of classical term rewrite systems. We define the addition on naturals as well as the function double with their usual meanings.

$r : +(n : 0, m : \bullet) \to r \gg m$
$r : +(n : succ(p : \bullet), m : \bullet) \to q : succ(k : +(p, m)); r \gg q$
$r : double(n : \bullet) \to q : +(n, n); r \gg q$

3   Dynamic Logic

It is now time to define the syntax and the semantics of the logic of graph modifiers that will be used as a tool to talk about rooted termgraphs.

3.1   Syntax

Like the language of propositional dynamic logic, the language of the logic of graph modifiers is based on the idea of associating with each action α of an action language a modal connective $[\alpha]$. The formula $[\alpha]\phi$ is read "after every terminating execution of α, ϕ is true". Consider, as in section 2, a countable set $\mathcal{F}$ (with typical members denoted a, b, etc) of edge labels and a countable set Ω (with typical members denoted ω, π, etc) of node labels. These labels are formulas defined below. A node labeled by π is called a π *node*.

Formally we define the set of all actions (with typical members denoted α, β, etc) and the set of all formulas (with typical members denoted ϕ, ψ, etc) as follows:

- $\alpha ::= a \mid U \mid n \mid \boldsymbol{n} \mid \phi? \mid (\omega :=_g \phi) \mid (\omega :=_l \phi) \mid (a + (\phi, \psi)) \mid (a - (\phi, \psi)) \mid$
  $(\alpha; \beta) \mid (\alpha \cup \beta) \mid \alpha^\star$
- $\phi ::= \omega \mid \bot \mid \neg\phi \mid (\phi \vee \psi) \mid [\alpha]\phi$

We adopt the standard abbreviations for the other Boolean connectives. Moreover, for all actions α and for all formulas ϕ, let $\langle\alpha\rangle\phi$ be $\neg[\alpha]\neg\phi$. As usual, we follow the standard rules for omission of the parentheses. An atomic action is either an edge label a in $\mathcal{F}$, the universal action U, a test $\phi?$ or an update action

n, $\boldsymbol{n}$, $\omega :=_g \phi$, $\omega :=_l \phi$, $a + (\phi, \psi)$ or $a - (\phi, \psi)$. U reads "go anywhere", n reads "add some new node", $\boldsymbol{n}$ reads "add some new node and go there", $\omega :=_g \phi$ reads "assign to ω nodes the truth value of ϕ everywhere (globally)", $\omega :=_l \phi$ reads "assign to ω the truth value of ϕ here (locally)", $a + (\phi, \psi)$ reads "add a edges from all ϕ nodes to all ψ nodes", and $a - (\phi, \psi)$ reads "delete a edges from all ϕ nodes to all ψ nodes". Complex actions are built by means of the regular operators ";", "$\cup$" and "*". An update action is an action that is not an edge label and in which no U occurs. An update action is $:=_l$-free if no local assignment $\omega :=_l \phi$ occurs in it.

3.2   Semantics

Like the truth-conditions of the formulas of ordinary modal logics, the truth-conditions of the formulas of the logic of graph modifiers is based on the idea of interpreting, within a rooted termgraph $G = (\mathcal{N}, \mathcal{E}, \mathcal{L}^n, \mathcal{L}^e, \mathcal{S}, \mathcal{T}, n_0)$, edge labels in $\mathcal{F}$ by sets of edges and node labels in Ω by sets of nodes. In this section, we consider a more general notion of node labeling functions $\mathcal{L}^n$ of termgraphs such that nodes can have several labels (propositions). In this case the labeling function has the following profile $\mathcal{L}^n : \mathcal{N} \rightarrow \mathcal{P}(\Omega)$. Node labeling functions considered in section 2 where a node can have at most one label is obviously a particular case. Let I_G be the interpretation function in G of labels defined as follows:

- $I_G(a) = \{e \in \mathcal{E}: \mathcal{L}^e(e) = a\}$,
- $I_G(\omega) = \{n \in \mathcal{N}: \omega \in \mathcal{L}^n(n)\}$.

For all abstract actions a, let $R_G(a) = \{(n_1, n_2): \text{there exists an edge } e \in I_G(a) \text{ such that } \mathcal{S}(e) = n_1 \text{ and } \mathcal{T}(e) = n_2\}$ be the binary relation interpreting the abstract action a in G. The truth-conditions of the formulas of the logic of graph modifiers are defined by induction as follows:

- $G \models \omega$ iff $n_0 \in I_G(\omega)$,
- $G \not\models \perp$,
- $G \models \neg\phi$ iff $G \not\models \phi$,
- $G \models \phi \vee \psi$ iff $G \models \phi$ or $G \models \psi$,
- $G \models [\alpha]\phi$ iff for all rooted termgraphs $G' = (\mathcal{N}', \mathcal{E}', \mathcal{L}^{n'}, \mathcal{L}^{e'}, \mathcal{S}', \mathcal{T}', n_0')$, if $G \longrightarrow_\alpha G'$ then $G' \models \phi$

where the binary relations $\longrightarrow_\alpha$ are defined by induction as follows:

- $G \longrightarrow_a G'$ iff $\mathcal{N}' = \mathcal{N}$, $\mathcal{E}' = \mathcal{E}$, $\mathcal{L}^{n'} = \mathcal{L}^n$, $\mathcal{L}^{e'} = \mathcal{L}^e$, $\mathcal{S}' = \mathcal{S}$, $\mathcal{T}' = \mathcal{T}$ and $(n_0, n_0') \in R_G(a)$,
- $G \longrightarrow_{\phi?} G'$ iff $\mathcal{N}' = \mathcal{N}$, $\mathcal{E}' = \mathcal{E}$, $\mathcal{L}^{n'} = \mathcal{L}^n$, $\mathcal{L}^{e'} = \mathcal{L}^e$, $\mathcal{S}' = \mathcal{S}$, $\mathcal{T}' = \mathcal{T}$, $n_0' = n_0$ and $G' \models \phi$,
- $G \longrightarrow_U G'$ iff $\mathcal{N}' = \mathcal{N}$, $\mathcal{E}' = \mathcal{E}$, $\mathcal{L}^{n'} = \mathcal{L}^n$, $\mathcal{L}^{e'} = \mathcal{L}^e$, $\mathcal{S}' = \mathcal{S}$ and $\mathcal{T}' = \mathcal{T}$,
- $G \longrightarrow_n G'$ iff $\mathcal{N}' = \mathcal{N} \cup \{n_1\}$ where n_1 is a new node, $\mathcal{E}' = \mathcal{E}$, $\mathcal{L}^{n'}(m) = \mathcal{L}^n(m)$ if $m \neq n_1$, $\mathcal{L}^{n'}(n_1) = \emptyset$, $\mathcal{L}^{e'} = \mathcal{L}^e$, $\mathcal{S}' = \mathcal{S}$, $\mathcal{T}' = \mathcal{T}$ and $n_0' = n_0$,

- $G \longrightarrow_n G'$ iff $\mathcal{N}' = \mathcal{N} \cup \{n_1\}$ where n_1 is a new node, $\mathcal{E}' = \mathcal{E}$, $\mathcal{L}^{n'}(m) = \mathcal{L}^n(m)$ if $m \neq n_1$, $\mathcal{L}^{n'}(n_1) = \emptyset$, $\mathcal{L}^{e'} = \mathcal{L}^e$, $\mathcal{S}' = \mathcal{S}$, $\mathcal{T}' = \mathcal{T}$ and $n_0' = n_1$,
- $G \longrightarrow_{w:=_g\phi} G'$ iff $\mathcal{N}' = \mathcal{N}$, $\mathcal{E}' = \mathcal{E}$, $\mathcal{L}^{n'}(m) = $ if $(\mathcal{N}, \mathcal{E}, \mathcal{L}^n, \mathcal{L}^e, \mathcal{S}, \mathcal{T}, m) \models \phi$ then $\mathcal{L}^n(m) \cup \{w\}$ else $\mathcal{L}^n(m) \setminus \{w\}$, $\mathcal{L}^{e'} = \mathcal{L}^e$, $\mathcal{S}' = \mathcal{S}$, $\mathcal{T}' = \mathcal{T}$ and $n_0' = n_0$,
- $G \longrightarrow_{w:=_l\phi} G'$ iff $\mathcal{N}' = \mathcal{N}$, $\mathcal{E}' = \mathcal{E}$, $\mathcal{L}^{n'}(n_0) = $ if $(\mathcal{N}, \mathcal{E}, \mathcal{L}^n, \mathcal{L}^e, \mathcal{S}, \mathcal{T}, n_0) \models \phi$ then $\mathcal{L}^n(n_0) \cup \{w\}$ else $\mathcal{L}^n(n_0) \setminus \{w\}$, $\mathcal{L}^{n'}(m) = \mathcal{L}^n(m)$ if $m \neq n_0$, $\mathcal{L}^{e'} = \mathcal{L}^e$, $\mathcal{S}' = \mathcal{S}$, $\mathcal{T}' = \mathcal{T}$ and $n_0' = n_0$,
- $G \longrightarrow_{a+(\phi,\psi)} G'$ iff $\mathcal{N}' = \mathcal{N}$, $\mathcal{E}' = \mathcal{E} \cup \{(n_1, a, n_2) : (\mathcal{N}, \mathcal{E}, \mathcal{L}^n, \mathcal{L}^e, \mathcal{S}, \mathcal{T}, n_1) \models \phi$ and $(\mathcal{N}, \mathcal{E}, \mathcal{L}^n, \mathcal{L}^e, \mathcal{S}, \mathcal{T}, n_2) \models \psi\}$, $\mathcal{L}^{n'} = \mathcal{L}^n$, $\mathcal{L}^{e'}(e) = $ if $e \in \mathcal{E}$ then $\mathcal{L}^e(e)$ else a, $\mathcal{S}'(e) = $ if $e \in \mathcal{E}$ then $\mathcal{S}(e)$ else e is of the form (n_1, a, n_2) and $\mathcal{S}'(e) = n_1$, $\mathcal{T}' = $ if $e \in \mathcal{E}$ then $\mathcal{T}(e)$ else e is of the form (n_1, a, n_2) and $\mathcal{T}'(e) = n_2$ and $n_0' = n_0$,
- $G \longrightarrow_{a-(\phi,\psi)} G'$ iff $\mathcal{N}' = \mathcal{N}$, $\mathcal{E}' = \mathcal{E} \setminus \{(n_1, a, n_2) : (\mathcal{N}, \mathcal{E}, \mathcal{L}^n, \mathcal{L}^e, \mathcal{S}, \mathcal{T}, n_1) \models \phi$ and $(\mathcal{N}, \mathcal{E}, \mathcal{L}^n, \mathcal{L}^e, \mathcal{S}, \mathcal{T}, n_2) \models \psi\}$, $\mathcal{L}^{n'} = \mathcal{L}^n$, $\mathcal{L}^{e'}(e) = \mathcal{L}^e(e)$, $\mathcal{S}' = \mathcal{S}$, $\mathcal{T}' = \mathcal{T}$ and $n_0' = n_0$,
- $G \longrightarrow_{\alpha;\beta} G'$ iff there exists a rooted termgraph $G'' = (\mathcal{N}'', \mathcal{E}'', \mathcal{L}^{n''}, \mathcal{L}^{e''}, \mathcal{S}'', \mathcal{T}'', n_0'')$ such that $G \longrightarrow_\alpha G''$ and $G'' \longrightarrow_\beta G'$,
- $G \longrightarrow_{\alpha \cup \beta} G'$ iff $G \longrightarrow_\alpha G'$ or $G \longrightarrow_\beta G'$,
- $G \longrightarrow_{\alpha^*} G'$ iff there exists a sequence $G^{(0)} \ldots, G^{(k)}$ of rooted termgraphs such that $G^{(i)} = (\mathcal{N}^{(i)}, \mathcal{E}^{(i)}, \mathcal{L}^{n(i)}, \mathcal{L}^{e(i)}, \mathcal{S}^{(i)}, \mathcal{T}^{(i)}, n_0^{(i)})$ for $i \in \{0, \ldots, k\}$, $G^{(0)} = G$, $G^{(k)} = G'$ and for all non-negative integers i, if $i < k$ then $G^{(i)} \longrightarrow_\alpha G^{(i+1)}$.

The above definitions of formulas reflect our intuitive understanding of the actions of the language of the logic of graph modifiers. Obviously, $G \models \langle\alpha\rangle\phi$ iff there exists a rooted termgraph $G' = (\mathcal{N}', \mathcal{E}', \mathcal{L}^{n'}, \mathcal{L}^{e'}, \mathcal{S}', \mathcal{T}', n_0')$ such that $G \longrightarrow_\alpha G'$ and $G' \models \phi$. The formula ϕ is said to be valid in class $\mathcal{C}$ of rooted termgraphs, in symbols $\mathcal{C} \models \phi$, iff $G \models \phi$ for each rooted termgraph $G = (\mathcal{N}, \mathcal{E}, \mathcal{L}^n, \mathcal{L}^e, \mathcal{S}, \mathcal{T}, n_0)$ in $\mathcal{C}$. The class of all rooted termgraphs will be denoted more briefly as $\mathcal{C}_{all}$.

3.3 Validities

Obviously, as in propositional dynamic logic, we have

- $\mathcal{C}_{all} \models [\phi?]\psi \leftrightarrow (\phi \rightarrow \psi)$,
- $\mathcal{C}_{all} \models [\alpha;\beta]\phi \leftrightarrow [\alpha][\beta]\phi$,
- $\mathcal{C}_{all} \models [\alpha \cup \beta]\phi \leftrightarrow [\alpha]\phi \wedge [\beta]\phi$,
- $\mathcal{C}_{all} \models [\alpha^*]\phi \leftrightarrow \phi \wedge [\alpha][\alpha^*]\phi$.

If α is a $:=_l$-free update action then

- $\mathcal{C}_{all} \models [\alpha]\bot \leftrightarrow \bot$,
- $\mathcal{C}_{all} \models [\alpha]\neg\phi \leftrightarrow \neg[\alpha]\phi$,
- $\mathcal{C}_{all} \models [\alpha](\phi \vee \psi) \leftrightarrow [\alpha]\phi \vee [\alpha]\psi$. .

These equivalences are the case because such actions are interpreted as total functions.

The next series of equivalences guarantees that each of our $:=_l$-free update actions can be moved across the abstract actions of the form a or U:

- $\mathcal{C}_{all} \models [n][a]\phi \leftrightarrow [a][n]\phi$,
- $\mathcal{C}_{all} \models [n][U]\phi \leftrightarrow [n]\phi \wedge [U][n]\phi$,
- $\mathcal{C}_{all} \models [\boldsymbol{n}][a]\phi \leftrightarrow \top$,
- $\mathcal{C}_{all} \models [\boldsymbol{n}][U]\phi \leftrightarrow [\boldsymbol{n}]\phi \wedge [U][\boldsymbol{n}]\phi$,
- $\mathcal{C}_{all} \models [\omega :=_g \phi][a]\psi \leftrightarrow [a][\omega :=_g \phi]\phi$,
- $\mathcal{C}_{all} \models [\omega :=_g \phi][U]\psi \leftrightarrow [U][\omega :=_g \phi]\psi$,
- $\mathcal{C}_{all} \models [a+(\phi,\psi)][b]\chi \leftrightarrow [b][a+(\phi,\psi)]\chi$ if $a \neq b$ and $\mathcal{C}_{all} \models [a+(\phi,\psi)][b]\chi \leftrightarrow [b][a+(\phi,\psi)]\chi \wedge (\phi \rightarrow [U](\psi \rightarrow [a+(\phi,\psi)]\chi))$ if $a = b$,
- $\mathcal{C}_{all} \models [a+(\phi,\psi)][U]\chi \leftrightarrow [U][a+(\phi,\psi)]\chi$,
- $\mathcal{C}_{all} \models [a-(\phi,\psi)][b]\chi \leftrightarrow [b][a-(\phi,\psi)]\chi$ if $a \neq b$ and $\mathcal{C}_{all} \models [a-(\phi,\psi)][b]\chi \leftrightarrow (\neg\phi \wedge [b][a-(\phi,\psi)]\chi) \vee (\phi \wedge [b](\neg\psi \rightarrow [a-(\phi,\psi)]\chi))$ if $a = b$,
- $\mathcal{C}_{all} \models [a-(\phi,\psi)][U]\chi \leftrightarrow [U][a-(\phi,\psi)]\chi$.

Finally, once we have moved each of our $:=_l$-free update actions across the abstract actions of the form a or U, these update actions can be eliminated by means of the following equivalences:

- $\mathcal{C}_{all} \models [n]\omega \leftrightarrow \omega$,
- $\mathcal{C}_{all} \models [\boldsymbol{n}]\omega \leftrightarrow \bot$,
- $\mathcal{C}_{all} \models [\omega :=_g \phi]\pi \leftrightarrow \pi$ if $\omega \neq \pi$ and $\mathcal{C}_{all} \models [\omega :=_g \phi]\pi \leftrightarrow \phi$ if $\omega = \pi$,
- $\mathcal{C}_{all} \models [a+(\phi,\psi)]\omega \leftrightarrow \omega$,
- $\mathcal{C}_{all} \models [a-(\phi,\psi)]\omega \leftrightarrow \omega$.

Proposition 1. *For all $:=_l$-free *-free formulas ϕ, there exists a $:=_l$-free *-free formula ψ without update actions such that $\mathcal{C}_{all} \models \phi \leftrightarrow \psi$.*

Proof. See the above discussion.

Just as for $:=_l$-free update actions, we have the following equivalences for the update actions of the form $\omega :=_l \phi$:

- $\mathcal{C}_{all} \models [\omega :=_l \phi]\bot \leftrightarrow \bot$,
- $\mathcal{C}_{all} \models [\omega :=_l \phi]\neg\psi \leftrightarrow \neg[\omega :=_l \phi]\psi$,
- $\mathcal{C}_{all} \models [\omega :=_l \phi](\psi \vee \chi) \leftrightarrow [\omega :=_l \phi]\psi \vee [\omega :=_l \phi]\chi$,
- $\mathcal{C}_{all} \models [\omega :=_l \phi]\pi \leftrightarrow \pi$ if $\omega \neq \pi$ and $\mathcal{C}_{all} \models [\omega :=_g \phi]\pi \leftrightarrow \phi$ if $\omega = \pi$.

But it is not possible to formulate reduction axioms for the cases $[\omega :=_l \phi][a]\psi$ and $[\omega :=_l \phi][U]\psi$. More precisely,

Proposition 2. *There exists a *-free formula ϕ such that for all *-free formulas ψ without update actions, $\mathcal{C}_{all} \not\models \phi \leftrightarrow \psi$.*

Proof. Take the *-free formula $\phi = [\omega :=_g \bot][U][\omega :=_l \top][a]\neg\omega$. The reader may easily verify that for all rooted termgraphs $G = (\mathcal{N}, \mathcal{E}, \mathcal{L}^n, \mathcal{L}^e, \mathcal{S}, \mathcal{T}, n_0)$, $G \models \phi$ iff $R_G(a)$ is irreflexive. Seeing that the binary relation interpreting an abstract action of the form a is irreflexive cannot be modally defined in propositional dynamic logic, then for all formulas ψ without update actions, $\mathcal{C}_{all} \not\models \phi \leftrightarrow \psi$.

3.4 Decidability, Axiomatization and a Link with Hybrid Logics

Firstly, let us consider the set L of all $:=_l$-free *-free formulas ϕ such that $\mathcal{C}_{all}$ $\models \phi$. Together with a procedure for deciding membership in *-free propositional dynamic logic, the equivalences preceding proposition 1 provide a procedure for deciding membership in L. Hence, membership in L is decidable.

Secondly, let us consider the set $L(:=_l)$ of all *-free formulas ϕ such that $\mathcal{C}_{all}$ $\models \phi$. Aucher *et al.* [3] have defined a recursive translation from the language of hybrid logic [2] into the set of all our *-free formulas that preserves satisfiability. It is known that the problem of deciding satisfiability of hybrid logic formulas is undecidable [1, Section 4.4]. The language of hybrid logic has formulas of the form $@_i\phi$ ("ϕ is true at i"), $@_x\phi$ ("ϕ is true at x") and $\downarrow x.\phi$ ("ϕ holds after x is bound to the current state"), where $NOM = \{i_1, \ldots\}$ is a set of nominals, and $SVAR = \{x_1, \ldots\}$ is a set of state variables. The (slightly adapted) translation of a given hybrid formula ϕ_0 is recursively defined as follows.

$$
\begin{array}{ll}
\tau(\omega) & = \omega \\
\tau(i) & = \omega_i \quad \text{where } \omega_i \text{ does not occur in } \phi_0 \\
\tau(x) & = \omega_x \quad \text{where } \omega_x \text{ does not occur in } \phi_0 \\
\tau(\neg\phi) & = \neg\tau(\phi) \\
\tau(\phi \vee \psi) & = \tau(\phi) \vee \tau(\psi) \\
\tau([a]\phi) & = [a]\tau(\phi) \\
\tau([U]\phi) & = [U]\tau(\phi) \\
\tau(@_i\phi) & = \langle U \rangle(\omega_i \wedge \tau(\phi)) \\
\tau(@_x\phi) & = \langle U \rangle(\omega_x \wedge \tau(\phi)) \\
\tau(\downarrow x.\phi) & = [\omega_x :=_g \bot][\omega_x :=_l \top]\tau(\phi)
\end{array}
$$

As the satisfiability problem is undecidable in hybrid logic, membership in $L(:=_l)$ is undecidable, too.

Thirdly, let us consider the set $L(^*)$ of all $:=_l$-free formulas ϕ such that $\mathcal{C}_{all}$ $\models \phi$. It is still an open problem whether membership in $L(^*)$ is decidable or not: while the update actions can be eliminated from $:=_l$-free formulas, it is not clear whether this can be done for formulas in which e.g. iterations of assignments occur.

As for the axiomatization issue, the equivalences preceding proposition 1 provide a sound and complete axiom system of L, whereas no axiom system of $L(:=_l)$ and $L(^*)$ is known to be sound and complete.

4 Definability of Classes of Termgraphs

For all abstract actions a, by means of the update actions of the form $\omega :=_l \phi$, we can express the fact that the binary relation interpreting an abstract action of the form a is deterministic, irreflexive or locally reflexive. More precisely, for all rooted termgraphs $G = (\mathcal{N}, \mathcal{E}, \mathcal{L}^n, \mathcal{L}^e, \mathcal{S}, \mathcal{T}, n_0)$,

- $G \models [\omega :=_g \bot][\pi :=_g \bot][U][\omega :=_l \top][a][\pi :=_l \top][U](\omega \rightarrow [a]\pi)$ iff $R_G(a)$ is deterministic,

- $G \models [\omega :=_g \bot][U][\omega :=_l \top][a]\neg\omega$ iff $R_G(a)$ is irreflexive,
- $G \models [\omega :=_g \bot][\omega :=_l \top]\langle a\rangle\omega$ iff $R_G(a)$ is locally reflexive in n_0.

Together with the update actions of the form $\omega :=_l \phi$, the regular operation "$*$" enables us to define non-elementary classes of rooted termgraphs. As a first example, the class of all infinite rooted termgraphs cannot be modally defined in propositional dynamic logic but the following formula pins it down:

- $[\omega :=_g \top][(U; \omega?; \omega :=_l \bot)^*]\langle U\rangle\omega$.

As a second example, take the class of all a-cycle-free rooted termgraphs. It cannot be modally defined in propositional dynamic logic but the following formula pins it down:

- $[\omega :=_g \top][U][\omega :=_l \bot][a^+]\omega$.

As a third example, within the class of all a-deterministic rooted termgraphs, the class of all a-circular rooted termgraphs[1] cannot be modally defined in propositional dynamic logic but the following formula pins it down:

- $[\omega :=_g \bot][U][\omega :=_l \top]\langle a^+\rangle\omega$.

Now, within the class of all rooted termgraphs that are a- and b-deterministic, the class of all $(a \leq b)$ rooted termgraphs[2] cannot be modally defined in propositional dynamic logic but the following formula pins it down:

- $[\omega :=_g \bot][\omega :=_l \top][\pi :=_g \bot][\pi :=_l \top][((U; \omega?; a; \neg\omega?; \omega :=_l \top);$
$(U; \pi?; b; \neg\pi?; \pi :=_l \top))^*](\langle U\rangle(\pi \wedge [b]\bot) \rightarrow \langle U\rangle(\omega \wedge [a]\bot))$.

Within the class of all finite $(a \cup b)$-cycle-free (a,b)-deterministic rooted termgraphs, the class of all (a,b)-binary rooted termgraphs cannot be modally defined in propositional dynamic logic but the following formula pins it down:

- $[\omega :=_g \bot][U][\omega :=_l \top][a][\pi :=_g \top][(a \cup b)^*][\pi :=_l \bot][U](\omega \rightarrow [b][(a \cup b)^*]\pi)$.

Finally, it is well-known that Hamiltonian graphs cannot be expressed by means of monadic second-order logic formulas (not MS_1-expressible, see e.g. [7]). We recall that a graph is Hamiltonian iff there exists a cycle that visits each node exactly once. The following formula expresses the existence of such cycle.[3] In this formula, α stands for $a_1 \cup \ldots \cup a_n$, where the a_i's are the possible features used in the graph ($\mathcal{F} = \{a_1, \ldots, a_n\}$):

---

[1] In an a-circular rooted termgraph for every node n there is an i and there are a_1, $\ldots a_n$ such that $a = a_1 = a_n$ and n_k is related to n_{k+1} by an edge labelled a, for all $k \leq i$.

[2] $(a \leq b)$ rooted termgraphs are termgraphs where the path obtained by following feature b is longer than or equal to the path obtained by following feature a.

[3] This presupposes that a graph with a single node has to have a reflexive edge in order to be Hamiltonian. If we want to specify such graphs we have to replace our formula by the following one:

- $\langle \omega :=_g \top; \pi :=_g \bot; \omega :=_l \bot; \pi :=_l \top; (\alpha; \omega?; \omega :=_l \bot)^*\rangle(\pi \wedge [U]\neg\omega \wedge ([U]\pi \rightarrow \langle\alpha\rangle\top))$.

$- \langle \omega :=_g \top; \pi :=_g \bot; \omega :=_l \bot; \pi :=_l \top; (\alpha; \omega?; \omega :=_l \bot)^\star \rangle (\pi \wedge [U] \neg \omega).$

Most important of all is the ability of the language of the logic of graph modifiers to characterize finite graph homomorphisms.

Proposition 3. *Let $G = (\mathcal{N}, \mathcal{E}, \mathcal{L}^n, \mathcal{L}^e, \mathcal{S}, \mathcal{T}, n_0)$ be a finite rooted termgraph. There exists a $\star$-free action α_G and a $\star$-free formula ϕ_G such that for all finite rooted termgraphs $G' = (\mathcal{N}', \mathcal{E}', \mathcal{L}^{n'}, \mathcal{L}^{e'}, \mathcal{S}', \mathcal{T}', n_0')$, $G' \models \langle \alpha_G \rangle \phi_G$ iff there exists a graph homomorphism from G into G'.*

Proof. Let $G = (\mathcal{N}, \mathcal{E}, \mathcal{L}^n, \mathcal{L}^e, \mathcal{S}, \mathcal{T}, n_0)$ be a finite rooted termgraph. Suppose that $\mathcal{N} = \{0, \ldots, N-1\}$ and consider a sequence $(\pi_0, \ldots, \pi_{N-1})$ of pairwise distinct elements of Ω. Each π_i will identify exactly one node of $\mathcal{N}$, and π_0 will identify the root.

We define the action α_G and the formula ϕ_G as follows:

- $\beta_G = (\pi_0 :=_g \bot); \ldots; (\pi_{N-1} :=_g \bot),$
- for all non-negative integers i, if $i < N$ then $\gamma_G^i = (\neg \pi_0 \wedge \ldots \wedge \neg \pi_{i-1})?; (\pi_i :=_l \top); U,$
- $\alpha_G = \beta_G; \gamma_G^0; \ldots; \gamma_G^{N-1},$
- for all non-negative integers i, if $i < N$ then $\psi_G^i = $ if $\mathcal{L}^n(i)$ is defined then $\langle U \rangle (\pi_i \wedge \mathcal{L}^n(i))$ else $\top,$
- for all non-negative integers i, j, if $i, j < N$ then $\chi_G^{i,j} = $ if there exists an edge $e \in \mathcal{E}$ such that $\mathcal{S}(e) = i$ and $\mathcal{T}(e) = j$ then $\langle U \rangle (\pi_i \wedge \langle \mathcal{L}^e(e) \rangle \pi_j)$ else $\top,$
- $\phi_G = \psi_G^0 \wedge \ldots \wedge \psi_G^{N-1} \wedge \chi_G^{0,0} \wedge \ldots \wedge \chi_G^{N-1,N-1}.$

The reader may easily verify that for all finite rooted termgraphs $G' = (\mathcal{N}', \mathcal{E}', \mathcal{L}^{n'}, \mathcal{L}^{e'}, \mathcal{S}', \mathcal{T}', n_0')$, $G' \models \langle \alpha_G \rangle \phi_G$ iff there exists a graph homomorphism from G to G'.

5 Definability of Transformations of Termgraphs

In this section we show how elementary actions over termgraphs as defined in Section 2 can be encoded by means of formulas of our dynamic logic. Let α_a be the action defined as follows:

- $\alpha_a = (\omega :=_g \bot); (\omega :=_l \top); (\pi :=_g \bot); (\pi :=_g \langle a \rangle \omega); (a - (\top, \omega)); n; (\omega :=_g \bot); (\omega :=_l \top); (a + (\pi, \omega)).$

The reader may easily verify that for all rooted termgraphs $G = (\mathcal{N}, \mathcal{E}, \mathcal{L}^n, \mathcal{L}^e, \mathcal{S}, \mathcal{T}, n_0)$ and $G' = (\mathcal{N}', \mathcal{E}', \mathcal{L}^{n'}, \mathcal{L}^{e'}, \mathcal{S}', \mathcal{T}', n_0')$, $G \longrightarrow_{\alpha_a} G'$ iff G' is obtained from G by redirecting every a-edge pointing to the current root towards a freshly created new root. Hence, together with the update actions n, $\boldsymbol{n}$, $\omega :=_g \phi$, $\omega :=_l \phi$, $a + (\phi, \psi)$ and $a - (\phi, \psi)$, the regular operations ";", "$\cup$" and "$\star$" enable us to define the elementary actions of node labelling, local redirection and global redirection of Section 2. Let us firstly consider the elementary action of node labelling: $n : f(a_1 \Rightarrow n_1, \ldots, a_k \Rightarrow n_k)$. Applying this elementary action consists in redirecting towards nodes $n_1, \ldots, n_k$ the targets of a_1-, $\ldots$, a_k- edges starting from node n. It corresponds to the action $nl(n : f(a_1 \Rightarrow n_1, \ldots, a_k \Rightarrow n_k))$ defined as follows:

- $nl(n : f(a_1 \Rightarrow n_1, \ldots, a_k \Rightarrow n_k)) = U; \pi_n?; (f :=_l \top);$
  $(a_1 + (\pi_n, \pi_{n_1})); \ldots; (a_k + (\pi_n, \pi_{n_k})).$

where the π_i's are as in the proof of Proposition 3. The reader may easily verify that for all rooted termgraphs $G = (\mathcal{N}, \mathcal{E}, \mathcal{L}^n, \mathcal{L}^e, \mathcal{S}, \mathcal{T}, n_0)$ and $G' = (\mathcal{N}', \mathcal{E}', \mathcal{L}^{n'}, \mathcal{L}^{e'}, \mathcal{S}', \mathcal{T}', n_0')$, $G \longrightarrow_{nl(n:f(a_1 \Rightarrow n_1, \ldots, a_k \Rightarrow n_k))} G'$ iff G' is obtained from G by redirecting towards nodes $n_1, \ldots, n_k$ the targets of a_1-, ..., a_k- edges starting from node n. Let us secondly consider the elementary action of local redirection: $n \gg_a^l m$. Applying this elementary action consists in redirecting towards node m the target of an a-edge starting from node n. It corresponds to the action $lr(n, a, m)$ defined as follows:

- $lr(n, a, m) = (a - (\pi_n, \top)); (a + (\pi_n, \pi_m)).$

The reader may easily verify that for all rooted termgraphs $G = (\mathcal{N}, \mathcal{E}, \mathcal{L}^n, \mathcal{L}^e, \mathcal{S}, \mathcal{T}, n_0)$ and $G' = (\mathcal{N}', \mathcal{E}', \mathcal{L}^{n'}, \mathcal{L}^{e'}, \mathcal{S}', \mathcal{T}', n_0')$, $G \longrightarrow_{lr(n,a,m)} G'$ iff G' is obtained from G by redirecting towards node m the target of an a-edge starting from node n. Let us thirdly consider the elementary action of global redirection: $n \gg_a^g m$. Applying this elementary action consists in redirecting towards node n the target of every a-edge pointing towards node m. It corresponds to the action $gr(n, a, m)$ defined as follows:

- $gr(n, a, m) = (\lambda_a :=_g \bot); (\lambda_a :=_g \langle a \rangle \pi_n); (a - (\top, \pi_n)); (a + (\lambda_a, \pi_m)).$

The reader may easily verify that for all rooted termgraphs $G = (\mathcal{N}, \mathcal{E}, \mathcal{L}^n, \mathcal{L}^e, \mathcal{S}, \mathcal{T}, n_0)$, $G' = (\mathcal{N}', \mathcal{E}', \mathcal{L}^{n'}, \mathcal{L}^{e'}, \mathcal{S}', \mathcal{T}', n_0')$, $G \longrightarrow_{gr(n,a,m)} G'$ iff G' is obtained from G by redirecting towards node n the target of every a-edge pointing towards node m. To redirect towards n the target of all edges pointing towards m, the action $gr(n, a, m)$ can be performed for all $a \in \mathcal{F}$. We get $gr(n, m) = \bigwedge_{a \in \mathcal{F}} gr(n, a, m)$.

6   Translating Rewrite Rules into Dynamic Logic

Now we are ready to show how termgraph rewriting can be specified by means of formulas of our dynamic logic.

Let $G \to (a_1, \ldots, a_n)$ be a rewrite rule as defined in Section 2, i.e., $G = (\mathcal{N}, \mathcal{E}, \mathcal{L}^n, \mathcal{L}^e, \mathcal{S}, \mathcal{T}, n_0)$ is a finite rooted termgraph and $(a_1, \ldots, a_n)$ is a finite sequence of elementary actions. We have seen how to associate to G a *-free action α_G and a *-free formula ϕ_G such that for all finite rooted termgraphs $G' = (\mathcal{N}', \mathcal{E}', \mathcal{L}^{n'}, \mathcal{L}^{e'}, \mathcal{S}', \mathcal{T}', n_0')$, $G' \models \langle \alpha_G \rangle \phi_G$ iff there exists a graph homomorphism from G into G'. We have also seen how to associate to the elementary actions a_1, ..., a_n actions $\alpha_1, \ldots, \alpha_n$. In the following proposition we show how to formulate the fact that a normal form with respect to a rewrite rule (generalization to a set of rules is obvious) satisfies a given formula φ. A termgraph t is in normal form with respect to a rule R iff t cannot be rewritten by means of R. Such formulation may help to express proof obligations of programs specified as termgraph rewrite rules. Let $n_1, \ldots, n_k$ be the list of all nodes occurring in $a_1, \ldots, a_n$ but not occurring in G. The truth of the matter is that

Proposition 4. *Let φ be a modal formula. For all finite rooted termgraphs G' $= (\mathcal{N}', \mathcal{E}', \mathcal{L}^{n'}, \mathcal{L}^{e'}, \mathcal{S}', \mathcal{T}', n_0')$, every normal form of G' with respect to $G \to$ $(a_1, \ldots, a_n)$ satisfies φ iff $G' \models [(\alpha_G; \phi_G?; \boldsymbol{n}; (\pi_{n_1} :=_g \bot); (\pi_{n_1} :=_l \top); \ldots; \boldsymbol{n};$ $(\pi_{n_k} :=_g \bot); (\pi_{n_k} :=_l \top); \alpha_1; \ldots; \alpha_n)^{\star}]([\alpha_G; \phi_G?]\bot \to \varphi).$*

In other respects, the following proposition shows how an invariant φ of a rewrite rule can be expressed in the proposed logic.

Proposition 5. *Let φ be a modal formula. The rewrite rule $G \to (a_1, \ldots, a_n)$ strongly preserves φ iff $\models \varphi \to [\alpha_G; \phi_G?; \boldsymbol{n}; (\pi_{n_1} :=_g \bot); (\pi_{n_1} :=_l \top); \ldots; \boldsymbol{n};$ $(\pi_{n_k} :=_g \bot); (\pi_{n_k} :=_l \top); \alpha_1; \ldots; \alpha_n]\varphi.$*

7 Conclusion

We have defined a dynamic logic which can be used either (i) to describe data-structures which are possibly defined by means of pointers and considered as termgraphs in this paper, (ii) to specify programs defined as rewrite rules which process these data-structures, or (iii) to reason about data-structures themselves and about the behavior of the programs under consideration. The features of our logic are very appealing. They contribute to define a logic which faithfully captures the behavior of termgraph rewrite systems. They also open new perspectives for the verification of programs manipulating pointers.

Our logic is undecidable in general. This is not surprising at all given its expressive power. However, this logic is very promising in developing new proof procedures regarding properties of termgraph rewrite systems. For instance, we have discussed a first fragment of the logic, consisting of formulas without re-labelling actions, where validity is decidable. Future work include mainly the investigation of new decidable fragments of our logic and their application to program verification.

References

1. Areces, C., Blackburn, P., Marx, M.: A road-map on complexity for hybrid logics. In: Flum, J., Rodríguez-Artalejo, M. (eds.) CSL 1999. LNCS, vol. 1683, pp. 307–321. Springer, Heidelberg (1999)
2. Areces, C., ten Cate, B.: Hybrid logics. In: Blackburn, P., van Benthem, J., Wolter, F. (eds.) Handbook of Modal Logic, vol. 3. Elsevier Science, Amsterdam (2006)
3. Aucher, G., Balbiani, P., Fariñas Del Cerro, L., Herzig, A.: Global and local graph modifiers. In: Electronic Notes in Theoretical Computer Science (ENTCS), Special issue Proceedings of the 5th Workshop on Methods for Modalities (M4M5 2007), vol. 231, pp. 293–307 (2009)
4. Baader, F., Nipkow, T.: Term Rewriting and All That. Cambridge University Press, Cambridge (1998)
5. Barendregt, H., van Eekelen, M., Glauert, J., Kenneway, R., Plasmeijer, M.J., Sleep, M.: Term graph rewriting. In: de Bakker, J.W., Nijman, A.J., Treleaven, P.C. (eds.) PARLE 1987. LNCS, vol. 259, pp. 141–158. Springer, Heidelberg (1987)

6. Caferra, R., Echahed, R., Peltier, N.: A term-graph clausal logic: Completeness and incompleteness results. Journal of Applied Non-classical Logics 18, 373–411 (2008)
7. Courcelle, B.: The expression of graph properties and graph transformations in monadic second-order logic. In: Rozenberg, G. (ed.) Handbook of Graph Grammars and Computing by Graph Transformations. Foundations, vol. 1, pp. 313–400. World Scientific, Singapore (1997)
8. Echahed, R.: Inductively sequential term-graph rewrite systems. In: Ehrig, H., Heckel, R., Rozenberg, G., Taentzer, G. (eds.) ICGT 2008. LNCS, vol. 5214, pp. 84–98. Springer, Heidelberg (2008)
9. Ehrig, H., Engels, G., Kreowski, H.-J., Rozenberg, G. (eds.): Handbook of Graph Grammars and Computing by Graph Transformations. Applications, Languages and Tools, vol. 2. World Scientific, Singapore (1999)
10. Ehrig, H., Kreowski, H.-J., Montanari, U., Rozenberg, G. (eds.): Handbook of Graph Grammars and Computing by Graph Transformations. Concurrency, Parallelism and Distribution, vol. 3. World Scientific, Singapore (1999)
11. Harel, D., Kozen, D., Tiuryn, J.: Dynamic Logic. MIT Press, Cambridge (2000)
12. Plump, D.: Term graph rewriting. In: Ehrig, H., Engels, G., Kreowski, H.J., Rozenberg, G. (eds.) Handbook of Graph Grammars and Computing by Graph Transformation, vol. 2, pp. 3–61. World Scientific, Singapore (1999)
13. Rozenberg, G. (ed.): Handbook of Graph Grammars and Computing by Graph Transformations. Foundations, vol. 1. World Scientific, Singapore (1997)
14. Tiuryn, J.: Fixed-points and algebras with infinitely long expression, part 1, regular algebras. Fundamenta Informaticae 2, 103–127 (1978)
15. Tiuryn, J.: Fixed-points and algebras with infinitely long expression, part 2, μ-clones of regular algebras. Fundamenta Informaticae 2(3), 317–335 (1979)

A New Type of Behaviour-Preserving Transition Insertions in Unfolding Prefixes

Victor Khomenko*

School of Computing Science, Newcastle University, UK
Victor.Khomenko@ncl.ac.uk

Abstract. A new kind of behaviour-preserving insertions of new transitions in Petri nets is proposed, and a method for computing such insertions using a complete unfolding prefix of the Petri net is developed. Moreover, as several transformations often have to be applied one after the other, the developed theory allows one to avoid (expensive) re-unfolding after each transformation, and instead use local modifications on the existing complete prefix to obtain a complete prefix of the modified net.

Keywords: Transition insertions, transformations, Petri net unfoldings.

1 Introduction

Many design methods based on Petri nets modify the original specification by behaviour-preserving insertion of new transitions. For example, in the design flow for asynchronous circuits based on Signal Transition Graphs (STGs) [1, 11] — a class of labelled Petri nets widely used for specifying the behaviour of asynchronous control circuits — transition insertions are used at two different stages: for resolving state encoding conflicts and for logic decomposition of gates. (At these stages, new internal signals are created in the circuit, and the changes in their values have to be modelled by inserting new transitions into the STG.) In the discussion below, though this particular motivating application is intended, the developed techniques and algorithms are not specific to this application domain and suitable for generic Petri nets (e.g. one can envisage applications to action refinement).

This paper focuses primarily on *SB-preserving* transformations, i.e. ones preserving Safeness and Behaviour (in the sense that the original and the transformed STGs are weakly bisimilar, provided that the newly inserted transitions are considered silent) of the Petri net.

In previous work [6] several types of transition insertions were introduced, and they turned out to be successful in certain applications, e.g. for resolution of state encoding conflicts [7]. However, for other applications, in particular for

---

* V. Khomenko is a Royal Academy of Engineering/EPSRC Post-Doctoral Research Fellow. This research was supported by the RAEng/EPSRC post-doctoral research fellowship EP/C53400X/1 (DAVAC) and EPSRC grant EP/G037809/1 (VERDAD).

H. Ehrig et al. (Eds.): ICGT 2010, LNCS 6372, pp. 75–90, 2010.

logic decomposition of asynchronous circuits, those types of transition insertions appear to be insufficient. Intuitively, there are relatively few transformations of the types described in [6] (their total number is just quadratic in the size of the Petri net under some reasonable assumptions), whereas in logic decomposition a new transition has to be inserted at positions where sub-expressions of a certain Boolean expression change their values, and there could be exponentially many such positions. Hence, in most cases one will not be able to find a set of transition insertions of the listed types which would cater for a particular sub-expression. Therefore, a new type of transition insertions is required, for which exponential (in the size of the STG) number of insertions is possible. (The technical report [8] explains this issue in more detail and provides a practical example.)

In this paper, the framework developed in [6] is extended with another type of transition insertions, called *generalised transition insertions (GTIs)*, and it is demonstrated how GTIs can be employed in practical applications. In particular:

- It is shown how to check efficiently whether a GTI is SB-preserving using a complete unfolding prefix of the Petri net.
- As several insertions often have to be applied one after the other, a theory allowing to avoid (expensive) re-unfolding of the Petri net after each insertion, and instead use local modifications on the existing prefix to obtain a complete prefix of the modified net, is developed. This has an additional advantage, viz. the produced prefix is similar to the original one, which is useful for visualisation and allows one to transfer some information (like computed encoding conflicts in asynchronous circuit design [7]) from the original prefix to the modified one, rather than having to re-compute it from scratch.
- A method allowing to avoid enumerating all GTIs and compute only potentially useful ones is developed; note that unlike the transition insertions proposed in [6], whose number is relatively small, there can be exponentially many GTIs (in the size of the Petri net), and so limiting the number of GTIs that have to be considered is very important in practice.

The formal proofs of all the results, as well as further explanations, can be found in the technical report [8] available online.

2   Basic Notions

This section recalls the basic definitions concerning Petri nets and the notions related to net unfoldings and their canonical prefixes (see also [3, 4, 5, 9, 10]).

Petri Nets

A *net* is a triple $N \stackrel{\mathrm{df}}{=} (P, T, F)$ such that P and T are disjoint sets of respectively *places* and *transitions*, and $F \subseteq (P \times T) \cup (T \times P)$ is a *flow relation*. A *marking* of N is a multiset M of places, i.e. $M : P \to \mathbb{N} = \{0, 1, 2, \ldots\}$. The standard rules about drawing nets are adopted in this paper, viz. places are represented as circles, transitions as boxes, the flow relation by arcs, and markings are shown by placing tokens within circles. As usual, $^{\bullet}z \stackrel{\mathrm{df}}{=} \{y \mid (y, z) \in F\}$ and $z^{\bullet} \stackrel{\mathrm{df}}{=} \{y \mid$

$(z, y) \in F\}$ denote the *preset* and *postset* of $z \in P \cup T$, and $\bullet Z \stackrel{\mathrm{df}}{=} \bigcup_{z \in Z} \bullet z$ and $Z \bullet \stackrel{\mathrm{df}}{=} \bigcup_{z \in Z} z \bullet$, for all $Z \subseteq P \cup T$. It is assumed that $\bullet t \neq \emptyset$, for every $t \in T$. A net N is *finite* if $P \cup T$ is finite, and *infinite* otherwise.

A *net system* is a tuple $\Sigma \stackrel{\mathrm{df}}{=} (P_\Sigma, T_\Sigma, F_\Sigma, M_\Sigma)$ where $(P_\Sigma, T_\Sigma, F_\Sigma)$ is a finite net and M_Σ is an *initial* marking. A transition $t \in T_\Sigma$ is *enabled* at a marking M, if $M(p) \geq 1$ for every $p \in \bullet t$. Such a transition can be *fired*, leading to the marking $M' \stackrel{\mathrm{df}}{=} M - \bullet t + t \bullet$, where '$-$' and '$+$' stand for the multiset difference and sum respectively. A finite sequence $t_1 t_2 t_3 \ldots$ of transitions is an *execution from a marking* M if either it is empty or t_1 is enabled at M and $t_2 t_3 \ldots$ is an execution from the marking reached by firing t_1; moreover, it is an execution of Σ if it is an execution from M_Σ.

The set of *reachable* markings of Σ is the smallest (w.r.t. $\subset$) set containing M_Σ and such that if M is in this set then firing any enabled at M transition leads to a marking in this set. A transition is *dead* if no reachable marking enables it. A transition is *live* if from every reachable marking there is an execution containing it. (Note that being live is a stronger property than being non-dead.)

A net system Σ is *k-bounded* if $M(p) \leq k$ for each reachable marking M and each place $p \in P_\Sigma$; *safe* if it is 1-bounded; and *bounded* if it is k-bounded for some $k \in \mathbb{N}$. A finite Σ has finitely many reachable markings iff it is bounded.

Branching Processes and Canonical Prefixes

A *finite and complete unfolding prefix* of a Petri net Σ is a finite acyclic net which implicitly represents all the reachable states of Σ together with transitions enabled at those states. Intuitively, it can be obtained through *unfolding* Σ, by successive firings of transitions, under the following assumptions: (a) for each new firing a fresh transition (called an *event*) is generated; (b) for each newly produced token a fresh place (called a *condition*) is generated. The unfolding is infinite whenever Σ has an infinite run; however, if Σ has finitely many reachable states then the unfolding eventually starts to repeat itself and can be truncated (by identifying a set E_{cut} of *cut-off* events beyond which the prefix is not generated) without loss of information, yielding a finite and complete prefix. Due to its structural properties (such as acyclicity), the reachable markings of Σ can be represented using *configurations* of any of its complete prefixes. Intuitively, a configuration is a partially ordered execution, i.e. an execution where the order of firing of some of the events (viz. concurrent ones) is not important.

Efficient algorithms exist for building finite and complete prefixes [4,5], which ensure that the number of non-cut-off events in the resulting prefix never exceeds the number of reachable states of Σ. In fact, complete prefixes are often exponentially smaller than the corresponding state graphs, especially for highly concurrent Petri nets, because they represent concurrency directly rather than by multidimensional 'diamonds' as it is done in state graphs. For example, if the original Petri net consists of 100 transitions which can fire once in parallel, the state graph will be a 100-dimensional hypercube with 2^{100} vertices, whereas the complete prefix will coincide with the net itself. The experimental results in [5] demonstrate that high levels of compression can indeed be achieved in practice.

Formally, two nodes (places or transitions) y and y' of a net are *in conflict*, denoted by $y \# y'$, if there are distinct transitions t and t' such that ${}^\bullet t \cap {}^\bullet t' \neq \emptyset$ and (t, y) and (t', y') are in the reflexive transitive closure of the flow relation, denoted by $\preceq$. A node y is in *self-conflict* if $y \# y$.

An *occurrence net* is a net $ON \stackrel{\mathrm{df}}{=} (B, E, G)$, where B is the set of *conditions* (places) and E is the set of *events* (transitions), satisfying the following: ON is acyclic (i.e. $\preceq$ is a partial order); for every $b \in B$, $|{}^\bullet b| \leq 1$; no node $y \in B \cup E$ is in self-conflict; and there are finitely many y' such that $y' \prec y$, where $\prec$ denotes the transitive closure of G. $Min(ON)$ will denote the set of minimal (w.r.t. $\prec$) elements of $B \cup E$. The relation $\prec$ is the *causality relation*. Two nodes are *concurrent*, denoted $y \parallel y'$, if neither $y \# y'$ nor $y \preceq y'$ nor $y' \preceq y$.

A *homomorphism* from an occurrence net ON to a net system Σ is a mapping $h : B \cup E \to P_\Sigma \cup T_\Sigma$ such that: $h(B) \subseteq P_\Sigma$ and $h(E) \subseteq T_\Sigma$ (conditions are mapped to places, and events to transitions); for all $e \in E$, the restriction of h to ${}^\bullet e$ is a bijection between ${}^\bullet e$ and ${}^\bullet h(e)$ and the restriction of h to $e^\bullet$ is a bijection between $e^\bullet$ and $h(e)^\bullet$ (transition environments are preserved); the restriction of h to $Min(ON)$ is a bijection between $Min(ON)$ and M_Σ (minimal conditions correspond to the initial marking); and for all $e, f \in E$, if ${}^\bullet e = {}^\bullet f$ and $h(e) = h(f)$ then $e = f$ (there is no redundancy). A *branching process* of Σ is a tuple $\pi_\Sigma \stackrel{\mathrm{df}}{=} (B_{\pi_\Sigma}, E_{\pi_\Sigma}, G_{\pi_\Sigma}, h_{\pi_\Sigma})$ such that $(B_{\pi_\Sigma}, E_{\pi_\Sigma}, G_{\pi_\Sigma})$ is an occurrence net and h_{π_Σ} is a homomorphism from it to Σ. If a node $x \in B_{\pi_\Sigma} \cup E_{\pi_\Sigma}$ is such

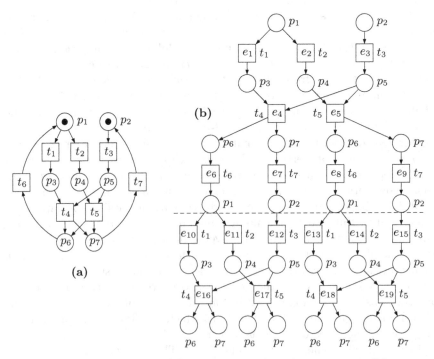

Fig. 1. A net system **(a)** and one of its branching processes **(b)**

that $h_{\pi_\Sigma}(x) = y \in P_\Sigma \cup T_\Sigma$, then x is often referred to as being y-*labelled* or as an *instance of* y.

A branching process $\pi'_\Sigma = (B_{\pi'_\Sigma}, E_{\pi'_\Sigma}, G_{\pi'_\Sigma}, h_{\pi'_\Sigma})$ of Σ is a *prefix* of π_Σ, denoted $\pi'_\Sigma \sqsubseteq \pi_\Sigma$, if $(B_{\pi'_\Sigma}, E_{\pi'_\Sigma}, G_{\pi'_\Sigma})$ is a subnet of $(B_{\pi_\Sigma}, E_{\pi_\Sigma}, G_{\pi_\Sigma})$ containing all minimal elements and such that: if $e \in E_{\pi'_\Sigma}$ and $(b, e) \in G_{\pi_\Sigma}$ or $(e, b) \in G_{\pi_\Sigma}$ then $b \in B_{\pi'_\Sigma}$; if $b \in B_{\pi'_\Sigma}$ and $(e, b) \in G_{\pi_\Sigma}$ then $e \in E_{\pi'_\Sigma}$; and $h_{\pi'_\Sigma}$ is the restriction of h_{π_Σ} to $B_{\pi'_\Sigma} \cup E_{\pi'_\Sigma}$. For each Σ there exists a unique (up to isomorphism) maximal (w.r.t. $\sqsubseteq$) branching process Unf_Σ, called the *unfolding* of Σ. To simplify the notation, h_Σ is used instead of h_{π_Σ}; this is justified since $h_{\pi_\Sigma}(x)$ is the same in any branching process of Σ containing x, in particular, one can always refer to Unf_Σ (branching processes of Σ are prefixes of Unf_Σ).

A set of events E' is *downward-closed* if all causal predecessors of the events in E' also belong to E'. Such a set *induces* a unique branching process $\pi_\Sigma \sqsubseteq Unf_\Sigma$ whose events are exactly those in E', and whose conditions are $Min(ON) \cup E'^\bullet$.

An example of a safe net system and one of its branching processes is shown in Fig. 1, where the homomorphism h_Σ is indicated by the labels of the nodes. In it, the set $\{e_1, \ldots, e_9\}$ induces a prefix shown above the dashed line.

Configurations and cuts. A *configuration* of a branching process π_Σ is a finite downward-closed set of events $C \subseteq E_{\pi_\Sigma}$ such that there are no $e, f \in C$ for which $e\#f$. For every event $e \in E_{\pi_\Sigma}$, the configuration $[e]_\Sigma \overset{\mathrm{df}}{=} \{f \mid f \preceq e\}$ is called the *local configuration* of e. Moreover, for a set of events E', $C \oplus E'$ denotes that $C \cup E'$ is a configuration and $C \cap E' = \emptyset$. Such a set is a *suffix* of C, and the configuration $C \oplus E'$ is an *extension* of C. For singleton suffixes, $C \oplus e$ is used instead of $C \oplus \{e\}$. For a transition $t \in T_\Sigma$ and a configuration C of π_Σ, $\#_t C$ denotes the number of instances of t in C.

A set of conditions B' such that for all distinct $b, b' \in B'$, $b \parallel b'$, is called a *coset*. A *cut* is a maximal (w.r.t. $\subset$) co-set. Every marking reachable from $Min(ON)$ is a cut. If C is a configuration of π_Σ then $Cut_\Sigma(C) \overset{\mathrm{df}}{=} (Min(ON) \cup C^\bullet) \setminus {}^\bullet C$ is a cut; moreover, the multiset of places $Mrk_\Sigma(C) \overset{\mathrm{df}}{=} h_\Sigma(Cut_\Sigma(C))$ is a reachable marking of Σ, called the *final marking* of C. A marking M of Σ is *represented* in π_Σ if there is a configuration C of π_Σ such that $M = Mrk_\Sigma(C)$. Every marking represented in π_Σ is reachable in the original net system Σ, and every reachable marking of Σ is represented in Unf_Σ.

Note that the notations $[\cdot]_\Sigma$, $Cut_\Sigma(\cdot)$ and $Mrk_\Sigma(\cdot)$ differ from the conventional ones by the presence of subscripts. They are necessary in the settings of this paper since the existing unfolding prefix is modified whenever the original Petri net is modified, i.e. the same event e may belong to the prefixes of two different Petri nets, viz. the original and modified ones, and the subscript is needed to distinguish between them. That Σ rather than π_Σ is used as the subscripts is justified in the view of the fact that the denoted objects are the same in any branching process of Σ containing the necessary events; in particular, one can always refer to Unf_Σ.

Cutting context. There exist different methods of truncating Petri net unfoldings. The differences are related to the kind of information about the original unfolding one wants to preserve in the prefix, as well as to the choice between using either only local configurations (which can improve the running time of an algorithm), or all configurations (which can result in a smaller prefix) for truncating the prefix.

In order to cope with different variants of the technique for truncating unfoldings, the abstract parametric model developed in [9] will be used. The main idea behind it is to speak about configurations of Unf_Σ rather than reachable markings of Σ. In this model, the first parameter determines the information one intends to preserve in the prefix (in the standard case, this is the set of reachable markings). Formally, this information corresponds to the equivalence classes of some equivalence relation $\approx$ on the configurations of Unf_Σ. The other parameters are more technical: they specify the circumstances under which an event can be designated as a cut-off event.

A triple $\Theta \stackrel{\text{df}}{=} \left(\approx, \lhd, \{\mathcal{C}_e\}_{e \in E_{Unf_\Sigma}}\right)$ is a *cutting context* if:

1. $\approx$ is an equivalence relation on the configurations of Unf_Σ.
2. $\lhd$, called an *adequate* order, is a strict well-founded partial order on the configurations of Unf_Σ refining $\subset$, i.e. $C' \subset C''$ implies $C' \lhd C''$.
3. $\approx$ and $\lhd$ are *preserved by finite extensions*, i.e. for every pair of configurations $C' \approx C''$, and for every finite suffix E'' of C'', there exists a finite suffix E' of C' such that
   (a) $C' \oplus E' \approx C'' \oplus E''$, and
   (b) if $C' \lhd C''$ then $C' \oplus E' \lhd C'' \oplus E''$.
4. For each event e of Unf_Σ, $\mathcal{C}_e$ is a set of configurations of Unf_Σ.

The main idea behind the adequate order is to specify which configurations will be preserved in the complete prefix; it turns out that all $\lhd$-minimal configurations in each equivalence class of $\approx$ will be preserved. The last parameter is needed to specify for each event e of Unf_Σ the set $\mathcal{C}_e$ of configurations which can be used as the corresponding configurations to declare e a cut-off event. For example, $\mathcal{C}_e$ may contain all configurations of Unf_Σ, or, as usually the case in practice, only the local ones.

For convenience, the domain of $\lhd$ is extended to the events of Unf_Σ as follows: $e \lhd f$ iff $[e]_\Sigma \lhd [f]_\Sigma$. Clearly, $\lhd$ is a well-founded partial order on the set of events. Hence, one can use Noetherian induction (see [2]) for definitions and proofs, i.e. it suffices to define or prove something for an event under the assumption that it has already been defined or proven for all its $\lhd$-predecessors.

In this paper it is assumed that the first component of the cutting context is the equivalence of final markings, defined as $C' \approx_{mar} C''$ iff $Mrk_\Sigma(C') = Mrk_\Sigma(C'')$. The first time the unfolding prefix is built, some standard cutting context, e.g. Θ_{ERV} corresponding to the framework proposed in [4], can be used. However, as transformations are applied to Σ and then mirrored in the corresponding unfolding prefix, the cutting context will change.

Completeness of branching processes. A branching process π_Σ is *complete* w.r.t. *a set* E_{cut} of its events if the following hold:
1. For each configuration C of Unf_Σ there is a configuration C' of π_Σ such that $C' \cap E_{cut} = \emptyset$ and $C \approx C'$.

2. For each configuration C of π_Σ such that $C \cap E_{cut} = \emptyset$, and for each event e of Unf_Σ such that $C \oplus e$ is a configuration of Unf_Σ, $C \oplus e$ is a configuration of π_Σ.

A branching process π_Σ is *complete* if it is complete w.r.t. some set E_{cut}.

Note that π_Σ remains complete after removing all causal successors of the events in E_{cut}, and so, w.l.o.g., one can assume that E_{cut} contains only causally maximal events of π_Σ. Note also that this definition depends only on the equivalence $\approx$, and not on the other components of the cutting context.

As an example, consider the net system in Fig. 1(a). If $\approx$ is taken to be $\approx_{mar}$ then the prefix in Fig. 1(b) is complete w.r.t. the set $E_{cut} = \{e_5, e_{16}, e_{17}\}$ (this choice is not unique: one could have chosen, e.g. $E_{cut} = \{e_4, e_{18}, e_{19}\}$). Notice that the events e_8, e_9, e_{13}–e_{15}, e_{18}, and e_{19} can be removed from the prefix without affecting its completeness.

Canonical prefix. Now one can define *static* cut-off events, without reference to any unfolding algorithm (hence the term 'static'), together with *feasible* events, which are precisely those events whose causal predecessors are not cut-off events, and as such must be included in the prefix determined by the static cut-off events.

The set of *feasible* events, denoted by $fsble_\Theta$, and the set of *static cut-off* events, denoted by cut_Θ, are two sets of events of Unf_Σ defined inductively, in the following way:

1. An event e is *feasible* if $([e]_\Sigma \setminus \{e\}) \cap cut_\Theta = \emptyset$.
2. An event e is a *static cut-off* if it is feasible, and there is a configuration $C \in \mathcal{C}_e$ such that $C \subseteq fsble_\Theta \setminus cut_\Theta$, $C \approx [e]_\Sigma$, and $C \lhd [e]_\Sigma$; any C satisfying these conditions will be called a *corresponding* configuration of e.

The sets $fsble_\Theta$ and cut_Θ are well-defined sets due to Noetherian induction [9].

Once the feasible events have been defined, the notion of the canonical prefix arises quite naturally, after observing that $fsble_\Theta$ is a downward-closed set of events. The branching process $Pref_\Sigma^\Theta$ induced by $fsble_\Theta$ is called the *canonical prefix* of Unf_Σ. Note that $Pref_\Sigma^\Theta$ is uniquely determined by the cutting context Θ, hence the term 'canonical'.

Several fundamental properties of $Pref_\Sigma^\Theta$ were proven in [9]. In particular, $Pref_\Sigma^\Theta$ is always complete w.r.t. cut_Θ, and it is finite if $\approx$ has finitely many equivalence classes (in the case of $\approx_{mar}$ the equivalence classes correspond to the reachable markings, i.e. this condition is equivalent to the boundedness of the Petri net) and, for each $e \in E_{Unf_\Sigma}$, $\mathcal{C}_e$ contains all the local configurations of Unf_Σ. Moreover, most unfolding algorithms proposed in literature, in particular those in [4,5], build precisely this canonical prefix.

3   Previous Work

In this section the transition insertions presented in [6] are briefly described. The theory developed in [6] allows one to check whether such transformations are SB-preserving using the canonical unfolding prefix, and to avoid re-unfolding the modified Petri net and instead use local modifications on the existing prefix to obtain the canonical prefix of the modified net.

Sequential pre-insertion. A sequential pre-insertion is essentially a generalised transition splitting, and is defined as follows. Given a transition t and a set of places $S \subseteq {}^\bullet t$, the sequential pre-insertion $S \wr t$ is the transformation inserting a new transition u (with an additional place) 'splitting off' the places in S from t. The picture below illustrates the sequential pre-insertion $\{p_1, p_2\} \wr t$.

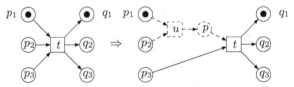

Sequential pre-insertions always preserve safeness and traces (i.e. executions with the newly inserted transition removed). However, in general, the behaviour is not preserved, and so a sequential pre-insertion is not guaranteed to be SB-preserving (in fact, it can introduce deadlocks). Given an unfolding prefix, it is quite easy to check whether a pre-insertion is SB-preserving [6].

Sequential post-insertion. Similarly to sequential pre-insertion, sequential post-insertion is also a generalisation of transition splitting, and is defined as follows. Given a transition t and a set of places $S \subseteq t^\bullet$, the sequential post-insertion $t \wr S$ is the transformation inserting a new transition u (with an additional place) 'splitting off' the places in S from t. Sequential post-insertions are always SB-preserving. The picture below illustrates the sequential post-insertion $t \wr \{q_1, q_2\}$.

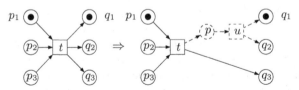

Concurrent insertion. Concurrent transition insertion can be advantageous for performance, since the inserted transition can fire in parallel with the existing ones. It is defined as follows. Given two distinct transitions, t' and t'', and an $n \in \{0, 1\}$, the concurrent insertion $t' {}^n{\mapsto} t''$ is the transformation inserting a new transition u (with a couple of additional places) between t' and t'', and putting n tokens in the place in its preset. The picture below illustrates the concurrent insertion $t_1 {}^1{\mapsto} t_3$ (note that the token in p is needed to prevent a deadlock).

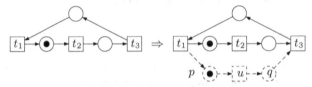

In general, concurrent insertions preserve neither safeness nor behaviour. In [6], an efficient test whether a concurrent insertion is SB-preserving, working on an unfolding prefix, has been developed.

4   Generalised Transition Insertions

Before introducing a new type of transition insertions, a few auxiliary notions are required. The idea of the following definition comes from [12], though the details are different.

Definition 1 (Lock relation). Two distinct transitions t' and t'' of a Petri net are in the *lock relation*, denoted $t' \circlearrowright t''$, if in any execution of the Petri net:

- the occurrences of t' and t'' alternate; and
- the first occurrence of t' precedes the first occurrence of t''.   ◇

It turns out that given a canonical unfolding prefix, the lock relation can be conservatively approximated.

Definition 2 (Approximated lock relation $\tilde{\circlearrowright}$). Let t' and t'' be two distinct transitions of a safe Petri net Σ, and $Tokens(t', t'', C) \overset{\text{df}}{=} \#_{t'}C - \#_{t''}C$ for any configuration C of Unf_Σ. Then for the given canonical prefix $Pref_\Sigma^\Theta$, the relation $\tilde{\circlearrowright}$ is defined as follows: $t' \tilde{\circlearrowright} t''$ iff

1. $Tokens(t', t'', [e]_\Sigma) = 1$ for each t'-labelled event e of the prefix; and
2. $Tokens(t', t'', [e]_\Sigma) = 0$ for each t''-labelled event e of the prefix; and
3. $Tokens(t', t'', [e]_\Sigma) = Tokens(t', t'', C^e)$ for each cut-off event e of the prefix with the corresponding configuration C^e.[1]   ◇

Observe that in contrast to $\circlearrowright$, the approximation $\tilde{\circlearrowright}$ is very easy to compute given $Pref_\Sigma^\Theta$.

Proposition 3 (Conservativeness of $\tilde{\circlearrowright}$). *For any pair of distinct transitions t' and t'' of a safe Petri net Σ, $t' \tilde{\circlearrowright} t''$ implies $t' \circlearrowright t''$.*   ◇

Note that the inverse of this property is, in general, not true, i.e. it can happen that $t' \circlearrowright t''$ but $t' \tilde{\not\circlearrowright} t''$. However, this is conservative, and the result below shows that in practically important cases $\circlearrowright$ and $\tilde{\circlearrowright}$ coincide.

Proposition 4 (Exactness of $\tilde{\circlearrowright}$ in the live case). *For any distinct transitions t' and t'' of a safe Petri net Σ such that at least one of them is live, $t' \tilde{\circlearrowright} t''$ iff $t' \circlearrowright t''$.*   ◇

The central notion of this paper, viz. *generalised transition insertion*, is now introduced (cf. Fig. 2). The inserted new transition will always be denoted by u.

Definition 5 (Generalised transition insertion). Let $\Sigma = (P_\Sigma, T_\Sigma, F_\Sigma, M_\Sigma)$ be a safe Petri net and $S, D \subseteq T_\Sigma$ be two disjoint non-empty sets of transitions called respectively *sources* and *destinations*, such that for every source $s \in S$ either $s \circlearrowright d$ for all $d \in D$ (in which case s is called *uninitiated*) or $d \circlearrowright s$ for all $d \in D$ (in which case s is called *initiated*). Then the *generalised transition insertion (GTI)* $S \rightarrowtail\!\!\!\rightarrow D$ is the transformation yielding the Petri net $\Sigma^u = (P_{\Sigma^u}, T_{\Sigma^u}, F_{\Sigma^u}, M_{\Sigma^u})$, where

---

[1] In general, a cut-off event e can have multiple corresponding configurations, but only one (any) of them is stored with e when the prefix is built.

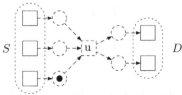

Fig. 2. Generalised transition insertion: the added nodes and arcs are shown in dashed lines; the dotted lines show the sets S and D. Some of the places in ${}^\bullet u$ can be initially marked, depending on the lock relation between the transitions in S and D.

- $P_{\Sigma^u} \stackrel{\mathrm{df}}{=} P_\Sigma \cup \{p_s \mid s \in S\} \cup \{p_d \mid d \in D\}$, where p_s and p_d are new places;
- $T_{\Sigma^u} \stackrel{\mathrm{df}}{=} T_\Sigma \cup \{u\}$, where u is a new transition;
- $F_{\Sigma^u} \stackrel{\mathrm{df}}{=} F_\Sigma \cup \{(s, p_s) \mid s \in S\} \cup \{(p_s, u) \mid s \in S\} \cup \{(u, p_d) \mid d \in D\} \cup \{(p_d, d) \mid d \in D\}$;
- $M_{\Sigma^u} \stackrel{\mathrm{df}}{=} M_\Sigma \cup \{p_s \mid s \in S$ and s is initiated$\}$.

A GTI is called *conservative* if $\tilde{\circlearrowleft}$ is used instead of $\circlearrowleft$ in this definition.[2]   ◊

Note that since by Prop. 3 $\tilde{\circlearrowleft}$ is a stricter relation than $\circlearrowleft$, conservative GTIs form a subset of all GTIs.

Proposition 6 (Validity of GTIs). *Generalised transition insertions are SB-preserving.*   ◇

5   GTIs in the Prefix

This section shows how to perform a GTI by locally modifying the existing prefix of the original net, avoiding thus re-unfolding.

The proposition below explains the natural correspondence between the configurations of Unf_Σ and Unf_{Σ^u}, assuming that Σ^u is obtained from Σ by applying a GTI. It turns out that any configuration C^u of Unf_{Σ^u} corresponds to the unique configuration $\psi(C^u)$ of Unf_Σ, and any configuration C of Unf_Σ corresponds to at most two configurations of Unf_{Σ^u}, denoted $\varphi(C)$ and $\overline{\varphi}(C)$, such that the latter is an extension of the former by a single u-labelled event.

Proposition 7 (Correspondence between configurations of Unf_Σ and Unf_{Σ^u}). *Let Σ^u is obtained from a safe Petri net Σ by applying a GTI, C be a configuration of Unf_Σ and C^u be a configuration of Unf_{Σ^u}. Then:*

1. *The set $\psi(C^u) \stackrel{\mathrm{df}}{=} \{e \in C^u \mid h_{\Sigma^u}(e) \neq u\}$ is a configuration of Unf_Σ.*
2. *There exists a unique configuration $\varphi(C)$ of Unf_{Σ^u} none of whose causally maximal events is u-labelled and such that $\psi(\varphi(C)) = C$. Moreover, there are at most two configurations in Unf_{Σ^u}, $\varphi(C)$ and $\varphi(C) \oplus e_u$, where e_u is u-labelled, such that $\psi(\varphi(C)) = \psi(\varphi(C \oplus e_u)) = C$. The latter configuration, if it exists, is denoted by $\overline{\varphi}(C)$; otherwise $\overline{\varphi}(C) \stackrel{\mathrm{df}}{=} \varphi(C)$.*

---

[2] Note that whether a GTI is conservative or not depends on a particular complete prefix for which $\tilde{\circlearrowleft}$ was computed; however, in the settings of this paper this prefix is fixed, and so there is no need to complicate the notation by parameterising the notion of conservativeness with it.

3. *Either $\varphi(\psi(C^u)) = C^u$ or $\varphi(\psi(C^u)) \oplus e_u = C^u$, and either $\overline{\varphi}(\psi(C^u)) = C^u$
   or $\overline{\varphi}(\psi(\overline{C^u})) = C^u \oplus e_u$, for some instance e_u of u.*
4. *If C is local then $\varphi(C)$ is local.* ◇

Note that in general, if C is local then $\overline{\varphi}(C)$ is not necessarily local, and if C^u
is local then $\psi(C^u)$ is not necessarily local.

The following algorithm, given a canonical prefix and a conservative GTI (the
conservativeness of which refers to the approximated lock relation $\check{\eth}$ computed
on this prefix), builds a canonical (w.r.t. some different cutting context) prefix
of the modified net.

Algorithm 8 (GTI in the prefix).

1 *For each initiated source $s \in S$ create a new initial p_s-labelled condition $c_{p_s}^{init}$.*
2 *For each instance e_s of each source transition $s \in S$ create a new p_s-labelled
  condition c_{p_s} and an arc (e_s, c_{p_s}).*
3 *For each co-set X such that $^\bullet X$ contains no cut-offs and $h_\Sigma(X) = \{p_s \mid s \in S\}$:*
  a. *create an instance e_u of the new transition u;*
  b. *for all $x \in X$, create the arcs (x, e_u);*
  c. *for each transition $d \in D$:*
     i) *create a new instance c_{p_d} of place p_d and the arc (e_u, c_{p_d});*
     ii) *for each instance e_d of d such that $[^\bullet X] \subset [e_d]_\Sigma$ and $[e_d]_\Sigma \setminus [^\bullet X]_\Sigma$ does
         not contain instances of d other than e_d, create an arc (c_{p_d}, e_d).*
4 *For each cut-off event e with a corresponding configuration C^e, change the
  corresponding configuration to $\varphi(C^e)$.* ◇

As was already mentioned, the prefix built by this algorithm is not canonical
w.r.t. the cutting context used to obtain the original prefix. Hence, a different
cutting context is defined below, w.r.t. which the resulting prefixes turns out to
be canonical.

Definition 9 (Cutting context Θ^u). Let Σ be a safe Petri net, Σ^u be the
Petri net obtained from Σ by applying a GTI, and $\Theta = \left(\approx_{mar}, \lhd, \{\mathcal{C}_e\}_{e \in E_{Unf_\Sigma}} \right)$
be the cutting context with which the canonical prefix $Pref_\Sigma^\Theta$ was built. Then
the cutting context $\Theta^u \overset{df}{=} \left(\approx_{mar}^u, \lhd^u, \{\mathcal{C}_e^u\}_{e \in E_{Unf_{\Sigma^u}}} \right)$ is defined as follows:

- $C' \approx_{mar}^u C''$ iff $Mrk_{\Sigma^u}(C') = Mrk_{\Sigma^u}(C'')$;
- $C' \lhd^u C''$ iff either $\psi(C') \lhd \psi(C'')$ or $\psi(C') = \psi(C'')$ and $\#_u C' < \#_u C''$;
- $\mathcal{C}_e^u \overset{df}{=} \emptyset$ if $h_{\Sigma^u}(e) = u$ and $\mathcal{C}_e^u \overset{df}{=} \{\varphi(C) \mid C \in \mathcal{C}_e\}$ if $h_{\Sigma^u}(e) \neq u$. ◇

To prove that this is indeed a cutting context, it is enough to show that $\lhd^u$ is an
adequate order on the configurations of Unf_{Σ^u}. (Note that $\approx_{mar}^u$ is the standard
equivalence of final markings on the configurations of Unf_{Σ^u}.)

Proposition 10 ($\lhd^u$ is adequate). *Let Σ^u be the Petri net obtained from a
safe Petri net Σ by a GTI $S \rightarrowtail\!\!\!\rightarrow D$ and $\lhd$ be an adequate order on the configurations of Unf_Σ. Then $\lhd^u$ is an adequate order on the configurations of Unf_{Σ^u}.
Moreover, $\lhd^u$ is total if $\lhd$ is total.* ◇

The following proposition states the correctness of Alg. 8, i.e. that the computed object $Pref^{alg}_{\Sigma^u}$ coincides with $Pref^{\Theta^u}_{\Sigma^u}$. Note that in this result it is essential that the GTI is conservative (with $\check{\circ}$ computed for $Pref^{\Theta}_{\Sigma}$); otherwise the final markings of the local configurations of cut-off events and their corresponding configurations might differ in $Pref^{alg}_{\Sigma^u}$.

Proposition 11 (Correctness of Alg. 8). *Let Σ be a safe Petri net and $Pref^{\Theta}_{\Sigma}$ be its canonical prefix w.r.t. a cutting context Θ. If the Petri net Σ^u is obtained from Σ by applying a conservative GTI $S \rightarrowtail\!\!\!\rightarrow D$ then the object $Pref^{alg}_{\Sigma^u}$ computed by Alg. 8 coincides with $Pref^{\Theta^u}_{\Sigma^u}$.*

Proof (Sketch). The idea of the proof in [8] is as follows. One can show that:

(*) $Pref^{alg}_{\Sigma^u}$ is a branching process of Σ^u (by checking the definition in Sect. 2).
(**) If e is a cut-off event of $Pref^{alg}_{\Sigma^u}$ with a corresponding configuration C^e then e is causally maximal, $C^e \approx^u_{mar} [e]_{\Sigma^u}$, $C^e \lhd^u [e]_{\Sigma^u}$ and $C^e \in \mathcal{C}^u_e$. Moreover, $Pref^{alg}_{\Sigma^u}$ cannot be extended without consuming a condition from the postset of some cut-off event. (The conservativeness of the GTI is essential here, as Def. 2(3) is needed for the proof in [8] to proceed.)

Due to (*), it remains to show is that the sets of feasible and cut-off events in $Pref^{alg}_{\Sigma^u}$ and $Pref^{\Theta^u}_{\Sigma^u}$ are the same, i.e.

(i) e is an event of $Pref^{alg}_{\Sigma^u}$ iff e is an event of $Pref^{\Theta^u}_{\Sigma^u}$; and
(ii) e is cut-off in $Pref^{alg}_{\Sigma^u}$ iff e is cut-off in $Pref^{\Theta^u}_{\Sigma^u}$.

Since $\lhd^u$ is an adequate order (Prop. 10), it is well-founded, and so one can use Noetherian induction on $\lhd^u$. That is, (i)&(ii) are proved assuming that (i)&(ii) holds for every $f \lhd^u e$ (note that Noetherian induction does not require the base case).

Suppose e is in one of $Pref^{alg}_{\Sigma^u}$, $Pref^{\Theta^u}_{\Sigma^u}$, but not in the other. Then, due to (**) and the completeness of $Pref^{\Theta^u}_{\Sigma^u}$, e is a causal successor of a cut-off event in one of them, but not in the other, i.e. there exists an event $f \lhd^u e$ which is cut-off in one of them, but not in the other, which contradicts the induction hypothesis. Hence (i) holds.

Now suppose e is in both prefixes and it is cut-off with a corresponding configuration C^e in one of them, but not cut-off in the other. Due to the induction hypothesis and $C^e \lhd^u [e]_{\Sigma^u}$, in both prefixes C^e contains no cut-off events or their causal successors. In what follows, two cases are considered.

First, suppose e is cut-off in $Pref^{alg}_{\Sigma^u}$ but not in $Pref^{\Theta^u}_{\Sigma^u}$. Since the algorithm never declares a u-labelled event cut-off in $Pref^{alg}_{\Sigma^u}$, $h_{\Sigma^u}(e) \neq u$, and so e was a cut-off event in $Pref^{\Theta}_{\Sigma}$ with a corresponding configuration $\psi(C^e)$. By (**), e satisfies the criteria of a cut-off event in $Pref^{\Theta^u}_{\Sigma^u}$ with a corresponding configuration C^e, i.e. $C^e \approx^u_{mar} [e]_{\Sigma^u}$, $C^e \lhd^u [e]_{\Sigma^u}$ and $C^e \in \mathcal{C}^u_e$, a contradiction.

Second, suppose e is cut-off in $Pref^{\Theta^u}_{\Sigma^u}$ but not in $Pref^{alg}_{\Sigma^u}$, i.e. $C^e \approx^u_{mar} [e]_{\Sigma^u}$, $C^e \lhd^u [e]_{\Sigma^u}$ and $C^e \in \mathcal{C}^u_e$. Since $h_{\Sigma^u}(e) \neq u$ (as otherwise $\mathcal{C}^u_e \stackrel{df}{=} \emptyset$ and so e

cannot be cut-off in $Pref_{\Sigma^u}^{\Theta^u}$), e was a cut-off in $Pref_{\Sigma}^{\Theta}$ with a corresponding configuration $\psi(C^e)$, because

- $C^e \approx_{mar}^{u} [e]_{\Sigma^u}$ implies $\psi(C^e) \approx_{mar} [e]_{\Sigma}$;
- $C^e \lhd^u [e]_{\Sigma^u}$ implies $\psi(C^e) \lhd [e]_{\Sigma}$, as $\psi(C^e) \neq [e]_{\Sigma}$ due to $h_{\Sigma^u}(e) \neq u$;
- $\psi(C^e) \in \mathcal{C}_e$ as $h_{\Sigma^u}(e) \neq u$.

Hence, the algorithm would have left e as a cut-off in $Pref_{\Sigma^u}^{alg}$, a contradiction.

Therefore, (ii) also holds, which completes the proof.   □

6   Computing Useful GTIs

As was already mentioned in the introduction, there are relatively few possible sequential and concurrent insertions, which makes it unlikely that insertions needed for logic decomposition exist. In contrast, in highly concurrent Petri nets the number of possible valid GTIs is usually exponential in the size of the net, and so there is a good chance that a suitable GTI can be found. On the flipside, though, it is no longer practical to enumerate all GTIs due to their large number. Hence, a method is needed that would allow one to reduce the number of GTIs that have to be considered, i.e. generate only potentially useful GTIs. Of course, which GTIs are useful depends on the application; however, there are some general techniques which can help here.

The purpose of this section is to demonstrate how GTIs that are potentially useful for logic decomposition of asynchronous circuits can be derived. The computation is performed in two steps. First, the possible sources are computed; this step is application specific, though the idea of using the incremental SAT technique for this is likely to be useful for other applications. Then, for a given set of sources, the possible destinations are computed; this step is relatively independent on the application.

Computing sources. In logic decomposition, given a Boolean expression implementing some local signal, its sub-expression $\mathcal{E}$ is selected, and a new internal signal is added in such a way that its implementation is $\mathcal{E}$. For this, several transitions of this signal have to be inserted into the STG (recall that STGs are a class of labelled Petri nets used to specify asynchronous circuits; in STGs, transitions model changes in values of the circuit signals). It turns out that such insertions have to be performed at the positions where $\mathcal{E}$ changes its value [8].

On the unfolding prefix, such a position can be formalised as a configuration C with each of its maximal events e labelled by a transition of some signal in the support of $\mathcal{E}$, and such that the values of $\mathcal{E}$ at the final states of $C \setminus \{e\}$ and C are different. This problem can be reduced to SAT and solved using an efficient off-the-shelf solver. Note that all such positions have to be computed, and so after one solution is returned by the solver, a new clause is added that rules out all the solutions for which the transitions corresponding to the causally maximal events of C are the same. Then the solver is executed again, giving another solution, and the process is continued until the added clauses make the SAT instance unsatisfiable. Note that many existing SAT solvers can take advantage of the similarity of this family of SAT instances (the technique that is called *incremental SAT*).

By considering all the computed configurations (more precisely, the transitions corresponding to their causally maximal events), the transition sets that can be used as sources of potentially useful GTIs are obtained.

Computing destinations. The above approach for computing sources yields several sets of transitions, and for each such a set $S \neq \emptyset$, one now has to compute all sets $D \neq \emptyset$ such that $S \rightarrowtail\twoheadrightarrow D$ is an SB-preserving GTI (only conservative GTIs are considered, which is a minor restriction in practice due to Prop. 4). Furthermore, the following minimality property is assumed: for all distinct $s, s' \in S$, $s \, \tilde{\emptyset} \, s'$, as otherwise s can be removed from S without any change in the resulting behaviour. This property is already guaranteed for the sources produced by the above approach, as the maximal events of any configuration are concurrent with each other and hence the corresponding transitions cannot be locked; for any other approach, this property can easily be enforced by removing some transitions from S.

Given such a set S, one can compute the set $L_S \overset{\mathrm{df}}{=} \{d \mid \forall s \in S : s \, \tilde{\circlearrowright} \, d \vee d \, \tilde{\circlearrowright} \, s\}$ of transitions locked with each $s \in S$. According to Def. 5, only such transitions can be in D, i.e. $D \subseteq L_S$. The binary *compatibility relation* $\bowtie$ on L_S introduced below specifies which transitions in L_S can be in the same set D. One of the requirements is the consistent locking with sources (see Def. 5), and the other requirement is the following minimality condition similar to that formulated above for sources: for all distinct transitions d and d', $d \, \tilde{\emptyset} \, d'$, as otherwise d' can be removed from the destinations without any change in the resulting behaviour.

Definition 12 (Compatibility relation $\bowtie$). The *compatibility relation* $\bowtie$ on the set of transitions L_S is defined as follows. Let $t', t'' \in L_S$ be a pair of distinct transitions. Then $t' \bowtie t''$ holds iff

- for each $s \in S$, $(t' \, \tilde{\circlearrowright} \, s \wedge t'' \, \tilde{\circlearrowright} \, s) \vee (s \, \tilde{\circlearrowright} \, t' \wedge s \, \tilde{\circlearrowright} \, t'')$; and
- $t' \, \tilde{\emptyset} \, t'' \wedge t'' \, \tilde{\emptyset} \, t'$. $\diamond$

The compatibility relation can be viewed as a graph with the vertex set L_S and the edges given by $\bowtie$. Now, any non-empty clique (including non-maximal ones) D in this graph forms a valid set of destinations, i.e. $S \rightarrowtail\twoheadrightarrow D$ is a valid conservative GTI. Hence, it is enough to enumerate all non-empty cliques and generate the corresponding GTIs.

In practice, further restrictions specific to the application domain can be incorporated into the definitions of L_S and $\bowtie$ in order to reduce the graph size. For example, in logic decomposition the newly inserted transitions always correspond to internal signals, and as such must never 'trigger' an input transition (as the inputs are controlled by the environment, which does not 'see' internal transitions and so cannot wait for them to occur); hence, all the input transitions should be removed from L_S before building the compatibility relation $\bowtie$.

Furthermore, if the number of cliques is still too large, the process of their enumeration can be stopped at any time, and one can continue to work with the curtailed set of GTIs, especially if the enumeration is implemented in such a way that the most useful (e.g. smallest or largest) cliques are produced first.

7   Conclusions

In this paper, a new type of transition insertions has been proposed, and the corresponding theory and a suite of algorithms have been developed to integrate it into the transformation framework developed in [6]. In particular, the contributions include:

- A method for computing the approximated lock relation $\tilde{\circlearrowleft}$ using a complete unfolding prefix (Def. 2). This approximation is always conservative (Prop. 3), and exact in the practical case of a live Petri net (Prop. 4).
- A new kind of transition insertion, called generalised transition insertion (Def. 5), that preserves safeness and behaviour of the Petri net (Prop. 6).
- An algorithm for efficient conversion of a given canonical prefix of a Petri net into a canonical prefix of the Petri net obtained by applying a conservative GTI (Alg. 8), avoiding thus (expensive) re-unfolding. Interestingly, a different cutting context has to be used for the resulting prefix (Prop. 11). (In general, re-unfolding the modified Petri net with the original cutting context can yield a very different prefix from that returned by Alg. 8.) As an auxiliary result, the correspondence between the configurations of the unfoldings of the original and modified Petri nets was established (Prop. 7).

  An additional advantage of this approach (besides avoiding re-unfolding) is that it yields a prefix very similar to the original one, which is useful for visualisation and allows one to transfer some information from the original prefix to the modified one (e.g. [7] used this feature, albeit for the transformations explained in Sect. 3, to transfer the yet unresolved encoding conflicts into the new prefix, avoiding thus re-computing them from scratch).
- Since the number of all possible GTIs grows exponentially with the size of the Petri net, their straightforward enumeration is impractical. Hence, a method for computing only potentially useful GTIs in the context of logic decomposition of asynchronous circuits was developed (Sect. 6); however, some parts of this method (viz. computing possible destination for the given set of sources) are relatively independent on the application domain.

These contributions form a complete framework for efficient use of GTIs together with the transformations developed earlier (see Sect. 3).

Currently, the developed theory (though it is quite generic) is applied to logic decomposition of asynchronous circuits. In future work, it would be interesting to integrate further transformations into the developed framework.

References

1. Chu, T.A.: Synthesis of Self-Timed VLSI Circuits from Graph-Theoretic Specifications. Ph.D. thesis, Lab. for Comp. Sci. MIT (1987)
2. Cohn, P.: Universal Algebra, 2nd edn. Reidel, Dordrecht (1981)

3. Engelfriet, J.: Branching processes of Petri nets. Acta Inf. 28, 575–591 (1991)
4. Esparza, J., Römer, S., Vogler, W.: An improvement of McMillan's unfolding algorithm. FMSD 20(3), 285–310 (2002)
5. Khomenko, V.: Model Checking Based on Prefixes of Petri Net Unfoldings. Ph.D. thesis, School of Comp. Sci., Newcastle Univ. (2003)
6. Khomenko, V.: Behaviour-preserving transition insertions in unfolding prefixes. In: Kleijn, J., Yakovlev, A. (eds.) ICATPN 2007. LNCS, vol. 4546, pp. 204–222. Springer, Heidelberg (2007)
7. Khomenko, V.: Efficient automatic resolution of encoding conflicts using STG unfoldings. IEEE Trans. VLSI Syst. 17(7), 855–868 (2009); Special section on asynchronous circuits and systems
8. Khomenko, V.: A new type of behaviour-preserving transition insertions in unfolding prefixes. Tech. Rep. CS-TR-1189, School of Comp. Sci., Newcastle Univ. (2010), http://www.cs.ncl.ac.uk/publications/trs/papers/1189.pdf
9. Khomenko, V., Koutny, M., Vogler, V.: Canonical prefixes of Petri net unfoldings. Acta Inf. 40(2), 95–118 (2003)
10. Murata, T.: Petri nets: Properties, analysis and applications. Proc. of the IEEE 77(4), 541–580 (1989)
11. Rosenblum, L., Yakovlev, A.: Signal graphs: from self-timed to timed ones. In: Proc. Workshop on Timed Petri Nets, pp. 199–206. IEEE Comp. Soc. Press, Los Alamitos (1985)
12. Vanbekbergen, P., Catthoor, F., Goossens, G., De Man, H.: Optimized synthesis of asynchronous control circuits from graph-theoretic specifications. In: Proc. ICCAD 1990, pp. 184–187. IEEE Comp. Soc. Press, Los Alamitos (1990)

On the Computation of McMillan's Prefix
for Contextual Nets and Graph Grammars*

Paolo Baldan[1], Alessandro Bruni[1], Andrea Corradini[2],
Barbara König[3], and Stefan Schwoon[4]

[1] Dipartimento di Matematica Pura e Applicata, Università di Padova, Italy
[2] Dipartimento di Informatica, Università di Pisa, Italy
[3] Abteilung für Informatik und Angewandte Kognitionswissenschaft,
Universität Duisburg-Essen, Germany
[4] LSV, ENS Cachan & CNRS, INRIA Saclay, France

Abstract. In recent years, a research thread focused on the use of the
unfolding semantics for verification purposes. This started with a paper
by McMillan, which devises an algorithm for constructing a finite complete prefix of the unfolding of a safe Petri net, providing a compact
representation of the reachability graph. The extension to contextual
nets and graph transformation systems is far from being trivial because
events can have multiple causal histories. Recently, we proposed an abstract algorithm that generalizes McMillan's construction to bounded
contextual nets without resorting to an encoding into plain P/T nets.
Here, we provide a more explicit construction that renders the algorithm
effective. To allow for an inductive definition of concurrency, missing in
the original proposal and essential for an efficient unfolding procedure,
the key intuition is to associate histories not only with events, but also
with places. Additionally, we outline how the proposed algorithm can be
extended to graph transformation systems, for which previous algorithms
based on the encoding of read arcs would not be applicable.

1 Introduction

Partial-order semantics are used to alleviate the state-explosion problem when
model checking concurrent systems. A thread of research started by McMillan [10,11] proposes the *unfolding* semantics for the verification of finite-state
systems, modelled as Petri nets. The unfolding of a Petri net [12] is a nondeterministic process of the net that completely expresses its behaviour; it is an
acyclic but usually infinite net. McMillan's algorithm constructs a finite *prefix*
of the unfolding, which is *complete*, i.e., each marking reachable in the original
net is represented in the prefix. Computing a complete prefix can be seen as a
preprocessing step for reachability analysis of a Petri net: a complete prefix is
generally larger than the original Petri net, but smaller than the reachability

---

* Supported by the MIUR Project SisteR, by the project AVIAMO of the University of
Padova and by the DFG project SANDS.

H. Ehrig et al. (Eds.): ICGT 2010, LNCS 6372, pp. 91–106, 2010.

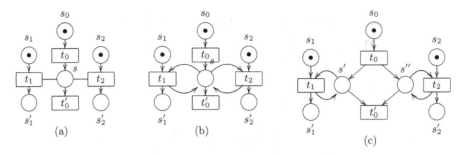

Fig. 1. (a) A safe contextual net; (b) its encoding by replacing read arcs with consume/produce loops; (c) its concurrency-preserving PR-encoding

graph, and reachability queries can be answered efficiently on the prefix. Other applications of unfoldings are, e.g. in fault diagnosis [7] and in planning [8].

The unfolding construction has been generalized to other rule-based formalisms such as contextual nets [14,4] and graph grammars [5]. However, concerning McMillan's construction, problems arise because these formalisms allow to preserve part of the state in a rewriting step. This has been observed originally for contextual nets by Vogler et al. [14]: they showed that for such nets the prefix generated by McMillan's algorithm might not be complete because events can have more than one causal history.

A solution to this problem is to encode contextual nets into an "equivalent" P/T net, to which McMillan's algorithm is then applied. Consider the net in Fig. 1 (a), where read arcs are drawn as undirected lines. Net (b) is obtained by replacing each read arc of (a) with a consume/produce loop, obtaining a net with the same reachability graph, but whose unfolding could grow exponentially due to the sequentialization imposed on readers. Net (c) is obtained like net (b), but first creating "private copies" of the read places for each reader: for safe nets this *Place-Replication (PR) encoding* preserves concurrency, and the size of the unfolding is comparable to that of the original net.

Unfortunately, such approaches are not viable for graph grammars nor, in general, for rewriting formalisms where states have a structure richer than multi-sets. Sticking to standard graph rewriting, if a rule preserves a node both encodings would transform it into a rule that deletes and creates again that node. Such two rules are not equivalent neither in the DPO approach (where because of the dangling condition, if there is an edge incident to the node then the second rule might not be applicable to a valid match for the first one) nor in SPO (where edges incident to the node would be deleted by the second rule as a side-effect), just to mention two of the most popular graph transformation approaches.

Another solution is to adapt McMillan's procedure; i.e., one generalizes it in such a way that the unfolding of a contextual net is itself a contextual net. In [14] this was done for the subclass of *read-persistent* contextual nets, for which a slight modification of McMillan's algorithm works, essentially because the restriction guarantees that each event has a single causal history. This approach has been successfully generalized to graph grammars in [2], by identifying a corresponding class of read-persistent, finite-state grammars, and showing how the finite complete prefix computed with a variation of McMillan's algorithm could be used for verifying properties of the original grammar.

However, read-persistency can be a strong restriction. Recently, we have proposed an algorithm that works for the whole class of bounded contextual nets [3]. The main idea is to equip events with causal histories. Instead of building the prefix by adding one event at a time, we add one pair *(event, history)* at a time.

While [3] provided the theoretical foundations of the finite complete prefixes of contextual nets, important practical concerns were left unresolved. E.g., [3] did not address how to actually compute the pairs that are added to the unfolding. Here, we refine the algorithm and design an effective, concrete procedure for computing the unfolding prefix. This is based on the idea of associating histories not only to events, but also to places. This eases the inductive computation of the relation of concurrency, and thus the construction of the prefix of the unfolding.

Moreover, we argue that the proposed algorithm can be adapted smoothly to graph grammars as well, even if, because of space constraints, the technical details are not worked out.

2   Contextual Nets and Their Unfolding

In this section we review contextual nets and their unfoldings. We refer to [3] for a fuller treatment with more examples illustrating the definitions.

2.1   Contextual Nets

Let A be a set; a *multiset* of A is a function $M : A \to \mathbb{N}$ where $\{a \in A : M(a) > 0\}$ is finite. The set of multisets of A is denoted by $A^\oplus$. Usual operations such as multiset union $\oplus$ or difference $\ominus$ are used. A function $f : A \to B$ induces a function on multisets denoted $f^\oplus : A^\oplus \to B^\oplus$. We write $M \le M'$ if $M(a) \le M'(a)$ for all $a \in A$, and $a \in M$ for $M(a) > 0$. For $n \in \mathbb{N}$, we denote by $\bar{n}$ the constant function that yields n for all arguments, i.e. $\bar{n}(a) = n$ for all a.

Definition 1 (contextual net). *A contextual Petri net (c-net) is a tuple* $N = \langle S, T, {}^\bullet(.), (.)^\bullet, (.), m \rangle$, *where S is a set of* places, *T is a set of* transitions, *and* ${}^\bullet(.), (.)^\bullet : T \to S^\oplus$, $(.) : T \to \mathcal{P}(S)$ *are functions which provide the pre-set, post-set, and context of a transition; $m \in S^\oplus$ is the* initial marking. *We assume* ${}^\bullet t \ne \bar{0}$ *for each transition $t \in T$.*

In the following when considering a c-net N, we will implicitly assume $N = \langle S, T, {}^\bullet(.), (.)^\bullet, (.), m \rangle$. Given a place $s \in S$ we define ${}^\bullet s = \{t \in T : s \in t^\bullet\}$, $s^\bullet = \{t \in T : s \in {}^\bullet t\}$, $\underline{s} = \{t \in T : s \in \underline{t}\}$.

An example of a contextual net, inspired by [14], is depicted in Fig. 2(a). For instance, referring to transition t_1 we have ${}^\bullet t_1 = \{s_1\}$, $t_1{}^\bullet = \{s_3\}$ and $\underline{t_1} = \{s_2\}$.

Definition 2 (firing). *Let N be a c-net. A transition $t \in T$ is enabled at a marking $M \in S^\oplus$ if ${}^\bullet t \oplus \underline{t} \le M$. In this case, its firing produces the marking $M' = M \ominus {}^\bullet t \oplus t^\bullet$, written as $M [t\rangle M'$.*

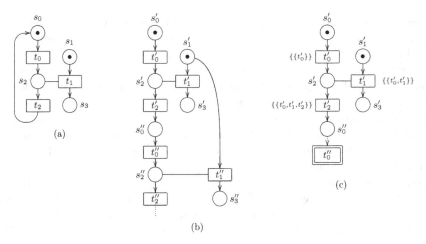

Fig. 2. (a) A c-net N_0, (b) its unfolding $\mathcal{U}_a(N_0)$ and (c) a complete enriched prefix

A marking M of a c-net N is called *reachable* if there is a finite sequence of firings leading from the initial marking to M, i.e., $m\,[t_1\rangle\,M_1\,[t_2\rangle\,M_2\ldots[t_n\rangle\,M$.

Definition 3 (bounded, safe and semi-weighted nets). *A c-net N is called n-bounded if every reachable marking M satisfies $M \leq \bar{n}$. It is called safe if it is 1-bounded and, for any $t \in T$, ${}^\bullet t$, $t^\bullet$ are sets (rather than general multisets). A c-net N is called semi-weighted if the initial marking m is a set and, for any $t \in T$, $t^\bullet$ is a set.*

We restrict to semi-weighted nets in order to simplify the presentation. The treatment of general nets would lead to some technical complications in the definition of the unfolding (Definition 9), related to the fact that an occurrence of a place would not be identified uniquely by its causal history.

2.2   Occurrence c-Nets

We will introduce two relations among transitions: causality and asymmetric conflict. Occurrence c-nets are safe c-nets where these relations satisfy certain acyclicity and well-foundedness requirements.

Causality is defined as for ordinary nets, with an additional clause stating that transition t causes t' if t generates a token in a context place of t'. Intuitively, $t < t'$ if t must happen before t' can happen.

Definition 4 (causality). *Let N be a safe c-net. The causality relation $<$ is the least transitive relation on $S \cup T$ such that (i) if $s \in {}^\bullet t$ then $s < t$; (ii) if $s \in t^\bullet$ then $t < s$; (iii) if $t^\bullet \cap \underline{t'} \neq \emptyset$ then $t < t'$. Given $x \in S \cup T$, we write $\lfloor x \rfloor$ for the set of causes of x in T, defined as $\lfloor x \rfloor = \{t \in T : t \leq x\} \subseteq T$, where $\leq_N$ is the reflexive closure of $<$.*

We say that a transition t is in *asymmetric conflict* with t', denoted $t \nearrow t'$, if whenever both t and t' fire in a computation, t fires before t'. The paradigmatic case is when transition t' consumes a token in the context of t, i.e., when $\underline{t} \cap {}^\bullet t' \neq \emptyset$, as for transitions t'_1 and t'_2 in Fig. 2(b) (see [13,9,14]).

Definition 5 (asymmetric conflict). *Let N be a safe c-net. The* asymmetric conflict relation $\nearrow$ *is the binary relation on T defined as*

$$t \nearrow t' \qquad if \qquad \underline{t} \cap {}^\bullet t' \neq \emptyset \quad or \quad (t \neq t' \wedge {}^\bullet t \cap {}^\bullet t' \neq \emptyset) \quad or \quad t < t'.$$

For $X \subseteq T$, $\nearrow_X$ denotes the restriction of $\nearrow$ to X, i.e., $\nearrow_X = \nearrow \cap (X \times X)$.

An occurrence c-net is a safe c-net that exhibits an acyclic behaviour, satisfying suitable conflict-freeness requirements.

Definition 6 (occurrence c-nets). *A c-net N is called* occurrence c-net *if*

- *N is safe, and for any $s \in S$, $|{}^\bullet s| \leq 1$ (no backward conflicts)*
- *$<$ is a strict partial order and $\lfloor t \rfloor$ is finite for any $t \in T$;*
- *$m = \{s \in S : {}^\bullet s = \emptyset\}$ (the initial marking is the set of minimal places);*
- *$\nearrow_{\lfloor t \rfloor}$ is acyclic for all $t \in T$.*

The last condition of the definition corresponds to the requirement of irreflexivity for the conflict relation in ordinary occurrence nets. An example of an occurrence c-net can be found in Fig. 2(b). From now on, consistently with the literature, we shall often call the transitions of an occurrence c-net *events*.

Definition 7 (configurations). *Let N be an occurrence c-net. A finite set of events $C \subseteq T$ is called a* configuration *if*

1. *$\nearrow_C$ is acyclic;*
2. *C is left-closed w.r.t. $\leq$, i.e. for all $t \in C$, $t' \in T$, $t' \leq t$ implies $t' \in C$.*

The marking produced by a configuration C is $C^\bullet = m \cup \bigcup_{t \in C} t^\bullet - \bigcup_{t \in C} {}^\bullet t$. A finite set M of places is called concurrent, *written $conc(M)$, if there exists a configuration C such that $M \subseteq C^\bullet$.*

We denote by $Conf(N)$ the set of all configurations of N. They are equipped with the ordering defined as $C_1 \sqsubseteq C_2$, if $C_1 \subseteq C_2$ and $\neg(t_2 \nearrow t_1)$ for all $t_1 \in C_1$, $t_2 \in C_2 \setminus C_1$. Furthermore two configurations C_1, C_2 are said to be in conflict, written $C_1 \# C_2$, when there is no $C \in Conf(N)$ such that $C_1 \sqsubseteq C$ and $C_2 \sqsubseteq C$.

Configurations characterise the possible (concurrent) computations of an occurrence c-net. The relation $\sqsubseteq$ is a computational order of configurations: $C \sqsubseteq C'$ if C can evolve and become C'. Differently from non-contextual occurrence nets this order is not simply subset inclusion among configurations.

Given a configuration C and an event $t \in C$, the *history of t in C* is the set of events that *must* precede t in the (concurrent) computation represented by C. For ordinary nets the history of an event t coincides with the set of causes $\lfloor t \rfloor$, independently of the configuration where t occurs. For c-nets, the presence of asymmetric conflicts implies that an event may have several histories.

Definition 8 (history). *Let N be an occurrence net. Given a configuration C and an event $t \in C$, the* history of t in C, *denoted by $C[\![t]\!]$, is defined as $C[\![t]\!] = \{t' \in C : t'(\nearrow_C)^* t\}$. The set of possible histories of an event t, namely $\{C[\![t]\!] : C \in Conf(N) \wedge t \in C\}$ is denoted by $Hist(t)$.*

For example, in the contextual net shown in Fig. 2(b), the event t'_2 has two possible histories: $\{t'_0, t'_2\}$ and $\{t'_0, t'_1, t'_2\}$.

2.3   Unfolding of Contextual Nets

Given a semi-weighted c-net N, an *unfolding* construction allows us to obtain an occurrence c-net $\mathcal{U}_a(N)$ that describes the behaviour of N [4,14]. The unfolding can be constructed inductively by starting from the initial marking of N and then by adding, at each step, an occurrence of each transition of N which is enabled by (the image of) a concurrent subset of the places already generated.

Definition 9 (unfolding). *Let $N = \langle S, T, {}^\bullet(.), (.)^\bullet, \underline{(.)}, m \rangle$ be a semi-weighted c-net. The unfolding $\mathcal{U}_a(N) = \langle S', T', {}^\bullet(.), (.)^\bullet, \underline{(.)}, m' \rangle$ of the net N is the occurrence c-net generated by the following inference rules, where $M_p, M_c \subseteq S'$; $M_p \cap M_c = \emptyset$; and $\pi_2(\langle x, y \rangle) = y$.*

$$\frac{s \in m}{\langle \emptyset, s \rangle \in S'} \qquad \frac{t' = \langle M_p, M_c, t \rangle \in T' \quad s \in t^\bullet}{\langle t', s \rangle \in S'}$$

$$\frac{t \in T \quad \pi_2^\oplus(M_p) = {}^\bullet t \quad \pi_2^\oplus(M_c) = \underline{t} \quad conc(M_p \cup M_c)}{\langle M_p, M_c, t \rangle \in T'}$$

The initial marking is $m' = \{\langle \emptyset, s \rangle : s \in m\}$, and given $t' = \langle M_p, M_c, t \rangle$

$$ {}^\bullet t' = M_p \qquad \underline{t'} = M_c \qquad t'^\bullet = \{\langle t', s \rangle : s \in t^\bullet\}$$

The folding morphism $f_N = \langle f_T, f_S \rangle : \mathcal{U}_a(N) \to N$ is a pair of mappings $f_T : T' \to T$ and $f_S : S' \to S$ defined by $f_T(t') = t$ for $t' = \langle M_p, M_c, t \rangle$ and $f_S(s') = s$ for $s' = \langle x, s \rangle$.

Places and events in the unfolding of a c-net represent respectively tokens and firing of transitions in the original net. Each place in the unfolding is a pair recording the "history" of the token and the corresponding place in the original net. Each event is a triple recording the precondition and context used in the firing, and the corresponding transition in the original net. A new place with empty history $\langle \emptyset, s \rangle$ is generated for each place s in the initial marking. Moreover, a new event $t' = \langle M_p, M_c, t \rangle$ is inserted in the unfolding whenever we can find a concurrent set of places (precondition M_p and context M_c) that corresponds, in the original net, to a marking that enables t. For each place s in the post-set of such t, a new place $\langle t', s \rangle$ is generated, belonging to the post-set of t'. The folding morphism f maps each place (event) of the unfolding to the corresponding place (transition) in the original net.

   An initial part of the unfolding of the net N_0 in Fig. 2(a) is represented in Fig. 2(b). The folding morphism from $\mathcal{U}_a(N_0)$ to N_0 is implicitly represented by the name of the items in the unfolding.

Proposition 1 (completeness of the unfolding). *Let N be a c-net and let $\mathcal{U}_a(N) = \langle S', T', {}^\bullet(.), (.)^\bullet, \underline{(.)}, m' \rangle$ be its unfolding. A marking $M \in S^\oplus$ is coverable in N iff there exists a concurrent subset $X \subseteq S'$ such that $M = f_S^\oplus(X)$.*

Proposition 1 captures the sense in which the unfolding is *complete* w.r.t. the original net. This notion of completeness is slightly weaker than that of [10,14], for example, as it is concerned with markings only, and not with transitions.

The result also suggests a method for checking reachability of a marking M of the original net: find the subsets X such that $M = f_S^\oplus(X)$ and check whether at least one is concurrent. Note that in (non-contextual) Petri net unfoldings this amounts to checking whether $\lfloor X \rfloor$ does not contain a (symmetric) conflicts, which takes linear time in $||\lfloor X \rfloor||$. For contextual nets, one needs to check the absence of cycles in the asymmetry relation, which is equally possible in linear time. Hence the complexity of checking the concurrency of a set X is the same in ordinary and contextual unfoldings, and since a contextual unfolding has in general fewer conditions and events than its PR-encoding, fewer concurrency checks need to be answered.

3   Computing the Prefix as an Enriched Occurrence Net

In this section we first recall the notion of enriched occurrence net, which is an occurrence net which records a subset of histories for each involved event. Next we describe an algorithm for computing a finite complete prefix of the full unfolding of a c-net N, which is a mild variation of that in [3]. The construction starts from the initial marking, and iteratively adds new extended events representing occurrences of transitions of N. The prefix will actually be an *enriched* occurrence net, where only histories which are considered "useful to produce new markings" are recorded.

Definition 10 (enriched occurrence net, extended event). *An enriched occurrence net is a pair $E = \langle N_E, \chi_E \rangle$, where $N_E = \langle S_E, T_E, {}^\bullet(.), (.)^\bullet, \underline{(.)}, m_E \rangle$ is an occurrence net and $\chi_E : T_E \to \mathcal{P}(\mathcal{P}(T_E))$ is a function such that*

- *for any $t \in T_E$, $\emptyset \neq \chi_E(t) \subseteq Hist(t)$*
- *for all $t, t' \in T_E$, for any $C \in \chi_E(t)$ if $t' \in C$ then $C[\![t']\!] \in \chi_E(t')$.*

An extended event for an occurrence net N_E is a pair $\epsilon = \langle t, H_t \rangle$, where $t \in T_E$ and $H_t \in Hist(t)$. We say that $\langle t, H_t \rangle$ covers another extended event $\langle t', H_{t'} \rangle$ when $H_{t'} \sqsubseteq H_t$. An enriched occurrence net $E = \langle N_E, \chi_E \rangle$ contains the extended event $\epsilon = \langle t, H_t \rangle$, written $\epsilon \in E$, if $t \in T_E$ and $H_t \in \chi_E(t)$.

From now on, $N = \langle S, T, {}^\bullet(.), (.)^\bullet, \underline{(.)}, m \rangle$ is a fixed semi-weighted c-net, $\mathcal{U}_a(N) = \langle S', T', {}^\bullet(.), (.)^\bullet, \underline{(.)}, m' \rangle$ is its unfolding, and $f_N : \mathcal{U}_a(N) \to N$ is the folding morphism.

Definition 11 (enriched prefix). *An enriched prefix of the unfolding $\mathcal{U}_a(N)$ is any enriched occurrence net E such that N_E is a prefix of $\mathcal{U}_a(N)$.*

An example of an enriched prefix of $\mathcal{U}_a(N_0)$ in Fig. 2(b) is given in Fig. 2(c). For any event t the set of histories $\chi_E(t)$ is written next to the event.

During the construction of a complete prefix of the unfolding, at each step we add a *possible extension* to the current prefix, that is, an extended event whose pre-set and context are already there, and whose history is compatible with the histories contained in the current prefix for the involved events.

Definition 12 (possible extension). *Let E be an enriched prefix of $\mathcal{U}_a(N)$. A possible extension of E is an extended event $\epsilon = \langle t, H_t \rangle$ of $\mathcal{U}_a(N)$ such that $\epsilon \notin E$ and for any $t' \in H_t - \{t\}$ it holds that $\langle t', H_t[\![t']\!] \rangle \in E$.*

Note that $\epsilon = \langle t, H_t \rangle$ is a possible extension when $E' \stackrel{def}{=} E \cup \epsilon$, obtained from E by inserting transition t and its post-set (if it is not already there) and by adding H_t to the set of histories of t, is an enriched prefix of $\mathcal{U}_a(N)$. E.g., for the prefix in Fig. 2(c) (assuming that t_0'' is not there) two possible extensions are $\epsilon = \langle t_0'', \{t_0', t_1', t_2', t_0''\} \rangle$ and $\epsilon' = \langle t_2', \{t_0', t_2'\} \rangle$. Instead, $\langle t_0'', \{t_0', t_2', t_0''\} \rangle$ is not a possible extension, since t_2' does not have the history $\{t_0', t_2'\}$.

A configuration of $\mathcal{U}_a(N)$ represents a computation in the unfolding, which in turn maps, via the folding morphism, to a computation of N. Hence we can define the marking of N after a configuration of the unfolding.

Definition 13 (marking after a configuration). *Let $C \in Conf(\mathcal{U}_a(N))$ be a configuration. The marking of N after C is defined as $mark(C) = f_S^\oplus(C^\bullet)$.*

In [10] a *cut-off* is defined as an event of the unfolding that can be omitted safely because there exists another event with a smaller causal history generating the same marking. In our setting a cut-off is defined as an *extended* event, thus taking histories explicitly into account.

Definition 14 (cut-off). *Let E be an enriched prefix of the unfolding. An extended event $\langle t, H_t \rangle$ in E is called a cut-off if either $mark(H_t) = m$, the initial marking of N, or there is another extended event $\langle t', H_{t'} \rangle$ of E satisfying (i) $mark(H_t) = mark(H_{t'})$ and (ii) $|H_{t'}| < |H_t|$.*

The algorithm shown in Fig. 3 computes a complete finite prefix. It is analogous to the standard algorithm for ordinary Petri nets, but applied to extended events: during the construction, each event t of Fin, the currently built part of the prefix, has associated also a set of histories $\chi_{Fin}(t)$, thus making the prefix under construction an enriched occurrence net.

The algorithm is similar to, but more abstract than, the algorithm in [3]. In fact, the present algorithm uses the function PE which, applied to the enriched prefix Fin and to an extended event ϵ, is expected to return a set of extended events. In [3], instead, the set of extended events added at each iteration was described (in a pretty abstract, non-constructive way) in the algorithm itself.

The main result shows that the algorithm terminates and correctly computes a complete finite prefix of the unfolding, provided that the set returned by $PE(Fin, \epsilon)$ contains at least all the possible extensions of Fin which cover ϵ.

$Fin := m'$, $\chi_{Fin} := \emptyset$
$pe := \{\langle t', \emptyset \rangle : t' \text{ enabled by } m'\}$
while $pe \neq \emptyset$ do

- take $\epsilon = \langle t, H \rangle \in pe$ such that $|H|$ is minimal.
- $pe = pe - \{\epsilon\}$
- if ϵ would be a cut-off in Fin, do nothing, else insert ϵ in Fin, i.e.,
  - if t is already present in Fin then add the history H to $\chi_{Fin}(t)$;
  - otherwise add t and the places in $t^\bullet$ to Fin, and set $\chi_{Fin}(t) := \{H\}$.
  - $pe := pe \cup PE(Fin, \epsilon)$

Fig. 3. Algorithm for computing the finite prefix

Theorem 1 (correctness and completeness). *If the net N is finite and n-bounded, and if function PE applied to any enriched prefix Fin and to an extended event ϵ returns a set of extended events containing at least all the possible extensions of Fin covering ϵ, then the algorithm in Fig. 3 terminates, and the resulting enriched prefix Fin_0 is complete.*

As an example, consider the net N_0 and its unfolding $\mathcal{U}_a(N_0)$ in Fig. 2. The algorithm (as detailed in the next section) would produce the enriched prefix depicted in Fig. 2(c). Note that it includes the event t'_2. In fact t'_2 has two possible histories: the minimal history $H_2 = \lfloor t'_2 \rfloor = \{t'_0, t'_2\}$ and $H'_2 = \{t'_0, t'_1, t'_2\}$. While $\langle t'_2, H_2 \rangle$ is a cut-off, the pair $\langle t'_2, H'_2 \rangle$ is not, and thus it is included.

4   Computing the Possible Extensions

During the construction of the complete prefix, we must compute the possible extensions of the current prefix. The key idea to do this efficiently and incrementally is to equip places with histories.

4.1   Extended Places

A first simple observation is that the histories of an event t in an enriched occurrence net E are obtained from the histories of the events that are in direct

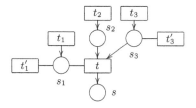

Fig. 4. Predecessors w.r.t. asymmetric conflict of an event t

asymmetric conflict with t. Referring to Fig. 4, the histories for t can be constructed by taking one history for every direct cause of t (i.e., t_1, t_2 and t_3), and possibly one for t_3' that is in direct asymmetric conflict with t. In contrast, a transition that merely reads from the context of t, such as t_1', is not in direct asymmetric conflict with t and therefore not considered. If these histories are "consistent" they can be joined to form a history of t. Once this history is added to the prefix, it can be used to generate new histories for the events (not depicted) that use the post-set s or that consume the context s_1. Therefore the generation of a new history for an event has to be "propagated" according to the structure of the net, because it can entail new histories for other events.

In order to concretely realize this sort of propagation in the algorithm we will rely on the notion of *extended place*, which is a place with an associated history. Hereafter, $E = \langle N, \chi \rangle$ denotes a fixed enriched occurrence net.

Definition 15 (histories for places). *The causal histories of a place $s \in S$, denoted $\chi_c(s)$, are the histories of the transition that generates s (i.e., $\chi_c(s) = \chi(\bullet s)$) if $s \notin m$. If $s \in m$, then $\chi_c(s) = \{\emptyset\}$. The read histories of s, denoted $\chi_r(s)$, are the histories of the events which read s, i.e., $\chi_r(s) = \bigcup_{t \in \underline{s}} \chi(t)$.*

In words, a place inherits histories from the (unique) event which generates the place (causal histories) and from events which read the place (read histories).

Definition 16 (extended places). *An extended place is a pair $\sigma = \langle s, H \rangle$ where $s \in S$ and $H \in \chi_c(s) \cup \chi_r(s)$. It is called a causal extended place if $H \in \chi_c(s)$ and a read extended place if $H \in \chi_r(s)$.*

Let $\langle s, H \rangle$ be a read extended place. It can be shown that there is a unique causal history $H' \in \chi_c(s)$ such that $H' \sqsubseteq H$. The extended place $\langle s, H' \rangle$ is denoted $\langle s, H \rangle^{\uparrow}$. We write π_H and π_S for the projections of an extended place to its components, i.e., given $\sigma = \langle s, H \rangle$, $\pi_H(\sigma) = H$ and $\pi_S(\sigma) = s$.

More intuition can be obtained from Fig. 5, where events and places are annotated with their histories. Place s_3 is associated with three extended places: one causal extended place $\langle s_3, \emptyset \rangle$ and two read extended places $\langle s_3, \{t_1\} \rangle$, $\langle s_3, \{t_2\} \rangle$; we do not consider $\langle s_3, \{t_1, t_2\} \rangle$ which would represent two readings of place s_3.

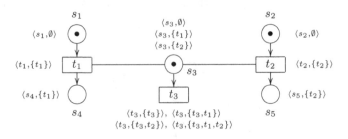

Fig. 5. Illustration of extended transitions and places

Now, according to Definition 8, there are four histories for transition t_3, i.e., $\{t_3\}$, $\{t_3, t_1\}$, $\{t_3, t_2\}$ and $\{t_3, t_1, t_2\}$. Interestingly, all of them can be obtained as a suitable combination of the extended places associated with s_3 (without considering the extended events corresponding to t_1 and t_2). Note that if we had n readers instead of just 2, we would have 2^n histories but only n read places. Thus, ignoring multiple readings of a place avoids a combinatorial explosion.

For each place we have, conceptually, one extended place for each possible history of the (only) event which generates the place, and one for each event that reads the place. In this way a new history for a transition t can be obtained by looking only at the extended places in its pre-set and context, without checking the transitions in asymmetric conflict with t, thus making the construction more local. The generation of a history for t in turn produces new histories (extended places) for the places in the post-set and in the context. This can be formalized elegantly by defining a notion of pre-set and post-set for extended events.

Definition 17 (pre-set and post-set of extended events). *Given an extended event* $\epsilon = \langle t, H \rangle$, *we define*

$$- \ {}^\bullet\epsilon = \{\langle s, H' \rangle \mid s \in {}^\bullet t \ \wedge \ H' \in \chi_c(s) \cup \chi_r(s) \ \wedge \ H' \sqsubseteq H\}$$
$$- \ \hat{\epsilon} = \{\langle s, H' \rangle \mid s \in \underline{t} \ \wedge \ H' \in \chi_c(s) \ \wedge \ H' \sqsubseteq H\}$$
$$- \ \epsilon^\bullet = \{\langle s, H \rangle \mid s \in t^\bullet\}$$
$$- \ \check{\epsilon} = \{\langle s, H \rangle \mid s \in \underline{t}\}$$

Note in particular, that an extended event has *two* "context sets": $\hat{\epsilon}$, the extended places which are read (these must be *causal* extended places, because reading a place should be causally unrelated to other concurrent events reading the same place) and $\check{\epsilon}$, the extended places which are generated by the readings of ϵ (these are read extended places). For instance, referring to Fig. 5, we can consider the extended event $\epsilon = \langle t_1, \{t_1\} \rangle$. Then we have ${}^\bullet\epsilon = \langle s_1, \emptyset \rangle$, $\hat{\epsilon} = \langle s_3, \emptyset \rangle$, $\epsilon^\bullet = \langle s_4, \{t_1\} \rangle$, and $\check{\epsilon} = \langle s_3, \{t_1\} \rangle$.

In order to formalize the intuition that the extended places used when generating a new extended event are "consistent", we introduce two relations on extended places: concurrency and subsumption.

Definition 18 (concurrency). *Let* $\sigma_1 = \langle s_1, H_1 \rangle$, $\sigma_2 = \langle s_2, H_2 \rangle$ *be extended places,* $\sigma_1 \neq \sigma_2$. *We say that they are* concurrent, *written* $\sigma_1 \frown \sigma_2$, *if* $\neg(H_1 \# H_2)$ *(thus* $H_1 \cup H_2$ *is a configuration) and* $s_1, s_2 \in (H_1 \cup H_2)^\bullet$.

In words, σ_1 and σ_2 are concurrent when the histories H_1 and H_2 associated with the two places are compatible, hence their union is a configuration C, and after executing C both places are marked.

We will see in Lemma 1 that, as expected, each pair of extended places in the pre-set of an extended event is concurrent. But the pre-set of an extended event also satisfies an important closure property:

Consider the net to the right. After the execution of t_1, for s we have one causal history $\langle s, \emptyset \rangle$ and one read history, namely $\langle s, \{t_1\} \rangle$. Now, transition t_2 could, in principle, be fired using the causal histories $\langle s_2, \{t_1\} \rangle$ and $\langle s, \emptyset \rangle$ and no read history. However, observe that the extended place $\langle s, \{t_1\} \rangle$ is implicitly there, since the inclusion of $\langle s_2, \{t_1\} \rangle$ implies that s has been read by t_1: we will say that $\langle s_2, \{t_1\} \rangle$ *subsumes* $\langle s, \{t_1\} \rangle$. Notice that the fact that $\langle s, \{t_1\} \rangle$ is not mentioned explicitly does not affect the new history that we are building for t_2, but it causes serious problems when computing the concurrency relation.

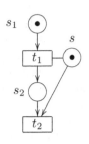

Definition 19 (subsumption relation). *For $\sigma = \langle s, H \rangle$, $\sigma' = \langle s', H' \rangle$ extended places, we write $\sigma \propto \sigma'$ (σ subsumes σ') when $s' \in H^\bullet$ and there exists $t \in H$ such that $s' \in \underline{t}$ and $H' = H[\![t]\!]$.*

In the example above $\langle s_2, \{t_1\} \rangle \propto \langle s, \{t_1\} \rangle$. Note that if $\sigma \propto \sigma'$, then σ' is necessarily a read extended place.

Lemma 1 provides a characterisation of extended events which will be used in the algorithm when looking for possible extensions of the current prefix. It says that an extended event $\epsilon = \langle t, H \rangle$ can be generated from a set of extended places which compose the pre-set and the context of t, and which have "consistent" histories, whose union will be H. Then ϵ generates a causal extended place for each event in the post-set of t and one read extended place for each event in the context of t. In particular, we note that the pre-set of an extended event is automatically closed under the subsumption relation. For this reason, in the algorithm, when looking for possible extensions we will take a set of extended places X which "enables" a transition and which satisfies a closure condition w.r.t. subsumption analogous to that in item 4 below.

Lemma 1 (events). *Let $\epsilon = \langle t, H \rangle$ be an extended event. If we let $X_p = {}^\bullet \epsilon$ and $X_c = \hat{\epsilon}$, and $X = X_p \cup X_c$, then*

1. *$\pi_S(X_p) = {}^\bullet t$ and $\pi_S(X_c) = \underline{t}$;*
2. *for any place $s \in \pi_S(X)$ there is a unique causal history $H' \in \chi_c(s)$ such that $\langle s, H' \rangle \in X$;*
3. *for any place $s \in \pi_S(X)$, if there is a read history $H' \in \chi_r(s)$ such that $\langle s, H' \rangle \in X$ then $s \in {}^\bullet t$;*
4. *for any $\sigma \in X$, if $\sigma \propto \sigma'$ and $\sigma'^\uparrow \in X_p$ then $\sigma' \in X_p$ (subsumption-closedness)*
5. *X is pairwise concurrent*

and $H = \bigcup \pi_H(X) \cup \{t\}$.

The next two lemmata provide an inductive characterisation of the concurrency and subsumption relations that will be pivotal for identifying the possible extensions of the prefix.

Lemma 2 (concurrency). *Let $\sigma = \langle s, H \rangle$ and $\sigma' = \langle s', H' \rangle$ be extended places in E, with $|H| \geq |H'|$. Then $\sigma \frown \sigma'$ if and only if $H = H' = \emptyset$ or $\sigma \in \epsilon^\bullet \cup \underline{\epsilon}$ for some ϵ and one of the following holds*

1. $\sigma' \in \epsilon^{\bullet} \cup \check{\underline{\epsilon}}$, $\sigma \neq \sigma'$
2. $\sigma' \in \hat{\underline{\epsilon}}$
3. (a) for all $\sigma'' \in {}^{\bullet}\epsilon \cup \hat{\underline{\epsilon}}$, it holds $\sigma'' \frown \sigma'$ and (b) if $\sigma' \propto \sigma'''$, with $\sigma'''^{\uparrow} \in {}^{\bullet}\epsilon$ then $\sigma''' \in {}^{\bullet}\epsilon$.

Lemma 3 (subsumption). *Let σ and σ' be extended places in E. Then $\sigma \propto \sigma'$ if and only if $\sigma \in \epsilon^{\bullet} \cup \check{\underline{\epsilon}}$ for some ϵ and one of the following holds*

1. $\sigma' \in \check{\underline{\epsilon}}$ or
2. there is $\sigma'' \in {}^{\bullet}\epsilon \cup \hat{\underline{\epsilon}}$ such that $\sigma'' \propto \sigma'$ and $\sigma \frown \sigma'$.

4.2 Computing the Possible Extensions

With the notions introduced so far, finally we can show how to generate the possible extensions of a given enriched prefix. We will refer to the construction of the prefix of the unfolding of a fixed semi-weighted c-net $N = \langle S, T, {}^{\bullet}(.), (.)^{\bullet}, \underline{(.)}, m \rangle$.

According to Lemma 1, in order to generate a new extended event, we can choose a concurrent set of extended places which "enables" the event. For any place in the pre-set and context we take a single causal history (condition 2 below). Only for places in the pre-set we can additionally take some read histories (condition 3 below). The fact that read histories can be consumed but not read corresponds to the fact that, referring to Fig. 4, to build a history for t we take additional histories only for transitions which read from consumed places (such as t'_3) and not for those that read from context places (such as t'_1).

Proposition 2 (possible extensions). *A possible extension of an enriched prefix Fin is an extended event $\epsilon = \langle t, H_t \rangle$ obtained as follows. Find sets of extended places X_p, X_c such that, if we denote by $X = X_p \cup X_c$ then*

1. *$f(\pi_S(X_p)) = {}^{\bullet}t_N$ and $f(\pi_S(X_c)) = \underline{t_N}$ for a transition t_N of the original net; the corresponding event in the unfolding is thus $t = \langle \pi_S(X_p), \pi_S(X_c), t_N \rangle$;*
2. *for any place $s \in \pi_S(X)$ there is exactly a single causal history $H \in \chi_c(s)$ such that $\langle s, H \rangle \in X$;*
3. *for any $s \in \pi_S(X)$ if there is a read history $\langle s, H \rangle \in X$ for some $H \in \chi_r(s)$ then $s \in {}^{\bullet}t$;*
4. *for any extended place $\sigma \in X$, if there is a σ' such that $\sigma \propto \sigma'$ and $\sigma'^{\uparrow} \in X_p$ then $\sigma' \in X$.*
5. *X is pairwise concurrent*

The history for t is defined as $H_t = \{t\} \cup \bigcup \pi_H(X)$.

Points 4 and 5 of Proposition 2 require to check the subsumption and the concurrency relations between certain pairs of extended places. The next result, that can be proved directly from the characterization of the two relations given in Lemmata 2 and 3, shows that both the concurrency and the subsumption relations can be computed inductively during the construction of the prefix.

Proposition 3 (concurrency and subsumption, inductively). *Let Fin be a finite enriched prefix of the unfolding obtained, according to the algorithm of Fig. 3, by starting with the initial marking m' and adding extended events $\epsilon_0, \ldots, \epsilon_n$. Then for each pair σ and σ' of extended places of Fin, $\sigma \frown \sigma'$ ($\sigma \propto \sigma'$, respectively) if and only if this can be deduced using the following rules:*

1. *[Base case]: For all $s_1, s_2 \in m'$ (initial marking) $\langle s_1, \emptyset \rangle \frown \langle s_2, \emptyset \rangle$.*
2. *[Inductive case]: Assume that the extended event $\epsilon_i = \langle t, H \rangle$ has been added, using the set $X = X_p \cup X_c$ of extended places, thus $t = \langle \pi_S(X_p), \pi_S(X_c), t_N \rangle$. Adding ϵ_i produces a set of extended places $Y = Y_p \cup Y_c$, where $Y_p = t^\bullet \times \{H\}$ are causal extended places and $Y_c = \underline{t} \times \{H\}$ are read extended places.*
   Then the concurrency relation can be extended to pairs including at least one new extended place using the following inference rules:

$$\frac{\sigma, \sigma' \in Y \quad \sigma \neq \sigma'}{\sigma \frown \sigma'} \qquad \frac{\sigma \in Y \quad \sigma' \in X_c}{\sigma \frown \sigma'}$$

$$\frac{\sigma \in Y \quad \forall \sigma'' \in X. \sigma'' \frown \sigma' \quad \forall \sigma'' \in X_p. (\sigma'' = \sigma'''^{\uparrow} \wedge \sigma' \propto \sigma''' \to \sigma''' \in X_p)}{\sigma \frown \sigma'}$$

Similarly, the subsumption relation can be extended to pairs including at least one new extended place, using rule (new) for the subsumptions induced by ϵ_i, and rule (inh) for inheriting the subsumptions of the premises, if not consumed:

$$\frac{\sigma \in Y \quad \sigma' \in Y_c}{\sigma \propto \sigma'} \ (new) \qquad \frac{\sigma \in Y \quad \sigma'' \in X \quad \sigma'' \propto \sigma' \quad \sigma \frown \sigma''}{\sigma \propto \sigma'} \ (inh)$$

We conclude by stating the correctness of the definition of the function PE based on Propositions 2 and 3.

Corollary 1 (definition and correctness of PE). *Let PE be the function that, when applied to an enriched prefix Fin and to an extended event ϵ, returns all the possible extensions of Fin, as defined in Proposition 2, built using at least one extended place generated by ϵ (i.e., using a set of extended places X such that $X \cap (\epsilon^\bullet \cup \underline{\epsilon}) \neq \emptyset$). Then PE satisfies the requirements of Theorem 1.*

5   Finite Prefix for Graph Transformation Systems

As mentioned in the introduction, the adequate treatment of read arcs and of the "reading" of items in general, is particularly important for graph transformation systems (GTSs). In fact, for GTSs—due to the presence of dangling conditions—the preservation of a node cannot be simulated by its deletion and subsequent re-creation. This means that the use of a "direct" algorithm which is not based on an encoding of read arcs as consume-create loops, while being an option for contextual nets, becomes mandatory for GTSs.

It has been observed earlier that in the absence of inhibiting conditions, which occur in DPO, the unfolding of graph transformation systems can be defined analogously to the unfolding of contextual nets. In [2,1] we have given the relevant definitions of occurrence graph grammars and we have generalized relations such as causality and asymmetric conflict to that setting. The role of places is played by the type graph, the initial graph represents the initial marking of the net and the rules correspond to transitions. In [2] the absence of inhibiting conditions was ensured by forbidding the deletion of nodes, whereas in [1] we used a more general setting, namely the SPO approach. Nevertheless, the more constrained format of rules allowed in [2] is balanced by the fact that one can take isomorphisms of graphs "up to isolated nodes", which leads to a more liberal notion of a "finite-state" graph transformation system.

Space limitations do not allow a detailed exposition of this development, but all constructions of the paper and especially the technical issues of Section 4—including the characterisation of extended events and the computation of the causality and subsumption relations—can be performed also for single-pushout rewriting. This would no longer be true if we switched to double-pushout rewriting where the presence of inhibiting conditions ("a node can not be deleted unless all adjacent edges are deleted") leads to a formalism which generalizes inhibitor nets, for which the unfolding semantics becomes much more involved.

In order to obtain *finite* complete McMillan prefixes for GTSs we clearly need to restrict to finite-state GTSs, where the number of reachable graphs is finite (up to isomorphism). More technically, this amounts to requiring that for every reachable graph, typed over the type graph T, the size of the preimage of an element of T is bounded by a constant k (this is analogous to the boundedness condition for Petri where one requires that for any reachable marking the number of tokens in each place is bounded by a fixed constant).

We also believe that this work can be generalized to rewriting in adhesive categories if we use a sesqui-pushout-like setting, as described in [6]. This would require to single out for a given rewriting system, a set of "atomic" subobjects from which any rewritten object can be built, playing the role of places.

6   Conclusions

We have described a worked-out procedure for computing McMillan prefixes for (bounded) contextual Petri nets and we argued that it can be extended to the computation of the prefix of the unfolding of finite-state graph transformation systems. Such prefixes are very valuable for the partial order verification of highly concurrent systems, where the use of unfolding techniques sometimes leads to an exponential gain in efficiency. While a part of the theoretical basis was already described in [3] it turned out that for aiming at an efficient implementation we needed more concepts such as extended places and the subsumption relation. A first prototype implementation for contextual nets has been realised. Although not yet optimised, it provides reassuring results on the feasibility of the approach.

As shown in this paper, the computation of finite complete prefixes for contextual nets is quite involved, but it is unclear whether it is possible to avoid the high complexity. The proposed technique becomes particularly interesting for graph transformation systems and related formalisms (with a structured state and preservation of items in a computational step), where approaches based on the encoding of item preservation by deletion and re-creation are not viable.

References

1. Baldan, P., Chatain, T., Haar, S., König, B.: Unfolding-based diagnosis of systems with an evolving topology. In: van Breugel, F., Chechik, M. (eds.) CONCUR 2008. LNCS, vol. 5201, pp. 203–217. Springer, Heidelberg (2008)
2. Baldan, P., Corradini, A., König, B.: Verifying finite-state graph grammars: an unfolding-based approach. In: Gardner, P., Yoshida, N. (eds.) CONCUR 2004. LNCS, vol. 3170, pp. 83–98. Springer, Heidelberg (2004)
3. Baldan, P., Corradini, A., König, B., Schwoon, S.: McMillan's complete prefix for contextual nets. In: Jensen, K., van der Aalst, W.M.P., Billington, J. (eds.) Transactions on Petri Nets and Other Models of Concurrency I. LNCS, vol. 5100, pp. 199–220. Springer, Heidelberg (2008)
4. Baldan, P., Corradini, A., Montanari, U.: An event structure semantics for P/T contextual nets: Asymmetric event structures. In: Nivat, M. (ed.) FOSSACS 1998. LNCS, vol. 1378, pp. 63–80. Springer, Heidelberg (1998)
5. Baldan, P., Corradini, A., Montanari, U., Ribeiro, L.: Unfolding Semantics of Graph Transformation. Information and Computation 205, 733–782 (2007)
6. Baldan, P., Corradini, A., Heindel, T., König, B., Sobociński, P.: Unfolding grammars in adhesive categories. In: Kurz, A., Lenisa, M., Tarlecki, A. (eds.) CALCO 2009. LNCS, vol. 5728, pp. 350–366. Springer, Heidelberg (2009)
7. Benveniste, A., Haar, S., Fabre, E., Jard, C.: Distributed monitoring of concurrent and asynchronous systems. In: Amadio, R.M., Lugiez, D. (eds.) CONCUR 2003. LNCS, vol. 2761, pp. 1–26. Springer, Heidelberg (2003)
8. Bonet, B., Haslum, P., Hickmott, S.L., Thiébaux, S.: Directed unfolding of Petri nets. In: Jensen, K., van der Aalst, W.M.P., Billington, J. (eds.) Transactions on Petri Nets and Other Models of Concurrency I. LNCS, vol. 5100, pp. 172–198. Springer, Heidelberg (2008)
9. Langerak, R.: Transformation and Semantics for LOTOS. Ph.D. thesis, Department of Computer Science, University of Twente (1992)
10. McMillan, K.: Using unfoldings to avoid the state explosion problem in the verification of asynchronous circuits. In: Probst, D.K., von Bochmann, G. (eds.) CAV 1992. LNCS, vol. 663, pp. 164–174. Springer, Heidelberg (1993)
11. McMillan, K.: Symbolic Model Checking. Kluwer Academic Publishers, Dordrecht (1993)
12. Nielsen, M., Plotkin, G., Winskel, G.: Petri Nets, Event Structures and Domains, Part 1. Theoretical Computer Science 13, 85–108 (1981)
13. Pinna, G.M., Poigné, A.: On the nature of events: another perspective in concurrency. Theoretical Computer Science 138(2), 425–454 (1995)
14. Vogler, W., Semenov, A., Yakovlev, A.: Unfolding and finite prefix for nets with read arcs. In: Sangiorgi, D., de Simone, R. (eds.) CONCUR 1998. LNCS, vol. 1466, pp. 501–516. Springer, Heidelberg (1998)

Verification of Graph Transformation Systems with Context-Free Specifications[*]

Barbara König[1] and Javier Esparza[2]

[1] Abteilung für Informatik und Angewandte Kognitionswissenschaft, Universität Duisburg-Essen, Germany
[2] Fakultät für Informatik, Technische Universität München, Germany

Abstract. We introduce an analysis method for graph transformation systems which checks that certain forbidden graphs are not reachable from the start graph. These forbidden graphs are specified by a context-free graph grammar. The technique is based on the approximation of graph transformation systems by Petri nets and on semilinear sets of markings. Especially we exploit Parikh's theorem which says that the Parikh image of a context-free grammar is semilinear. An important application is deadlock analysis for interaction nets and we specifically show how to apply the technique to an infinite-state dining philosopher's system.

1 Introduction

In recent years there have been several approaches directed specifically at the verification of graph transformation systems, for instance [21,10,24,1,2,4,22]. They usually address the following problem: Given a graph transformation system T generating a set $reach(T)$ of reachable graphs, and a set F of forbidden graphs, is the intersection $reach(T) \cap F$ empty? This problem is undecidable even for the case in which the set F contains one single graph, and so all approaches approximate the set $reach(T)$ in some way. In this paper we further develop the approach of [1,2], which—given T—constructs a Petri net whose reachable markings encode a set L of graphs satisfying $L \supseteq reach(T)$. So emptiness of $L \cap F$ implies emptiness of $reach(T) \cap F$.

The decidability, complexity, and practicality of checking emptiness of $L \cap F$ depend on the characteristics of the set F. If F is defined as the set of all graphs containing a given subgraph, then emptiness reduces to the coverability problem of Petri nets. In [3] we studied the case in which F is the set of models of a formula in first-order or second-order monadic logic on graphs.

Unfortunately, many forbidden sets of interest for applications are not expressible in first-order or monadic second-order logic. This follows from the fact that all expressible sets are recognizable in the sense of Courcelle [7], while the forbidden sets are often not. For instance counting constraints usually specify non-recognizable languages. These sets can, however, often be described by (a

---

[*] Research supported by DFG project SANDS.

H. Ehrig et al. (Eds.): ICGT 2010, LNCS 6372, pp. 107–122, 2010.
© Springer-Verlag Berlin Heidelberg 2010

slight variant of) context-free graph grammars. In this paper we propose an extended emptiness check for this class.

The main ideas behind the new check can already be introduced for transformation systems acting on words. Assume that T is such a system. The approximation of the word language T can be intuitively seen as equivalent to a finite automaton A and a *counting constraint* C.[1] The approximation $\mathcal{L}$ is given by $\mathcal{L} = L(A) \cap L(C)$, i.e., the words of $\mathcal{L}$ are those accepted by A and satisfying C. Later such counting constraints will be specified via the reachable markings of a Petri net.

Assume further that $\mathcal{F}$ is the language of forbidden words generated by a context-free word grammar G, i.e., $\mathcal{F} = L(G)$. Our task is to check the emptiness of $\mathcal{F} \cap \mathcal{L} = L(G) \cap L(A) \cap L(C)$. For this, we proceed in three steps:

(1) Compute a grammar G' accepting $L(G) \cap L(A)$ (using the well-known fact that the intersection of a context-free and a regular language is context-free). If $L(G') = \emptyset$, then terminate and conclude $\mathcal{F} \cap \mathcal{L} = \emptyset$, otherwise proceed with (2).

(2) Compute the strongest counting constraint C' satisfied by $L(G')$, specifically $L(G') \subseteq L(C')$. This is possible by Parikh's theorem, which states that the Parikh image of a context-free grammar is semilinear set, which can be formulated as a counting constraint.

(3) Check the emptiness of $L(C) \cap L(C')$ by deciding whether the conjunction $C \wedge C'$ is unsatisfiable. If it is unsatisfiable we have $\mathcal{F} \cap \mathcal{L} = \emptyset$, otherwise the intersection is non-empty.

The first contribution of the paper is an extension of this procedure to graph transformation systems. T is now a graph transformation system, and $\mathcal{F}$ is the set of graphs generated by a context-free graph grammar. The decidability of the satisfiability of $C \wedge C'$ is proved using results from Petri net theory.

While the decidability has some theoretical interest, the complexity of the resulting procedure is too high for practical purposes. Using techniques developed in [11], it is possible to over-approximate C' by a system of linear constraints C''. The satisfiability of $C \wedge C''$ can then be checked with appropriate solvers for linear programming (such as lp_solve).

An important application is to verify the absence of deadlocks, which often manifest themselves as "vicious cycles" (cycles of processes waiting for an action from another process). We show how to detect such cycles in the general setting of interaction nets and treat—as a concrete running example—an infinite-state version of the dining philosophers.

A simplified version of our results appeared as a workshop paper in [14]. That paper uses regular expressions instead of the more powerful context-free graph grammars employed here. Furthermore, in the current paper the theory is for the first time applied to the general setting of interaction nets.

---

[1] A constraint on the number of times that each letter can appear in a word of L; for instance, $(\#a - 2 \cdot \#b) \geq 3$ is a counting constraint for the alphabet $\{a, b\}$.

2   Preliminaries

2.1   Monoids and Semilinear Sets

Let M be a *monoid* with an associative binary operation written $m_1 m_2$ for $m_1, m_2 \in M$. Furthermore there exists a unit $1 \in M$. In this paper we will mainly consider the *free monoid* over a finite set S, denoted by S^*, and the *free commutative monoid* over S, denoted by $S^{\oplus}$. The former monoid consists of all *words* over S whereas the latter is isomorphic to the set of all mappings from S to $\mathbb{N}$, denoted by $\mathbb{N}^S$. Its elements will also be called *multisets*.

The *rational sets* of a monoid M are inductively defined as follows:

– Every finite subset of M is rational.
– Whenever A and B are rational sets, then $A \cup B$ and $AB = \{ab \mid a \in A, b \in B\}$ are rational sets.
– Whenever A is a rational set, then $A^* = \{a_1 \ldots a_n \mid n \geq 0,\, a_1, \ldots, a_n \in A\}$ is a rational set. Alternatively one can define A^* as the intersection of all sub-monoids of M containing A.

A *semilinear set* of a monoid M is of the form $\bigcup_{i \in I} \{a_i\} B_i^*$, where $a_i \in M$ and the B_i are finite subsets of M and I is a finite index set. Sets of the form $\{a_i\} B_i^*$ are called *linear*. For commutative monoids, every rational set is also semilinear. Furthermore semilinear sets are closed under intersection and complement [12].

In the following we will denote the monoid operation and the Kleene closure by $\oplus$ whenever we are working in a commutative monoid. Furthermore for a function $f \colon S \to T$ we will denote by $f^* \colon S^* \to T^*$ and $f^{\oplus} \colon S^{\oplus} \to T^{\oplus}$ its obvious extensions to words and multisets.

Whenever $\mu \colon S^* \to S^{\oplus}$ is the canonical monoid morphism and $A \subseteq S^*$, then $\mu(A)$ is called the *Parikh* image of A in $S^{\oplus}$. The Parikh image of a rational set is always semilinear.

2.2   Petri Nets

A *Petri net* is given by a tuple $N = (S_N, T_N, {}^{\bullet}(), ()^{\bullet}, p_N, m_0)$ where S_N is a set of places, T_N is a set of transitions; ${}^{\bullet}(), ()^{\bullet} \colon T_N \to S_N^{\oplus}$ assign to each transition its pre-set and post-set (where both are multisets), $p_N \colon T_N \to \Lambda$ assigns an action label to each transition and $m_0 \in S_N^{\oplus}$ is the initial marking.

Reachability and coverability in Petri nets is defined as usual. The set of all reachable markings of a net N is denoted by $reach(N)$.

Proposition 1. *Let N be a Petri net where S_N is the set of places. Given a semilinear set A of $S_N^{\oplus}$ it is decidable whether there exists a reachable marking m of N that is contained in A.*

The proof is based on a reduction to the reachability problem for Petri nets. But although this problem is decidable it is not known to be primitive recursive and quite impossible to solve in practice. Hence, we aim at solving the reachability problem in an approximative way by solving systems of linear constraints, as detailed at the end of Section 3.

2.3   Hypergraphs

We assume that Λ is a fixed and finite set of labels and that each label $\ell \in \Lambda$ is equipped with an arity $ar(\ell) \in \mathbb{N}$.

Definition 1 (Hypergraph, hypergraph morphism). *A (Λ-)hypergraph G is a tuple (V_G, E_G, c_G, l_G), where V_G is a finite set of nodes, E_G is a finite set of edges, $c_G : E_G \to V_G^*$ is a connection function and $l_G : E_G \to \Lambda$ is the labelling function for edges satisfying $ar(l_G(e)) = |c_G(e)|$ for every $e \in E_G$. Nodes are not labelled.*

Let G, G' be (Λ-)hypergraphs. A hypergraph morphism $\varphi : G \to G'$ consists of a pair of total functions $\langle \varphi_V : V_G \to V_{G'}, \varphi_E : E_G \to E_{G'} \rangle$ such that for every $e \in E_G$ it holds that $l_G(e) = l_{G'}(\varphi_E(e))$ and $\varphi_V^*(c_G(e)) = c_{G'}(\varphi_E(e))$. The morphism φ is called edge-bijective *if φ_E is bijective.*

We will now introduce the language of all graphs that can be mapped homomorphically to a given graph. This is related to regular (word) languages, which intuitively contain all graphs consisting of a single path that can be mapped homomorphically to a finite automaton (taking into account also initial and final states). However note that the recognizable graph languages of Courcelle are more general.

Definition 2 (Language induced by a graph). *Let G be a hypergraph. Then we denote by $\mathcal{I}_G$ the language induced by G, defined by*

$$\mathcal{I}_G = \{G' \mid \exists \varphi : G' \to G\}.$$

For later use we need the following notion of node fusion.

Definition 3 (Closure under node fusion). *Let $\mathcal{L}$ be a graph language. By $nf(\mathcal{L})$ we denote the language obtained from $\mathcal{L}$ by fusing arbitrary nodes, i.e.,*

$$nf(\mathcal{L}) = \{G/{\equiv} \mid G \in \mathcal{L} \text{ and } \equiv \text{ is an equivalence on } V_G\}$$
$$= \{G' \mid \varphi : G \to G', G \in \mathcal{L}, \varphi \text{ edge-bijective}\}.$$

2.4   Graph Transformation Systems

We now define the notion of a graph transformation system. In order to simplify the analysis we focus on restricted rewriting rules which can not delete nodes (but nodes may eventually become isolated).

Definition 4 (Graph transformation system). *A graph transformation system (GTS) $\mathcal{T} = (G_0, \mathcal{R})$ consists of an initial graph G_0 and a set $\mathcal{R}$ of rewriting rules of the form $r = (L, R, \alpha)$, where L, R are graphs, called* left-hand side *and* right-hand side, *respectively, and $\alpha : V_L \to V_R$ is a function.*

A match *of a rewriting rule r in a graph G is a morphism $\varphi : L \to G$ which is injective on edges. We can apply r to a match in G obtaining a new graph H, written $G \overset{r}{\Rightarrow} H$, where the target graph H is defined as follows. Let $\equiv$ be the smallest equivalence on V_R satisfying $\alpha(w_1) \equiv \alpha(w_2)$ whenever $\varphi(w_1) = \varphi(w_2)$*

for two nodes $w_1, w_2 \in V_L$. *Furthermore let* $\overline{\varphi}\colon V_G \uplus V_R \to V_H$ *be defined as follows:*

$$\overline{\varphi}(v) = \begin{cases} v & \text{if } v \in V_G - \varphi(V_L) \\ [\alpha(w)]_\equiv & \text{if } v \in \varphi(V_L),\ v = \varphi(w) \\ [v]_\equiv & \text{if } v \in V_R \end{cases}$$

Then we can define

$$V_H = (V_G - \varphi(V_L)) \uplus (V_R/\equiv) \qquad E_H = (E_G - \varphi(E_L)) \uplus E_R$$
$$e \in E_G - \varphi(E_L) \quad \Rightarrow \quad c_H(e) = \overline{\varphi}(c_G(e)) \quad l_H(e) = l_G(e)$$
$$e \in E_R \quad \Rightarrow \quad c_H(e) = \overline{\varphi}(c_R(e)), \quad l_H(e) = l_R(e)$$

The set of all reachable graphs, i.e., graphs that can be derived from G_0 *via repeated rule application, will be denoted by* $reach(\mathcal{T})$.

The application of r to G at the match φ first removes from G the image of the nodes and edges of L. Then the graph G is extended by adding the new nodes and edges in R, which are connected accordingly. Observe that the nodes of R have to be grouped into equivalence classes, i.e. merged, depending on whether they are assigned to the same nodes in G. The notion of graph rewriting that we use is a special case of the double-pushout approach.

Example 1. As a running example we consider an infinite-state dining philosphers system. The start graph of the system is depicted in Figure 1 where we see three philosophers (indicated by label H, since they are hungry) and three forks (indicated by labels F or F'). Some edges will be pointing clockwise along the cycle, others (indicated by a prime) counter-clockwise. Note that for binary edges, i.e., for edges connected to exactly two nodes, the first node is indicated by an arrowhead.

In order to avoid deadlocks we use a technique similar to the one presented in [6] where a dining philosophers system is seen as a directed acyclic graph with philosophers as nodes and forks as directed edges. In our setting only a philosopher which is maximal wrt. the partial order established by the forks, i.e., a philosopher which has only forks pointing towards it, may start eating.

Figure 2 depicts the rules governing the behaviour of the dining philosophers. A hungry philosopher may take up a fork pointing towards it (rule *(Wait)*) and

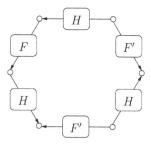

Fig. 1. Start graph of the infinite-state dining philosophers system

become a waiting philosopher (labelled W'). Subsequently a waiting philosopher may pick up another fork also pointing towards it and become an eating philosopher (rule *(Eat)*). The reason for modelling an eating philosopher by two edges is that we will later view this GTS as an instance of an interaction net system, where a left-hand side must always consist of exactly two edges. Since eating philosophers may react of their own impulse it is here necessary to represent them by two edges forming a valid redex. An eating philosopher may then either give back its forks in such a way that they are pointing away from it (rule *(Hungry)*) or replicate itself in order to create another hungry philosopher with another fork (rule *(Rep)*). This last rule makes the system infinite-state and hence non-trivial to analyze. Our aim is to show that no deadlock – which in this case manifests itself as a vicious cycle – will ever occur.

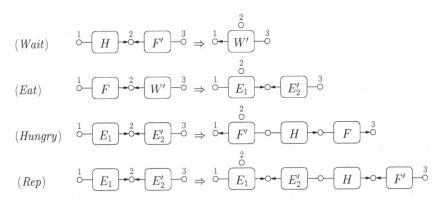

Fig. 2. Productions of the infinite-state dining philosophers system

2.5   Context-Free Graph Grammars

In order to characterize graph languages more complex than the ones of Definition 2, we introduce context-free graph grammars, also called hyperedge replacement grammars [13]. Similar to the case of word languages, the left-hand side of a context-free rule consists of a single hyperedge, whereas right-hand sides can be of arbitrary shape.

Definition 5 (Single hyperedge). *Let $\ell \in \Lambda$ be a label of arity $ar(\ell)$. Then $edge(\ell)$ is a hypergraph H which consists of a single edge e labelled ℓ and $ar(\ell)$ nodes such that the nodes in $c_H(e)$ are all pairwise different.*

Definition 6 (Context-free graph grammar). *Assume that the set of labels Λ is the disjoint union of Λ_T (the set of terminal labels) and Λ_N (the set of non-terminal labels). A context-free graph grammar $\mathcal{G}$ is a special graph transformation system where*

- *for every rule $r = (L, R, \alpha)$ it holds that $L = edge(A)$ for some $A \in \Lambda_N$ and α is injective.*
- *the start graph G_0 is of the form $edge(S)$ for some $S \in \Lambda_N$ with $ar(S) = 0$.*

Definition 7 (Context-free graph language). *Let $\mathcal{G}$ be a context-free graph grammar. Then the* language generated by $\mathcal{G}$, *denoted by $\mathcal{L}(\mathcal{G})$, is defined as*

$$\mathcal{L}(\mathcal{G}) = \{G \in reach(\mathcal{G}) \mid all\ labels\ of\ G\ are\ contained\ in\ \Lambda_T\}.$$

In analogy to word languages, $\mathcal{L}(\mathcal{G})$ is also called a context-free (graph) language.

Note that not every language induced by a graph according to Definition 2 is context-free. Especially the language of all hypergraphs is induced by the final hypergraph (i.e., a graph consisting of one node and one edge for every label), but it is not context-free (cf. [13]).

Furthermore, given a graph language $\mathcal{L}$ over a label set Λ_T, its *Parikh image* is $\{l_G^{\oplus}(E_G) \mid G \in \mathcal{L}\} \subseteq \Lambda_T^{\oplus}$, i.e., for every graph in the language take the multiset of its edge labels.

2.6 Approximating Graph Transformation Systems by Petri Nets

In this section we sketch the algorithm, introduced in [1], for the construction of a finite approximation of the unfolding of a graph transformation system. Of course there is a straightforward counting abstraction via a Petri net which just counts the number of occurrences of each edge, regardless of structure. However, this will be too coarse for our purposes and we use Petri nets over a hypergraph structure, so-called *Petri graphs*.

Definition 8 (Petri graphs). *Let $\mathcal{T} = (G_0, \mathcal{R})$ be a GTS. A* Petri graph P *(over $\mathcal{T}$) is a tuple (G, N) where*

- *G is a graph;*
- *$N = (E_G, T_N, {}^{\bullet}(), ()^{\bullet}, p_N, m_0)$ is an $\mathcal{R}$-labelled Petri net, where the set of places is E_G, i.e., the edge set of G;*
- *$m_0 = \iota^{\oplus}(E_{G_0})$ for a graph morphism $\iota: G_0 \to G$ (i.e., m_0 must properly correspond to the initial state of the GTS $\mathcal{T}$).*

A marking $m \in E_G^{\oplus}$ will be called reachable (coverable) *in P if it is reachable (coverable) from the initial marking in the Petri net underlying P.*

A marking m of a Petri graph can be seen as an abstract representation of a graph in the following sense.

Definition 9. *Let (G, N) be a Petri graph and let $m \in E_G^{\oplus}$ be a marking of N. The graph H generated by m, denoted by $graph(m)$, is defined as follows: $V_H = \{v \in V_G \mid \exists e \in m\ \exists i: (v = [c_G(e)]_i)\}$, $E_H = \{(e, i) \mid e \in m \wedge 1 \leq i \leq m(e)\}$, $c_H((e, i)) = c_G(e)$ and $l_H((e, i)) = l_G(e)$.*[2]

That is, we take only the nodes adjacent to a marked edge and make parallel copies of all edges according to their multiplicity (see also Example 2 below).

---

[2] Note that $[\tilde{s}]_i$ denotes the i-th element of the string $\tilde{s}$ and that $m(e)$ denotes the multiplicity of e in the multiset m.

Given a GTS $T = (G_0, \mathcal{R})$, with some additional minor constraints on the format of rewriting rules (see [1,2]), we can construct a Petri graph approximation of T, called *covering* and denoted by $\mathcal{U}(T)$. The (0-)covering is produced by the following (terminating) algorithm which generates a sequence $P_i = (G_i, N_i)$ of Petri graphs. (There is also an extension which produces k-coverings, having a better precision.)

1. $P_0 = (G_0, N_0)$, where the net N_0 contains no transitions and $m_0 = E_{G_0}$.
2. As long as one of the following steps is applicable, transform P_i into P_{i+1}, giving precedence to folding steps.

   Unfolding. Find a rule $r = (L, R, \alpha) \in \mathcal{R}$ and a match $\varphi \colon L \to G_i$ such that $\varphi^{\oplus}(E_L)$ is coverable in P_i. Then extend P_i by "attaching" R to G_i according to α and add a transition t, labelled by r, describing the application of rule r.

   Folding. Find a rule $r = (L, R, \alpha) \in \mathcal{R}$ and two matches $\varphi, \varphi' \colon L \to G_i$ such that $\varphi^{\oplus}(E_L)$ and $\varphi'^{\oplus}(E_L)$ are coverable in N_i and the second match is causally dependent on the transition unfolding the first match. Then merge the two matches by setting $\varphi(e) \equiv \varphi'(e)$ for each $e \in E_L$ and factoring through the resulting equivalence relation $\equiv$.

The first construction above is an unfolding construction that "unwinds" a system, while still preserving its concurrent behaviour. Hence unfoldings can lead to very compact descriptions of the state space of a concurrent system. Here, we add so called folding steps to the unfolding construction in order to obtain a finite over-approximation, even if the GTS has an infinite state space. We lose information by merging nodes that would actually be distinct. The covering $\mathcal{U}(T)$ is an abstraction of the original GTS $(G_0, \mathcal{R})$ in the following sense.

Proposition 2 (Abstraction [3]). *Let $T = (G_0, \mathcal{R})$ be a graph transformation system and let $\mathcal{U}(T) = (G, N)$ be its covering where N has initial marking m_0. Then there exists a simulation $\mathcal{S} \subseteq reach(T) \times reach(N)$ with the following properties:*

- $(G_0, m_0) \in \mathcal{S}$;
- *whenever $(G, m) \in \mathcal{S}$ and $G \overset{r}{\Rightarrow} G'$, then there exists a marking m', obtained from m by firing a transition labelled r, and $(G', m') \in \mathcal{S}$;*
- *for every $(G, m) \in \mathcal{S}$ there is an edge-bijective morphism $\varphi \colon G \to graph(m)$.*

From this one can easily deduce that the set of all reachable graphs is over-approximated by the set of all graphs generated by reachable markings where edges can be "pulled apart" at the nodes in an arbitrary way, where by "pulling apart" we mean the reverse of node folding introduced earlier:

$$nf^{-1}(\{graph(m) \mid m \in reach(N)\}) \supseteq reach(T) \tag{1}$$

Example 2. With the approximated unfolding technique sketched above we obtain the Petri graph in Figure 3. It has been computed automatically using the tool AUGUR 2 [16]. Note that edges of the graph and places of the net coincide and that arcs between places and transitions are drawn with dashed lines. The marking shown is the initial marking m_0 and the corresponding graph $graph(m_0)$ has two nodes and six edges. Two parallel F'-labelled and three parallel H-labelled edges are going from right to left whereas one F-labelled edge is pointing from left to right. This graph is an over-approximation of G_0 in the sense described above.

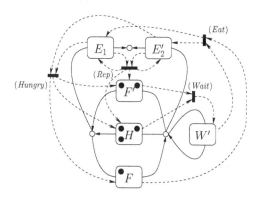

Fig. 3. A Petri graph approximating the dining philosophers system

3   Specifying Sets of Forbidden Graphs by Semilinear Sets

We assume that we are provided with a context-free graph grammar $\mathcal{G}$ that specifies the set of all "bad" or forbidden graphs, i.e., graphs representing error states. The set of forbidden graphs will not be the language generated by $\mathcal{G}$, but also all graphs G' that arise from graphs of $\mathcal{L}(\mathcal{G})$ by node fusion, i.e., we are interested in $nf(\mathcal{L}(\mathcal{G}))$. The introduction of node fusion is due to two reasons: first, the operator nf provides us with the possibility of specifying languages that can otherwise not be specified with a context-free grammar. For instance, the language of all graphs is not context-free (see [13]), but it can be obtained from a context-free language via node fusion. (Of course, this also means that other languages are excluded.) Second, our approximation method over-approximates with respect to the identity of nodes (see equation (1)), so if we find out—via the approximation—that a graph G is not reachable, the same is true for every preimage of G wrt. node fusion.

Given a graph transformation system $\mathcal{T}$, an over-approximating Petri graph $P = (G, N)$, i.e., a Petri graph satisfying equation (1), and a context-free graph grammar $\mathcal{G}$, our task is to construct a set S of markings of N which satisfies: *For a marking m of N it holds that $m \in S$ if and only if $graph(m) \in nf(\mathcal{L}(\mathcal{G}))$.*

If we can then show that no marking $m \in S$ is reachable in N, we know for sure that no graph $G \in nf(\mathcal{L}(\mathcal{G}))$ is reachable in $\mathcal{T}$. To prove this, assume that such a graph is reachable in $\mathcal{T}$. Then it holds that there exists a reachable marking m and an edge-bijective morphism $\varphi: G \to graph(m)$, i.e., $graph(m)$ is a "folded" variant of G. Furthermore $graph(m) \in nf(\mathcal{L}(\mathcal{G}))$ and hence as required above $m \in S$. But this is a contradiction, since we have checked that no marking of S is reachable in the net.

It will turn out that S is semilinear and it corresponds to the strongest counting constraint mentioned in the introduction. As a first step towards the construction of S we determine a context-free graph grammar that generates the intersection of a context-free graph language with a language induced by a graph, i.e., the set of graphs that can be mapped to a fixed graph G. Here G will usually be the graph component of a Petri graph.

As far as we know the underlying construction is original. It has similarities to the construction of the "cross-product" of a context-free grammar with a finite automaton. Furthermore there is a related result by Courcelle [7] who shows that the intersection of recognizable set and a equational set of graphs is always equational. Equational sets correspond to context-free graph languages.

Proposition 3. *For a context-free graph grammar $\mathcal{G}$ and a graph G there exists a context-free graph grammar $\mathcal{G}'$ such that $\mathcal{L}(\mathcal{G}') = \mathcal{L}(\mathcal{G}) \cap \mathcal{I}_G$.*

The construction used in the proof of the previous proposition is now the key to the construction of the required semilinar set S. Specifically, we can derive S from the grammar $\mathcal{G}'$ of Proposition 3, which implies that S is semilinear.

Proposition 4. *For a context-free graph grammar $\mathcal{G}$ and a graph G the following set $S_G^{\mathcal{G}}$ is a semilinear set of $E_G^{\oplus}$: $S_G^{\mathcal{G}} = \{m \in E_G^{\oplus} \mid graph(m) \in nf(\mathcal{L}(\mathcal{G}))\}$.*

Finally it is well-known that $S_G^{\mathcal{G}}$ is effectively computable and furthermore it follows from Proposition 1 that it is decidable whether there exists a reachable marking of a net that is contained in a given semilinear set. This would allow us to automatically verify whether $graph(m) \notin nf(\mathcal{L}(\mathcal{G}))$ for all reachable markings m, which—as detailed above—implies that no graph contained in $nf(\mathcal{L}(\mathcal{G}))$ is reachable in the graph transformation system $\mathcal{T}$. Figure 4 summarizes the technique.

However, checking whether no marking of a semilinear set is reachable is at least as costly as the reachability problem for Petri nets [19]. Since this problem is not known to be primitive recursive and all known algorithms are very complex,

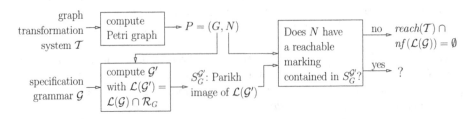

Fig. 4. Summary of the technique

it will usually be necessary to resort to approximate reachability methods based on the marking equation and traps [11]. These techniques are based on the efficient solving of linear constraints over natural numbers, which can be handled using a tool such as lp_solve.

4 Characterizing Deadlocks of Interaction Nets

Interaction nets [17] are a special form of graph transformation systems with strong restrictions on the start graph and on the form of rules. In this setting one can ensure confluence and furthermore a deadlock (i.e., non-applicability of any rule) always manifests itself structurally in the presence of vicious cycles. We will here present a slight adaptation of interaction nets, especially we will differ by allowing isolated nodes, which however have no influence on further reductions.

In the following, the first node attached to a hyperedge will be called *principal node* and denoted by an arrow head. The *degree* of a node v is the number of "tentacles" attached to a node, i.e., $\deg(v) = |\{(e, i) \mid [c_G(e)]_i = v\}|$.

Definition 10 (Interaction net (system)). *A graph transformation system* T *is called an interaction net system whenever the following conditions hold:*

- *Every node of the start graph is either isolated or has degree 2. (Any such graph is called* interaction net.*)*
- *For a rewriting rule* (L, R, α) *the left-hand side* L *has the following form:*

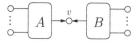

We require that $\deg(\alpha(v)) = 0$ *and that* v *is the only node that* α *maps to* $\alpha(v)$. *For every other node* v' *in the right-hand side* R *it holds that:* $\deg(v') = 2 - |\alpha^{-1}(v')|$. *(That is, nodes which are not in the image of* α *have degree 2 and whenever two nodes of the left-hand side are merged, the resulting node has degree 0. The remaining nodes have degree 1.)*
Finally, for every pair $A, B \in \Lambda$ *of labels there exists a rule with a left-hand side as depicted above.*

Note that the second condition above is required in order to make sure that interaction nets are closed under rewriting.

All nodes in the left-hand side which are different from v and their images in the right-hand side are called *external*.

Example 3. The example graph transformation system introduced in Section 2.4 satisfies the conditions imposed by Definition 10, apart from the last. This last condition can be satisfied by adding dummy rules for the remaining left-hand sides. These dummy rules have no effect, i.e., a left-hand side with two edges (e.g., F, F') is replaced by the same left-hand side plus an isolated node.

It can then be shown that in an interaction net system every reachable graph has only nodes which are either isolated or have degree 2. Furthermore it can be shown (see [17]) that a graph that does not allow the application of a rule has a vicious cycle. In order to formally characterize vicious cycles we first need the following definition.

Definition 11 (C-path). *Let C be a set of triples (i, A, j), where $A \in \Lambda$ and $i, j \in \{1, \ldots, ar(A)\}$, $i \neq j$. A C-path is a sequence $v_0, e_1, v_1, \ldots, e_n, v_n$ of nodes and edges such that two nodes v_k, v_{k+1} are connected by the edge e_k where $[c_G(e_k)]_i = v_{k-1}$, $[c_G(e_k)]_j = v_k$ and $(i, l_G(e_i), j) \in C$. We also require that all edges e_k are distinct. A C-path is a C-cycle if additionally $v_n = v_0$.*

We call a C-path a principal *C-path whenever for all $(i, A, j) \in C$ it holds that either $i=1$ or $j=1$. It is a* clockwise *C-path if always $j=1$ and a* counterclockwise *C-path if always $i=1$. (Similar for C-cycles.)*

A clockwise C-cycle where $C = \{(i, A, 1) \mid A \in \Lambda, i \in \{2, \ldots, ar(A)\}\}$ (i.e., C contains all possible triples) is also called vicious cycle.

Proposition 5 (Vicious cycles [17]). *An interaction net in which no reduction is possible is either the empty graph or it contains at least one vicious cycle.*

Note that this proposition depends heavily on the satisfaction of the last item of Definition 10 and hence, in our setting, on the existence of dummy rules.

Graphs not containing vicious cycles can be specified using hyperedge replacement grammars (and node fusion). However, if we do not have additional information about the potential form of the vicious cycles, a better and more local condition can be used in order to check for the absence of deadlocks. It is enough to check that for each reachable marking m there exists a node v such that the number of tokens placed on edges having v as principal node is strictly larger than the number of tokens placed on edges having v as non-principal node. The node v represents several nodes and the condition implies that at least one of them is the principal node of two edges. The negation of this requirement can be expressed as a semilinear set of markings, or, even simpler, directly encoded into a linear constraint and passed to a linear constraint solver.

Example 4. Note that the technique described above does not work for the running example with the Petri graph depicted in Figure 3. Specifically, it discovers a possible vicious cycle with labels W', W', H, F that can be reached after firing the transition representing rule (*Wait*) twice. However, this cycle is not among the reachable graphs and a modification of the technique is necessary.

5   A Refined Deadlock Analysis

If we know that all reachable interaction nets satisfy a certain invariant, we can omit deadlock configurations which are known to be unreachable and only specify the remaining situations via a context-free graph grammar. An invariant that works well in a number of examples is being $\{C_i\}_i$-*cycle-covered.*, i.e., a graph is covered by cycles $k_1, \ldots, k_n$, where k_i is a C_i-cycle (for a set C_i, see

Definition 11), and all other cycles are trivial, which means that they consist of only one node.

It is possible to define sufficient conditions implying this invariant. Furthermore it follows from the invariant that an arising vicious cycle must be one of the C_i-cycles, and that the remaining edges are from the other cycles. This stronger non-local property can be described by a hyperedge replacement grammar.

A context-free grammar $\mathcal{G}_\ell$ specifying the forbidden graphs with a vicious C_ℓ-cycle either generates the empty graph or a clockwise C_ℓ-cycle or a counterclockwise C_ℓ-cycle for some index ℓ. In addition it must generate all edges present in C_m for $m \neq \ell$. This can be done with the context-free graph grammar, having nonterminals S, CW, CC, given in Figure 5. Then it is necessary to check that there is an empty intersection of the reachable graphs with each language $\mathcal{L}(\mathcal{G}_\ell)$.

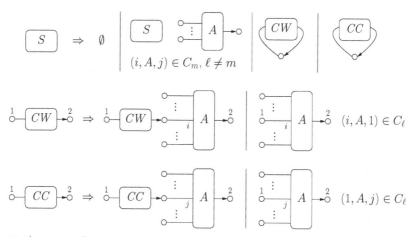

Fig. 5. A context-free graph grammar $\mathcal{G}_\ell$ for the generation of vicious C_ℓ-cycles

Example 5. In our running example we can easily check the following invariant, by inspection of the rules and the initial graph: every reachable graph is a C-cycle for

$$C = \{(1, H, 2), (1, F, 2), (1, E_1, 2), (2, F', 1), (2, W', 1), (2, E'_2, 1)\}.$$

(So, in particular, every reachable graph is $\{C\}$-cycle-covered.) It follows that a vicious cycle of a reachable graph must necessarily be equal to the graph itself.

In this case, the context-free graph grammar generating all vicious C-cycles either generates a counterclockwise cycle consisting only of edge labels H, F, E_1 or a clockwise cycle consisting only of edge labels F', W', E'_2. Applying the procedure for computing semilinear sets of Section 3 to this grammar and the Petri graph of Figure 3 we obtain $S = \{H \oplus F\}^\oplus \cup \{W'\}^\oplus$ (where edges are denoted by their labels, since every label occurs at most once in our Petri graph). The first set corresponds to a situation where only H- and F-edges are present and their numbers are the same, i.e., to a vicious cycle of the form $H, F, H, F, \ldots$. The second set corresponds to the cycle $W', W', \ldots$.

No marking of S is reachable. This can be shown by setting up systems of linear constraints, one for each linear set. In our case it is sufficient to use the marking equation and the equations describing the semilinear set S above. No solution can be found, and so that this dining-philosophers system will never reach a deadlock state.

Finally, we can also specify as error graphs all graphs that allow only the application of the dummy rules introduced above, a more useful property to verify. Since C-cycles have been established as invariant, all reachable cycles can be represented as words over the alphabet C by cutting the cycle at an arbitrary point. The following regular expression r describes the language of all "good" cycles allowing the application of at least one non-trivial rule as shown in Figure 2:

$$r = C^*(2, H, 1)(1, F', 2)C^* + C^*(2, F, 1)(1, W', 2)C^* + C^*(2, E_1, 1)(1, E_2', 2)C^*$$
$$+ (1, F', 2)C^*(2, H, 1) + (1, W', 2)C^*(2, F, 1) + (1, E_2', 2)C^*(2, E_1, 1)$$

Using standard techniques one can compute a regular expression for the complement language (C-cycles allowing only the application of trivial rules) and turn it into a context-free graph grammar. Applying our techniques we obtain the same semilinear set S as above and the same reasoning applies.

6   Conclusion

We have shown how to analyze a graph transformation system where the set of forbidden graphs is specified by a context-free graph grammar. This is done by using Parikh's theorem which states that the Parikh image of a context-free language is semi-linear. Semilinear sets can then be expressed using linear constraints. Similarly, the specification of the error graphs can be written as linear constraints. If the resulting constraint system has no solution, we can infer that no error graph is reachable.

We use graph transformation to model and analyze systems with dynamically evolving communication structures. Several papers in this area concentrate on identifying decidable classes. Lammich et al. study dynamic pushdown networks, in which communication between parallel threads is either abstracted away or limited (see e.g. [18]), and Meyer describes a decidable fragment of the π-calculus in [20], subsuming decidability results of previous papers. None of these formalisms, however, can model the examples of this paper. Results on parametrized verification cannot be directly applied either, because in those approaches the number of processes stays fixed during the computation, and the communication structure is restricted to trees or other structures.

We follow the line of overapproximating the set of reachable states, which in our case are graphs. Apart from the work on the verification of graph transformation mentioned in the introduction, related work along this line is shape analysis [23] and, more recently, separation logic [9]. Both, however, are mainly specialized towards the verification of specific pointer structures, such as singly-linked or doubly-linked lists, rather than arbitrary graphs.

The idea of using Parikh's theorem for overapproximating context-free structures has been used by Deutsch [8] and by Bouajjani et al. [5] in settings different from ours.

The method described in the paper has been partially implemented in the tool AUGUR 2 [16]. Especially, we have implemented the computation of the over-approximating Petri graph and the computation of the semi-linear set from a context-free grammar. We plan to investigate techniques to derive linear constraints directly from the context-free grammar, without going via semi-linear sets, and techniques to automatically generate invariants like the ones obtained here through vicious cycles.

Finally, verification may also be unsuccessful because the Petri net over-approximates too coarsely. We have developed a counterexample-guided abstraction refinement technique [15] that could be adapted to the setting of this paper.

Acknowledgements. We would like to thank Volker Diekert, Nicolas Relange, Paolo Baldan, Arwed von Merkatz, Stefan Kiefer, Vitali Kozioura and Salil Joshi for helpful discussions on relevant topics, for earlier related work and for their help with the implementation.

References

1. Baldan, P., Corradini, A., König, B.: A static analysis technique for graph transformation systems. In: Larsen, K.G., Nielsen, M. (eds.) CONCUR 2001. LNCS, vol. 2154, pp. 381–395. Springer, Heidelberg (2001)
2. Baldan, P., König, B.: Approximating the behaviour of graph transformation systems. In: Corradini, A., Ehrig, H., Kreowski, H.-J., Rozenberg, G. (eds.) ICGT 2002. LNCS, vol. 2505, pp. 14–29. Springer, Heidelberg (2002)
3. Baldan, P., König, B., König, B.: A logic for analyzing abstractions of graph transformation systems. In: Cousot, R. (ed.) SAS 2003. LNCS, vol. 2694, pp. 255–272. Springer, Heidelberg (2003)
4. Bauer, J., Wilhelm, R.: Static analysis of dynamic communication systems by partner abstraction. In: Riis Nielson, H., Filé, G. (eds.) SAS 2007. LNCS, vol. 4634, pp. 249–264. Springer, Heidelberg (2007)
5. Bouajjani, A., Esparza, J., Touili, T.: A generic approach to the static analysis of concurrent programs with procedures. In: Proc. of POPL 2003, pp. 62–73. ACM, New York (2003)
6. Chandy, K.M., Misra, J.: The drinking philosophers problem. ACM Transactions on Programming Languages and Systems 6(4), 632–646 (1984)
7. Courcelle, B.: The expression of graph properties and graph transformations in monadic second-order logic. In: Rozenberg, G. (ed.) Handbook of Graph Grammars and Computing by Graph Transformation, ch. 5. Foundations, vol. 1. World Scientific, Singapore (1997)
8. Deutsch, A.: Interprocedural may-alias analysis for pointers: Beyond k-limiting. In: Proc. of PLDI 1994. SIGPLAN Notices, vol. 29(6), pp. 230–241. ACM, New York (1994)
9. Distefano, D., O'Hearn, P.W., Yang, H.: A local shape analysis based on separation logic. In: Hermanns, H., Palsberg, J. (eds.) TACAS 2006. LNCS, vol. 3920, pp. 287–302. Springer, Heidelberg (2006)

10. Dotti, F.L., Foss, L., Ribeiro, L., Marchi Santos, O.: Verification of distributed object-based systems. In: Najm, E., Nestmann, U., Stevens, P. (eds.) FMOODS 2003. LNCS, vol. 2884, pp. 261–275. Springer, Heidelberg (2003)
11. Esparza, J., Melzer, S.: Verification of safety properties using integer programming: Beyond the state equation. Formal Methods in System Design 16, 159–189 (2000)
12. Ginsburg, S., Spanier, E.H.: Semigroups, Presburger formulas, and languages. Pacific Journal of Mathematics 16(2), 285–296 (1966)
13. Habel, A.: Hyperedge Replacement: Grammars and Languages. In: Habel, A. (ed.) Hyperedge Replacement: Grammars and Languages. LNCS, vol. 643. Springer, Heidelberg (1992)
14. König, B.: Graph transformation systems, Petri nets and semilinear sets: Checking for the absence of forbidden paths in graphs. In: Proc. of PNGT 2006. Electronic Communications of the EASST, vol. 2 (2007)
15. König, B., Kozioura, V.: Counterexample-guided abstraction refinement for the analysis of graph transformation systems. In: Hermanns, H., Palsberg, J. (eds.) TACAS 2006. LNCS, vol. 3920, pp. 197–211. Springer, Heidelberg (2006)
16. König, B., Kozioura, V.: AUGUR 2—a new version of a tool for the analysis of graph transformation systems. In: Proc. of GT-VMT 2006. ENTCS, vol. 211, pp. 201–210. Elsevier, Amsterdam (2006)
17. Lafont, Y.: Interaction nets. In: Proc. of POPL 1990, pp. 95–108. ACM Press, New York (1990)
18. Lammich, P., Müller-Olm, M., Wenner, A.: Predecessor sets of dynamic pushdown networks with tree-regular constraints. In: Bouajjani, A., Maler, O. (eds.) Computer Aided Verification. LNCS, vol. 5643, pp. 525–539. Springer, Heidelberg (2009)
19. Mayr, E.W.: An algorithm for the general Petri net reachability problem. SIAM J. Comput. 13(3), 441–460 (1984)
20. Meyer, R.: On boundedness in depth in the pi-calculus. In: Proc. of IFIP TCS 2008. IFIP, vol. 273, pp. 477–489. Springer, Heidelberg (2008)
21. Rensink, A., Distefano, D.: Abstract graph transformation. In: Proc. of SVV 2005. ENTCS, vol. 157.1, pp. 39–59 (2005)
22. Rieger, S., Noll, T.: Abstracting complex data structures by hyperedge replacement. In: Ehrig, H., Heckel, R., Rozenberg, G., Taentzer, G. (eds.) ICGT 2008. LNCS, vol. 5214, pp. 69–83. Springer, Heidelberg (2008)
23. Sagiv, M., Reps, T., Wilhelm, R.: Parametric shape analysis via 3-valued logic. TOPLAS (ACM Transactions on Programming Languages and Systems) 24(3), 217–298 (2002)
24. Varró, D.: Towards symbolic analysis of visual modeling languages. In: Proc. of GT-VMT 2002. ENTCS, vol. 72. Elsevier, Amsterdam (2002)

Saturated LTSs for Adhesive Rewriting Systems*

Filippo Bonchi[1], Fabio Gadducci[2],
Giacoma Valentina Monreale[2], and Ugo Montanari[2]

[1] INRIA Saclay and LIX, École Polytechnique
[2] Dipartimento di Informatica, Università di Pisa

Abstract. G-Reactive Systems (GRSs) are a framework for the derivation of labelled transition systems (LTSs) from a set of unlabelled rules. A label for a transition from A to B is a context $C[-]$ such that $C[A]$ may perform a reaction and reach B. If either all contexts, or just the "minimal" ones, are considered, the resulting LTS is called saturated (GIPO, respectively). The borrowed contexts (BCs) technique addresses the issue in the setting of the DPO approach. Indeed, from an adhesive rewriting system (ARS) a GRS can be defined such that DPO derivations correspond to reactions, and BC derivations to transitions of the GIPO LTS. This paper extends the BCs technique in order to derive saturated LTSs for ARSs, applying it to capture bisimilarity for asynchronous calculi.

1 Introduction

The complexity of the formalisms adopted for the specification of open-ended systems (let them be either Web Services description languages [26], or process calculi for biological systems [11], or...) is putting novel emphasis on the use of reduction semantics for modelling their dynamics. Roughly, the modelling technique is based on a reduction system: a set of states of the device equipped with a binary relation, representing the possible evolutions of the device. The system itself is presented inductively, by instantiating a few reduction rules.

The ease of use of the approach based on reductions led to its increasing adoption, most often in presence of specification languages offering a complex interaction between their operators (i.e., between the components of a compound system). However, a main drawback of reduction-based solutions is poor compositionality, since the dynamic behaviour of arbitrary stand-alone terms can be interpreted only by inserting them in appropriate contexts, where a reduction may take place. It would then be preferable to equip such formalisms with a labelled semantics, providing them with a specification of their dynamics which is based on a set of labelled rules, each one compactly describing some behavioural information/features of the component it is applied to. This is the first step leading to the definition of suitable observational equivalences, abstractly characterising when two systems have the same behaviour, thus allowing the possibility of verifying properties of system compositions.

---

* Supported by the MIUR Project SisteR, and carried out during the first author's tenure of an ERCIM "Alain Bensoussan" Fellowship Programme.

H. Ehrig et al. (Eds.): ICGT 2010, LNCS 6372, pp. 123–138, 2010.

The identification of the "right" labels is a difficult task: the assessment of their validity may vary for each formalism, and it is usually left to the ingenuity of the researcher. A case at hand is the calculus of mobile ambients [10]: despite its rapid acceptance in the process calculi literature, and the intense scrutiny it was subject to, the development of a suitable labelled semantics defied the researchers until quite recently [23,30], and it might not be fully settled.

G-Reactive Systems (GRSs) [22,31] are a successful meta-framework addressing the need of observational equivalences for systems specified by unlabelled rules. The key idea for getting an LTS is simple: a label for a transition from state A to state B is a context $C[-]$ such that the composed state $C[A]$, obtained by inserting A into $C[-]$, may perform a reaction and reach B. If either all possible contexts are considered, or just the "minimal" ones, the resulting LTS is called saturated or GIPO, respectively.

The framework proved quite general, covering a wide range of applications [25,16,9], as well as theoretically rich enough to be largely extended over the years. Various notions of equivalence were defined for the resulting LTS, such as saturated [8] and barbed [18,7]. Moreover, the possibility of obtaining *sorted* contexts was investigated [3], adding flexibility to the minimality requirement, for a favourite instance of GRSs, bigraphical reactive systems [25,17].

The borrowed contexts (BCs) technique allows the derivation of LTSs in the setting of the DPO approach. What is noteworthy is that it represents a constructive presentation of GRSs. Indeed, consider a widely adopted generalisation of graph transformation, *adhesive rewriting systems* (ARSs) [21]: for every ARS a GRS can be defined such that DPO derivations correspond to reactions, and BC derivations correspond to transitions of the GIPO LTS.

Despite validating case studies [28], so far the boundaries of the BC framework were less tested and extended. The resulting lack of flexibility is clearly a drawback in terms of BCs usability, yet it casts a shadow also on the novel notions introduced for GRSs: indeed, these are of an eminently prescriptive nature, and they would then benefit (as a sanity check, or for clarifying their meaning) from the constructive recasting offered by the BC setting. Our paper aims at addressing this sort of "technology transfer" among the frameworks: we introduce a technique based on BCs which allows to derive saturated LTSs for ARSs, exploiting it to propose an equivalence based on a notion of barb, for those (saturated) LTSs derived by the BC mechanism. Our proposal is then tested on a case study, drawn from the mold of the visual specification of process calculi.

We believe that interesting challenges, as well as one of the most successful application of the BC synthesis mechanism, arise in the derivation of LTSs for process calculi. The modelling approach is straightforward [6,5]: given a calculus, a graphical encoding is found such that structurally congruent processes are mapped into isomorphic graphs, and the reduction semantics is captured by a set of DPO rules, along the lines of [15,13]. These graphs are amenable to the BC mechanism, and a GRS on processes can be obtained such that the resulting GIPO semantics (hopefully) captures the standard bisimilarity for the calculus at hand, as it happens for strong bisimilarity in synchronous CCS [5].

Unfortunately, for most calculi strong bisimilarity is not the preferred equivalence, since it is too discriminating: this fact holds true especially for the asynchronous ones, such as *asynchronous* CCS. Similarly, the GIPO semantics is often unsatisfactory: as proved e.g. for mobile ambients in [7], a suitable notion of saturated (and barbed) GRS has to be taken into account. The reasoning in the above papers was modelled along the following pattern: after using the BC mechanism to derive the labels, a GRS is reverse-engineered, and the larger array of tools of the GRS framework exploited in order to obtain an LTS and a suitable equivalence for the calculus at hand. The results of our paper allows to skip the derivation of the GRS, and to constructively reason directly on ARSs.

Synopsis. Section 2 surveys with some care the main results on ARSs and GRSs, showing that the former are an instance of the latter: DPO derivations in ARSs correspond to reductions in GRSs, and BC derivations to GIPO transitions. The paper extends this correspondence: Section 4 introduces saturated transitions for ARSs, allowing for an ARS modelling asynchronous CCS (presented in Section 3). The equivalence thus derived with the BC mechanism is strictly contained into asynchronous bisimilarity, hence Section 5 proposes a notion of barb for ARSs: barbs and saturated transitions are able to capture asynchronous bisimilarity. The concluding Section 6 discusses future works.

2   On G-Reactive Systems and Borrowed Contexts

The purpose of this section is to recall, as briefly as possible, the main notions concerning G-reactive systems and borrowed contexts, including a sketch of their relationship, in order to fully understand the developments proposed later on.

2.1   G-Reactive Systems

This section introduces *G-reactive systems* (GRSs) [33,32], an extension of *reactive systems* [22] to *groupoidal enriched categories* (*G-categories*). For our purposes, it suffices to know that (a) *2-categories* are categories equipped with *2-cells* (intuitively, "arrows between arrows") and (b) G-categories are 2-categories where all 2-cells are isomorphisms: we refer the reader to the references above.

A *G*-category **B** models the syntax of a formalism. Objects represent system *interfaces* (with 0 the empty one). A *system* with interface I_1 is an arrow $A : 0 \to I_1$: it can be plugged into the *context* $B : I_1 \to I_2$ via arrow composition $A; B$. Given arrows $A, B : I_1 \to I_2$, a 2-cell $\alpha : A \Rightarrow B$ represents an isomorphism (intuitively, a proof of equivalence) between contexts A and B. The semantics is instead given via *reduction rules*: pairs of systems $\langle L, R \rangle$ with the same interface.

Definition 1 (G-Reactive System). *A* G-reactive system $\mathbb{C}$ *consists of*

1. *a G-category* **B***;*
2. *a distinguished object* 0*;*
3. *a set* **D** $\subseteq$ **B** *of 2-cells closed, composition-reflecting reactive contexts;*
4. *a set* $\mathfrak{R} \subseteq \bigcup_{I \in |\mathbf{C}|} \mathbf{B}(0, I) \times \mathbf{B}(0, I)$ *of reduction rules.*

Intuitively, reactive contexts are those arrows inside which a reduction can occur. By composition-reflecting we mean that $D'; D \in \mathbf{D}$ implies $D, D' \in \mathbf{D}$.

The reduction relation is generated by closing the reduction rules under all reactive contexts and 2-cells. Formally, the *reduction relation* is defined by taking $A \rightsquigarrow B$ if there exist $\langle L, R \rangle \in \mathfrak{R}$, $D \in \mathbf{D}$, $\alpha : A \Rightarrow L; D$ and $\alpha' : B \Rightarrow R; D$.

The behaviour of a GRS is given by an unlabelled transition system. In order to obtain a labelled one, we plug a system A into a context C and observe if a reduction occurs. Categorically, this means that $A; C$ is isomorphic to $L; D$ (there is a 2-cell $\alpha : A; C \Rightarrow L; D$) for rule $\langle L, R \rangle$ and reactive context D. This situation is depicted in the diagram in Fig. 1, a *redex square*.

Definition 2 (Saturated Transition System). *Let $\mathbb{C}$ be a GRS and $\mathbf{B}$ its underlying G-category. The* saturated transition system $\mathrm{SLTS}(\mathbb{C})$ *is defined as*

- *states: arrows $A : 0 \to I$ in $\mathbf{B}$, for arbitrary I;*
- *transitions: $A \xrightarrow{C}_{SAT} B$ if $A; C \rightsquigarrow B$.*

Bisimilarity on $\mathrm{SLTS}(\mathbb{C})$ *is referred to as* saturated bisimilarity $(\sim^S)$.

Clearly $\sim^S$ is a congruence, and it is the coarsest bisimulation relation that is so. Unfortunately, $\mathrm{SLTS}(\mathbb{C})$ is often infinite-branching, since all contexts allowing reductions may occur as labels. Moreover, it has redundant transitions: the CCS process like $a.0$ would have both transitions $a.0 \xrightarrow{\bar{a}.0|-}_{SAT} 0$ and $a.0 \xrightarrow{P|\bar{a}.0|-}_{SAT} P$, yet P does not "concur" to the reduction.

The notion of "minimal context allowing a reduction" is modelled by groupoidal-idem pushouts (GIPOs) in G-categories. We refer the reader to [31]: for our purposes, it suffices to know that such "minimality" can be derived: it will become clearer when introducing its constructive version, *borrowed contexts*.

Definition 3 (GIPO Transition System). *Let $\mathbb{C}$ be a GRS and $\mathbf{B}$ its underlying G-category. The* GIPO transition system $\mathrm{GLTS}(\mathbb{C})$ *is defined as*

- *states: $A : 0 \to I$ in $\mathbf{B}$, for arbitrary I;*
- *transitions: $A \xrightarrow{C}_{GIPO} A'$ if there is $D \in \mathbf{D}$, rule $\langle L, R \rangle \in \mathfrak{R}$, and 2-cell $\alpha : A; C \Rightarrow L; D$ making the diagram in Fig. 1 a GIPO with A' iso to $R; D$.*

Bisimilarity on $\mathrm{GLTS}(\mathbb{C})$ *is referred to GIPO bisimilarity $(\sim^G)$.*

Under certain conditions (see [33,32] for details), $\sim^G$ is a congruence. Unfortunately, it turns out that in many interesting cases $\sim^G$ is too strict [4]. The first and the last authors together with König introduced *semi-saturated bisimilarity*: an alternative, (in some cases) finitary characterisation of saturated bisimilarity. The theory was developed [8] for standard RSs [22], but it is easily lifted to GRSs.

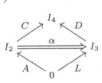

Fig. 1. A GIPO

Definition 4 (Semi-Saturated Bisimulation). *A symmetric relation $\mathcal{R}$ is a* semi-saturated bisimulation *if whenever $A \mathcal{R} B$ then*

- *if $A \xrightarrow{C}_{GIPO} A'$, then $B \xrightarrow{C}_{SAT} B'$ and $A' \mathcal{R} B'$.*

Semi-saturated bisimilarity is shown [8] to coincide with $\sim^S$.

2.2 DPO Rewriting with Borrowed Contexts

This section introduces *double-pushout* (DPO) rewriting and its interactive extension with *borrowed contexts* (BCs) [12]. We present them by relying on adhesive categories [21] as in [33]. In order to uniformly introduce DPO and BCs, we consider DPO derivations for *systems with interface*: morphisms $J \to G$ where G represents a system and J its interface. A *production* or *rewrite rule* is a span $L \hookleftarrow I \to R$ in an adhesive category $\mathbf{A}$ where the left-hand side $I \rightarrowtail L$ is monic. A *DPO adhesive rewriting system* (ARS) is a pair $\langle \mathbf{A}, P \rangle$ where P is a set of productions. In the definitions below, we refer to a chosen ARS $S = \langle \mathbf{A}, P \rangle$.

Definition 5 (DPO Derivation for Systems with Interfaces). *Let $J \to G$ and $J \to H$ be two systems with interface and $p : (L \hookleftarrow I \to R)$ a production. A match of p in G is a morphism $m : L \to G$. A direct derivation from $J \to G$ to $J \to H$ via p and m is a commuting diagram as in Fig. 2, such that two squares are pushouts (PO), denoted $J \to G \Longrightarrow J \to H$.*

The morphism $k : J \to C$ (making the left triangle commute) is unique, whenever it exists. If such a morphism does not exist, the rewriting step is not feasible.

In these derivations, the left-hand side L of a production must then occur completely in G. In a BC derivation L might occur partially in G, since the latter may interact with the environment through the interface J in order to exactly match L. Those BCs are the "smallest" contexts needed to obtain the image of L in G, and they may be used as suitable labels. Given an ARS S, $\mathrm{BC}(S)$ denotes the LTS derived via the BC mechanism defined below.

$$
\begin{array}{ccccc}
L & \longleftarrow\!\!\!\leftarrow I & \longrightarrow R \\
m \downarrow & PO \quad \downarrow & PO \quad \downarrow \\
G & \longleftarrow\!\!\!\leftarrow C & \longrightarrow H
\end{array}
$$

Fig. 2. A direct derivation

Definition 6 (Rewriting with Borrowed Contexts). *Given a production $p : L \hookleftarrow I \to R$, a system with interface $J \to G$ and a span of monos $d : G \hookleftarrow D \rightarrowtail L$, we say that $J \to G$ reduces to $K \to H$ with label $J \rightarrowtail F \leftarrow K$ via p and d if there are objects G^+, C and additional morphisms such that the diagram in Fig. 3 commutes and the squares are either pushouts (PO) or pullbacks (PB). We write $J \to G \xrightarrow{J \rightarrowtail F \leftarrow K} K \to H$, called* rewriting step with borrowed context.

The upper left-hand square of the diagram in Fig. 3 merges the left-hand side L and the object G to be rewritten according to a partial match $G \hookleftarrow D \rightarrowtail L$. The resulting G^+ contains a total match of L and is rewritten as in the DPO approach, producing the two other squares in the upper row. The pushout in the lower row gives the BC F which is missing for obtaining a total match of L, along with a morphism $J \rightarrowtail F$ indicating how F should be pasted to G. Finally, the interface for H is obtained by "intersecting" F and C via a pullback.

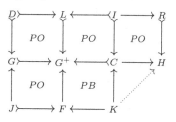

Fig. 3. A BC derivation

Note that two pushout complements that are needed in Definition 6, namely C and F, may not exist. In this case, the rewriting step is not feasible.

2.3   Relating Borrowed Contexts and G-Reactive Systems

We are now ready for showing that ARSs are instances of GRSs [32]. We consider cospans as contexts, and for this reason we need to work in bicategories [2] (with iso 2-cells) instead of G-categories. For the aim of this paper is indeed enough to know that a bicategory can be described, roughly, as a 2-category where associative and identity laws of composition hold up to isomorphism. In order to transfer the notions of GIPOs and GRPOs (in Section 2.1) to bicategories, it suffices to introduce the coherent associativity isomorphisms where necessary.

Bicategories of Cospans. Let $\mathbf{A}$ be an adhesive category with chosen pushouts. The bicategory of cospans of $\mathbf{A}$ has the same objects as $\mathbf{A}$ and morphism pairs $I_1 \xrightarrow{i_C} C \xleftarrow{o_C} I_2$ as arrows from I_1 to I_2, denoted $C_{i_C}^{o_C} : I_1 \to I_2$. Objects I_1 and I_2 are thought of as the input and the output interface of $C_{i_C}^{o_C}$.

Given the cospans $C_{i_C}^{o_C} : I_1 \to I_2$ and $D_{i_D}^{o_D} : I_2 \to I_3$, their composition $C_{i_C}^{o_C}; D_{i_D}^{o_D} : I_1 \to I_3$ is the cospan obtained by taking the chosen pushout of o_C and i_D, as depicted in Fig. 4. Note that, since arrows composition is a chosen pushout, it is associative only up to isomorphism.

A 2-cell $h : C_{i_C}^{o_C} \Rightarrow D_{i_D}^{o_D}$: $I_1 \to I_2$ is an arrow $h : C \to D$ in $\mathbf{A}$ satisfying $i_C; h = i_D$ and $o_C; h = o_D$, and it is *isomorphic* if h is an isomorphism in $\mathbf{A}$.

A cospan $C_{i_C}^{o_C}$ is *input linear* when i_C is mono in $\mathbf{A}$, and the composition of two input linear

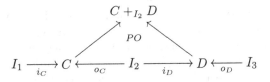

Fig. 4. Cospan composition

cospans yields another input linear cospan. For this reason, we can define the *input linear cospans bicategories* over $\mathbf{A}$, denoted by $ILC(\mathbf{A})$, as the bicategory consisting of input linear cospans and isomorphic 2-cells.

From ARSs to GRSs. Consider an ARS $S = \langle \mathbf{A}, P \rangle$, where the adhesive category $\mathbf{A}$ has initial object 0. This can be seen as a GRS where

- the base category is $ILC(\mathbf{A})$,
- the distinguished object is 0 (the initial object),
- all arrows in $ILC(\mathbf{A})$ are reactive,
- rules are pairs $\langle 0 \to L \hookleftarrow I, 0 \to R \leftarrow I \rangle$ for any $L \hookleftarrow I \to R$ rule in P.

For an ARS S, $\mathbb{C}_S$ denotes its associated GRS. A system with interface $J \to G$ in S can be thought as the arrow $0 \to G \leftarrow J$ of $ILC(\mathbf{A})$.

Proposition 1 ([14]). $J \to G \Longrightarrow J \to H$ in S iff $J \to G \rightsquigarrow J \to H$ in $\mathbb{C}_S$.

The above result allows for stating the correspondence between ARSs and BCs: GIPOs for GRSs over input linear cospans are equivalent to BCs for ARSs.

Proposition 2 ([32]). BC(S) = GLTS($\mathbb{C}_S$).

Sections 4 and 5 present the main results of the paper, strengthening the correspondence between the theories. We introduce (barbed) saturated semantics for BCs and we show how it relates to (barbed) saturated semantics for GRSs.

$$p ::= \bar{a}, \; p_1 \mid p_2, \; (\nu a)p, \; m \qquad m ::= \mathbf{0}, \; \alpha.p, \; m_1 + m_2 \qquad \alpha ::= a, \; \tau$$

$p \mid q \equiv q \mid p$	$(p \mid q) \mid r \equiv p \mid (q \mid r)$	$p \mid \mathbf{0} \equiv p$
$m + n \equiv n + m$	$(m + n) + o \equiv m + (n + o)$	$m + \mathbf{0} \equiv m$
$(\nu a)(\nu b)p \equiv (\nu b)(\nu a)p$	$(\nu a)(p \mid q) \equiv p \mid (\nu a)q \quad$ if $a \notin fn(p)$	$(\nu a)\mathbf{0} \equiv \mathbf{0}$
	$(\nu a)p \equiv (\nu b)(p\{^b/_a\}) \;$ if $b \notin fn(p)$	

$$\bar{a} \mid (a.p + m) \to p \qquad \tau.p + m \to p \qquad \frac{p \to q}{(\nu a)p \to (\nu a)q} \qquad \frac{p \to q}{p \mid r \to q \mid r}$$

Fig. 5. Syntax, structural congruence and reduction relation of ACCS

$$a.p + m \xrightarrow{a} p \qquad \tau.p + m \xrightarrow{\tau} p \qquad \bar{a} \xrightarrow{\bar{a}} \mathbf{0}$$

$$\frac{p \xrightarrow{\mu} q \quad a \notin n(\mu)}{(\nu a)p \xrightarrow{\mu} (\nu a)q} \qquad \frac{p \xrightarrow{\mu} q}{p \mid r \xrightarrow{\mu} q \mid r} \qquad \frac{p \xrightarrow{a} p_1 \quad q \xrightarrow{\bar{a}} q_1}{p \mid q \xrightarrow{\tau} p_1 \mid q_1}$$

Fig. 6. Labelled semantics of ACCS

3 Process Semantics via a Graphical Encoding

This section considers asynchronous CCS (ACCS). We present an encoding for the finite fragment of the calculus (adapting the one for the synchronous version in [5]), and we model the reduction semantics of the calculus via a set of DPO rules. Finally, we show that the resulting GIPO semantics is too strict.

Asynchronous CCS. The syntax of ACCS is shown in Fig. 5: $\mathcal{N}$ is a set of *names*, ranged over by $a, b, \ldots$, with $\tau \notin \mathcal{N}$. We let $p, q, \ldots$ range over the set $\mathcal{P}$ of processes and $m, n, \ldots$ over the set $\mathcal{S}$ of summations. With respect to synchronous CCS, the calculus lacks output prefixes: process $\bar{a}$ is thought of as a message, available on a communication media named a, that disappears after its reception. The *free names* $fn(p)$ of a process p are defined as usual.

Processes are taken up to a *structural congruence* (Fig. 5), denoted by $\equiv$. The *reduction relation*, denoted by $\to$, describes process evolution: the least relation $\to \subseteq \mathcal{P} \times \mathcal{P}$, closed under $\equiv$, inductively generated by the rules in Fig. 5. The interactive semantics for ACCS is given by the relation over processes up to $\equiv$, obtained by the rules in Fig. 6. We let μ range over the set of labels $\{\tau, a, \bar{a} \mid a \in \mathcal{N}\}$: the names of μ, denoted by $n(\mu)$, are defined as usual. Differently from synchronous calculi, sending messages is non-blocking. Hence, an observer might send messages without knowing about their reception, and inputs are thus deemed as unobservable. This is mirrored in the chosen bisimilarity [1].

Definition 7 (Asynchronous Bisimulation). *A symmetric relation $\mathcal{R}$ is an asynchronous bisimulation if whenever $p \, \mathcal{R} \, q$ then*

- *if $p \xrightarrow{\tau} p'$ then $q \xrightarrow{\tau} q'$ and $p' \, \mathcal{R} \, q'$,*
- *if $p \xrightarrow{\bar{a}} p'$ then $q \xrightarrow{\bar{a}} q'$ and $p' \, \mathcal{R} \, q'$,*
- *if $p \xrightarrow{a} p'$ then either $q \xrightarrow{a} q'$ and $p' \, \mathcal{R} \, q'$ or $q \xrightarrow{\tau} q'$ and $p' \, \mathcal{R} \, q' \mid \bar{a}$.*

Asynchronous bisimilarity $\sim^A$ is the largest asynchronous bisimulation.

For example, the processes $a.\bar{a} + \tau.\mathbf{0}$ and $\tau.\mathbf{0}$ are asynchronous bisimilar. If $a.\bar{a} + \tau.\mathbf{0} \xrightarrow{a} \bar{a}$, then $\tau.\mathbf{0} \xrightarrow{\tau} \mathbf{0}$ and clearly $\bar{a} \sim^A \mathbf{0} \mid \bar{a}$.

Graphical Encoding for ACCS. For the sake of space, we do not present the formal definition of the graphical encoding of ACCS processes. Indeed, it is analogous to the encoding for processes of the synchronous calculus into typed graphs with interfaces, presented in [5, Definition 9]: it differs only for the choice of the typed graph T_A, depicted in Fig. 7. We remark that choosing a graph typed over T_A means to consider graphs where each node (edge) is labelled by a node (edge) of T_A, and the incoming and outcoming tentacles are preserved.

Intuitively, a graph having as root a node of type • (⋄) corresponds to a process (respectively a summation), while each node of type ∘ basically represents a name. Indeed, even if the encoding could be defined by means of operators on typed graphs with interfaces, for ACCS the situation is summed up by saying that a typed graph with interfaces is the encoding of a process P if its underlying graph is almost the syntactic tree of P: each internal node of type • has exactly one incoming edge, except for root, to which an edge labelled go is attached.

Going back to the type graph, the edge rcv (snd) simulates the input prefix (output operator, respectively), while there is no edge for the parallel composition, non-deterministic choice and restriction operators. Edge c is a syntactical device for "coercing" the occurrence of a summation inside a process context, while similarly edge go detects the "entry" point of the computation, thus avoiding to perform any reaction below the outermost prefix operators: the latter is needed to properly simulate the reduction semantics of the calculus.

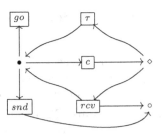

Fig. 7. Type graph T_A

The encoding of a process P, with respect to a set of names Γ including the free names of P, is a graph with interfaces $(\{p\} \cup \Gamma, \emptyset)$. It is sound and complete with respect to structural congruence of the calculus, that is, two processes are equivalent if and only if they are mapped into isomorphic graphs.

Fig. 8 depicts the graph encoding for process $P = (\nu b)(\bar{b} \mid b.\bar{a} + a)$. The two leftmost edges labelled c and snd have the same root, into which the node p of the interface is mapped. They are the top edges of the two subgraphs representing the parallel components of the process. In particular, the edge labelled snd represents

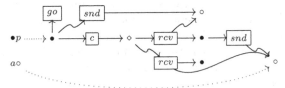

Fig. 8. Encoding for process $(\nu b)(\bar{b} \mid b.\bar{a} + a)$

the output over the restricted channel b, namely $\bar{b}$, while the c edge is the syntactical operator denoting that its subgraph represents a summation, that is, $b.\bar{a} + a$. The two leftmost edges of this last subgraph, both labelled rcv, model the two input prefixes b and a, while the rightmost snd edge represent the operator $\bar{a}$. Note that channel name a is in the interface since it is free in P, while instead the bound name b does not belong to the interface.

An ARS for ACCS. Fig. 9 presents the DPO rules which simulate the reduction semantics for ACCS. The two rules p_{com} and p_τ mirror the two axioms of the reductions relation in Fig. 5, and a soundness and completeness result of our encoding with respect to reductions is easily obtained (see [5, Proposition 2]).[1]

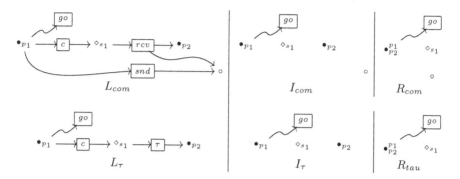

Fig. 9. The productions $p_{com} : L_{com} \hookleftarrow I_{com} \to R_{com}$ and $p_\tau : L_\tau \hookleftarrow I_\tau \to R_\tau$

Relevant here is that this graphical encoding is amenable to BC mechanism. By exploiting the pruning techniques of [5], we can show that the derived GIPO transition system is substantially equivalent to the one in Fig. 6: $\xrightarrow{a}$ corresponds to $\to_{GIPO}^{-|\bar{a}}$, $\xrightarrow{\bar{a}}$ to $\to_{GIPO}^{-|a}$ and $\xrightarrow{\tau}$ to $\to_{GIPO}^{-}$, for $-$ the identity context.

However, GIPO bisimilarity is too fine grained for ACCS. Consider the asynchronously bisimilar processes $a.\bar{a} + \tau.\mathbf{0}$ and $\tau.\mathbf{0}$. It is easy to verify that they are not GIPO bisimilar, since the former can perform the GIPO transition $a.\bar{a} + \tau.\mathbf{0} \to_{GIPO}^{-|\bar{a}} \bar{a} \mid \mathbf{0}$ in Fig. 16 (corresponding to the transition $a.\bar{a} + \tau.\mathbf{0} \xrightarrow{a} \bar{a}$ in the ordinary semantics of Fig. 6), while the latter has no such transition.

4   Saturated Rewriting with Borrowed Contexts

This section introduces a technique based on BCs which allows us to derive saturated LTSs for ARSs. As the BC technique offers a constructive solution for calculating the minimal contexts enabling a rule, the approach we present represents a constructive solution for calculating all contexts enabling a rule.

Definition 8 (Saturated Rewriting). *Given a production* $p : L \hookleftarrow I \to R$ *and a system with interface* $J \to G$, *we say that* $J \to G$ *reduces to* $K \to H$ *with label* $J \rightarrowtail F \leftarrow K$ *via* p *if there are objects* G^+, C *and additional morphisms such that the diagram in Fig. 10 commutes and three squares are pushouts (PO). We write* $J \to G \xrightarrow{J \rightarrowtail F \leftarrow K}_{BCSAT} K \to H$, *called* saturated rewriting step.

---

[1] The correspondence accounts for the discarding of sub-processes, due to the solving of non-deterministic choices: after a DPO derivation there can be parts of the graph (representing the discarded components) that are not reachable from the root.

In the diagram on Fig. 10, G^+ is any object containing a total match of both the object G and the left-hand side L. Therefore it can be rewritten as in the standard DPO approach, producing the two squares in the upper row. The pushout in the lower row gives the context F which we need to add to G in order to obtain G^+, along with a morphism $J \rightarrowtail F$ indicating how F should be pasted to G. Fi-

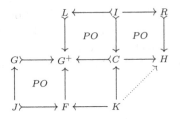

Fig. 10. A saturated derivation

nally, the interface for the resulting object H is any object K such that the lower right-hand square commutes. If the two pushout complements, which are needed in Definition 8, namely C and F, do not exist, the rewriting step is not feasible.

Note that in this case the object G^+ might not be the minimal object containing both G and L: it could also contain something else. Therefore, the context F needed to extend G to G^+ might be not the minimal context allowing the DPO derivation for G. Fig. 11 shows an example of saturated rewriting for the graphical encoding of $\tau.0$ with respect to the set of names $\Gamma = \{a\}$. Note that $-|\bar{a}$ is not the minimal context allowing the reaction: the parallel component $\bar{a}$ would not be necessary since it does non concur to the transition.

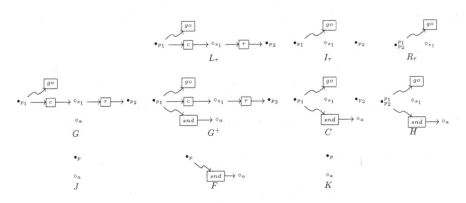

Fig. 11. The saturated transition $\tau.0 \xrightarrow{-|\bar{a}}_{SAT} 0|\bar{a}$.

Given an ARS S, we denote by $\text{SBC}(S)$ the LTS derived via the saturated BC mechanism as defined above. Mirroring the correspondence between BCs and GIPOs, the proposition below states that this LTS coincides with the saturated LTS for the GRS $\mathbb{C}_S$ associated to S.

Proposition 3. $\text{SBC}(S) = \text{SLTS}(\mathbb{C}_S)$.

Semi-Saturated Borrowed Contexts LTS. Adopting the semi-saturation point of view, it is possible to show that saturated rewriting steps can be deduced in a uniform fashion from standard rewriting steps with borrowed context.

Definition 9. *Let S be an ARS. The* semi-saturated *LTS* $\mathrm{SSBC}(S)$ *is defined as the set of transitions* $J \to G \xrightarrow{J \rightarrowtail F \leftarrow K}_{SSAT} K \to H$ *such that there are a BC transition* $J \to G \xrightarrow{J \rightarrowtail F' \leftarrow K'}_{BC} K' \to H'$ *and a graphical context* $K' \rightarrowtail F'' \leftarrow K$ *making the diagrams below commute and two squares pushouts (PO).*

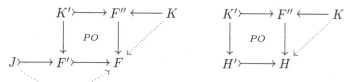

We now draw the connection with saturated BCs.

Proposition 4. $\mathrm{SBC}(S) = \mathrm{SSBC}(S)$.

Consider again the saturated transition introduced in Fig. 11. It can be derived from the minimal transition depicted in Fig. 13. More precisely, it is obtained as described in Definition 9, by considering the graphical context $K' \rightarrowtail F'' \leftarrow K$ represented in Fig. 12.

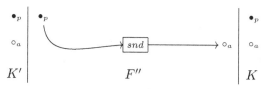

Fig. 12. The graphical context $- \mid \bar{a}$

We can exploit the above observation in order to show that the graphical encoding of $a.\bar{a} + \tau.\mathbf{0}$ and of $\tau.\mathbf{0}$ are saturated bisimilar. Consider $\sim^S$ as a semi-saturated bisimulation. If $a.\bar{a} + \tau.\mathbf{0} \xrightarrow{-|\bar{a}}_{GIPO} \bar{a} \mid \mathbf{0}$ (Fig. 16, corresponding to transition $a.\bar{a} + \tau.\mathbf{0} \xrightarrow{a} \bar{a}$), then $\tau.\mathbf{0} \xrightarrow{-}_{GIPO} \mathbf{0}$ (Fig. 13): from the latter (by Definition 9) $\tau.\mathbf{0} \xrightarrow{-|\bar{a}}_{SAT} \mathbf{0} \mid \bar{a}$ is derived (Fig. 11). Note that this is analogous to the one of asynchronous bisimulation (Definition 7). If $p \xrightarrow{a} p'$ (corresponding to $\xrightarrow{-|\bar{a}}_{GIPO}$) then $q \xrightarrow{\tau} q'$ (corresponding to $\xrightarrow{-}_{GIPO}$) with $p'Rq' \mid \bar{a}$.

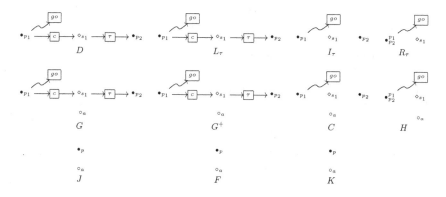

Fig. 13. The BC transition corresponding to $\tau.\mathbf{0} \xrightarrow{-}_{GIPO} \mathbf{0}$

However, $\sim^A$ is strictly included into $\sim^S$. In order to provide the simplest example possible, take into account the full process syntax, including recursion. Then, consider the processes $p = rec_X.(\tau.X + a.X)$ and $q = rec_X.\tau.X$: the latter just offers continuously an unobservable action τ, while the former may at any time perform either τ or a. Note that p is not asynchronous bisimilar to $q \mid \bar{a}$: then the same follows for p and q, since q can not simulate the reduction $p \xrightarrow{a} p$. Instead, p and $q \mid \bar{a}$ are saturated bisimilar: indeed, $\bar{a}$ remains idle in the latter process, and no parallel composition $- \mid r$ may distinguish between p and q. As it is standard for process calculi, this kind of problem can be solved by considering *barbs*, adding in the bisimulation game the check of suitable predicates over the states of a system. Next section introduces the notion of barbs for ARS.

5   Barbs for Adhesive Rewriting Systems

We have shown that neither $\sim^G$ nor $\sim^S$ capture asynchronous bisimilarity for ACCS: the former is too strict, while the latter is too coarse. The same problem arises for many formalisms (as discussed in [4]), so that the coincidence of standard bisimilarity for the synchronous CCS with $\sim^G$ appears the exception [5]. This fact lead the first three authors to introduce *barbed saturated bisimilarity* for GRSs [6]: it refines $\sim^S$ by adding state observations, called *barbs* [24]. In this section we fix a set of barbs O and write $A \downarrow_o$ if $o \in O$ holds in A.

Definition 10 (Barbed Saturated Bisimulation). *A symmetric relation* $\mathcal{R}$ *is a* barbed saturated bisimulation *if whenever* $A \mathcal{R} B$ *then*

- $\forall C,\ A; C \mathcal{R} B; C,$
- *if* $A \downarrow_o$ *then* $B \downarrow_o,$
- *if* $A \rightsquigarrow A'$ *then* $B \rightsquigarrow B'$ *and* $A' \mathcal{R} B'.$

Barbed saturated bisimilarity $\sim^{BS}$ *is the largest barbed saturated bisimulation.*

The above definition is general enough to capture the abstract semantics of many important formalisms such as mobile ambients, CCS, π-calculus and their asynchronous variants [6]. However, it is parametric with respect to the choice of the set of barbs O and defining the "right" barbs is not a trivial task, as witnessed by several papers about this topic (e.g. [29,20]).

We now introduce a novel notion of barb for ARSs, trying to keep in line with the constructive nature of the BC mechanism. To this end, our intuition is driven by the graphical encodings of calculi, and by the nature of barbs in most examples from that setting, there basically (a) barbs check the presence of some suitable subsystem, such that (b) this subsystem is needed to perform an interaction with the environment. For instance, in ACCS, barbs are parallel outputs [1], formally (a) $p \downarrow_{\bar{a}}$ if $p \equiv p_1 \mid \bar{a}$ and (b) these outputs can interact with the environment through the rule $\bar{a} \mid a.q + m \rightsquigarrow q$. In mobile ambients, barbs are ambients at the topmost level [23], formally (a) $p \downarrow_m$ if $p \equiv m[p_1] \mid p_2$ and (b) these ambients can be interact with the environment via the rule $open\ m.q_1 \mid m[q_2] \rightsquigarrow q_1 \mid q_2$.

Concerning ARSs, in order to respect intuition (a), we think of a barb D for a system G as a sub-object $D \rightarrowtail G$. Then, we assume that what can be observed on G depends on the interface J and thus we require barbs to be parametric with respect to the interfaces. Thus, a barb D for the interface J is a span $b : J \leftarrowtail J_D \rightarrowtail D$. In order to respect intuition (b), we further require that the barbs occurs in the left-hand side of some production. Formally, that D is a sub-object of L (there exists $D \rightarrowtail L$) for some production $L \leftarrowtail I \rightarrow R$.

Definition 11 (Barbs for Borrowed Contexts). *Given a production $p :$ $L \leftarrowtail I \rightarrow R$, a system with interface $J \rightarrow G$ and spans of monos $d : G \leftarrowtail D \rightarrowtail L$ and $b : J \leftarrowtail J_D \rightarrowtail D$, we say that $J \rightarrow G$ satisfies the barb b via p and d if there is an object F_D and additional morphisms such that the diagram in Fig. 14 commutes and one square is a pushout (PO). We write $(J \rightarrow G) \downarrow_b$.*

Put differently, a barb $J \leftarrowtail J_D \rightarrowtail D$ is satisfied by a system with interface $J \rightarrowtail G$ if it at first ensures that a suitable BC derivation can potentially be performed, even if it might not be feasible (see Definition 6 and Fig. 3). Additionally, the interface J_D is required to involve all the components of D shared with other components occurring in L (the pushout condition), while the factorisation through J guarantees that these components are uniquely identified in G.

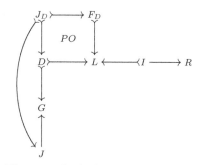

Fig. 14. The barb $J \leftarrowtail J_D \rightarrowtail D$

Consider again our running example: in ACCS we have that $\forall a \in \mathcal{N}$ we state $p \downarrow \bar{a}$ if $p \equiv \bar{a} \mid q$. The same requirement can be enforced in the graphical encoding of a process with the barb $\bar{a}_{J_D}$ shown in Fig. 15, with respect to the interface $J_D = \bullet_p \quad \circ_a$. For any larger interface J (i.e., such that there is $J \rightarrowtail J'$), the barb $\bar{a}_J$ is the obvious span $J \leftarrowtail J_D \rightarrowtail D$.

Note that the occurrence of the name node in J_D guarantees that we are observing an output on a non-restricted channel. Instead, the go edge in the graph D guarantees that the output observed is at

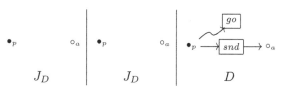

Fig. 15. Graphical barb for ACCS

the top level of the process. These output barbs are precisely those that are needed to capture $\sim^A$ via $\sim^{BS}$.

Proposition 5. *Let p and q be two ACCS processes. Let Γ be a set of names such that $fn(p) \cup fn(q) \subseteq \Gamma$. Let $[\![p]\!]_\Gamma$ and $[\![q]\!]_\Gamma$ be the graphical encodings of p and q with respect to Γ as described in Section 3, and let the barbs $\bar{a}_J$ be as defined above. Then $p \sim^A q$ iff $[\![p]\!]_\Gamma \sim^{BS} [\![q]\!]_\Gamma$.*

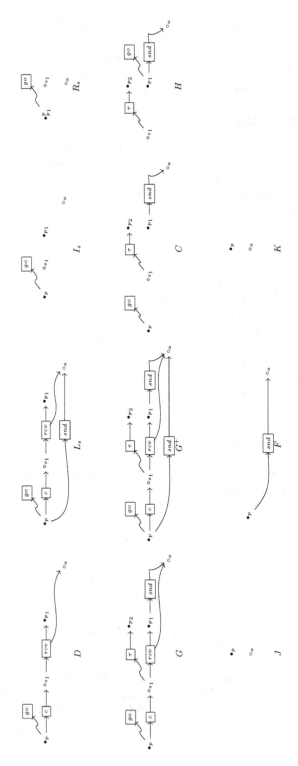

Fig. 16. The BC transition corresponding to the GIPO transition $a.\bar{a} + \tau.0 \xrightarrow{-|\bar{a}|}_{GIPO} \bar{a}|0$

6   Conclusions and Further Works

The aim of the paper is to add flexibility to the BC mechanism. To this end, we introduced saturated LTSs for ARSs, proving their correspondence to a constructive presentation of the analogous notion for GRSs. We also presented barbed semantics for ARSs: they are needed to address the semantics of many process calculi, such as asynchronous CCS and mobile ambients, but their application could be wider. For instance, [28] studies model refactoring that preserves GIPO bisimilarity: the latter could be safely replaced by barbed saturated bisimilarity.

Our work also opens new venues of investigation. Indeed, barbs in the GRS setting are quite general, hence quite poor from e.g. the point of view of its modularity properties. On the contrary, the simpler notion of barbs that we devised for BCs, besides its constructive features, may represent a major step towards solving the problem of automatically deriving suitable barbs for GRSs.

Finally, we believe that the modelling of calculi is one of the strongest assets of the BC mechanism: graphical encodings for processes can be easily provided, and their reduction semantics simulated by DPO rules. By integrating our results with *negative application conditions* [19] (adapted to the BC framework in [27]) we might provide suitable behavioural theories for larger classes of calculi.

References

1. Amadio, R., Castellani, I., Sangiorgi, D.: On bisimulations for the asynchronous π-calculus. Theoretical Computer Science 195(2), 291–324 (1998)
2. Bénabou, J.: Introduction to bicategories. In: Midwest Category Seminar I. LNM, vol. 47, pp. 1–77. Springer, Heidelberg (1967)
3. Birkedal, L., Debois, S., Hildebrandt, T.: On the construction of sorted reactive systems. In: van Breugel, F., Chechik, M. (eds.) CONCUR 2008. LNCS, vol. 5201, pp. 218–232. Springer, Heidelberg (2008)
4. Bonchi, F.: Abstract Semantics by Observable Contexts. Ph.D. thesis, Department of Informatics, University of Pisa (2008)
5. Bonchi, F., Gadducci, F., König, B.: Synthesising CCS bisimulation using graph rewriting. Information and Computation 207(1), 14–40 (2009)
6. Bonchi, F., Gadducci, F., Monreale, G.V.: Labelled transitions for mobile ambients (as synthesized via a graphical encoding). In: Hildebrandt, T., Gorla, D. (eds.) EXPRESS 2008. ENTCS, vol. 242(1), pp. 73–98. Elsevier, Amsterdam (2009)
7. Bonchi, F., Gadducci, F., Monreale, G.V.: Reactive systems, barbed semantics, and the mobile ambients. In: de Alfaro, L. (ed.) FOSSACS 2009. LNCS, vol. 5504, pp. 272–287. Springer, Heidelberg (2009)
8. Bonchi, F., König, B., Montanari, U.: Saturated semantics for reactive systems. In: LICS 2006, pp. 69–80. IEEE Computer Society, Los Alamitos (2006)
9. Bonchi, F., Brogi, A., Corfini, S., Gadducci, F.: On the use of behavioural equivalences for web services' development. Fundamenta Informaticae 89(4), 479–510 (2008)
10. Cardelli, L., Gordon, A.: Mobile ambients. Theoretical Computer Science 240(1), 177–213 (2000)
11. Cardelli, L.: Brane calculi. In: Danos, V., Schachter, V. (eds.) CMSB 2004. LNCS (LNBI), vol. 3082, pp. 257–278. Springer, Heidelberg (2005)
12. Ehrig, H., König, B.: Deriving bisimulation congruences in the DPO approach to graph rewriting with borrowed contexts. Mathematical Structures in Computer Science 16(6), 1133–1163 (2006)

13. Gadducci, F.: Graph rewriting for the π-calculus. Mathematical Structures in Computer Science 17(3), 407–437 (2007)
14. Gadducci, F., Heckel, R.: An inductive view of graph transformation. In: Parisi-Presicce, F. (ed.) WADT 1997. LNCS, vol. 1376, pp. 219–233. Springer, Heidelberg (1998)
15. Gadducci, F., Montanari, U.: A concurrent graph semantics for mobile ambients. In: Brookes, S., Mislove, M. (eds.) MFPS 2001. ENTCS, vol. 45. Elsevier, Amsterdam (2001)
16. Gianantonio, P.D., Honsell, F., Lenisa, M.: RPO, second-order contexts, and lambda-calculus. Logical Methods in Computer Science 5(3) (2009)
17. Grohmann, D., Miculan, M.: Reactive systems over directed bigraphs. In: Caires, L., Vasconcelos, V.T. (eds.) CONCUR 2007. LNCS, vol. 4703, pp. 380–394. Springer, Heidelberg (2007)
18. Grohmann, D., Miculan, M.: Deriving barbed bisimulations for bigraphical reactive systems. In: Corradini, A., Tuosto, E. (eds.) ICGT 2008 - Doctoral Symposium. Electronic Communications of the EASST, vol. 16. EASST (2009)
19. Habel, A., Heckel, R., Taentzer, G.: Graph grammars with negative application conditions. Fundamenta Informaticae 26(3/4), 287–313 (1996)
20. Honda, K., Yoshida, N.: On reduction-based process semantics. Theoretical Computer Science 151(2), 437–486 (1995)
21. Lack, S., Sobocinski, P.: Adhesive and quasiadhesive categories. Theoretical Informatics and Applications 39(3), 511–545 (2005)
22. Leifer, J., Milner, R.: Deriving bisimulation congruences for reactive systems. In: Palamidessi, C. (ed.) CONCUR 2000. LNCS, vol. 1877, pp. 243–258. Springer, Heidelberg (2000)
23. Merro, M., Zappa Nardelli, F.: Behavioral theory for mobile ambients. Journal of the ACM 52(6), 961–1023 (2005)
24. Milner, R., Sangiorgi, D.: Barbed bisimulation. In: Kuich, W. (ed.) ICALP 1992. LNCS, vol. 623, pp. 685–695. Springer, Heidelberg (1992)
25. Milner, R.: Bigraphical reactive systems. In: Larsen, K.G., Nielsen, M. (eds.) CONCUR 2001. LNCS, vol. 2154, pp. 16–35. Springer, Heidelberg (2001)
26. OWL-S Coalition: OWL-S for services, www.ai.sri.com/daml/services/owl-s/
27. Rangel, G., König, B., Ehrig, H.: Deriving bisimulation congruences in the presence of negative application conditions. In: Amadio, R.M. (ed.) FOSSACS 2008. LNCS, vol. 4962, pp. 413–427. Springer, Heidelberg (2008)
28. Rangel, G., Lambers, L., König, B., Ehrig, H., Baldan, P.: Behavior preservation in model refactoring using DPO transformations with borrowed contexts. In: Ehrig, H., Heckel, R., Rozenberg, G., Taentzer, G. (eds.) ICGT 2008. LNCS, vol. 5214, pp. 242–256. Springer, Heidelberg (2008)
29. Rathke, J., Sassone, V., Sobociński, P.: Semantic barbs and biorthogonality. In: Seidl, H. (ed.) FOSSACS 2007. LNCS, vol. 4423, pp. 302–316. Springer, Heidelberg (2007)
30. Rathke, J., Sobociński, P.: Deriving structural labelled transitions for mobile ambients. In: van Breugel, F., Chechik, M. (eds.) CONCUR 2008. LNCS, vol. 5201, pp. 462–476. Springer, Heidelberg (2008)
31. Sassone, V., Sobocinski, P.: Deriving bisimulation congruences using 2-categories. Nordic Journal of Computing 10(2), 163–183 (2003)
32. Sassone, V., Sobociński, P.: Reactive systems over cospans. In: LICS 2005, pp. 311–320. IEEE Computer Society Press, Los Alamitos (2005)
33. Sobociński, P.: Deriving bisimulation congruences from reduction systems. Ph.D. thesis, BRICS, Department of Computer Science, University of Aaurhus (2004)

A Hoare Calculus for Graph Programs

Christopher M. Poskitt and Detlef Plump

Department of Computer Science
The University of York, UK

Abstract. We present Hoare-style axiom schemata and inference rules for verifying the partial correctness of programs in the graph programming language GP. The pre- and postconditions of this calculus are the nested conditions of Habel, Pennemann and Rensink, extended with expressions for labels in order to deal with GP's conditional rule schemata and infinite label alphabet. We show that the proof rules are sound with respect to GP's operational semantics.

1 Introduction

Recent years have seen an increased interest in formally verifying properties of graph transformation systems, motivated by the many applications of graph transformation to specification and programming. Typically, this work has focused on verification techniques for sets of graph transformation rules or graph grammars, see for example [16,2,10,3,5].

Graph transformation languages and systems such as PROGRES [17], AGG [18], Fujaba [12] and GrGen [4], however, allow one to use control constructs on top of graph transformation rules for solving graph problems in practice. The challenge to verify programs in such languages has, to the best of our knowledge, not yet been addressed.

A first step beyond the verification of plain sets of rules has been made by Habel, Pennemann and Rensink in [6], by providing a construction for weakest preconditions of so-called high-level programs. These programs allow one to use constructs such as sequential composition and as-long-as-possible iteration over sets of conditional graph transformation rules. The verification method follows Dijkstra's approach to program verification: one calculates the weakest precondition of a program and then needs to prove that it follows from the program's precondition. High-level programs fall short of practical graph transformation languages though, in that their rules cannot perform computations on labels (or attributes).

In this paper, we present an approach for verifying programs in the graph programming language GP [13,11]. Rather than adopting a weakest precondition approach, we follow Hoare's seminal paper [9] and devise a calculus of proof rules which are directed by the syntax of the language's control constructs. Similar to classical Hoare logic, the calculus aims at human-guided verification and allows the compositional construction of proofs.

H. Ehrig et al. (Eds.): ICGT 2010, LNCS 6372, pp. 139–154, 2010.

The pre- and postconditions of our calculus are nested conditions [5], extended with expressions for labels and so-called assignment constraints; we refer to them as *E-conditions*. The extension is necessary for two reasons. Firstly, when a label alphabet is infinite, it is impossible to express a number of simple properties with finite nested conditions. For example, one cannot express with a finite nested condition that a graph over the set of integers is non-empty, since it is impossible to finitely enumerate every integer. Secondly, the conditions in [5] cannot express relations between labels such as "x and y are integers and $x^2 = y$". Such relations can be expressed, however, in GP's rule schemata.

We briefly review the preliminaries in Section 2 and graph programs in Section 3. Following this, we present E-conditions in Section 4, and then use them to define a proof system for GP in Section 5, where its use will be demonstrated by proving a property of a graph colouring program. In Section 6, we formally define the two transformations of E-conditions used in the proof system, before proving the axiom schemata and inference rules sound in the sense of partial correctness, with respect to GP's operational semantics [13,14]. Finally, we conclude in Section 7. A long version of this paper with the abstract syntax and operational semantics of GP, as well as detailed proofs of results, is available online [15].

2   Graphs, Assignments, and Substitutions

Graph transformation in GP is based on the double-pushout approach with relabelling [8]. This framework deals with partially labelled graphs, whose definition we recall below. We deal with two classes of graphs, "syntactic" graphs labelled with expressions and "semantic" graphs labelled with (sequences of) integers and strings. We also introduce assignments which translate syntactic graphs into semantic graphs, and substitutions which operate on syntactic graphs.

A *graph* over a label alphabet $\mathcal{C}$ is a system $G = (V_G, E_G, s_G, t_G, l_G, m_G)$, where V_G and E_G are finite sets of *nodes* (or *vertices*) and *edges*, $s_G, t_G \colon E_G \to V_G$ are the *source* and *target* functions for edges, $l_G \colon V_G \to \mathcal{C}$ is the partial node labelling function and $m_G \colon E_G \to \mathcal{C}$ is the (total) edge labelling function. Given a node v, we write $l_G(v) = \bot$ to express that $l_G(v)$ is undefined. Graph G is *totally labelled* if l_G is a total function.

Unlabelled nodes will occur only in the interfaces of rules and are necessary in the double-pushout approach to relabel nodes. There is no need to relabel edges as they can always be deleted and reinserted with changed labels.

A *graph morphism* $g \colon G \to H$ between graphs G and H consists of two functions $g_V \colon V_G \to V_H$ and $g_E \colon E_G \to E_H$ that preserve sources, targets and labels; that is, $s_H \circ g_E = g_V \circ s_G$, $t_H \circ g_E = g_V \circ t_G$, $m_H \circ g_E = m_G$, and $l_H(g(v)) = l_G(v)$ for all v such that $l_G(v) \neq \bot$. Morphism g is an *inclusion* if $g(x) = x$ for all nodes and edges x. It is *injective* (*surjective*) if g_V and g_E are injective (surjective). It is an *isomorphism* if it is injective, surjective and satisfies $l_H(g_V(v)) = \bot$ for all nodes v with $l_V(v) = \bot$. In this case G and H are *isomorphic*, which is denoted by $G \cong H$.

We consider graphs over two distinct label alphabets. Graph programs and E-conditions contain graphs labelled with expressions, while the graphs on which programs operate are labelled with (sequences of) integers and character strings. We consider graphs of the first type as syntactic objects and graphs of the second type as semantic objects, and aim to clearly separate the levels of syntax and semantics.

Let $\mathbb{Z}$ be the set of integers and Char be a finite set of characters (that can be typed on a keyboard). We fix the label alphabet $\mathcal{L} = (\mathbb{Z} \cup \text{Char}^*)^+$ of all non-empty sequences over integers and character strings, and denote by $\mathcal{G}(\mathcal{L})$ the set of all graphs over $\mathcal{L}$.

The other label alphabet we are using consists of expressions according to the EBNF grammar of Figure 1, where VarId is a syntactic class[1] of variable identifiers. We write $\mathcal{G}(\text{Exp})$ for the set of all graphs over the syntactic class Exp.

Exp	::=	(Term | String) ['_' Exp]
Term	::=	Num | VarId | Term ArithOp Term
ArithOp	::=	'+' | '-' | '*' | '/'
Num	::=	['-'] Digit {Digit}
String	::=	' " ' {Char} ' " '

Fig. 1. Syntax of expressions

Each graph in $\mathcal{G}(\text{Exp})$ represents a possibly infinite set of graphs in $\mathcal{G}(\mathcal{L})$. The latter are obtained by instantiating variables with values from $\mathcal{L}$ and evaluating expressions. An *assignment* is a mapping $\alpha \colon \text{VarId} \to \mathcal{L}$. Given an expression e, α is *well-typed* for e if for every term $t_1 \oplus t_2$ in e, with $\oplus \in \text{ArithOp}$, we have $\alpha(\mathbf{x}) \in \mathbb{Z}$ for all variable identifiers $\mathbf{x}$ in $t_1 \oplus t_2$. In this case we inductively define the value $e^\alpha \in \mathcal{L}$ as follows. If e is a numeral or a sequence of characters, then e^α is the integer or character string represented by e. If e is a variable identifier, then $e^\alpha = \alpha(e)$. Otherwise, if e has the form $t_1 \oplus t_2$ with $\oplus \in \text{ArithOp}$ and $t_1, t_2 \in \text{Term}$, then $e^\alpha = t_1^\alpha \oplus_\mathbb{Z} t_2^\alpha$ where $\oplus_\mathbb{Z}$ is the integer operation represented by $\oplus$. Finally, if e has the form t_e_1 with $t \in \text{Term} \cup \text{String}$ and $e_1 \in \text{Exp}$, then $e^\alpha = t^\alpha e_1^\alpha$ (the concatenation of t^α and e_1^α).

Given a graph G in $\mathcal{G}(\text{Exp})$ and an assignment α that is well-typed for all expressions occurring in G, we write G^α for the graph in $\mathcal{G}(\mathcal{L})$ that is obtained from G by replacing each label e with e^α. If $g \colon G \to H$ is a graph morphism between graphs in $\mathcal{G}(\text{Exp})$, then g^α denotes the morphism $\langle g_V^\alpha, g_E^\alpha \rangle \colon G^\alpha \to H^\alpha$.

A *substitution* is a mapping $\sigma \colon \text{VarId} \to \text{Exp}$. Given an expression e, σ is *well-typed* for e if for every term $t_1 \oplus t_2$ in e, with $\oplus \in \text{ArithOp}$, we have $\sigma(\mathbf{x}) \in \text{Term}$ for all variable identifiers $\mathbf{x}$ in $t_1 \oplus t_2$. In this case the expression e^σ is obtained from e by replacing every occurrence of a variable $\mathbf{x}$ with $\sigma(\mathbf{x})$. Given a graph G in $\mathcal{G}(\text{Exp})$, we write G^σ for the graph that is obtained by replacing each label e

---

[1] For simplicity, we use the non-terminals of our grammars to denote the syntactic classes of strings that can be derived from them.

with e^σ. If $g\colon G \to H$ is a graph morphism between graphs in $\mathcal{G}(\mathrm{Exp})$, then g^σ denotes the morphism $\langle g_V^\sigma, g_E^\sigma \rangle\colon G^\sigma \to H^\sigma$.

Given an assignment α, the substitution σ_α *induced* by α maps every variable x to the expression that is obtained from $\alpha(\mathbf{x})$ by replacing integers and strings with their syntactic counterparts. For example, if $\alpha(\mathbf{x})$ is the sequence $56, a, bc$, where 56 is an integer and a and bc are strings, then $\sigma_\alpha(\mathbf{x}) = 56\_"a"\_"bc"$.

3   Graph Programs

We briefly review GP's conditional rule schemata and discuss an example program. Technical details (including an operational semantics later used in our soundness proof) and further examples can be found in [13,14].

3.1   Conditional Rule Schemata

Conditional rule schemata are the "building blocks" of graph programs: a program is essentially a list of declarations of conditional rule schemata together with a command sequence for controlling the application of the schemata. Rule schemata generalise graph transformation rules in the double-pushout approach with relabelling [8], in that labels can contain expressions over parameters of type integer or string. Figure 2 shows a conditional rule schema consisting of the identifier bridge followed by the declaration of formal parameters, the left and right graphs of the schema which are graphs in $\mathcal{G}(\mathrm{Exp})$, the node identifiers 1, 2, 3 specifying which nodes are preserved, and the keyword where followed by a rule schema condition.

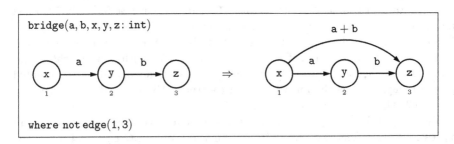

Fig. 2. A conditional rule schema

In the GP programming system [11], rule schemata are constructed with a graphical editor. Labels in the left graph comprise only variables and constants (no composite expressions) because their values at execution time are determined by graph matching. The condition of a rule schema is a Boolean expression built from arithmetic expressions and the special predicate edge, where all variables occurring in the condition must also occur in the left graph. The predicate edge demands the (non-)existence of an edge between two nodes in the graph to which the rule schema is applied. For example, the expression not edge(1, 3) in the condition of Figure 2 forbids an edge from node 1 to node 3 when the left graph is matched. The grammar of Figure 3 defines the syntax of rule schema conditions, where Term is the syntactic class defined in Figure 1.

BoolExp ::= **edge** '(' Node ',' Node ')' | Term RelOp Term
          | **not** BoolExp | BoolExp BoolOp BoolExp
Node   ::= Digit {Digit}
RelOp   ::= '=' | '\=' | '>' | '<' | '>=' | '<='
BoolOp ::= **and** | **or**

Fig. 3. Syntax of rule schema conditions

Conditional rule schemata represent possibly infinite sets of conditional graph transformation rules over graphs in $\mathcal{G}(\mathcal{L})$, and are applied according to the double-pushout approach with relabelling. A rule schema $L \Rightarrow R$ with condition Γ represents conditional rules $\langle\langle L^\alpha \leftarrow K \rightarrow R^\alpha\rangle, \Gamma^{\alpha,g}\rangle$, where K consists of the preserved nodes (which are unlabelled) and $\Gamma^{\alpha,g}$ is a predicate on graph morphisms $g\colon L^\alpha \rightarrow G$ (see [13,14]).

3.2   Programs

We discuss an example program to familiarise the reader with GP's features. This program will be a running example throughout the remainder of the paper.

Example 1 (Colouring). A *colouring* for a graph is an assignment of colours (integers) to nodes such that the source and target of each non-looping edge have different colours. The program **colouring** in Figure 4 produces a colouring for every integer-labelled input graph, recording colours as so-called tags. In general, a tagged label is a sequence of expressions separated by underscores.

   The program initially colours each node with 1 by applying the rule schema **init** as long as possible, using the iteration operator '!'. It then repeatedly increments the target colour of edges with the same colour at both ends. Note that this process is highly nondeterministic: Figure 4 shows an execution producing a colouring with two colours, but a colouring with three colours could have been produced for the same input graph.

   It is easy to see that whenever **colouring** terminates, the resulting graph is a correctly coloured version of the input graph. This is because the output cannot contain an edge with the same colour at both incident nodes, as then **inc** would have been applied at least one more time. Also, it can be shown that every execution of the program terminates after at most a quadratic number of rule schema applications [13].

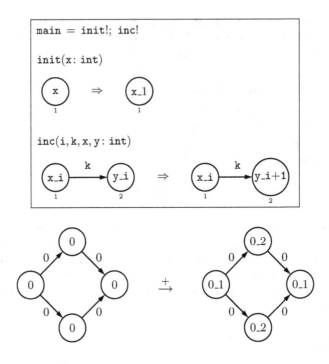

Fig. 4. The program `colouring` and one of its executions

4   Nested Graph Conditions with Expressions

We introduce nested graph conditions with expressions (or E-conditions) to specify graph properties in the pre- and postconditions of graph programs. E-conditions extend the nested conditions of [5] with expressions for labels, and assignment constraints which restrict the values that can be assigned to variables. The resulting conditions can be considered as representations of possibly infinite sets of ordinary nested conditions.

Definition 1 (Assignment constraint). An *assignment constraint* is a Boolean expression conforming to the grammar in Figure 5. We require that the arguments of the operators $>$, $<$, $\geq$ and $\leq$ belong to the syntactic class Term and that the arguments of $=$ and $\neq$ belong both to either Term, String or Exp $-$ (Term $\cup$ String). (See Figure 1 for the definition of Term, String and Exp.)

Given an assignment constraint γ and an assignment $\alpha\colon \text{VarId} \to \mathcal{L}$, the *value* γ^α in $\mathbb{B} = \{\mathbf{tt}, \mathbf{ff}\}$ is inductively defined[2]. If $\gamma = \text{true}$, then $\gamma^\alpha = \mathbf{tt}$. If γ has the form $e_1 \bowtie e_2$ with $\bowtie \in \text{ACRelOp}$ and $e_1, e_2 \in \text{Exp}$, then $\gamma^\alpha = \mathbf{tt}$ if and only if $e_1^\alpha \bowtie_{\mathcal{L}} e_2^\alpha$ where $\bowtie_{\mathcal{L}}$ is the obvious relation on $\mathcal{L}$ represented by $\bowtie$. If $\gamma = \neg \gamma_1$ with $\gamma_1 \in \text{ACBoolExp}$, then $\gamma^\alpha = \mathbf{tt}$ if and only if $\gamma_1^\alpha = \mathbf{ff}$. If $\gamma = \gamma_1 \oplus \gamma_2$ with

---

[2] We assume that α is well-typed for γ, which is defined in a similar way to before.

ACBoolExp ::= Exp ACRelOp Exp | '¬' ACBoolExp
          | ACBoolExp ACBoolOp ACBoolExp
          | 'type' '(' VarId ')' '=' Type | 'true'
ACRelOp   ::= '=' | '≠' | '>' | '<' | '≥' | '≤'
ACBoolOp ::= '∧' | '∨'
Type   ::= 'int' | 'string' | 'tagged'

Fig. 5. Syntax of assignment constraints

$\gamma_1, \gamma_2 \in$ ACBoolExp and $\oplus \in$ ACBoolOp, then $\gamma^\alpha = \gamma_1^\alpha \oplus_\mathbb{B} \gamma_2^\alpha$ where $\oplus_\mathbb{B}$ is the Boolean operation on $\mathbb{B}$ represented by $\oplus$. Finally, if γ has the form type$(\mathbf{x}) = t$ with $\mathbf{x} \in$ VarId and $t \in$ Type, then $\gamma^\alpha = \mathtt{tt}$ if and only if $type(\alpha(\mathbf{x})) = t$, where the function $type \colon \mathcal{L} \to$ Type is defined by

$$type(l) = \begin{cases} \text{int} & \text{if } l \in \mathbb{Z}, \\ \text{string} & \text{if } l \in \text{Char}^*, \\ \text{tagged} & \text{otherwise.} \end{cases}$$

Example 2 (Assignment constraint). Consider the assignment constraint $\gamma = \mathtt{a} > \mathtt{b} \wedge \mathtt{b} \neq 0 \wedge$ type$(\mathtt{a}) = $ int. Let $\alpha_1 = (\mathtt{a} \mapsto 5, \mathtt{b} \mapsto 1)$ and $\alpha_2 = (\mathtt{a} \mapsto 3, \mathtt{b} \mapsto 0)$. Then $\gamma^{\alpha_1} = \mathtt{tt}$ and $\gamma^{\alpha_2} = \mathtt{ff}$.

Note that variables in assignment constraints do not have a type per se, unlike the variables in GP rule schemata. Rather, the operator 'type' can be used to constrain the type of a variable. Note also that an assignment constraint such as $\mathtt{a} > 5 \wedge$ type$(\mathtt{a}) = $ string evaluates under every assignment to $\mathtt{ff}$, because we assume that assignments are well-typed.

A substitution $\sigma \colon$ VarId $\to$ Exp is well-typed for an assignment constraint γ if the replacement of every occurrence of a variable $\mathbf{x}$ in γ with $\sigma(\mathbf{x})$ results in an assignment constraint. In this case the resulting constraint is denoted by γ^σ.

Definition 2 (E-condition). An *E-condition* c over a graph P is of the form true or $\exists(a|\gamma, c')$, where $a \colon P \hookrightarrow C$ is an injective[3] graph morphism with $P, C \in \mathcal{G}(\text{Exp})$, γ is an assignment constraint, and c' is an E-condition over C. Moreover, Boolean formulas over E-conditions over P yield E-conditions over P, that is, $\neg c$ and $c_1 \wedge c_2$ are E-conditions over P, where c_1 and c_2 are E-conditions over P.

For brevity, we write false for $\neg$true, $\exists(a|\gamma)$ for $\exists(a|\gamma, \text{true})$, $\exists(a, c')$ for $\exists(a|\text{true}, c')$, and $\forall(a|\gamma, c')$ for $\neg\exists(a|\gamma, \neg c')$. In examples, when the domain of morphism $a \colon P \hookrightarrow C$ can unambiguously be inferred, we write only the codomain C. For instance, an E-condition $\exists(\emptyset \hookrightarrow C, \exists(C \hookrightarrow C'))$ can be written as $\exists(C, \exists(C'))$, where the domain of the outermost morphism is the empty graph, and the domain of the nested morphism is the codomain of the encapsulating E-condition's

---

[3] For simplicity, we restrict E-conditions to injective graph morphisms since this is sufficient for GP.

morphism. An E-condition over a graph morphism whose domain is the empty graph is referred to as an *E-constraint*. We later refer to E-conditions over left- and right-hand sides of rule schemata as *E-app-conditions*.

Example 3 (E-condition). The E-condition $\forall(\,$ ⓧ→ⓨ $\,|\,$ x > y, $\exists(\,$ ⓧ→ⓨ $\,))$ (which is an E-constraint) expresses that every pair of adjacent integer-labelled nodes with the source label greater than the target label has a loop incident to the source node. The unabbreviated version of the condition is as follows:

$$\neg\exists(\emptyset \hookrightarrow \text{ⓧ→ⓨ} \,|\, x > y,\ \neg\exists(\text{ⓧ→ⓨ} \hookrightarrow \text{ⓧ→ⓨ} \,|\, \text{true, true})).$$

The *satisfaction* of E-conditions by injective graph morphisms over $\mathcal{L}$ is defined inductively. Every such morphism satisfies the E-condition true. An injective graph morphism $s\colon S \hookrightarrow G$ with $S, G \in \mathcal{G}(\mathcal{L})$ satisfies the condition $c = \exists(a\colon P \hookrightarrow C|\gamma, c')$, denoted $s \models c$, if there exists an assignment α such that $S = P^\alpha$ and there is an injective graph morphism $q\colon C^\alpha \hookrightarrow G$ with $q \circ a^\alpha = s$, $\gamma^\alpha = \mathtt{tt}$, and $q \models (c')^{\sigma_\alpha}$, where σ_α is the substitution induced by α.

A graph G in $\mathcal{G}(\mathcal{L})$ satisfies an E-condition c, denoted $G \models c$, if the morphism $\emptyset \hookrightarrow G$ satisfies c.

The application of a substitution σ to an E-condition c is defined inductively, too. We have $\mathrm{true}^\sigma = \mathrm{true}$ and $\exists(a|\gamma, c')^\sigma = \exists(a^\sigma|\gamma^\sigma, (c')^\sigma)$, where we assume that σ is well-typed for all components it is applied to.

5   A Hoare Calculus for Graph Programs

We present a system of partial correctness proof rules for GP, in the style of Hoare [1], using E-constraints as the assertions. We demonstrate the proof system by proving a property of our earlier `colouring` graph program, and sketch a proof of the rules' soundness according to GP's operational semantics [13,14].

Definition 3 (Partial correctness). A graph program P is *partially correct* with respect to a precondition c and a postcondition d (both of which are E-constraints), if for every graph $G \in \mathcal{G}(\mathcal{L})$, $G \models c$ implies $H \models d$ for every graph H in $[\![P]\!]G$.

Here, $[\![\_]\!]$ is GP's semantic function (see [14]), and $[\![P]\!]G$ is the set of all graphs resulting from executing program P on graph G. Note that partial correctness of a program P does not entail that P will actually terminate on graphs satisfying the precondition.

Given E-constraints c, d and a program P, a triple of the form $\{c\}\ P\ \{d\}$ expresses the claim that whenever a graph G satisfies c, then any graphs resulting from the application of P to G will satisfy d. Our proof system in Figure 6 operates on such triples. As in classical Hoare logic [1], we use the proof system to construct proof trees, combining axiom schemata and inference rules (an example will follow). We let c, d, e, inv range over E-constraints, P, Q over arbitrary command sequences, r, r_i over conditional rule schemata, and $\mathcal{R}$ over sets of conditional rule schemata.

$$[\text{rule}] \; \frac{}{\{\text{Pre}(r,c)\} \; r \; \{c\}} \qquad\qquad [\text{ruleset}_1] \; \frac{}{\{\neg\text{App}(\mathcal{R})\} \; \mathcal{R} \; \{\text{false}\}}$$

$$[\text{ruleset}_2] \; \frac{\{c\} \; r_1 \; \{d\} \; \ldots \; \{c\} \; r_n \; \{d\}}{\{c\} \; \{r_1,\ldots,r_n\} \; \{d\}} \qquad\qquad [!] \; \frac{\{inv\} \; \mathcal{R} \; \{inv\}}{\{inv\} \; \mathcal{R}! \; \{inv \wedge \neg\text{App}(\mathcal{R})\}}$$

$$[\text{comp}] \; \frac{\{c\} \; P \; \{e\} \qquad \{e\} \; Q \; \{d\}}{\{c\} \; P; \; Q \; \{d\}} \qquad\qquad [\text{cons}] \; c \Longrightarrow c' \; \frac{\{c'\} \; P \; \{d'\}}{\{c\} \; P \; \{d\}} \; d' \Longrightarrow d$$

$$[\text{if}] \; \frac{\{c \wedge \text{App}(\mathcal{R})\} \; P \; \{d\} \qquad \{c \wedge \neg\text{App}(\mathcal{R})\} \; Q \; \{d\}}{\{c\} \; \text{if } \mathcal{R} \text{ then } P \text{ else } Q \; \{d\}}$$

Fig. 6. Partial correctness proof system for GP

Two transformations — App and Pre — are required in some of the assertions. Intuitively, App takes as input a set $\mathcal{R}$ of conditional rule schemata, and transforms it into an E-condition specifying the property that a rule in $\mathcal{R}$ is applicable to the graph. Pre constructs the weakest precondition such that if $G \models \text{Pre}(r,c)$, and the application of r to G results in a graph H, then $H \models c$. The transformation Pre is informally described by the following steps: (1) form a disjunction of right E-app-conditions for the possible overlappings of c and the right-hand side of the rule schema r, (2) convert the right E-app-condition into a left E-app-condition (i.e. over the left-hand side of r), (3) nest this within an E-condition that is quantified over every L and also accounts for the applicability of r.

Note that two of the proof rules deal with programs that are restricted in a particular way: both the condition C of a branching command if C then P else Q and the body P of a loop $P!$ must be sets of conditional rule schemata. This restriction does not affect the computational completeness of the language, because in [7] it is shown that a graph transformation language is complete if it contains single-step application and as-long-as-possible iteration of (unconditional) sets of rules, together with sequential composition.

Example 4 (Colouring). Figure 7 shows a proof tree for the colouring program of Figure 4. It proves that if colouring is executed on a graph in which the node labels are exclusively integers, then any graph resulting will have the property that each node label is an integer with a colour attached to it, and that adjacent nodes have distinct colours. That is, it proves the triple $\{\neg\exists(\;\text{ⓐ}\;|\; \text{type}(a) \neq \text{int})\}$ init!; inc! $\{\forall(\;\text{ⓐ}_1\;, \exists(\;\text{ⓐ}_1\;|\; a = b\_c \wedge \text{type}(b,c) = \text{int})) \wedge \neg\exists(\;\text{(x.i)} \xrightarrow{k} \text{(y.i)}\;|\; \text{type}(i,k,x,y) = \text{int})\}$. For conciseness, we abuse our notation (in this, and later examples), and allow $\text{type}(x_1,\ldots,x_n) = \text{int}$ to represent $\text{type}(x_1) = \text{int} \wedge \ldots \wedge \text{type}(x_n) = \text{int}$.

$$[\text{rule}] \; \cfrac{\{\text{Pre}(\text{init}, e)\} \; \text{init} \; \{e\}}{\cfrac{\cfrac{[\text{cons}]}{[!]} \; \cfrac{\{e\} \; \text{init} \; \{e\}}{\{e\} \; \text{init!} \; \{e \wedge \neg\text{App}(\{\text{init}\})\}}}{[\text{cons}] \; \cfrac{\{c\} \; \text{init!} \; \{d\}}{}}} \qquad [\text{rule}] \; \cfrac{\{\text{Pre}(\text{inc}, d)\} \; \text{inc} \; \{d\}}{\cfrac{[\text{cons}]}{[!]} \; \cfrac{\{d\} \; \text{inc} \; \{d\}}{\{d\} \; \text{inc!} \; \{d \wedge \neg\text{App}(\{\text{inc}\})\}}}$$

$$[\text{comp}] \; \cfrac{}{\{c\} \; \text{init!}; \; \text{inc!} \; \{d \wedge \neg\text{App}(\{\text{inc}\})\}}$$

$$c = \neg\exists(\; \text{ⓐ} \mid \text{type}(a) \neq \text{int})$$
$$d = \forall(\; \text{ⓐ}_1, \exists(\; \text{ⓐ}_1 \mid a = b\_c \wedge \text{type}(b, c) = \text{int}))$$
$$e = \forall(\; \text{ⓐ}_1, \exists(\; \text{ⓐ}_1 \mid \text{type}(a) = \text{int}) \vee \exists(\; \text{ⓐ}_1 \mid a = b\_c \wedge \text{type}(b, c) = \text{int}))$$
$$\neg\text{App}(\{\text{init}\}) = \neg\exists(\; \text{ⓧ} \mid \text{type}(x) = \text{int})$$
$$\neg\text{App}(\{\text{inc}\}) = \neg\exists(\; \text{x\_i} \xrightarrow{k} \text{y\_i} \mid \text{type}(i, k, x, y) = \text{int})$$
$$\text{Pre}(\text{init}, e) = \forall(\; \text{ⓧ}_1 \; \text{ⓐ}_2 \mid \text{type}(x) = \text{int}, \exists(\; \text{ⓧ}_1 \; \text{ⓐ}_2 \mid \text{type}(a) = \text{int})$$
$$\vee \; \exists(\; \text{ⓧ}_1 \; \text{ⓐ}_2 \mid a = b\_c \wedge \text{type}(b, c) = \text{int}))$$
$$\text{Pre}(\text{inc}, d) = \forall(\; \text{x\_i}_1 \xrightarrow{k} \text{y\_i}_2 \; \text{ⓐ}_3 \mid \text{type}(i, k, x, y) = \text{int},$$
$$\exists(\; \text{x\_i}_1 \xrightarrow{k} \text{y\_i}_2 \; \text{ⓐ}_3 \mid a = b\_c \wedge \text{type}(b, c) = \text{int}))$$

Fig. 7. A proof tree for the program **colouring** of Figure 4

6   Transformations and Soundness

We provide full definitions of the transformations App and Pre in this section. In order to define Pre, it is necessary to first define the intermediary transformations A and L, which are adapted from basic transformations of nested conditions [5]. Following this, we will show that our proof system is sound according to the operational semantics of GP.

Proposition 1 (Applicability of a set of rule schemata). *For every set $\mathcal{R}$ of conditional rule schemata, there exists an E-constraint $\text{App}(\mathcal{R})$ such that for every graph $G \in \mathcal{G}(\mathcal{L})$,*

$$G \models \text{App}(\mathcal{R}) \iff G \in \text{Dom}(\Rightarrow_\mathcal{R}),$$

where $G \in \text{Dom}(\Rightarrow_\mathcal{R})$ if there is a direct derivation $G \Rightarrow_\mathcal{R} H$ for some graph H.

The transformation App gives an E-constraint that can only be satisfied by a graph G if at least one of the rule schemata from $\mathcal{R}$ can directly derive a graph H from G. The idea is to generate a disjunction of E-constraints from the left-hand sides of the rule schemata, with nested E-conditions for handling restrictions on the application of the rule schemata (such as the dangling condition when deleting nodes).

Construction. Define $\text{App}(\{\}) = \text{false}$ and $\text{App}(\{r_1, \ldots, r_n\}) = \text{app}(r_1) \vee \ldots \vee \text{app}(r_n)$. For a rule schema $r_i = \langle L_i \hookleftarrow K_i \hookrightarrow R_i \rangle$ with rule schema condition Γ_i, define $\text{app}(r_i) = \exists(\emptyset \hookrightarrow L_i | \gamma_{r_i}, \neg\text{Dang}(r_i) \wedge \tau(L_i, \Gamma_i))$ where γ_{r_i} is a conjunction of expressions constraining the types of variables in r_i to the corresponding types in the declaration of r_i. For example, if r_i corresponds to the declaration of $\texttt{inc}$ (Figure 4), then γ_{r_i} would be the Boolean expression $\text{type}(\texttt{i}) = \text{int} \wedge \text{type}(\texttt{k}) = \text{int} \wedge \text{type}(\texttt{x}) = \text{int} \wedge \text{type}(\texttt{y}) = \text{int}$.

Define $\text{Dang}(r_i) = \bigvee_{a \in A} \exists a$, where the index set A ranges over all[4] injective graph morphisms $a : L_i \hookrightarrow L_i^\oplus$ such that the pair $\langle K_i \hookrightarrow L_i, a \rangle$ has no natural pushout[5] complement, and each $L_i^\oplus$ is a graph that can be obtained from L_i by adding either (1) a loop, (2) a single edge between distinct nodes, or (3) a single node and a non-looping edge incident to that node. All items in $L_i^\oplus - L_i$ are labelled with single variables, distinct from each other, and distinct from those in L_i. If the index set A is empty, then $\text{Dang}(r_i) = \text{false}$.

We define $\tau(L_i, \Gamma_i)$ inductively (see Figure 3 for the syntax of rule schema conditions). If there is no rule schema condition, then $\tau(L_i, \Gamma_i) = \text{true}$. If Γ_i has the form $t_1 \bowtie t_2$ with t_1, t_2 in Term and $\bowtie$ in RelOp, then $\tau(L_i, \Gamma_i) = \exists(L_i \hookrightarrow L_i | t_1 \bowtie_{\text{ACRelOp}} t_2)$ where $\bowtie_{\text{ACRelOp}}$ is the symbol in ACRelOp that corresponds to the symbol $\bowtie$ from RelOp. If Γ_i has the form $\textbf{not } b_i$ with b_i in BoolExp, then $\tau(L_i, \Gamma_i) = \neg\tau(L_i, b_i)$. If Γ_i has the form $b_1 \oplus b_2$ with b_1, b_2 in BoolExp and $\oplus$ in BoolOp, then $\tau(L_i, \Gamma_i) = \tau(L_i, b_1) \oplus_{\wedge,\vee} \tau(L_i, b_2)$ where $\oplus_{\wedge,\vee}$ is $\wedge$ for $\textbf{and}$ and $\vee$ for $\textbf{or}$. Finally, if Γ_i is of the form $\textbf{edge}(n_1, n_2)$ with n_1, n_2 in Node, then $\tau(L_i, \Gamma_i) = \exists(L_i \hookrightarrow L_i')$ where L_i' is a graph isomorphic to L_i, except for an additional edge whose source is the node with identifier n_1, whose target is the node with identifier n_2, and whose label is a variable distinct from all others in use.

Proposition 2 (From E-constraints to E-app-conditions). *There is a transformation* A *such that, for all E-constraints* c, *all rule schemata* $r : L \Rightarrow R$ *sharing no variables with* c[6], *and all injective graph morphisms* $h : R^\alpha \hookrightarrow H$ *where* $H \in \mathcal{G}(\mathcal{L})$ *and* α *is a well-typed assignment,*

$$h \models \text{A}(r, c) \Longleftrightarrow H \models c.$$

The idea of A is to consider a disjunction of all possible overlappings of R and the graphs of the E-constraint. Substitutions are used to replace label variables in c with portions of labels from R, facilitating the overlappings.

Construction. All graphs used in the construction of the transformation belong to the class $\mathcal{G}(\text{Exp})$. For E-constraints $c = \exists(a : \emptyset \hookrightarrow C | \gamma, c')$ and rule schemata r, define $\text{A}(r, c) = \text{A}'(i_R : \emptyset \hookrightarrow R, c)$. For injective graph morphisms $p : P \hookrightarrow P'$, and E-conditions over P,

---

[4] We equate morphisms with isomorphic codomains, so A is finite.
[5] A pushout is *natural* if is simultaneously a pullback [8].
[6] It is always possible to replace the label variables in c with new ones that are distinct from those in r.

$$A'(p, \text{true}) = \text{true},$$
$$A'(p, \exists(a|\gamma, c')) = \bigvee_{\sigma \in \Sigma} \bigvee_{e \in \varepsilon_\sigma} \exists(b|\gamma^\sigma, A'(s, (c')^\sigma)).$$

Construct the pushout (1) of p and a leading to injective graph morphisms $a' : P' \hookrightarrow C'$ and $q : C \hookrightarrow C'$. The finite double disjunction $\bigvee_{\sigma \in \Sigma} \bigvee_{e \in \varepsilon_\sigma}$ ranges first over substitutions from Σ, which have the special form $(a_1 \mapsto \beta_1, \ldots, a_k \mapsto \beta_k)$ where each a_i is a distinct label variable from C that is not also in P, and each β_i is a portion (or the entirety) of some label from P'. For each $\sigma \in \Sigma$, the double disjunction then ranges over every surjective graph morphism $e : (C')^\sigma \to E$ such that $b = e \circ (a')^\sigma$ and $s = e \circ q^\sigma$ are injective graph morphisms. The set ε_σ is the set of such sur-

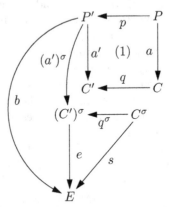

jective graph morphisms for a particular σ, the codomains of which we consider up to isomorphism. For a surjective graph morphism $e_1 : (C_1')^{\sigma_1} \to E_1$, E_1 is considered redundant and is excluded from the disjunction if there exists a surjective graph morphism, $e_2 : (C_2')^{\sigma_2} \to E_2$, such that $E_2 \not\cong E_1$, and there exists some $\sigma \in \Sigma$ such that $E_2^\sigma \cong E_1$.

The transformation A is extended for Boolean formulas over E-conditions in the same way as transformations over conditions (see [5]).

Example 5. Let r correspond to the rule schema inc (Figure 4), and E-constraint $c = \neg \exists(\text{ⓐ} \mid \text{type}(a) = \text{int})$. Then,

$$A(r, c) = \neg \exists(\boxed{x\_i} \xrightarrow{k} \boxed{y\_i+1} \hookrightarrow \boxed{x\_i} \xrightarrow{k} \boxed{y\_i+1} \, \text{ⓐ}) \mid \text{type}(a) = \text{int})$$

Proposition 3 (Transformation of E-app-conditions). *There is a transformation L such that, for every rule schema $r = \langle L \hookleftarrow K \hookrightarrow R \rangle$ with rule schema condition Γ, every right E-app-condition c for r, and every direct derivation $G \Rightarrow_{r,g,h} H$ with $g : L^\alpha \hookrightarrow G$ and $h : R^\alpha \hookrightarrow H$ where $G, H \in \mathcal{G}(\mathcal{L})$ and α is a well-typed assignment,*

$$g \models L(r, c) \Longleftrightarrow h \models c.$$

Construction. All graphs used in the construction of the transformation belong to the class $\mathcal{G}(\text{Exp})$. $L(r, c)$ is inductively defined as follows. Let $L(r, \text{true}) = \text{true}$ and $L(r, \exists(a|\gamma, c')) = \exists(b|\gamma, L(r^*, c'))$ if $\langle K \hookrightarrow R, a \rangle$ has a natural pushout complement (1) with $r^* = \langle Y \hookleftarrow Z \hookrightarrow X \rangle$ denoting the "derived" rule by constructing natural pushout (2). If $\langle K \hookrightarrow R, a \rangle$ has no natural pushout complement, then $L(r, \exists(a|\gamma, c')) = \text{false}$.

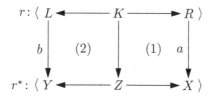

Example 6. Continuing from Example 5, we get $L(r, A(r, c)) = \neg\exists(\,\text{x\_i} \xrightarrow{\text{k}} \text{y\_i}\, \hookrightarrow$ $\text{x\_i} \xrightarrow{\text{k}} \text{y\_i}\ \text{a}\ |\text{type}(\text{a}) = \text{int})$.

Proposition 4 (Transformation of postconditions into preconditions).
There is a transformation Pre *such that, for every E-constraint c, every rule schema* $r = \langle L \hookleftarrow K \hookrightarrow R \rangle$ *with rule schema condition* Γ*, and every direct derivation* $G \Rightarrow_r H$,

$$G \models \text{Pre}(r, c) \implies H \models c.$$

Construction. Define $\text{Pre}(r, c) = \forall(\emptyset \hookrightarrow L|\gamma_r, (\neg\text{Dang}(r) \wedge \tau(L, \Gamma) \implies L(r, A(r, c))))$, where γ_r is as defined in Proposition 1.

Example 7. Continuing from Examples 5 and 6, we get $\text{Pre}(r, c) = \forall(\,\text{x\_i} \xrightarrow{\text{k}} \text{y\_i}\,|$ $\text{type}(\text{i}, \text{k}, \text{x}, \text{y}) = \text{int}, \neg\exists(\,\text{x\_i} \xrightarrow{\text{k}} \text{y\_i}\ \text{a}\ |\text{type}(\text{a}) = \text{int}))$. Since r does not delete any nodes, and does not have a rule schema condition, $\neg\text{Dang}(r) \wedge \tau(L, \Gamma) = \text{true}$, simplifying the nested E-condition generated by Pre.

Our main result is that the proof rules of Figure 6 are sound for proving partial correctness of graph programs. That is, a graph program P is partially correct with respect to a precondition c and a postcondition d (in the sense of Definition 3) if there exists a full proof tree whose root is the triple $\{c\}\ P\ \{d\}$.

Theorem 1. *The proof system of Figure 6 is sound for graph programs, in the sense of partial correctness.*

Proof. To prove soundness, we consider each proof rule in turn, appealing to the semantic function $[\![P]\!]G$ (defined in [13,14]). The result then follows by induction on the length of proofs.

Let c, d, e, inv be E-constraints, P, Q be arbitrary graph programs, $\mathcal{R}$ be a set of conditional rule schemata, r, r_i be conditional rule schemata, and $G, H, \overline{G}, G'$, $H' \in \mathcal{G}(\mathcal{L})$. $\to$ is a small-step transition relation on configurations of graphs and programs. We decorate the names of the semantic inference rules of [14] with "SOS", in order to fully distinguish them from the names in our Hoare calculus.

[rule]. Follows from Proposition 4.

[ruleset$_1$]. Suppose that $G \models \neg\text{App}(\mathcal{R})$. Proposition 1 implies that $G \notin \text{Dom}(\Rightarrow_\mathcal{R})$, hence from the inference rule [Call$_2$]$_\text{SOS}$ we obtain the transition $\langle\mathcal{R}, G\rangle \to \text{fail}$ (intuitively, this indicates that the program terminates but without returning a graph). No graph will result; this is captured by the postcondition false, which no graph can satisfy.

[ruleset$_2$]. Suppose that we have a non-empty set of rule schemata $\{r_1, \ldots, r_n\}$ denoted by $\mathcal{R}$, that $G \models c$, and that we have a non-empty set of graphs $\bigcup_{r \in \mathcal{R}} \{H \in \mathcal{G}(\mathcal{L}) | G \Rightarrow_r H\}$ such that each $H \models d$ (if the set was empty, then [ruleset$_1$] would apply). For the set to be non-empty, at least one $r \in \mathcal{R}$ must be applicable to G. That is, there is a direct derivation $G \Rightarrow_\mathcal{R} H$ for some graph H that satisfies d. From the inference rule [Call$_1$]$_{\text{SOS}}$ and the assumption, we get $[\![\mathcal{R}]\!]G = \{H \in \mathcal{G}(\mathcal{L}) | \langle \mathcal{R}, G \rangle \rightarrow H\}$ such that each $H \models d$.

[comp]. Suppose that $G \models c$, $[\![P]\!]G = \{G' \in \mathcal{G}(\mathcal{L}) | \langle P, G \rangle \rightarrow^+ G'\}$ such that each $G' \models e$, and $[\![Q]\!]G' = \{H \in \mathcal{G}(\mathcal{L}) | \langle Q, G' \rangle \rightarrow^+ H\}$ such that each $H \models d$. Then $[\![P; Q]\!]G = \{H \in \mathcal{G}(\mathcal{L}) | \langle P; Q, G \rangle \rightarrow^+ \langle Q, G' \rangle \rightarrow^+ H\}$ such that each $H \models d$ follows from the inference rule [Seq$_2$]$_{\text{SOS}}$.

[cons]. Suppose that $G' \models c'$, $c \Longrightarrow c'$, $d' \Longrightarrow d$, and $[\![P]\!]G' = \{H' \in \mathcal{G}(\mathcal{L}) | \langle P, G' \rangle \rightarrow^+ H'\}$ such that each $H' \models d'$. If $G \models c$, we have $G \models c'$ since $c \Longrightarrow c'$. The assumption then gives us an $H \in [\![P]\!]G$ such that $H \models d'$. From $d' \Longrightarrow d$, we get $H \models d$.

[if]. *Case One.* Suppose that $G \models c$, $[\![P]\!]G = \{H \in \mathcal{G}(\mathcal{L}) | \langle P, G \rangle \rightarrow^+ H\}$ such that each $H \models d$, and $G \models \text{App}(\mathcal{R})$. Then by Proposition 1, executing $\mathcal{R}$ on G will result in a graph. Hence by the assumption and the inference rule [If$_1$]$_{\text{SOS}}$, $[\![\text{if } \mathcal{R} \text{ then } P \text{ else } Q]\!]G = \{H \in \mathcal{G}(\mathcal{L}) | \langle \text{if } \mathcal{R} \text{ then } P \text{ else } Q, G \rangle \rightarrow \langle P, G \rangle \rightarrow^+ H\}$ such that each $H \models d$. *Case Two.* Suppose that $G \models c$, $[\![Q]\!]G = \{H \in \mathcal{G}(\mathcal{L}) | \langle Q, G \rangle \rightarrow^+ H\}$ such that each $H \models d$, and $G \models \neg\text{App}(\mathcal{R})$. Then by Proposition 1, executing $\mathcal{R}$ on G will not result in a graph. Hence by the assumption and the inference rule [If$_2$]$_{\text{SOS}}$, $[\![\text{if } \mathcal{R} \text{ then } P \text{ else } Q]\!]G = \{H \in \mathcal{G}(\mathcal{L}) | \langle \text{if } \mathcal{R} \text{ then } P \text{ else } Q, G \rangle \rightarrow \langle Q, G \rangle \rightarrow^+ H\}$ such that each $H \models d$.

[!]. We prove the soundness of this proof rule by induction over the number of executions of $\mathcal{R}$ that do not result in finite failure, which we denote by n. Assume that for any graph G' such that $G' \models inv$, $[\![\mathcal{R}]\!]G' = \{H' \in \mathcal{G}(\mathcal{L}) | \langle \mathcal{R}, G' \rangle \rightarrow^+ H'\}$ such that each $H' \models inv$. *Induction Basis (n = 0).* Suppose that $G \models inv$. Only the inference rule [Alap$_2$]$_{\text{SOS}}$ can be applied, that is, $[\![\mathcal{R}!]\!]G = \{G \in \mathcal{G}(\mathcal{L}) | \langle \mathcal{R}!, G \rangle \rightarrow G\}$. Since the graph is not changed, trivially, the invariant holds, i.e. $G \models inv$. Since the execution of $\mathcal{R}$ on G does not result in a graph, $G \models \neg\text{App}(\mathcal{R})$. *Induction Hypothesis (n = k).* Assume that there exists a configuration $\langle \mathcal{R}!, G \rangle$ such that $\langle \mathcal{R}!, G \rangle \rightarrow^* \langle \mathcal{R}!, H \rangle \rightarrow H$. Hence for $[\![\mathcal{R}!]\!]G = \{H \in \mathcal{G}(\mathcal{L}) | \langle \mathcal{R}!, G \rangle \rightarrow^* \langle \mathcal{R}!, H \rangle \rightarrow H\}$, we assume that if $G \models inv$, then each $H \models inv$ and $H \models \neg\text{App}(\mathcal{R})$. *Induction Step (n = k + 1).* We have $[\![\mathcal{R}!]\!]G = \{H \in \mathcal{G}(\mathcal{L}) | \langle \mathcal{R}!, G \rangle \rightarrow \langle \mathcal{R}!, \overline{G} \rangle \rightarrow^* \langle \mathcal{R}!, H \rangle \rightarrow H\}$. Let $G \models inv$, and $G \cong G'$. From the assumption, we get $H' \models inv$, $H' \cong \overline{G}$, and hence $\overline{G} \models inv$. It follows from the induction hypothesis that each $H \models inv$ and $H \models \neg\text{App}(\mathcal{R})$.

7   Conclusion

We have presented the first Hoare-style verification calculus for an implemented graph transformation language. This required us to extend the nested graph conditions of Habel, Pennemann and Rensink with expressions for labels and assignment constraints, in order to deal with GP's powerful rule schemata and infinite label alphabet. We have demonstrated the use of the calculus for proving the partial correctness of a highly nondeterministic colouring program, and have shown that our proof rules are sound with respect to GP's formal semantics.

Future work will investigate the completeness of the calculus. Also, we intend to add termination proof rules in order to verify the total correctness of graph programs. Finally, we will consider how the calculus can be generalised to deal with GP programs in which the conditions of branching statements and the bodies of loops can be arbitrary subprograms rather than sets of rule schemata.

Acknowledgements. We are grateful to the anonymous referees for their comments which helped to improve the presentation of this paper.

References

1. Apt, K.R., de Boer, F.S., Olderog, E.-R.: Verification of Sequential and Concurrent Programs, 3rd edn. Springer, Heidelberg (2009)
2. Baldan, P., Corradini, A., König, B.: A framework for the verification of infinite-state graph transformation systems. Information and Computation 206(7), 869–907 (2008)
3. Bisztray, D., Heckel, R., Ehrig, H.: Compositional verification of architectural refactorings. In: de Lemos, R. (ed.) Architecting Dependable Systems VI. LNCS, vol. 5835, pp. 308–333. Springer, Heidelberg (2009)
4. Geiß, R., Batz, G.V., Grund, D., Hack, S., Szalkowski, A.M.: GrGen: A fast SPO-based graph rewriting tool. In: Corradini, A., Ehrig, H., Montanari, U., Ribeiro, L., Rozenberg, G. (eds.) ICGT 2006. LNCS, vol. 4178, pp. 383–397. Springer, Heidelberg (2006)
5. Habel, A., Pennemann, K.-H.: Correctness of high-level transformation systems relative to nested conditions. Mathematical Structures in Computer Science 19(2), 245–296 (2009)
6. Habel, A., Pennemann, K.-H., Rensink, A.: Weakest preconditions for high-level programs. In: Corradini, A., Ehrig, H., Montanari, U., Ribeiro, L., Rozenberg, G. (eds.) ICGT 2006. LNCS, vol. 4178, pp. 445–460. Springer, Heidelberg (2006)
7. Habel, A., Plump, D.: Computational completeness of programming languages based on graph transformation. In: Honsell, F., Miculan, M. (eds.) FOSSACS 2001. LNCS, vol. 2030, pp. 230–245. Springer, Heidelberg (2001)
8. Habel, A., Plump, D.: Relabelling in graph transformation. In: Corradini, A., Ehrig, H., Kreowski, H.-J., Rozenberg, G. (eds.) ICGT 2002. LNCS, vol. 2505, pp. 135–147. Springer, Heidelberg (2002)
9. Hoare, C.A.R.: An axiomatic basis for computer programming. Communications of the ACM 12(10), 576–580 (1969)
10. König, B., Kozioura, V.: Towards the verification of attributed graph transformation systems. In: Ehrig, H., Heckel, R., Rozenberg, G., Taentzer, G. (eds.) ICGT 2008. LNCS, vol. 5214, pp. 305–320. Springer, Heidelberg (2008)
11. Manning, G., Plump, D.: The GP programming system. In: Proc. Graph Transformation and Visual Modelling Techniques (GT-VMT 2008). Electronic Communications of the EASST, vol. 10 (2008)
12. Nickel, U., Niere, J., Zündorf, A.: The FUJABA environment. In: Proc. International Conference on Software Engineering (ICSE 2000), pp. 742–745. ACM Press, New York (2000)

13. Plump, D.: The graph programming language GP. In: Bozapalidis, S., Rahonis, G. (eds.) Algebraic Informatics. LNCS, vol. 5725, pp. 99–122. Springer, Heidelberg (2009)
14. Plump, D., Steinert, S.: The semantics of graph programs. In: Proc. Rule-Based Programming (RULE 2009). EPTCS, vol. 21, pp. 27–38 (2010)
15. Poskitt, C.M., Plump, D.: A Hoare calculus for graph programs (long version) (2010), http://www.cs.york.ac.uk/plasma/publications/pdf/PoskittPlump.ICGT.10.Long.pdf
16. Rensink, A., Schmidt, Á., Varró, D.: Model checking graph transformations: A comparison of two approaches. In: Ehrig, H., Engels, G., Parisi-Presicce, F., Rozenberg, G. (eds.) ICGT 2004. LNCS, vol. 3256, pp. 226–241. Springer, Heidelberg (2004)
17. Schürr, A., Winter, A., Zündorf, A.: The PROGRES approach: Language and environment. In: Ehrig, H., Engels, G., Kreowski, H.-J., Rozenberg, G. (eds.) Handbook of Graph Grammars and Computing by Graph Transformation, ch. 13, vol. 2, pp. 487–550. World Scientific, Singapore (1999)
18. Taentzer, G.: AGG: A graph transformation environment for modeling and validation of software. In: Pfaltz, J.L., Nagl, M., Böhlen, B. (eds.) AGTIVE 2003. LNCS, vol. 3062, pp. 446–453. Springer, Heidelberg (2004)

Formal Analysis of Functional Behaviour for Model Transformations Based on Triple Graph Grammars

Frank Hermann[1], Hartmut Ehrig[1], Fernando Orejas[2], and Ulrike Golas[1]

[1] Institut für Softwaretechnik und Theoretische Informatik,
Technische Universität Berlin, Germany
{frank,ehrig,ugolas}@cs.tu-berlin.de
[2] Departament de Llenguatges i Sistemes Informàtics,
Universitat Politècnica de Catalunya, Barcelona, Spain
orejas@lsi.upc.edu

Abstract. Triple Graph Grammars (TGGs) are a well-established concept for the specification of model transformations. In previous work we have formalized and analyzed already crucial properties of model transformations like termination, correctness and completeness, but functional behaviour is missing up to now.

In order to close this gap we generate forward translation rules, which extend standard forward rules by translation attributes keeping track of the elements which have been translated already. In the first main result we show the equivalence of model transformations based on forward resp. forward translation rules. This way, an additional control structure for the forward transformation is not needed. This allows to apply critical pair analysis and corresponding tool support by the tool AGG. However, we do not need general local confluence, because confluence for source graphs not belonging to the source language is not relevant for the functional behaviour of a model transformation. For this reason we only have to analyze a weaker property, called translation confluence. This leads to our second main result, the functional behaviour of model transformations, which is applied to our running example, the model transformation from class diagrams to database models.

Keywords: Model Transformation, Triple Graph Grammars, Confluence, Functional Behaviour.

1 Introduction

Model transformations based on triple graph grammars (TGGs) have been introduced by Schürr in [18]. TGGs are grammars that generate languages of graph triples, consisting of source and target graphs, together with a correspondence graph "between" them. Since 1994, several extensions of the original TGG definitions have been published [19,13,8] and various kinds of applications have been presented [20,9,12]. For source-to-target model transformations, so-called

H. Ehrig et al. (Eds.): ICGT 2010, LNCS 6372, pp. 155–170, 2010.

forward transformations, we derive rules which take the source graph as input and produce a corresponding target graph. Major properties expected to be fulfilled for model transformations are termination, correctness and completeness, which have been analyzed in [1,3,4,6,7].

In addition to these properties, functional behaviour of model transformations is an important property for several application domains. Functional behaviour means that for each graph in the source language the model transformation yields a unique graph (up to isomorphism) in the target language. It is well-known that termination and local confluence implies confluence and hence functional behaviour. Since termination has been analyzed already in [4] the main aim of this paper is to analyze local confluence in the view of functional behaviour for model transformations based on general TGGs. Our new technique is implicitly based on our constructions in [4], where the "on-the-fly" construction uses source and forward rules, which can be generated automatically from the triple rules. In this paper, we introduce forward translation rules which combine the source and forward rules using additional translation attributes for keeping track of the source elements that have been translated already. The first main result of this paper shows that there is a bijective correspondence between model transformations based on source consistent forward sequences and those based on forward translation sequences. As shown in [11] the translation attributes can be separated from the source model in order to keep the source model unchanged.

In contrast to non-deleting triple rules, the corresponding forward translation rules are deleting and creating on the translation attributes. This means that some transformation steps can be parallel dependent. In this case we can apply the well-known critical pair analysis techniques to obtain local confluence. Since they are valid for all $\mathcal{M}$-adhesive systems (called weak adhesive HLR systems in [2]), they are also valid for typed attributed triple graph transformation systems. In fact, our model transformations based on forward translation rules can be considered as special case of the latter. However, we do not need general local confluence, because local confluence for transformations of all those source graphs, which do not belong to the source language, is not relevant for the functional bahaviour of a model transformation. In fact, we only analyze a weaker property, called translation confluence. This leads to our second main result, the functional behaviour of model transformations based on translation confluence. We have applied this technique for showing functional behaviour of our running example, the model transformation from class diagrams to database models, using our tool AGG [21] for critical pair analysis. Note that standard techniques are not applicable to show functional behaviour based on local confluence.

This paper is organized as follows: In Sec. 2 we review the basic notions of TGGs and model transformations based on forward rules. In Sec. 3 we introduce forward translation rules and characterize in our first main result model transformations in the TGG approach by forward translation sequences. In Sec. 4 we show in our second main result how functional behaviour of model transformations can be analyzed by translation confluence and we apply the technique to our running example. Related work and our conclusion - including a summary of our results and future work - is presented in Sections 5 and 6, respectively.

2   Review of Triple Graph Grammars

Triple graph grammars [18] are a well known approach for bidirectional model transformations. Models are defined as pairs of source and target graphs, which are connected via a correspondence graph together with its embeddings into these graphs. In [13], Königs and Schürr formalize the basic concepts of triple graph grammars in a set-theoretical way, which is generalized and extended by Ehrig et al. in [6] to typed, attributed graphs. In this section, we review main constructions and results of model transformations based on triple graph grammars [19,4].

A triple graph $G = (G_S \xleftarrow{s_G} G_C \xrightarrow{t_G} G_T)$ consists of three graphs G_S, G_C, and G_T, called source, correspondence, and target graphs, together with two graph morphisms $s_G : G_C \to G_S$ and $t_G : G_C \to G_T$. A triple graph morphism $m = (m_S, m_C, m_T) : G \to H$ consists of three graph morphisms $m_S : G_S \to H_S$, $m_C : G_C \to H_C$ and $m_T : G_T \to H_T$ such that $m_S \circ s_G = s_H \circ m_C$ and $m_T \circ t_G = t_H \circ m_C$. A typed triple graph G is typed over a triple graph TG by a triple graph morphism $type_G : G \to TG$.

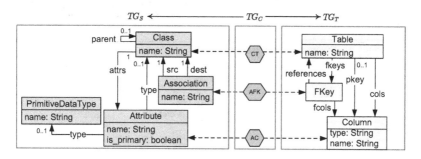

Fig. 1. Triple type graph for *CD2RDBM*

Example 1 (Triple Type Graph). Fig. 1 shows the type graph TG of the triple graph grammar TGG for our example model transformation *CD2RDBM* from class diagrams to database models. The source component TG_S defines the structure of class diagrams while in its target component the structure of relational database models is specified. Classes correspond to tables, attributes to columns, and associations to foreign keys. Throughout the example, originating from [6], elements are arranged left, center, and right according to the component types source, correspondence and target. Morphisms starting at a correspondence part are specified by dashed arrows. Furthermore, the triple rules of the grammar shown in Fig. 2 ensure several multiplicity constraints, which are denoted within the type graph. In addition, the source language $\mathcal{L}_S = CD$ contains only those class diagrams where classes have unique primary attributes and subclasses have no primary attributes to avoid possible confusion.

Note that the case study uses attributed triple graphs based on E-graphs as presented in [6] in the framework of $\mathcal{M}$-adhesive categories (called weak adhesive HLR in [2]).

Triple rules synchronously build up source and target graphs as well as their correspondence graphs, i.e. they are non-deleting. A triple rule tr is

$$L = (L_S \xleftarrow{s_L} L_C \xrightarrow{t_L} L_T) \qquad L \xhookrightarrow{tr} R$$
$$tr\downarrow \ tr_S\downarrow \qquad tr_C\downarrow \qquad tr_T\downarrow \qquad m\downarrow \ (PO) \ \downarrow n$$
$$R = (R_S \xleftarrow{s_R} R_C \xrightarrow{t_R} R_T) \qquad G \xhookrightarrow{t} H$$

an injective triple graph morphism $tr = (tr_S, tr_C, tr_T) : L \to R$ and w.l.o.g. we assume tr to be an inclusion. Given a triple graph morphism $m : L \to G$, a triple graph transformation (TGT) step $G \xrightarrow{tr,m} H$ from G to a triple graph H is given by a pushout of triple graphs with comatch $n : R \to H$ and transformation inclusion $t : G \hookrightarrow H$. A grammar $TGG = (TG, S, TR)$ consists of a triple type graph TG, a triple start graph S and a set TR of triple rules.

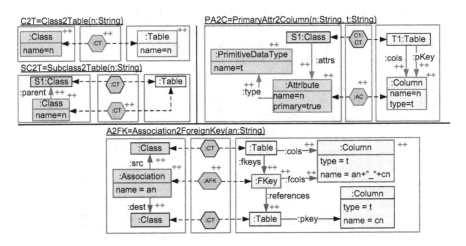

Fig. 2. Rules for the model transformation *Class2Table*

Example 2 (Triple Rules). The triple rules in Fig. 2 are part of the rules of the grammar TGG for the model transformation *CD2RDBM*. They are presented in short notation, i.e. left and right hand sides of a rule are depicted in one triple graph. Elements, which are created by the rule, are labeled with green "++" and marked by green line colouring. The rule "*Class2Table*" synchronously creates a class in a class diagram with its corresponding table in the relational database. Accordingly, subclasses are connected to the tables of its super classes by rule "*Subclass2Table*". Attributes are created together with their corresponding columns in the database component. The depicted rule "*Primary-Attr2Column*" concerns primary attributes with primitive data types for which an edge of type "pKey" is inserted that points to the column in the target component. This additional edge is not created for standard attributes, which are created by the rule *"Attr2Column"*, which is not depicted. Finally, the rule "Association2ForeignKey" creates associations between two classes together with their corresponding foreign keys and an additional column that specifies the relation between the involved tables.

$$
\begin{array}{ccc}
(L_S \longleftarrow \varnothing \longrightarrow \varnothing) & \qquad (R_S \xleftarrow{\ tr_S \circ s_L\ } L_C \xrightarrow{\ t_L\ } L_T) \\
tr_S \downarrow \qquad \downarrow \qquad \downarrow & \qquad id \downarrow \qquad tr_C \downarrow \qquad \downarrow tr_T \\
(R_S \longleftarrow \varnothing \longrightarrow \varnothing) & \qquad (R_S \xleftarrow{\ s_R\ } R_C \xrightarrow{\ t_R\ } R_T) \\
\text{source rule } tr_S & \qquad \text{forward rule } tr_F
\end{array}
$$

The operational rules for model transformations are automatically derived from the set of triple rules TR. From each triple rule tr we derive a forward rule tr_F for forward transformation sequences and a source rule tr_S for the construction resp. parsing of a model of the source language. By TR_S and TR_F we denote the sets of all source and forward rules derived from TR. Analogously, we derive a target rule tr_T and a backward rule tr_B for the construction and transformation of a model of the target language leading to the sets TR_T and TR_B.

A set of triple rules TR and the start graph $\varnothing$ generate a visual language VL of integrated models, i.e. models with elements in the source, target and correspondence component. The source language VL_S and target language VL_T are derived by projection to the triple components, i.e. $VL_S = proj_S(VL)$ and $VL_T = proj_T(VL)$. The set VL_{S0} of models that can be generated resp. parsed by the set of all source rules TR_S is possibly larger than VL_S and we have $VL_S \subseteq VL_{S0} = \{G_S \mid \varnothing \Rightarrow^* (G_S \leftarrow \varnothing \rightarrow \varnothing) \text{ via } TR_S\}$. Analogously, we have $VL_T \subseteq VL_{T0} = \{G_T \mid \varnothing \Rightarrow^* (\varnothing \leftarrow \varnothing \rightarrow G_T) \text{ via } TR_T\}$.

As introduced in [6,4] the derived operational rules provide the basis to define model transformations based on source consistent forward transformations $G_0 \Rightarrow^* G_n$ via $(tr_{1,F}, \ldots, tr_{n,F})$, short $G_0 \xRightarrow{tr_F^*} G_n$. A forward sequence $G_0 \xRightarrow{tr_F^*} G_n$ is source consistent, if there is a source sequence $\varnothing \xRightarrow{tr_S^*} G_0$ such that the sequence $\varnothing \xRightarrow{tr_S^*} G_0 \xRightarrow{tr_F^*} G_n$ is match consistent, i.e. the S-component of each match $m_{i,F}$ of $tr_{i,F}(i = 1 \ldots n)$ is uniquely determined by the comatch $n_{i,S}$ of $tr_{i,S}$, where $tr_{i,S}$ and $tr_{i,F}$ are source and forward rules of the same triple rules tr_i. Thus, source consistency is a control condition for the construction of the forward sequence.

Definition 1 (Model Transformation based on Forward Rules). *A model transformation sequence* $(G_S, G_0 \xRightarrow{tr_F^*} G_n, G_T)$ *consists of a source graph* G_S, *a target graph* G_T, *and a source consistent forward* TGT-*sequence* $G_0 \xRightarrow{tr_F^*} G_n$ *with* $G_S = G_{0,S}$ *and* $G_T = G_{n,T}$.

A model transformation $MT : VL_{S0} \Rightarrow VL_{T0}$ *is defined by all model transformation sequences* $(G_S, G_0 \xRightarrow{tr_F^*} G_n, G_T)$ *with* $G_S \in VL_{S0}$ *and* $G_T \in VL_{T0}$. *All the corresponding pairs* (G_S, G_T) *define the* model transformation relation *$MTR_F \subseteq VL_{S0} \times VL_{T0}$.*

In [6,4] we have proved that source consistency ensures completeness and correctness of model transformations based on forward rules with respect to the language *VL* of integrated models. Moreover, source consistency is the basis for the on-the-fly construction defined in [4].

3   Model Transformations Based on Forward Translation Rules

Model transformations as defined in the previous section are based on source consistent forward sequences. In order to analyze functional behaviour, we present in this section a characterizion by model transformations based on forward translation rules, which integrate the control condition source consistency using additional attributes (see Thm. 1). For each node, edge and attribute of a graph a new attribute is created and labeled with the prefix "*tr*". If this prefix is used already for an existing attribute, then a unique extended prefix is chosen.

Definition 2 (Graph with Translation Attributes). *Given an attributed graph $AG = (G, D)$ and a subgraph $G_0 \subseteq G$ we call AG' a graph with translation attributes over AG if it extends AG with one boolean-valued attribute tr_x for each element x (node or edge) in G_0 and one boolean-valued attribute tr_x_a for each attribute associated to such an element x in G_0. The set of all these additional translation attributes is denoted by Att_{G_0}. $Att_{G_0}^v$, where $v = \mathbf{T}$ or $v = \mathbf{F}$, denotes a translation graph where all the attributes in Att_{G_0} are set to v. By $AG' = AG \oplus Att_{G_0}$ we specify that AG is extended by the attributes in Att_{G_0}. Moreover, we define $Att^v(AG) := AG \oplus Att_G^v$.*

The extenstion of forward rules to forward translation rules ensures that the effective elements of the rule may only be matched to those elements that have not been translated so far. A first intuitive approach would be to use NACs on the correspondence component of the forward rule in order to check that the effective elements are unrelated. However, this approach is too restrictive, because e.g. edges and attributes in the source graph cannot be checked separately, but only via their attached nodes. Moreover, the analysis of functional behaviour of model transformations with NACs is general more complex compared to using boolean valued translation attributes instead. Thus, the new concept of forward translation rules extends the construction of forward rules by additional translation attributes, which keep track of the elements that have been translated at any point of the transformation process. This way, each element in the source graph cannot be translated twice, which is one of the main aspects of source consistency. For that reason, all translation attributes of the source model of a model transformation are set to false and in the terminal graph we expect that all the translation attributes are set to true. Moreover, also for that reason, the translation rules set to true all the elements of the source rule that would be generated by the corresponding source rule. This requires that the rules are deleting on the translation attributes and we extend a transformation step from a single (total) pushout to the classical double pushout (DPO) approach [2]. Thus, we can ensure source consistency by merely using attributes in order to completely translate a model. Therefore, we call these rules forward translation

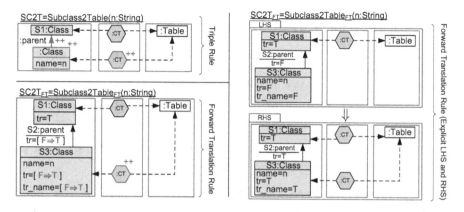

Fig. 3. Forward translation rule $Subclass2Table_{FT}(n : String)$

rules, while pure forward rules need to be controlled by the source consistency condition. Note that the extension of a forward rule to a forward translation rule is unique.

Definition 3 (Forward Translation Rule). *Given a triple rule* $tr = (L \to R)$, *the* forward translation rule *of tr is given by* $tr_{FT} = (L_{FT} \xleftarrow{l_{FT}} K_{FT} \xrightarrow{r_{FT}} R_{FT})$ *defined as follows using the forward rule* $(L_F \xrightarrow{tr_F} R_F)$ *and the source rule* $(L_S \xrightarrow{tr_S} R_S)$ *of tr, where we assume w.l.o.g. that tr is an inclusion:*

$$- K_{FT} = L_F \oplus Att_{L_S}^{\mathbf{T}},$$
$$- L_{FT} = L_F \oplus Att_{L_S}^{\mathbf{T}} \oplus Att_{R_S \setminus L_S}^{\mathbf{F}},$$
$$- R_{FT} = R_F \oplus Att_{L_S}^{\mathbf{T}} \oplus Att_{R_S \setminus L_S}^{\mathbf{T}} = R_F \oplus Att_{R_S}^{\mathbf{T}},$$
$$- l_{FT} \text{ and } r_{FT} \text{ are the induced inclusions.}$$

Example 3 (Derived Forward Translation Rules). Figure 3 shows the derived forward translation rule *"Subclass2Table $_{FT}$"* for the triple rule *"Subclass2Table"* in Fig. 2. Note that we abbreviate *"tr_x"* for an item (node or edge) x by *"tr"* and *"tr_x_a"* by *"tr_type(a)"* in the figures to increase readability. The compact notation of forward translation rules specifies the modification of translation attributes by "$[\mathbf{F} \Rightarrow \mathbf{T}]$", meaning that the attribute is matched with the value "$\mathbf{F}$" and set to "$\mathbf{T}$" during the transformation step.

From the application point of view model transformation rules should be applied along matches that do not identify structural elements. But it would be too restrictive to require injectivity of the matches also on the data part, because the matching should allow to match two different variables in the left hand side of a rule to the same data value in the host graph of a transformation step. This requirement applies to all model transformations based on abstract syntax graphs with attribution. For this reason we introduce the notion of almost injective matches, which requires that matches are injective except for the data value nodes. This way, attribute values can still be specified as terms within a rule and matched non-injectively to the same value.

Definition 4 (Almost Injective Match and Completeness). *An attributed triple graph morphism* $m : L \to G$ *is called* almost injective, *if it is non-injective at most for the set of variables and data values in* L_{FT}. *A forward translation sequence* $G_0 \xrightarrow{tr^*_{FT}} G_n$ *with almost injective matches is called* complete *if* G_n *is completely translated, i.e. all translation attributes of* G_n *are set to true (*"$\mathbf{T}$"*).*

Now we are able to show the equivalence of complete forward translation sequences with source consistent forward sequences.

Fact 1 (Complete Forward Translation Sequences) *Given a triple graph grammar* $TGG = (TG, \varnothing, TR)$ *and a triple graph* $G_0 = (G_S \leftarrow \varnothing \to \varnothing)$ *typed over* TG. *Let* $G'_0 = (Att^{\mathbf{F}}(G_S) \leftarrow \varnothing \to \varnothing)$. *Then, the following are equivalent for almost injective matches:*

1. $\exists$ *a source consistent* TGT-*sequence* $G_0 \xrightarrow{tr^*_F} G$ *via forward rules and* $G = (G_S \leftarrow G_C \to G_T)$.

2. $\exists$ *a complete* TGT-*sequence* $G'_0 \xrightarrow{tr^*_{FT}} G'$ *via forward translation rules and* $G' = (Att^{\mathbf{T}}(G_S) \leftarrow G_C \to G_T)$.

Proof (Sketch). Using Thm. 1 in [4] we know that in sequence 1 all matches are forward consistent as defined for the on-the-fly construction in [4]. This allows to show for each step $G_{i-1} \Rightarrow G_i$ starting with $i = 1$ that there is a transformation step $G_{i-1} \xrightarrow{tr_{i,F}} G_i$ iff there is a transformation step $G'_{i-1} \xrightarrow{tr_{i,FT}} G'_i$ leading to the complete equivalence of both sequences as shown in detail in [11]. □

Now, we define model transformations based on forward translation rules in the same way as for forward rules in Def. 1, where source consistency of the forward sequence is replaced by completeness of the forward translation sequence. Note that we can separate the translation attributes from the source model as shown in [11] in order to keep the source model unchanged.

Definition 5 (Model Transformation Based on Forward Translation Rules). *A* model transformation sequence $(G_S, G'_0 \xrightarrow{tr^*_{FT}} G'_n, G_T)$ *based on forward translation rules consists of a source graph* G_S, *a target graph* G_T, *and a complete* TGT-*sequence* $G'_0 \xrightarrow{tr^*_{FT}} G'_n$ *with almost injective matches,* $G'_0 = (Att^{\mathbf{F}}(G_S) \leftarrow \varnothing \to \varnothing)$ *and* $G'_n = (Att^{\mathbf{T}}(G_S) \leftarrow G_C \to G_T)$.
  A model transformation $MT : VL_{S0} \Rightarrow VL_{T0}$ *based on forward translation rules is defined by all model transformation sequences* $(G_S, G'_0 \xrightarrow{tr^*_{FT}} G'_n, G_T)$ *based on forward translation rules with* $G_S \in VL_{S0}$ *and* $G_T \in VL_{T0}$. *All these pairs* (G_S, G_T) *define the* model transformation relation $MTR_{FT} \subseteq VL_{S0} \times VL_{T0}$. *The model transformation is* terminating *if there are no infinite* TGT-*sequences via forward translation rules and almost injective matches starting with* $G'_0 = (Att^{\mathbf{F}}(G_S) \leftarrow \varnothing \to \varnothing)$ *for some source graph* G_S.

The main result of this section in Thm. 1 below states that model transformations based on forward translation rules are equivalent to those based on forward rules.

Theorem 1 (Equivalence of Model Transformation Concepts). *Given a triple graph grammar, then the model transformation* $MT_F : VL_{S0} \Rightarrow VL_{T0}$ *based on forward rules and the model transformation* $MT_{FT} : VL_{S0} \Rightarrow VL_{T0}$ *based on forward translation rules, both with almost injective matches, define the same model transformation relation* $MTR_F = MTR_{FT} \subseteq VL_{S0} \times VL_{T0}$.

Proof. The theorem follows directly from Def. 1, Def. 5 and Fact 1.   □

Remark 1. It can be shown that the model transformation relation MTR defined by the triple rules TR coincides with the relations MTR_F and MTR_{FT} of the model transformations based on forward and forward translation rules TR_F and TR_{FT}, respectively.

The equivalence of model transformations in Thm. 1 above directly implies Thm. 2 beneath, because we already have shown the results for model transformations based on forward rules in [4]. Note that the provided condition for termination is sufficient and in many cases also necessary. The condition is not necessary only for the case that there are some source identic triple rules, but none of them is applicable to any integrated model in the triple language VL.

Theorem 2 (Termination, Correctness and Completeness). *Each model transformation* $MT : VL_{S0} \Rightarrow VL_{T0}$ *based on forward translation rules is*

- terminating, *if each forward translation rule changes at least one translation attribute,*
- correct, *i.e. for each model transformation sequence* $(G_S, G'_0 \xrightarrow{tr^*_{FT}} G'_n, G_T)$ *there is* $G \in VL$ *with* $G = (G_S \leftarrow G_C \rightarrow G_T)$*, and it is*
- complete, *i.e. for each* $G_S \in VL_S$ *there is* $G = (G_S \leftarrow G_C \rightarrow G_T) \in VL$ *with a model transformation sequence* $(G_S, G'_0 \xrightarrow{tr^*_{FT}} G'_n, G_T)$.

Proof (Sketch). By Def. 3 we have that a rule changes the translation attributes iff the source rule of the original triple rule is creating, which is a sufficient criteria for termination by Thm. 3 in [4]. The correctness and completeness are based on Thm. 1 above and the proof of Thm. 3 in [3].   □

Example 4 (Model Transformation). Figure 4 shows a triple graph $G \in VL$. By Thm. 1 and Thm. 2 we can conclude that the class diagram G_S of the source language can be translated into the relation database model G_T by the application of the forward translation rules, i.e. there is a forward translation sequence $G_0 \xrightarrow{tr^*_{FT}} G_n$ starting at the source model with translation attributes $G_0 = (Att^{\mathbf{F}}(G_S) \leftarrow \varnothing \rightarrow \varnothing)$ and ending at a completely translated model $G_n = (Att^{\mathbf{T}}(G_S) \leftarrow G_C \rightarrow G_T)$. Furthermore, any other complete translation sequence leads to the same target model G_T (up to isomorphism). We show in Ex. 5 in Sec. 4 that the model transformation has this functional behaviour for each source model.

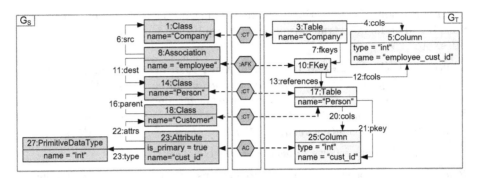

Fig. 4. Result of a model transformation after removing translation attributes

4   Analysis of Functional Behaviour

When a rewriting or transformation system describes some kind of computational process, it is often required that it shows a functional behaviour, i.e. every object can be transformed into a unique (terminal) object that cannot be transformed anymore. One way of ensuring this property is proving termination and confluence of the given transformation system. Moreover, if the system is ensured to be terminating, then it suffices to show local confluence according to Newman's Lemma [14]. However, an extension of the notion of critical pairs to encompass the additional control condition source consistency directly would be quite complex. Indeed, it would need to cover the interplay between a pair of forward steps and its corresponding pair of source steps at the same time including the relating morphism between them. As a consequence, an extension of the implemented critical pair analysis of AGG would probably be significantly less efficient, because of the generation of more possible overlappings.

We now show, how the generation and use of forward translation rules enables us to ensure termination and to integrate the control structure source consistency in the analysis of functional behaviour, such that we can apply the existing results [2] for showing local confluence of the transformation system leading to functional behaviour of the model transformation. The standard approach to check local confluence is to check the confluence of all *critical pairs* $(P_1 \Leftarrow K \Rightarrow P_2)$, which represent the minimal objects where a confluence conflict may occur. The technique is based on two results. On one hand, the *completeness* of critical pairs implies that every confluence conflict $(G_1 \Leftarrow G \Rightarrow G_2)$ embeds a critical pair $(P_1 \Leftarrow K \Rightarrow P_2)$. On the other hand, it is also based on the fact that the transformations $(P_1 \overset{*}{\Rightarrow} K' \overset{*}{\Leftarrow} P_2)$ obtained by confluence of the critical pair can be embedded into transformations $(G_1 \overset{*}{\Rightarrow} G' \overset{*}{\Leftarrow} G_2)$ that solve the original confluence conflict. However, as shown by Plump [16,17] confluence of critical pairs is not sufficient for this purpose, but a slightly stronger version, called strict confluence, which additionally requires that the preserved elements of the given steps are preserved in the merging steps. This result is also valid for typed attributed graph transformation systems [2] and we apply them to show functional behaviour of model transformations in the following sense, where the source language $\mathcal{L}_S$ may be a subset of VL_S derived from the triple rules.

Definition 6 (Functional Behaviour of Model Transformations). *A model transformation has* functional behaviour *if each model G_S of the source language $\mathcal{L}_S \subseteq VL_S$ is transformed into a unique terminal model G_T and, furthermore, G_T belongs to the target language VL_T.*

Model transformations based on forward translation rules are terminating, if each rule rewrites at least one translation attribute from "**F**" to "**T**". In contrast to that, termination of model transformations based on forward rules is ensuresed by an additional control structure – either source consistency in [4] or a controlling transformation algorithm as e.g. in [19]. A common alternative way of ensuring termination is the extension of rules by NACs that prevent an application at the same match. However, termination is only one aspect and does not ensure correctness and completeness of the model transformation. In particular, this means that matches must not overlap on effective elements, i.e. elements that are created by the source rule, because this would mean to translate these elements twice. But matches are allowed to overlap on other elements. Since the forward rules are identic on the source part there is no general way to prevent a partial overlapping of the matches by additional NACs and even nested application conditions [10] do not suffice. Nevertheless, in our case study *CD2RDBM* partial overlapping of matches can be prevented by NACs using the created correspondence nodes, but this is not possible for the general case with more complex rules.

Therefore, an analysis of functional behaviour based on the results for local confluence strictly depends on the generation of the system of forward translation rules. This means that, in principle, to prove functional behaviour of a model transformation, it is enough to prove local confluence of the forward translation rules. However, local confluence or confluence may be too strong to show functional behavior in our case. In particular, a model transformation system has a functional behavior if each source model, G_S, can be transformed into a unique target model, G_T. Or, more precisely, that $(Att^{\mathbf{F}}(G_S) \leftarrow \varnothing \rightarrow \varnothing)$ can be transformed into a unique completely translated graph $(Att^{\mathbf{T}}(G_S) \leftarrow G_C \rightarrow G_T)$. However, this does not preclude that it may be possible to transform $(Att^{\mathbf{F}}(G_S) \leftarrow \varnothing \rightarrow \varnothing)$ into some triple graph $(G'_S \leftarrow G'_C \rightarrow G'_T)$ where not all translation attributes in G'_S are set to true and no other forward translation rule is applicable. This means that, to show the functional behaviour of a set of forward translation rules, it is sufficient to use a weaker notion of confluence, called translation confluence.

Definition 7 (Translation Confluence). *Let TR_{FT} be a set of forward translation rules for the source language $\mathcal{L}_S \subseteq VL_S$. Then, TR_{FT} is translation confluent if for every triple graph $G = (Att^{\mathbf{F}}(G_S) \leftarrow \varnothing \rightarrow \varnothing)$ with $G_S \in \mathcal{L}_S \subseteq VL_{S0}$, we have that if $G \overset{*}{\Rightarrow} G_1$ and $G \overset{*}{\Rightarrow} G_2$ and moreover G_1 and G_2 are completely translated graphs, then the target components of G_1 and G_2 are isomorphic, i.e. $G_{1,T} \cong G_{2,T}$.*

The difference between confluence with terminal graphs and translation confluence is that, given $G_1 \overset{*}{\Leftarrow} G \overset{*}{\Rightarrow} G_2$, we only have to care about the confluence of these two transformations if both graphs, G_1 and G_2 can be transformed into completely translated graphs and furthermore, that they do not necessarily coincide on the correspondence part. This concept allows us to show the second main result of this paper in Thm. 3 that characterizes the analysis of functional behaviour of model transformations based on forward translation rules by the analysis of translation confluence, which is based on the analysis of critical pairs.

In Ex. 5 we will show that the set of forward translation rules of our model transformation *CD2RDBM* is translation confluent and hence, we have functional behaviour according to the following Thm. 3. In future work we will give sufficient conditions in order to ensure translation confluence, which will lead to a more efficient analysis technique for functional behaviour.

Theorem 3 (Functional Behaviour). *A model transformation based on forward translation rules has functional behaviour, iff the corresponding system of forward translation rules is translation confluent.*

Proof. "**if**": For $G_S \in \mathcal{L}_S \subseteq VL_S$, there is a transformation $\varnothing \xrightarrow{tr_S^*} (G_S \leftarrow \varnothing \rightarrow \varnothing) = G_0$ via source rules leading to a source consistent transformation $G_0 \xrightarrow{tr_F^*} G_n = (G_S \leftarrow G_C \rightarrow G_T)$ (see [4,6]). Using Fact 1, there is also a complete transformation $G_0' = (Att^{\mathbf{F}}(G_S) \leftarrow \varnothing \rightarrow \varnothing) \xrightarrow{tr_{FT}^*} (Att^{\mathbf{T}}(G_S) \leftarrow G_C \rightarrow G_T) = G_n'$ leading to $(G_S, G_T) \in MTR_{FT}$. For any other complete transformation via forward translation rules $\overline{tr}_{FT}^*$ we have $G_0' \xrightarrow{\overline{tr}_{FT}^*} (Att^{\mathbf{T}}(G_S) \leftarrow G_C' \rightarrow G_T')$. Translation confluence implies that $G_T \cong G_T'$, i.e. G_T is unique up to isomorphism.

"**only if**": For $G_S \in \mathcal{L}_S \subseteq VL_S$, suppose $G \overset{*}{\Rightarrow} G_1$ and $G \overset{*}{\Rightarrow} G_2$ with $G = (Att^{\mathbf{F}}(G_S) \leftarrow \varnothing \rightarrow \varnothing)$ and G_1, G_2 are completely translated. This means that $(G_S, G_{1,T}), (G_S, G_{2,T}) \in MTR_{FT}$, and the functional behaviour of the model transformation implies that $G_{1,T} \cong G_{2,T}$. □

The flattening construction presented in [3] for triple graph grammars enables us to use the tool AGG [21] for typed attributed graph grammars for generating and analyzing the critical pairs of the system of forward translation rules. The construction requires that the correspondence component TG_C of the type graph TG is discrete, i.e. has no edges. This condition is fulfilled for our case study and many others as well.

Using Thm. 1 we know that the system of forward translation rules has the same behaviour as the system of forward rules controlled by the source consistency condition. Therefore, it suffices to analyze the pure transformation system of forward translation rules without any additional control condition. This allows us furthermore, to transfer the analysis from a triple graph transformation system to a plain graph transformation system using Thm. 2 in [3], which states that there is a one-to-one correspondence between a triple graph transformation sequence and its flattened plain transformation sequence. Hence, we can analyze confluence, in particular critical pairs, of a set of triple rules by analyzing the corresponding set of flattened rules. This allows us to use the tool AGG for the generation of critical pairs for the flattened forward translation rules. The additional translation attributes increase the size of the triple rules by one attribute for each element in the source model. However, in order to improve scalability, they can be reduced by grouping those which always occur in the same combination within the rules.

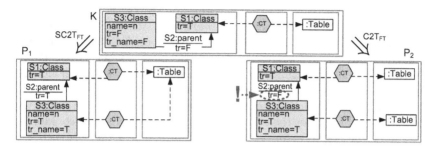

Fig. 5. Critical pair for the rules $Subclass2Table_{FT}$ and $Class2Table_{FT}$

Example 5 (Functional Behaviour). We show functional behaviour of the model transformation $CD2RDBM$, which is terminating, because all rules are source creating, but the system is not confluent w.r.t. terminal graphs. The tool AGG generates four critical pairs respecting the maximum multiplicity constraints according to Fig. 1. We explain why they can be neglected and refer to [11] for further details. The first two pairs can be neglected, because they refer to class diagrams containing a class with two primary attributes, which is not allowed for the source language $\mathcal{L}_S = CD$ (see Ex. 1). The two remaining critical pairs are identical – only the order is swapped. One pair is shown in Fig. 5. The edge "S2" in the graph P_2 is labeled with "**F**" but its source node is labeled with "**T**". Only the rule "$SC2T_{FT}$" can change the translation attribute of a "parent"-edge, but it requires that the source node is labeled with "**F**". Thus, no forward translation sequence where rule "$C2T_{FT}$" is applied to a source node of a parent edge, will lead to a completely translated graph. Assuming that our system is not translation confluent by two diverging complete forward translation sequences $s_1 = (G \overset{*}{\Rightarrow} G_1)$ and $s_2 = (G \overset{*}{\Rightarrow} G_2)$ leads to a contradiction. If the first diverging pair of steps in s_1 and s_2 is parallel dependent we can embed the critical pair and have that one sequence is incomplete, because the particular edge "S2" remains untranslated. Otherwise, we can merge them using the Local Church Rosser (LCR) Thm. leading to possibly two new diverging pairs of steps. If they are dependent we can embed the critical pair and conclude by LCR that the problematic step of the critical pair was also applied in the first diverging situation leading to incompleteness of s_1 or s_2. This procedure is repeated till the end of the sequences and we can conclude that there is no parallel dependent situation and by LCR we have that $G_1 \cong G_2$, because we have termination. The system is translation confluent and we can apply Thm. 3 to show the functional behaviour of the model transformation $CD2RDBM$.

5   Related Work

As pointed out in the introduction our work is based on triple graph grammars presented by Schürr et.el. in [19,18,13] with various applications in [8,9,12,13,20]. The formal approach to TGGs has been developed in [1,3,4,5,6,7]. In [6] it is shown how to analyze bi-directional model transformations based on TGGs with respect to information preservation, which is based on a decomposition and composition result for triple graph transformation sequences.

As shown in [1] and [7], the notion of *source consistency* ensures correctness and completeness of model transformations based on TGGs. A construction technique for correct and complete model transformation sequences *on-the-fly* is presented in [4], i.e. correctness and completeness properties of a model transformation do not need to be analyzed after completion, but are ensured by construction. In this construction, source consistency is checked on-the-fly, which means during and not after the construction of the forward sequence. Moreover, a strong sufficient condition for termination is given. The main construction and results are used for the proof of Fact 1 and hence, also for our first main result in Thm. 1. Similarly to the generated forward translation rules in this paper, the generated operational rules in [15] also do not need an additional control condition. However, the notion of correctness and completeness is much more relaxed, because it is not based on a given triple graph grammar, but according to a pattern specification, from which usually many triple rules are generated.

A first approach to analyze functional behaviour for model transformations based on TGGs was already given in [5] for triple rules with distinguished kernel typing. This strong restriction requires e.g. that there is no pair of triple rules handling the same source node type - which is, however, not the case for the first two rules in our case study *CD2RDBM*. The close relationship between model transformations based on TGGs and those on "plain graph transformations" is discussed in [3], but without considering the special control condition source consistency. The treatment of source consistency based on translation attributes is one contribution of this paper in order to analyze functional behaviour. As explained in Sec. 3 additional NACs are not sufficient to obtain this result. Functional behaviour for a case study on model transformations based on "plain graphs" is already studied in [2] using also critical pair analysis in order to show local confluence. But the additional main advantage of our TGG-approach in this paper is that we can transfer the strong results concerning termination, correctness and completeness from previous TGG-papers [3,4] based on source consistency to our approach in Thm. 2 by integrating the control structure source consistency in the analysis of functional behaviour. Finally there is a strong relationship with the model transformation algorithm in [19], which provides a control mechanism for model transformations based on TGGs by keeping track of the elements that are translated so far. In [4] we formalized the notion of elements that are translated at a current step by so-called effective elements. In this paper we have shown that the new translation attributes can be used to automatically keep track of the elements that have been translated so far.

6   Conclusion

In this paper we have analyzed under which conditions a model transformation based on triple graph grammars (TGGs) has functional behaviour. For this purpose, we have shown how to generate automatically forward translation rules from a given set of triple rules, such that model transformations can be defined equivalently by complete forward translation sequences. The main result shows that a terminating model transformation has functional behaviour if the set of forward translation rules is translation confluent. This allows to apply the well-known critical pair analysis techniques for typed attributed graph transformations with support from the tool AGG to the system of forward translation rules, which was not possible before, because the control condition source consistency could not be integrated in the analysis. These techniques have been applied to show functional behaviour of our running example, the model transformation from class diagrams to data base models. In order to keep the source model unchanged during the transformation the translation attributes can be separated from the source model as presented in [11]. Alternatively, the model transformation can be executed using the on-the-fly construction in [4], which is shown to be equivalent by Thm. 1. In future work we give sufficient conditions in order to check translation confluence, which will further improve the analysis techniques. Moreover, we will extend the results to systems with control structures like negative application conditions (NACs), rule layering and amalgamation. In order to extend the main result concerning functional behaviour to the case with NACs, we have to extend the generation of forward translation rules by extending the NACs with translation attributes and we have to prove the equivalence of the resulting model transformation with the on-the-fly construction in [7].

References

1. Ehrig, H., Ehrig, K., Hermann, F.: From Model Transformation to Model Integration based on the Algebraic Approach to Triple Graph Grammars. In: Ermel, C., de Lara, J., Heckel, R. (eds.) Proc. GT-VMT 2008, EC-EASST, EASST, vol. 10 (2008)
2. Ehrig, H., Ehrig, K., Prange, U., Taentzer, G.: Fundamentals of Algebraic Graph Transformation. EATCS Monographs. Springer, Heidelberg (2006)
3. Ehrig, H., Ermel, C., Hermann, F.: On the Relationship of Model Transformations Based on Triple and Plain Graph Grammars. In: Karsai, G., Taentzer, G. (eds.) Proc. GraMoT 2008. ACM, New York (2008)
4. Ehrig, H., Ermel, C., Hermann, F., Prange, U.: On-the-Fly Construction, Correctness and Completeness of Model Transformations based on Triple Graph Grammars. In: Schürr, A., Selic, B. (eds.) MODELS 2009. LNCS, vol. 5795, pp. 241–255. Springer, Heidelberg (2009)
5. Ehrig, H., Prange, U.: Formal Analysis of Model Transformations Based on Triple Graph Rules with Kernels. In: Ehrig, H., Heckel, R., Rozenberg, G., Taentzer, G. (eds.) ICGT 2008. LNCS, vol. 5214, pp. 178–193. Springer, Heidelberg (2008)
6. Ehrig, H., Ehrig, K., Ermel, C., Hermann, F., Taentzer, G.: Information preserving bidirectional model transformations. In: Dwyer, M.B., Lopes, A. (eds.) FASE 2007. LNCS, vol. 4422, pp. 72–86. Springer, Heidelberg (2007)

7. Ehrig, H., Hermann, F., Sartorius, C.: Completeness and Correctness of Model Transformations based on Triple Graph Grammars with Negative Application Conditions. In: Heckel, R., Boronat, A. (eds.) Proc. GT-VMT 2009, EC-EASST, EASST, vol. 18 (2009)

8. Guerra, E., de Lara, J.: Attributed typed triple graph transformation with inheritance in the double pushout approach. Tech. Rep. UC3M-TR-CS-2006-00, Universidad Carlos III, Madrid, Spain (2006)

9. Guerra, E., de Lara, J.: Model view management with triple graph grammars. In: Corradini, A., Ehrig, H., Montanari, U., Ribeiro, L., Rozenberg, G. (eds.) ICGT 2006. LNCS, vol. 4178, pp. 351–366. Springer, Heidelberg (2006)

10. Habel, A., Pennemann, K.H.: Correctness of high-level transformation systems relative to nested conditions. Mathematical Structures in Computer Science 19, 1–52 (2009)

11. Hermann, F., Ehrig, H., Golas, U., Orejas, F.: Formal Analysis of Functional Behaviour for Model Transformations Based on Triple Graph Grammars - Extended Version. Tech. Rep. TR 2010-8, TU Berlin (2010),
http://www.eecs.tu-berlin.de/menue/forschung/forschungsberichte/2010

12. Kindler, E., Wagner, R.: Triple graph grammars: Concepts, extensions, implementations, and application scenarios. Tech. Rep. TR-ri-07-284, Department of Computer Science, University of Paderborn, Germany (2007)

13. Königs, A., Schürr, A.: Tool Integration with Triple Graph Grammars - A Survey. In: Proc. SegraVis School on Foundations of Visual Modelling Techniques. ENTCS, vol. 148, pp. 113–150. Elsevier Science, Amsterdam (2006)

14. Newman, M.H.A.: On theories with a combinatorial definition of "equivalence". Annals of Mathematics 43(2), 223–243 (1942)

15. Orejas, F., Guerra, E., de Lara, J., Ehrig, H.: Correctness, completeness and termination of pattern-based model-to-model transformation. In: Kurz, A., Lenisa, M., Tarlecki, A. (eds.) CALCO 2009. LNCS, vol. 5728, pp. 383–397. Springer, Heidelberg (2009)

16. Plump, D.: Hypergraph rewriting: Critical pairs and undecidability of confluence. In: Term Graph Rewriting: Theory and Practice, pp. 201–213. John Wiley, Chichester (1993)

17. Plump, D.: Confluence of graph transformation revisited. In: Middeldorp, A., van Oostrom, V., van Raamsdonk, F., de Vrijer, R. (eds.) Processes, Terms and Cycles: Steps on the Road to Infinity. LNCS, vol. 3838, pp. 280–308. Springer, Heidelberg (2005)

18. Schürr, A.: Specification of Graph Translators with Triple Graph Grammars. In: Mayr, E.W., Schmidt, G., Tinhofer, G. (eds.) WG 1994. LNCS, vol. 903, pp. 151–163. Springer, Heidelberg (1995)

19. Schürr, A., Klar, F.: 15 years of triple graph grammars. In: Ehrig, H., Heckel, R., Rozenberg, G., Taentzer, G. (eds.) ICGT 2008. LNCS, vol. 5214, pp. 411–425. Springer, Heidelberg (2008)

20. Taentzer, G., Ehrig, K., Guerra, E., de Lara, J., Lengyel, L., Levendovsky, T., Prange, U., Varro, D., Varro-Gyapay, S.: Model Transformation by Graph Transformation: A Comparative Study. In: Proc. MoDELS 2005 Workshop MTiP 2005 (2005)

21. TFS-Group, TU Berlin: AGG (2009), http://tfs.cs.tu-berlin.de/agg

Conflict Detection for Model Versioning Based on Graph Modifications

Gabriele Taentzer[1], Claudia Ermel[2],
Philip Langer[3], and Manuel Wimmer[4]

[1] Philipps-Universität Marburg, Germany
taentzer@mathematik.uni-marburg.de
[2] Technische Universität Berlin, Germany
claudia.ermel@tu-berlin.de
[3] Johannes-Kepler-Universität Linz, Austria
philip.langer@jku.at
[4] Technische Universität Wien, Austria
wimmer@big.tuwien.ac.at

Abstract. In model-driven engineering, models are primary artifacts and can evolve heavily during their life cycle. Therefore, versioning of models is a key technique which has to be offered by an integrated development environment for model-driven engineering. In contrast to text-based versioning systems we present an approach which takes abstract syntax structures in model states and operational features into account. Considering the abstract syntax of models as graphs, we define model revisions as graph modifications which are not necessarily rule-based. Building up on the DPO approach to graph transformations, we define two different kinds of conflict detection: (1) the check for operation-based conflicts, and (2) the check for state-based conflicts on results of merged graph modifications.

1 Introduction

A key benefit of model-driven engineering is the management of the complexity of modern systems by abstracting its compelling details using models. Like source code, models may heavily evolve during their life cycle and, therefore, they have to be put under version control. Especially optimistic versioning is of particular importance because it allows for concurrent modifications of one and the same artifact performed by multiple modelers at the same time. When concurrent modifications are endorsed, contradicting and inconsistent changes, and therewith versioning conflicts, might occur. Traditional version control systems for code usually work on file-level and perform conflict detection by line-oriented text comparison. When applied to the textual serialization of models, the result is unsatisfactory because the information stemming from the graph-based structure is destroyed and associated syntactic and semantic information is lost.

To tackle this problem, dedicated model versioning systems have been proposed [1,2,3,4]. However, a uniform and effective approach for precise conflict

H. Ehrig et al. (Eds.): ICGT 2010, LNCS 6372, pp. 171–186, 2010.
© Springer-Verlag Berlin Heidelberg 2010

detection and supportive conflict resolution in model versioning still remains an open problem. For the successful establishment of dedicated model versioning systems, a profound understanding by means of formal definitions of potentially occurring kinds of conflicts is indispensable, but yet missing.

Therefore, we present a formalization of two different kinds of conflicts based on graph modifications. We introduce this new notion of graph modifications to generalize graph transformations. Graph modifications are not necessarily rule-based, but just describe changes in graphs. Two kinds of conflict detection are defined based on graph modifications: (1) operation-based conflicts and (2) state-based conflicts. The specification of operations is based on rules. Therefore, we extract minimal rules from graph modifications and/or select suitable pre-defined operations and construct graph transformations in that way. Conflict detection is then based on parallel dependence of graph transformations and the extraction of critical pairs as presented in [5]. State-based conflicts are concerned with the well-formedness of the result after merging graph modifications. To detect state-based conflicts, two graph modifications are merged and the result is checked against pre-defined constraints. The proposed critical pair extraction is not complex and corresponds to the procedure of textual versioning systems, i.e. they are applied whenever graph modifications have occurred and are being checked in. For each check-in, all those modifications (check-ins) are checked for conflicts that have taken place since the check-out of the model performed by the user who is performing the current check-in.

The paper is structured as follows: In Section 2, we introduce the concept of graph modification and recall the notion of graph transformation. While Section 3 is concerned with the detection of operation-based conflicts, Section 4 presents the detection of state-based conflicts. Sections 5 and 6 discuss implementation issues and related work and finally, Section 7 concludes this paper.

2   Graph Modifications and Graph Transformations

Throughout this paper, we describe the underlying structure of a model by a graph. To capture all important information, we use typed, attributed graphs and graph morphisms as presented in [5]. In the following, we omit the terms "typed" and "attributed" when mentioning graphs and graph morphisms.

All model modifications are considered on the level of the abstract syntax where we deal with graphs. We introduce graph modifications generalizing the concept of graph transformation to graph changes which are not necessarily rule-based. A graph modification is a partial injective graph mapping being defined by a span of injective graph morphisms.

Definition 1 (Graph modification). *Given two graphs G and H, a direct graph modification $G \Longrightarrow H$ is a span of injective morphisms $G \xleftarrow{g} D \xrightarrow{h} H$. A sequence $G = G_0 \Longrightarrow G_1 \Longrightarrow \ldots \Longrightarrow G_n = H$ of direct graph modifications is called* graph modification *and is denoted by $G \overset{*}{\Longrightarrow} H$.*

Graph D characterizes an intermediate graph where all deletion actions have been performed but nothing has been added yet.

Example 1 (Graph and graph revision). Consider the following model versioning scenario for statecharts. The abstract syntax of a statechart can be defined by a typed, attributed graph, as e.g. shown in Fig. 1 (a). The node type is given in the top compartment of a node. The name of each node of type State is written in the attribute compartment below the type name. We model hierarchical statecharts by using containment edges. For instance, in Fig. 1 (a), state S0 contains S1 and S2 as substates. (In contrast to UML state machines, we distinguish these edges which present containment links by composition decorators.) Note that we abstract from transition events, guards and actions, as well as from other statechart features. Furthermore, from now on we use a compact notation of the abstract syntax of statecharts, where we draw states as nodes (rounded rectangles with their names inside) and transitions as directed arcs between state nodes. The compact notation of the statechart in Fig. 1 (a) is shown in Fig. 1 (b). Fig. 1 (c) shows the statechart in the well-known concrete syntax.

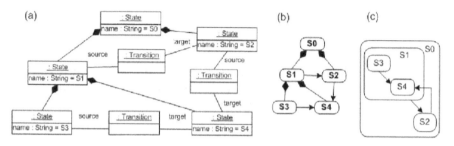

Fig. 1. Sample statechart: abstract syntax graph (a), compact notation (b) and concrete syntax (c)

In our scenario for model versioning, three users check out the current statechart shown in Fig. 1 and change it in three different ways. User A intends to perform a refactoring operation on it. She moves state S3 up in the state hierarchy (cf. Fig. 2).

User B refines the statechart by adding a new state S5 inside superstate S0 and connects this newly added state S5 to state S2 in the same superstate by drawing a new transition between them. Moreover, the transition connecting S2 to S4 is deleted in this refinement step. This graph modification is shown in Fig. 3. Finally, user C deletes state S3 together with its adjacent transition to state S4 (cf. Fig. 4).

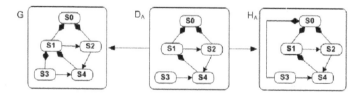

Fig. 2. Refactoring step as graph modification gm_A

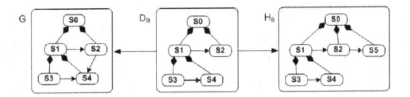

Fig. 3. Refinement step as graph modification gm_B

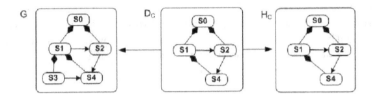

Fig. 4. Deletion step as graph modification gm_C

Obviously, conflicts occur when these users try to check in their changes: state S3 is deleted by user C but is moved to another container by user A. Furthermore, user B and user C delete different transitions adjacent to state S4. This may lead to a problem if the statechart language forbids isolated states (not adjacent to any transition), although each single change does not create a forbidden situation.

Since we will use rules to detect conflicts between graph modifications, we recall the notions of graph rule and transformation here. We use the DPO approach for our work, since its comprehensive theory as presented in [5] is especially useful to formalize conflict detection in model versioning.

Definition 2 (Graph rule and transformation). *A graph rule* $p = L \xleftarrow{l} K \xrightarrow{r} R$ *consists of graphs* L, K *and* R *and injective graph morphisms* l *and* r. *Given a match* $m : L \to G$, *graph rule* p *can be applied to* G *if a double-pushout (DPO) exists as shown in the diagram below with pushouts* (PO_1) *and* (PO_2) *in the category of typed, attributed graphs. Graph* D *is the intermediate graph after removing* $m(L)$, *and* H *is constructed as gluing of* D *and* R *along* K *(see* *[5]).* $G \stackrel{p,m}{\Longrightarrow} H$ *is called* graph transformation.

$$
\begin{array}{ccccc}
L & \xleftarrow{\ l\ } & K & \xrightarrow{\ r\ } & R \\
{\scriptstyle m}\downarrow & (PO_1) & \downarrow{\scriptstyle k} & (PO_2) & \downarrow{\scriptstyle m'} \\
G & \xleftarrow{\ g\ } & D & \xrightarrow{\ h\ } & H
\end{array}
$$

Obviously, each graph transformation can be considered as graph modification by forgetting about the rule and its match. If the rule and its match are given, the pushout (PO_1) has to be constructed as pushout complement. We recall the definition of a pushout complement as presented in [5]. From an operational point of view, a pushout complement determines the part in graph G that does not come from L, but includes K.

Definition 3 (PO complement). *Given morphisms* $l\colon K \to L$ *and* $m\colon L \to G$, *then* $k\colon K \to D$ *and* $g\colon D \to G$ *is the pushout complement (POC) of* l *and* m, *if* (PO_1) *in Def. 2 is a pushout.*

3   Detection of Operation-Based Conflicts

For the detection of operation-based conflicts, we have to find out the operation resulting in a particular graph modification. Thus, we have to find a corresponding rule and match, i.e. a corresponding graph transformation to the given graph modification. An approach which is always possible is to extract a minimal rule [6], i.e. a minimal operation which contains all atomic actions (i.e. creation and deletion of nodes, edges, and attribute values) that are performed by the given graph modification. Thus, the extraction of a minimal rule together with its match leads to a minimal graph transformation performing a given graph modification.

If the operation which led to a graph modification is not known but can be specified by a graph rule, a suitable method to identify the right operation is to extract again the minimal rule and to find the corresponding operation (out of a set of pre-defined operations) by comparing it with the minimal rule.

After having specified graph modifications by graph transformations, the parallel independence of transformations can be checked and critical situations are identified as conflicts. These conflicts can be specified as critical pairs [5].

3.1   Extraction of a Minimal Rule

As first step, we use the construction of minimal rules by Bisztray et al. [6]. This construction yields in a natural way a minimal DPO rule for a graph modification. Minimal rules contain the proper atomic actions on graphs with minimal contexts. Bisztray et al. have shown that this construction is unique, i.e. no smaller rule than the minimal rule can be constructed for a given graph modification.

Definition 4 (Minimal graph rule and transformation). *Rule* $p = (L \xleftarrow{l} K \xrightarrow{r} R$ *is* minimal *over direct graph modification* $G \xleftarrow{g} D \xrightarrow{h} H$ *if for each rule* $L' \xleftarrow{l'} K' \xrightarrow{r'} R'$ *with injective morphism* $K' \to D$ *and pushouts (3) and (4), there are unique morphisms* $L \to L'$, $K \to K'$, *and* $R \to R'$ *such that the following diagram commutes and (1), (2), (1) + (3), and (2) + (4) are pushouts. Graph transformation* $G \overset{p}{\Longrightarrow} H$ *is also called* minimal.

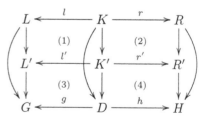

The following minimal rule construction extracts all deletion and creation actions from a given transformation in graphs L_1 and R_1 by constructing so-called initial pushouts. Roughly speaking, an initial pushout extracts a graph morphism consisting of the changing part of the given graph morphism, i.e. the non-injective mapping part as well as the codomain part that is not in the image of the morphism (see [5]). This is done for both sides of a graph modification, leading to the left and the right-hand sides of the minimal rule. In the middle, two gluing graphs are constructed which have to be glued together, and the left and the right-hand sides are potentially extended by further necessary context.

Definition 5 (Initial pushout). *Let* $g: D \to G$ *be a graph morphism, an initial pushout* *over* g *consists of graph morphisms* $l_1: L_1 \to G$, $b_1: B_1 \to L_1$, *and* $d_1: B_1 \to D$ *(cf. diagram below) such that* g *and* l_1 *are a pushout over* b_1 *and* d_1. *For every other pushout over* g *consisting of* $l'_1: L'_1 \to G$, $b'_1: B'_1 \to L'_1$, *and* $d'_1: B'_1 \to D$, *there are unique graph morphisms* $b: B_1 \to B'_1$ *and* $l: L_1 \to L'_1$ *such that* $l'_1 \circ l = l_1$ *and* $b'_1 \circ b = b_1$. *Moreover,* (l, b'_1) *is a pushout over* (b_1, b).

Definition 6 (Minimal rule construction).
Given a direct graph modification $G \xleftarrow{g} D \xrightarrow{h} H$, *we construct a rule* $L \xleftarrow{l} K \xrightarrow{r} R$ *which shall be minimal.*

$$L_1 \xleftarrow{b_1} B_1 \qquad\qquad B_2 \longrightarrow R_1$$
$$l_1 \downarrow \;(IPO_1) \qquad\qquad (IPO_2) \downarrow$$
$$G \xleftarrow{\quad g \quad} D \xrightarrow{\quad h \quad} H$$

1. *Construct the initial pushouts* (IPO_1) *over* $g: D \to G$ *and* (IPO_2) *over* $h: D \to H$.
2. *Define* $B_1 \longleftarrow P \longrightarrow B_2$ *as pullback of* $B_1 \longrightarrow D \longleftarrow B_2$ *and* $B_1 \longrightarrow K \longleftarrow B_2$ *as pushout* (PO_4) *of* $B_1 \longleftarrow P \longrightarrow B_2$ *with induced morphism* $K \to D$.
3. *Construct* $L_1 \longrightarrow L \longleftarrow K$ *as pushout* (PO_3) *of* $L_1 \longleftarrow B_1 \longrightarrow K$ *with induced morphism* $L \to G$. *Similarly, construct* $R_1 \longrightarrow R \longleftarrow K$ *as pushout* (PO_5) *of* $R_1 \longleftarrow B_2 \longrightarrow K$ *with induced morphism* $R \to H$.
4. *Since* (IPO_1) *and* (PO_3) *are pushouts,* $(IPO_1) + (PO_3)$ *is also a pushout, due to pushout composition properties [5]. Similarly, since* (IPO_2) *and* (PO_5) *are pushouts,* $(IPO_2) + (PO_5)$ *is also a pushout.*

Proposition 1 (Minimal rule). *Given a graph modification* $G \xleftarrow{g} D \xrightarrow{h} H$, *rule* $L \xleftarrow{l} K \xrightarrow{r} R$ *constructed as in Definition 6, is minimal.*

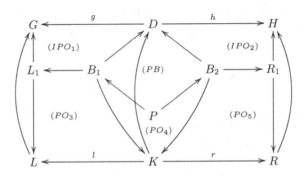

Fig. 5. Minimal rule construction

The proof of Prop. 1 is given in [6]. Note that after Step 1, the minimal rule extraction may also be considered as E-concurrent rule constructed from a deletion rule on the left and a creation rule on the right.

Example 2 (Minimal rule construction). The construction of the minimal rule p_A for graph modification gm_A (the refactoring performed by user A in Fig. 2) is depicted in Fig. 6. Note that the minimal rule does not contain any attributes, since they are not changed within the graph modification. The minimal rules p_B for gm_B (the refinement performed by user B in Fig. 3) and p_C for gm_C (the deletion by user C in Fig. 4) are constructed analogously. The results are depicted in Figs. 7 and 8. Note that the initial pushout construction leads to variables as attribute values of deleted and newly created nodes.

Comparing the applications of minimal rules p_A and p_C, we see that minimal rule p_C deletes state S3 that is used by minimal rule p_A to perform its refactoring.

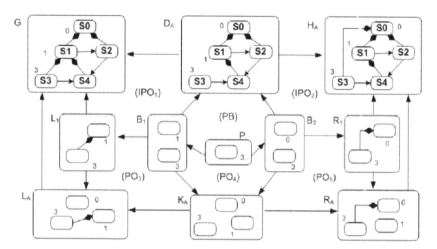

Fig. 6. Construction of minimal rule $p_A = (L_A \leftarrow K_A \rightarrow R_A)$ for gm_A

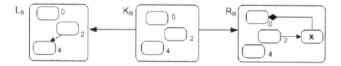

Fig. 7. Minimal rule p_B for graph modification gm_B

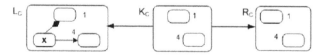

Fig. 8. Minimal rule p_C for graph modification gm_C

Such conflicts are called Delete/Use-conflicts and are defined below. They can be automatically detected using the graph transformation tool AGG [7].

3.2   Identification of Operations

Up to now, we investigated the actual actions performed by different users which we extracted in minimal rules. This approach does not take any predefined operations into account. Given a set of change operations defined by graph rules, we can identify the right operation that has been performed for a graph modification gm by the following method: we extract again the minimal rule for gm and find the corresponding operation (out of a set of pre-defined operations) by comparing with the minimal rule. An operation o is *executable* wrt. a given graph modification $G \longleftarrow D \longrightarrow H$ if the extracted minimal rule is a subrule of operation o and o is applicable to the original graph G in a compatible way.

Definition 7 (Subrule). *A minimal rule* $L_s \xleftarrow{l_s} K_s \xrightarrow{r_s} R_s$ *is a subrule of rule* $L \xleftarrow{l} K \xrightarrow{r} R$ *if morphisms* $i_l\colon L_s \to L$, $i_k\colon K_s \to K$, *and* $i_r\colon R_s \to R$ *exist and the diagram on the right commutes and* (PO_1) *and* (PO_2) *are pushouts.*

$$
\begin{array}{ccccc}
L_s & \xleftarrow{l_s} & K_s & \xrightarrow{r_s} & R_s \\
\downarrow{i_l} & (PO_1) & \downarrow{i_k} & (PO_2) & \downarrow{i_r} \\
L & \xleftarrow{l} & K & \xrightarrow{r} & R
\end{array}
$$

Definition 8 (Minimal rule-related operation execution). *Given a minimal rule* $s = L_s \xleftarrow{l_s} K_s \xrightarrow{r_s} R_s$ *applicable to graph* G *by match* $m_s\colon L_s \to G$, *an operation given by rule* $o = L \xleftarrow{l} K \xrightarrow{r} R$ *is executable on graph* G *wrt. rule* s *if* s *is a subrule of* o *(as defined in Def. 7) and if* o *is applicable to* G *by a match* $m\colon L \to G$ *such that* $m \circ i_l = m_s$ *with* $i_l\colon L_s \to L$.

Note that Def. 8 is useful for identifying single operations per minimal rule. It cannot be used to identify sequences of operations which relate to one minimal rule. This problem is left to future work.

Example 3. We define an operation enabling the user to move a state s to another container only if the new container state contains the previous container state of s. This avoids producing cyclic containments. The operation rule o_A used for graph modification gm_A has more context than its minimal rule (cf. Fig. 9 for the relation between these two rules where the minimal rule p_A is shown in the upper half and the operation rule o_A in the lower half).

Once the executed operations have been identified, we can start the conflict detection also for these rules. Since they can come with more context, more conflicts might occur compared to the conflict detection based on minimal rules.

3.3   Detection of Operation-Based Conflicts

After having constructed graph transformations from graph modifications by minimal rule extraction and/or operation rule selection, we can use critical pairs as presented in [5] to define operation-based conflicts.

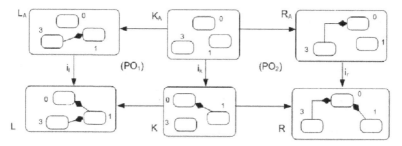

Fig. 9. *MoveState* operation

First we check if two graph transformations are parallel independent which means that rule matches overlap in preserved items only.

Definition 9 (Parallel independent transformations). *Two direct graph transformations* $G \overset{p_1,m_1}{\Longrightarrow} H_1$ *and* $G \overset{p_2,m_2}{\Longrightarrow} H_2$ *being applications of rules* $p_1 = (L_1 \overset{l_1}{\longleftarrow} K_1 \overset{r_1}{\longrightarrow} R_1)$ *and* $p_2 = (L_2 \overset{l_2}{\longleftarrow} K_2 \overset{r_2}{\longrightarrow} R_2)$ *at matches* $m_1 : L_1 \rightarrow G$ *and* $m_2 : L_2 \rightarrow G$ *are* parallel independent *if the transformations preserve all items in the intersection of both matches, i.e.* $m_1(L_1) \cap m_2(L_2) \subseteq m_1(l_1(K_1)) \cap m_2(l_2(K_2))$.

To concentrate on the proper conflict, we abstract from unnecessary context of identified parallel dependent transformations. This leads to the notion of a *critical pair* which is defined below. A critical pair consists of two parallel dependent transformations starting from a smaller graph K now. K can be considered as a suitable gluing of left-hand sides L_1 and L_2 of corresponding rules. For each two parallel dependent transformations $G \overset{p_1,m_1}{\Longrightarrow} H_1$ and $G \overset{p_2,m_2}{\Longrightarrow} H_2$ a corresponding critical pair can be found. We consider critical pairs as operation-based conflicts.

Definition 10 (Critical pair). *A* critical pair *consists of two parallel dependent graph transformations* $K \overset{p_1,o_1}{\Longrightarrow} P_1$ *and* $K \overset{p_2,o_2}{\Longrightarrow} P_2$ *with matches* $o_1 : L_1 \rightarrow K$ *and* $o_2 : L_2 \rightarrow K$ *being jointly surjective.*

Proposition 2 (Completeness of critical pairs). *Given two direct graph transformations* $G \overset{p_1,m_1}{\Longrightarrow} H_1$ *and* $G \overset{p_2,m_2}{\Longrightarrow} H_2$ *which are parallel dependent, there is a critical pair* $K \overset{p_1,o_1}{\Longrightarrow} P_1$ *and* $K \overset{p_2,o_2}{\Longrightarrow} P_2$ *such that there is a graph morphism* $o : K \rightarrow G$ *with* $o \circ o_1 = m_1$ *and* $o \circ o_2 = m_2$.

The proof can be found in [5]. Note that this proposition is very useful to find operation-based conflicts. If the pair of transformations considered is parallel dependent, we get a critical pair, i.e. an operation-based conflict. By proposition 2 we know that we get all operation-based conflicts that way.

This formalization enables us to detect so-called *Delete/Use conflicts:* one transformation deletes a graph item while the other one reads it. Note that a particular kind of delete/use conflicts are sometimes called *Change/Use conflicts.* Here, the first transformation changes the value of an attribute, while the second one either changes it too, or just checks its value. Since attribute value bindings are modeled by edges, attribute value changes involve the deletion of edges (cf. [5]).

In case we find critical pairs, we can identify the conflict in form of the minimal context of the critical match overlappings. Note that since we have given the overlapping of the left-hand sides of the minimal rules already, we need to check only this overlapping situation for a conflict. This procedure needs much less effort than the normal critical pair analysis in AGG which computes all possible contexts.

Example 4. Considering the minimal rules p_A and p_C in Figs. 6 and 8 applied to graph G in Figs. 2 - 4 with the obvious matches, we get a Delete/Use-conflict based on deletion and usage of state S3.

4   Detection of State-Based Conflicts

Besides *operation-based* conflicts, we want to detect *state-based* conflicts which can occur in merged modification results. These conflicts occur if a merged modification result shows some abnormality not present in the modification results before merging. Detection of state-based conflicts is done by constraint checking. The constraints may be language-specific, i.e. potentially induced by the corresponding graph language definition.

In the following, we present a procedure to merge two different graph modifications to find state-based conflicts. We show that this merge procedure yields the expected result if there are no operation-based conflicts. Otherwise, the procedure cannot be performed. Thus, a natural ordering of conflict detection is

1. to extract minimal rules and analyze minimal transformations for operation-based conflicts,
2. to find further operation-based conflicts by analyzing operations,
3. to check for state-based conflicts after all operation-based conflicts have been resolved.

4.1   Merging of Graph Modifications

To determine the merge graph of two graph modifications, the "least common denominator" of both modifications is determined. It is called D in the construction below. Considering both modifications, the least common denominator is extended by all the changes of modifications 1 and 2, first separately in C_1 and C_2, and finally glued together in graph X. (Compare the diagram in Def. 11.)

Definition 11 (Merging of graph modifications). *Given two modifications* $G \xleftarrow{g_1} D_1 \xrightarrow{h_1} H_1$ *and* $G \xleftarrow{g_2} D_2 \xrightarrow{h_2} H_2$, *the merged graph* X *and the merged graph modification* $G \longleftarrow D \longrightarrow X$ *can be constructed as follows:*

1. Construct $D_1 \xleftarrow{d_1} D \xrightarrow{d_2} D_2$ as pullback of $D_1 \xrightarrow{g_1} G \xleftarrow{g_2} D_2$.
2. Construct PO-complements $d_i' : D \to C_i$ and $c_i : C_i \to H_i$ from morphisms $d_i : D \to D_i$ and $h_i : D_i \to H_i$ for $i = 1, 2$.
3. Construct $C_1 \xrightarrow{x_1} X \xleftarrow{x_2} C_2$ as pushout of $C_1 \xleftarrow{d_1'} D \xrightarrow{d_2'} C_2$.
4. The merged graph modification is $G \xleftarrow{g_1 \circ d_1} D \xrightarrow{x_1 \circ d_1'} X$.

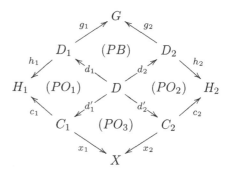

Proposition 3 (Existence and uniqueness of merging). *The construction in Def. 11 leads to a unique graph X and furthermore, unique graph modification $G \xleftarrow{g_1 \circ d_1} D \xrightarrow{x_1 \circ d_1'} X$ up to isomorphism, if graph transformations $G \xRightarrow{p_1, m_1} H_1$ and $G \xRightarrow{p_2, m_2} H_2$ with minimal rules $p_1 = L_1 \xleftarrow{l_1} K_1 \xrightarrow{r_1} R_1$ and $p_2 = L_2 \xleftarrow{l_2} K_2 \xrightarrow{r_2} R_2$ uniquely extracted from graph modifications $G \xleftarrow{g_1} D_1 \xrightarrow{h_1} H_1$ and $G \xleftarrow{g_2} D_2 \xrightarrow{h_2} H_2$ are parallel independent.*

Proof Idea: While pullback and pushouts always exist uniquely in the category of graph and graph morphisms, this is not the case for pushout complements. We show the existence of pushout complements in (PO_1) and (PO_2), using the assumption that the graph transformations $G \xRightarrow{p_1, m_1} H_1$ and $G \xRightarrow{p_2, m_2} H_2$ are independent. The full proof is given in [8].

Example 5 (Merging of graph modifications). The merging construction applied to gm_B and gm_C (the refinement and deletion modifications by users B and C) is depicted in Fig. 10. We see that the merged graph X at the bottom of Fig. 10 contains a forbidden situation: state S4 is isolated, i.e. it is not adjacent to a transition anymore. We want to find such a situation automatically by checking a graph constraint forbidding any situation where a state is isolated.

4.2   Detection of State-Based Conflicts

Using graph conditions as defined in [9], we can specify well-formedness constraints to be checked on the merged graph.

Definition 12 (Graph condition and graph constraint). *A graph condition over graph G is of the form* true *or* $\exists(a, c)$ *where* $a : P \to C$ *is a graph morphism and c is a condition over C. Moreover, Boolean formulas over conditions over P yield conditions over P, i.e. $\neg c$ and $\wedge_{j \in J} c_j$ are (Boolean) conditions over P where J is an index set and $c, (c_j)_{j \in J}$ are conditions over P. Additionally, $\exists a$ abbreviates $\exists(a, true)$, $\forall(a, c)$ abbreviates $\neg \exists(a, \neg c)$, false abbreviates $\neg true$, $\vee_{j \in J} c_j$ abbreviates $\neg \wedge_{j \in J} \neg c_j$, and $c \implies d$ abbreviates $\neg c \vee d$.*

Every graph morphism satisfies true*. A morphism $p : P \to G$ satisfies condition $\exists(a, c)$ if there is an injective graph morphism $q : C \to G$ such that $q \circ a = p$*

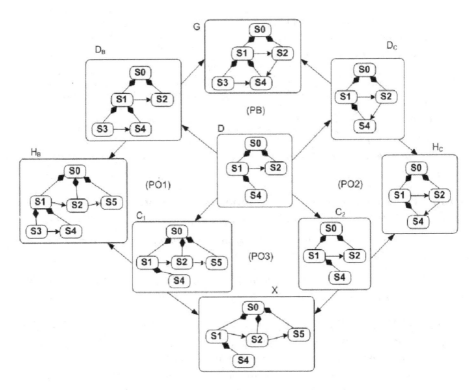

Fig. 10. Merging of graph modifications gm_B and gm_C

and q satisfies c. A graph G satisfies a condition $\exists(a, c)$ if this condition is satisfied by graph morphism $\emptyset \to G$. In the context of graphs, graph conditions are called graph constraints. The satisfaction of conditions by graphs and morphisms is extended to Boolean conditions in the usual way.

The notation of graph constraints of the form $\exists(a : \emptyset \to G, c)$ can be shortened to $\exists(G, c)$ without loss of information. A rule application condition of the form $\neg \exists a$ is usually called *negative application condition* (see [9]).

Definition 13 (State-based conflict). *Given a merged graph modification as in Def. 11, a state-based conflict $(C, H_1 \Longrightarrow X, H_2 \Longrightarrow X)$ consists of a graph constraint C, and graph modifications $H_1 \Longrightarrow X$ and $H_2 \Longrightarrow X$ such that C is satisfied by graphs H_1 and H_2 but not by X.*

Example 6. Fig. 10 shows a merged graph X that contains an isolated state which should not be allowed. This situation can be formalized by a graph constraint $C = \forall G_0((\exists a : G_0 \to G_1) \vee (\exists a : G_0 \to G_2))$ where G_0 consists of a state contained in some other state, and G_1 and G_2 show the alternative required contexts for G_0 in Fig. 11. C is satisfied by graphs H_B and H_C in Fig. 10, but not by graph X.

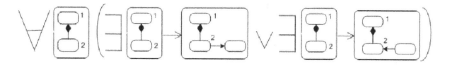

Fig. 11. Graph constraint forbidding isolated states

Further typical state-based conflicts can occur w.r.t. well-formedness constraints in meta models. Consider e.g. the constraint that a transition may have at most one event. If two graph modifications add an event to the same transition, then this leads to a state-based conflict after merging.

5 Implementation Issues

AGG. AGG [5] is an integrated development environment for algebraic graph transformation systems. It implements basic concepts and constructions such as graphs, graph morphisms, matching, and pushouts. On top of these, graph transformation, a restricted form of graph constraint checking as well as parallel dependency checking of rule applications are supported. This is a solid basis to implement initial pushouts, pullbacks, and pushout complements as basis for minimal rule extraction, merging and conflict detection with a convenient user interface in the near future.

AMOR. We now compare the model versioning system AMOR [1] with the formal definitions presented in this paper.

Model differencing. Before conflicts are detected, the concurrently performed changes have to be determined. In this paper, the modifications are extracted applying the construction of minimal rules (cf. Def. 5). To identify executions of predefined operations, it is checked whether the extracted minimal rule is a subrule of a specific predefined operation. In AMOR, all applied changes are derived by conducting EMF Compare [10]. The resulting difference model is conceptually equivalent to a minimal graph rule, because it consists of a match model explicating common elements (cf. D in a graph modification) and a set of atomic changes describing the differences between G and H. Executions of predefined operations are detected in AMOR by searching for their specific change patterns in the derived difference model and, subsequently for each match, by evaluating their pre- and postconditions. Again, this technique is compliant to the operation identification presented in this paper.

Operation-based conflicts. As proposed in Sect. 3, we check minimal rules and operations for critical pairs which are considered as operation-based conflicts. In AMOR, operation-based conflicts are detected by comparing each change applied by one user to each change applied by another. If two changes are contradicting, a conflict explicating the contradicting changes is reported. If further operation executions have been identified before, the preconditions of this operation are evaluated after all atomic changes of the opposite side are executed.

With this it is checked whether the operation may still be executed (according to its preconditions) to a model incorporating the opposite changes.

State-based conflicts. These conflicts occur if a merged model bears prohibited conditions. In this paper, such conditions are defined by graph constraints which are evaluated on the merged graph (cf. Sec. 4.1). According to Proposition 3, this is possible only, if there are no parallel dependent modifications. In AMOR, the merged model is validated against the metamodel and its OCL constraints to find state-based conflicts as well. However, the applied merge differs slightly from the merge presented in this paper, since a merged model is also created, if parallel dependent changes exist. In such cases, only those changes are propagated to the merged model which are not parallel dependent and the user has to manually resolve the operation-based conflicts.

6   Related Work

The main contribution of this work is a formal definition of model versioning conflicts as basis for automatic conflict detection by using the DPO approach to graph transformations. Therefore, we distinguish two kinds of related work. First, we discuss the state-of-the-art of current model versioning systems, and second, we compare our work to other approaches aiming at the formalization of model versioning conflicts.

Model Versioning Systems. In the last decades a lot of research has been conducted in the domain of software versioning which is profoundly outlined in [11,12]. Most of the approaches focus on source code versioning, others focus on two-way comparison of models [13], but there are also some dedicated approaches aiming at the versioning of models by a three-way merge. For example, Odyssey-VCS [2] supports the versioning of UML models. This system performs the conflict detection at a very fine-grained level, hence it is able to merge modifications concerning different model elements or even different attributes of one model element. EMF Compare [10] is an Eclipse plug-in, for comparing and merging models independently of the underlying meta model. CoObRA [4] is integrated in the Fujaba tool suite and logs the changes performed on a model. The modifications performed by the modeler who did the later commit are replayed on the updated version of the repository. Conflicts are reported if an operation may not be applied due to a violated precondition. Similar to CoObRA, Unicase [3] also provides three-way merging based on edit logs.

   Although *delete/use conflicts* and *change/use conflicts* are captured by all of these systems, they do not take predefined operations like refactorings and, consequently, their bigger contexts into account. EMF Compare and Unicase also miss to detect changes causing state-based conflicts. None of the four mentioned systems aims at providing a precise formalization of conflict detection.

Formalization of versioning conflicts. Another category-theoretical approach which formalizes model versioning is given in [14]. Similarly to our approach, modifications are considered as spans of morphisms to describe a partial mapping of models and model merging is defined by pushout constructions. Moreover, syntactic conflicts such as adding structure to an element which has been deleted, are identified. This kind of conflicts is very close to our delete/use-conflicts which we can identify after having extracted minimal rules. In contrast to [14], we refer to a formal analysis of operation-based conflicts. In addition, we consider state-based conflict detection. This has been indicated as future work in [14], where conflict detection based on user-specified operations are not mentioned at al.

Alanen and Porres [15] define a difference and merge operator for MOF-based models from a set-theoretical view. Differences are represented by atomic changes leading from a base version to the working copy. With their approach, they are able to detect *Delete/Use* and *Change/Use-conflicts*, also incorporating advanced concepts such as ordered features. However, conflicts going beyond atomic changes as well as state-based conflicts remain undetected.

The approach by Blanc et al. [16,17] detects state-based inconsistencies using Prolog empowered first-order logics. Structural and methodological constraints are formalized in consistency rules as logic formulae over a sequence of model construction operations. However, they do not consider to detect operations going beyond atomic changes as it is supported by our approach.

7 Conclusion

Two different kinds of conflicts in model versioning are defined in this paper based on the notion of graph modifications: *operation-based* and *state-based* conflicts. Graph modifications are not necessarily rule-based, but just describe changes in graphs. Operation-based conflicts are detected by extracting minimal rules from modifications first and selecting pre-defined operation rules thereafter if possible. As a consequence, we can use the well-known conflict characterization for graph transformations based on parallel dependence checking and extraction of critical pairs. The detection of state-based conflicts builds directly on merged graph modifications and constraint checking.

In this paper, operations are specified simply by rules without additional application conditions. Several extensions are imaginable here: If operations are specified by rules with negative application conditions, an additional kind of conflict can be identified namely *Produce/Forbid-conflicts*. New parallel independence results for rules with more complex application conditions and their applications are currently elaborated by Habel et al. Moreover, the detection of operation sequences for minimal rules is left to future work.

Throughout this paper, we concentrate on the formalization of *conflict detection*. What can *conflict resolution* mean in this setting? The resolution of an operation-based conflict means to show the confluence of the corresponding critical pair (see [5]), while state-based conflicts might be solved by the definition of repair actions. Usually, different conflict resolutions are possible and it is up to future work to develop adequate resolution strategies for this formal setting.

References

1. Brosch, P., Kappel, G., Langer, P., Seidl, M., Wieland, K., Wimmer, M., Kargl, H.: Adaptable Model Versioning in Action. In: Modellierung 2010. LNI, vol. 161. GI (2010)
2. Murta, L., Corrêa, C., Prudêncio, J.G., Werner, C.: Towards Odyssey-VCS 2: Improvements over a UML-based Version Control System. In: 2nd Int. Workshop on Comparison and Versioning of Software Models @ ICSE 2008 (2008)
3. Kögel, M.: Towards Software Configuration Management for Unified Models. In: Workshop on Comparison and Versioning of Software Models @ ICSE 2008 (2008)
4. Schneider, C., Zündorf, A., Niere, J.: CoObRA - A Small Step for Development Tools to Collaborative Environments. In: Workshop on Directions in Software Engineering Environments @ ICSE 2004 (2004)
5. Ehrig, H., Ehrig, K., Prange, U., Taentzer, G.: Fundamentals of Algebraic Graph Transformation. In: Monographs in Theoretical Computer Science. Springer, Heidelberg (2006)
6. Bisztray, D., Heckel, R., Ehrig, H.: Verification of architectural refactorings: Rule extraction and tool support. ECEASST 16 (2008)
7. TFS-Group, TU Berlin: AGG (2009), http://tfs.cs.tu-berlin.de/agg
8. Taentzer, G., Ermel, C., Langer, P., Wimmer, M.: Conflict detection for model versioning based on graph modfications: Long version. Technical Report 2010/09, Technische Universität Berlin (to appear 2010), http://www.eecs.tu-berlin.de/menue/forschung/forschungsberichte/2010
9. Habel, A., Pennemann, K.H.: Correctness of high-level transformation systems relative to nested conditions. Math. Struct. in Comp. Sci. 19(2), 245–296 (2009)
10. Brun, C., Pierantonio, A.: Model Differences in the Eclipse Modeling Framework. UPGRADE, The European Journal for the Informatics Professional (2008)
11. Conradi, R., Westfechtel, B.: Version Models for Software Configuration Management. ACM Computing Surveys 30(2), 232–282 (1998)
12. Mens, T.: A State-of-the-Art Survey on Software Merging. IEEE Transactions on Software Engineering 28(5), 449–462 (2002)
13. Kelter, U., Wehren, J., Niere, J.: A Generic Difference Algorithm for UML Models. In: Software Engineering 2005. LNI, vol. 64, pp. 105–116. GI (2005)
14. Rutle, A., Rossini, A., Lamo, Y., Wolter, U.: A Category-Theoretical Approach to the Formalisation of Version Control in MDE. In: Chechik, M., Wirsing, M. (eds.) FASE 2009. LNCS, vol. 5503, pp. 64–78. Springer, Heidelberg (2009)
15. Alanen, M., Porres, I.: Difference and union of models. In: Stevens, P., Whittle, J., Booch, G. (eds.) UML 2003. LNCS, vol. 2863, pp. 2–17. Springer, Heidelberg (2003)
16. Blanc, X., Mounier, I., Mougenot, A., Mens, T.: Detecting model inconsistency through operation-based model construction. In: ICSE 2008–30th Int. Conference on Software Engineering (2008)
17. Blanc, X., Mougenot, A., Mounier, I., Mens, T.: Incremental Detection of Model Inconsistencies Based on Model Operations. In: van Eck, P., Gordijn, J., Wieringa, R. (eds.) CAiSE 2009. LNCS, vol. 5565, pp. 32–46. Springer, Heidelberg (2009)

A Component Concept for Typed Graphs with Inheritance and Containment Structures

Stefan Jurack and Gabriele Taentzer

Philipps-Universität Marburg
Germany

Abstract. Model-driven development (MDD) has become a promising trend in software engineering. The model-driven development of highly complex software systems may lead to large models which deserve a modularization concept to enable their structured development in larger teams. Graphs are a natural way to represent the underlying structure of visual models. Typed graphs with inheritance and containment are well suited to describe the essentials of models based on the Eclipse Modeling Framework (EMF). EMF models already support the physical distribution of model parts. Based on the concept of distributed graphs, we propose typed composite graphs with inheritance and containment to specify logical distribution structures of EMF models. The category-theoretical foundation of this kind of composite graphs forms a solid basis for the precise definition of typed composite graph transformations obeying inheritance and containment conditions.

1 Introduction

Software engineers have to cope with a continuously increasing complexity in software development. A promising paradigm addressing this issue is model-driven software development. Here, the main artifacts are models being ideal means to master complexity by abstraction. All implementation details are added by a code generator which is a highly reused part of a model-driven infrastructure. Models enable a developer to focus on application-specific aspects of a software system.

Considering traditional software development, it is common practice to work in distributed development teams. The question arises how model-driven development can be performed by several distributed teams. An obvious idea is to set up a central repository for models which can be used by all teams. Central repositories support a physical distribution of models, while still dealing with logically undistributed models. This concept may not fit in every case of distributed development e.g. consider Open Source Development with distributed developer teams working in independent projects.

In [1] we propose a component concept for models based on the category of graphs and graph morphisms which allows the logical distribution of models. A composite graph is built up from graph parts inter-operating directly or via interfaces. An elaborated component concept consists of components where each one is equipped with an arbitrary number of export and import interfaces. Export interfaces specify model parts that are provided to the environment while import interfaces specify model parts being

H. Ehrig et al. (Eds.): ICGT 2010, LNCS 6372, pp. 187–202, 2010.

required. The explicit declaration of interfaces allows for an independent definition of component models such that they can be connected later.

In the modeling community, the Eclipse Modeling Framework (EMF) [2,3] has evolved to a well-known and widely used technology. EMF provides modeling and code generation capabilities based on so-called structural data models. As they describe structural aspects only, they are mainly used to specify domain-specific languages. EMF complies with Essential MOF (EMOF) as part of OMG's Meta Object Facility (MOF) 2.0 specification [4]. In EMF models, the containment concept plays an important role as it describes an ownership relation aiming at acyclic containment structures in instance models. Consequently, objects must belong to at most one container and cyclic containment is forbidden.

EMF supports a concept for physical model distribution. Each model element can refer directly to elements contained in remote resources, i.e. stored elsewhere. This concept is fine as long as a logically undistributed model is expected. The consequence is that each element can be referred to in an unstructured manner which is not always desired. Consider for example the development of two management tools. One is concerned with the management of departments in a company while the other is a typical project management tool. Assume that both tools should be developed independently of each other in a model-based or model-driven manner. The interaction of both developments shall be made explicit by import and export interfaces. For example, an integrating tool suite should offer the possibility to exchange information about employees working in projects.

In this paper, we consider typed graphs with node inheritance and containment edges as formal basis for EMF models. Needless to say that object nodes can be attributed. For the component concept presented later, attributes play a minor role and are not further considered. However, attributes can be formalized by attributed graphs as done in e.g. [5]. To summarize, we build up on the formalization approach for EMF models by Biermann et al. in [6] and extend it to partial graph morphisms and later towards a component concept for typed graphs with inheritance and containment, taking composite models presented in [1] into account. Component types can be used to identify a class of similar component instances. Import and export interfaces may have simpler inheritance structures and need not show the complete containment structures of their component bodies. Moreover, import interfaces may be served partially only by connected export interfaces. In this case, a sort of component inconsistency is exposed.

This paper is structured as follows: In Sec. 2 we consider typed graphs with inheritance and containment structures. All concepts are illustrated by a running example. In Sec. 3, we present a component concept for this kind of graphs leading to composite typed graphs with inheritance and containment structures. In Section 4, we sketch how far the results presented can be used to define composite transformations. Finally, we discuss related work (cf. Sec. 5) and conclude our work (cf. Sec. 6).

2   Typed Graphs with Inheritance and Containment Structures

In this section, we formally define typed graphs with inheritance and containment. For convenience, definitions are illustrated and explained by a running example containing models for a department management and project management component.

Graphs are a natural means to represent the underlying structure of visual models. They may be extended by attributes as e.g. presented by Ehrig et. al. in [5].

Definition 1 (Graph). *A graph* $G = (G_N, G_E, s_G, t_G)$ *consists of a set* G_N *of nodes, a set* G_E *of edges, as well as source and target functions* $s_G, t_G : G_E \rightarrow G_N$. *In the following we call* G simple graph.

Definition 2 (Partial Graph Morphism). *Given two graphs* G *and* H, *a pair of partial functions* (f_N, f_E) *with* $f_N : G_N \rightarrow H_N$ *and* $f_E : G_E \rightarrow H_E$ *forms a partial graph morphism* $f : G \rightarrow H$, *shortly* graph morphism *or* morphism, *if the following properties hold:*

1. *If* f_E *is defined for* $e \in G_E$, f_N *is defined for* $s_G(e)$ *and* $t_G(e)$,
2. $f_N \circ s_G = s_H \circ f_E$ *and*
3. $f_N \circ t_G = t_H \circ f_E$.

If f_N *and* f_E *are total,* f *is called* total graph morphism.
If both f_N *and* f_E *are inclusions,* f *is called* graph inclusion *written* $G \subseteq H$.
$dom(f)$ *is the domain of* f, *i.e. the subset of* G *where* f *is defined.*

Definition 3 (Relational Operators on Morphisms). *Given two partial graph morphisms* $f, g : G \rightarrow H$, *the following operators can be defined:*

1. $f \leq g : \forall x \in dom(f) : f(x) = g(x)$
2. $f = g : f \leq g \wedge g \leq f$

Graphs can be extended by inheritance and containment. Definition 4 below revises corresponding definitions in [6]. In the following Def. a revised definition is presented. The semantics of inheritance follows that in object-oriented programming languages i.e. features of parents are inherited by their children. Note that edges are the only features to inherit once we neglect attributes. An important constraint with regard to inheritance is that inheritance cycles are forbidden.

In our formalization we extend a simple graph with an inheritance graph, a set of abstract nodes and a set of containment edges. Furthermore, we define relations *clan* and *contains*. The former identifies all clan members of a given parent node i.e. children, grandchildren, etc. Note that the given node is part of its clan as well. We also show how simple graphs can be lifted to equivalent graphs with trivial inheritance. Relation *contains* identifies nodes which are connected by edges with containment property. This relation respects inherited containment and is transitively defined. In EMF it is a desired property for models that at least all non-abstract nodes are transitively contained in one root node. At the instance level, this enables the creation of a tree structure which can be persisted conveniently. Accordingly we define a property *rooted*. Further conceivable extensions as multiplicities and additional constraints are left out in order to concentrate on structural aspects.

Definition 4 (Graph with Inheritance and Containment). *A tuple* $G = (T, I, A, C)$, *called* graph with inheritance and containment *or short* IC-graph, *consists of a graph* $T = (N, E, s, t)$, *a graph* I, *called* inheritance graph, *with* $I_N = N$ *and* $I_E \cap E = \emptyset$, *a set* $A \subseteq N$ *of abstract nodes and a set* $C \subseteq E$ *of containment edges. In addition, we require:*

– *For each node $n \in I_N$ the inheritance* clan *is defined by* $clan_G(n) = \{n' \in I_N \mid \exists$ *path* $e_1, ..., e_i \in I_E$ *with* $i > 1$ *such that* $s_I(e_1) = n'$ *and* $s_I(e_k) = t_I(e_{k-1})$ *with* $1 < k \leq i$ *and* $t_I(e_i) = n\} \cup \{n' \in I_N \mid \exists e \in I_E : s_I(e) = n' \wedge t_I(e) = n\} \cup \{n\}$,
– $\forall n, m \in I_N : n \in clan_G(m) \wedge m \in clan_G(n) \Rightarrow n = m$ *(no inheritance cycles)*.

Furthermore the following is defined:

– $clan_G(M) = \bigcup_{n \in M} clan_G(n)$.
– $contains_G$ *defines a containment relation corresponding to* C:
   $contains'_G = \{(n, m) \in N \times N \mid \exists c \in C \wedge x, y \in N : s_T(c) = x$ *with* $n \in clan_G(x) \wedge t_T(c) = y$ *with* $m \in clan_G(y)\}$.
   $contains_G$ *is the transitive closure of* $contains'_G$.
– *Graph* G *is called* rooted, *if there is one non-abstract node* $r \in N - A$, *called* root *node, which contains all other non-abstract nodes transitively:* $\forall n \in N - A - \{r\}$: $(r, n) \in contains_G$.
– *For each simple graph* K *a straight inheritance extension with containment is defined as IC-graph* $KC = (K, I, \emptyset, \emptyset)$ *with* $I_E = \emptyset$.

Tuple elements T, I, A *and* C *of graph* G *can also be referred to by* $T(G), I(G), A(G)$ *and* $C(G)$ *respectively.*

Example 1. Figure 1 shows two IC-graphs each representing a software component. A containment relation is illustrated with a diamond at the container's side. Inheritance is visualized by a triangle at the parent's side. Both are UML compliant visualizations. Note that in this example attributes are shown just to motivate the setting. They are not part of the formalization given in this paper. Note furthermore, in terms of our formalization, source nodes of edges in set C are considered being the container of nodes on the opposite. In set I, a source node is interpreted as child while a target node is considered as parent.

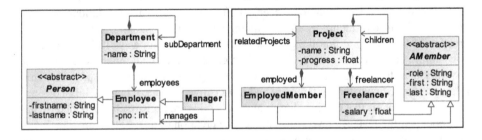

Fig. 1. IC-graphs modeling department management (left) and project management (right)

On the left-hand side, a department management component is depicted which consists of a named `Department` being constituted by a number of `Employees` and `Managers` while a manager as specialized employee can supervise other employees. Managers and employees have a full name due to their abstract parent `Person`. An employee is additionally equipped with a personal number. Furthermore, departments may have sub-departments. The resulting IC-graph is *rooted* since node `Department` can be considered as non-abstract root containing all remaining (at least non-abstract) nodes. Node `Manager` is contained indirectly due to its parent-child relation with `Employee`.

The illustration on the right-hand side depicts a project management component for companies. A named `Project` may refer to related projects as well as child projects. Projects are implemented by employees of a company (`EmployedMembers`) and `Freelancers`. Both nodes derive from abstract parent `AMember` which provides a number of useful attributes e.g. first name, last name and project role. Moreover, a freelancer has a salary attribute. The project management component is a rooted IC-graph as well, with `Project` considered as root.

To allow interrelated IC-graphs we define IC-graph morphisms below. They come with three essential constraints which preserve structural equivalence. However, they provide a much more flexible mapping than ordinary graph morphisms. Constraint (1) allows a source (target) node to be mapped either to the source (target) of the mapped edge directly or alternatively, to one of its children. Constraint (2) ensures that clans are mapped to correspondent clans. The length of inheritance paths does not matter. The preservation of containment properties is guaranteed by constraint (3).

Definition 5 (Graph Morphism with Inheritance and Containment). *Given two IC-graphs G and H, a morphism $f\colon G \to H$, called* graph morphism with inheritance and containment, *shortly* IC-graph morphism *or* IC-morphism, *is defined over graph morphism $f_T\colon T(G) \to T(H)$. The mapping between sets $A(G)$ and $A(H)$ is induced by $f_{T,N}$. The node mapping between $I(G)_N$ and $I(H)_N$ is induced by $f_{T,N}$ as well such that edges of $I(G)_E$ may be mapped to paths in $I(H)_E$. Mappings between sets $C(G)$ and $C(H)$ are induced by f_E. Consequently, from now on we use f as abbreviation of f_T, i.e. f_E and f_N relate to mapped edges and nodes instead of $f_{T,E}$ and $f_{T,N}$. Additionally, we require the following to hold:*

1. $f_N \circ s_{T(G)} \subseteq clan_{I(H)} \left(s_{T(H)} \circ f_E \right)$ *and*
   $f_N \circ t_{T(G)} \subseteq clan_{I(H)} \left(t_{T(H)} \circ f_E \right)$ *(clan-compatible source and target mappings)*
2. $f_N \left(clan_{I(G)} \right) \subseteq clan_{I(H)} \left(f_N \right)$ *(clan-compatible node mapping)*
3. $f_E(C(G)) \subseteq C(H)$ *(containment-compatible edge mapping)*

Proposition 1. *Given IC-graphs G, H and K as well as IC-morphisms $f\colon G \to H$ and $g\colon H \to K$, then $g \circ f\colon G \to K$ is an IC-graph morphism, too.*

Proof idea. *A general composition of morphisms is constructed and each property is checked, i.e. compatibility of edge mappings with source, target and node mappings, as well as preservation of the containment property of edges (see [7] for the complete proof).*

Proposition 2. *IC-graphs and IC-graph morphisms form a category, called* ICGRAPH.

Proof idea. *Identity morphism and composition of morphisms are shown (see [7] for the complete proof).*

Special IC-graph morphisms are used for typing of IC-graphs. A typing IC-graph morphism runs from an instance IC-graph, a simple graph with straight inheritance extension and induced containment edges, to its type IC-graph having a proper inheritance structure and an identified set of containment edges. Furthermore, instance nodes must have at most one container and cyclic containment is invalid. If the type graph is rooted, these constraints induce a tree structure. A corresponding definition is given in [6] and revised in Def. 6 below.

Definition 6 (Typed IC-Graph, Type IC-Graph and Typing IC-Graph Morphism).
Given a simple graph $G = (N, E, s, t)$ *with straight inheritance extension and containment* GC' *and IC-graph* TGC. *IC-graph* $GC = (G, I(GC'), A(GC'), C)$, *called* typed IC-graph *or* instance IC-graph, *is defined along total IC-morphism* $type_{GC}$: $GC \rightarrow TGC$, *called* typing IC-morphism, *with:*

1. $C = \{e \in E \mid type_{GC}(e) \in C(TGC)\}$ *i.e.* C *is induced by the typing morphism,*
2. $e1, e2 \in C : t_G(e1) = t_G(e2) \Rightarrow e1 = e2$ *(at most one container), and*
3. $(x, x) \notin contains_{GC}$ *for all* $x \in N$ *(no containment cycles).*

This morphism is called concrete typing IC-morphism, *if all nodes in* G_N *are typed over concrete types i.e.* $\forall n \in G_N : type_{GC}(n) \notin A(TGC)$.

Remark 1. Please note that typing IC-morphisms are IC-morphisms in the sense of Def. 5. While constraints 5.1 and 5.2 are obviously satisfied, constraint 5.3 holds as C(GC) is compatibly induced by the typing morphism.

Definition 7 (Typed IC-Morphism). *Considering IC-graphs and IC-morphisms only: Given graph* G *typed over graph* TG *by typing morphism* $type_G$, *and graph* H *typed over graph* TH *by typing morphism* $type_H$. *Furthermore, morphism* $g : TG \rightarrow TH$ *is given. An IC-morphism* $f : G \rightarrow H$, *called* typed IC-morphism, *is typed over* g *if the following holds:* $type_H \circ f \leq g \circ type_G$ *(cf. Fig. 2).*

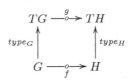

Fig. 2. Typed IC-graph morphism

Example 2. Figure 3 shows two instance IC-graphs typed over two type IC-graphs in analogy to the schema in Fig. 2.

In detail, two different department management component type graphs are shown at the top of Fig. 3 which are mapped by an IC-graph morphism denoted by dotted arrows. `Department` is mapped straightforward to `Department`. Furthermore, `Employee` and `Manager` on the left-hand side are both mapped to `Employee` on the right-hand side which is a valid clan-compatible node mapping. The containment edge `employees` between `Department` and `Employee` is mapped to a containment edge `persons` between `Department` and `Person` nodes. This is a valid IC-mapping, since the sources of both mapped edges are directly mapped. The target `Employee` is mapped to a child of `Person` which complies to the clan-compatible edge mapping constraint. Please note that this morphism is non-total.

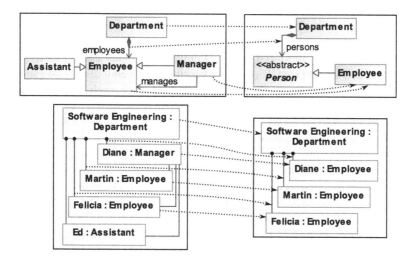

Fig. 3. Example for typed IC-graphs and a typed IC-graph morphism

For each type graph, an instance graph is shown in the lower part of Fig. 3. Small circles at edges denote containments at the instance level[1] and are located at the container's side equivalent to the diamonds on type level. To enable proper differentiation each instance node is equipped with a name and its type e.g. Diane : Manager. Corresponding to the given type, there exists a mapping to its type node. Please note that the types of instance edges are not explicitly shown but can be uniquely deduced from their contexts. Edge mappings are also not shown, since they can be uniquely determined, following Def. 5. On the upper left e.g., the edge between SoftwareEngineering : Department and Diane : Manager is mapped to edge type employees which runs between types Department and Employee. Since Manager is a child of Employee, this mapping is valid. Similarly, instances of edge manages are mapped validly, since the types of targeted instances :Employee and :Assistant are in the clan of type Employee. It can be argued analogously for the right-hand instance and type graphs.

It is obvious that the typed morphism between both instances in Fig. 3 is truly partial as the Assistant node and the manages edge are not mapped. The type graph morphism is partial as well, because node type Assistant and edge type manages are not mapped. Managers and employees are both mapped to employee instances analogously to the type level mapping. Since instance IC-graphs are straight inheritance extensions of simple graphs i.e. they have a trivial inheritance structure only and their containments are induced by the typing, mappings between instance graphs correspond to ordinary graph morphisms. However, according to Def. 7, a typed IC-morphism has to be typed over the morphism between related type graphs such that Fig. 2 commutes.

---

[1] At instance level there is no UML compliant and distinguishable visualization for instance edges typed over compositions/aggregations or ordinary associations.

3   Composite IC-Graphs

In this section, we recall and adapt the concept of composite graphs and its formalization which has been introduced in [1]. Mainly, such graphs are considered on the network layer and object layer. The former describes the overall distribution structure, the latter describes the structure of each local graph. This approach is very general and can be used to formalize different kinds of model compositions.

Here we concentrate on *composite graphs with explicit export and import interfaces*. Such a composite graph is constituted by a set of *components*[2] each consisting of a *body graph*, and a set of interface graphs, namely *export and import graphs*. While export interfaces contain graph elements that are provided to the environment, import interfaces specify required elements. Interface graph elements refer to elements of their component's body, thus there is a total morphisms from each interface graph to its body graph. In order to connect components, each import graph has to be connected to an export graph. This connection is realized by partial graph morphisms where a non-total, i.e. proper partial morphism, describes an import which is not fully served by the export connected. Such a case may be interpreted as some sort of inconsistency of a composite graph.

Composite graphs with explicit interfaces offer several advantages. On the one hand, syntactical compatibility between components is ensured by compatible interfaces, while on the other hand information hiding is realized by export interfaces which expose a subset of body graph elements only. Moreover, explicit declaration of interfaces in advance allows for independent development of graphs components to be connected later, if their corresponding interfaces fit together. We consider composite network graphs and composite instance graphs laying a basis for the subsequent definition of typed composite graphs. The composite graph concepts as presented in [1] are extended by node type inheritance and containment edges to provide a formal basis for composite EMF models.

In the following, we define typed composite IC-graphs with explicit interfaces consisting of interconnected components. First of all, a definition of composite network graphs is given (cf. [1]). Thereafter, IC-graph specific definitions follow.

Definition 8 (Composite Network Graph). *A composite network graph is a graph* $G = (N, E, s, t)$ *with the following structure: Let G_N consist of disjoint sets of body nodes, export nodes, and import nodes:* $G_N = G_{Bod} \uplus G_{Exp} \uplus G_{Imp}$ *and G_E consists of disjoint sets of edges:* $G_E = G_{EB} \uplus G_{IB} \uplus G_{IE}$. *An element of G_N is called* network node, *while an element of G_E is called* network edge. *Then, the following conditions have to be satisfied:*

- Body nodes *are not source of any network edge in G_E:* $\forall e \in G_E : \nexists n \in G_{Bod}$ *with $s_G(e) = n$.*
- *Each* export node *is source of exactly one network edge with its target being a body node:* $\forall n \in G_{Exp} : \exists e \in G_E$ *with $s_G(e) = n \wedge t_G(e) \in G_{Bod}$ and $\forall e_1, e_2 \in G_E :$* $s_G(e_1) \in G_{Exp} \wedge s_G(e_2) \in G_{Exp} \wedge s_G(e_1) = s_G(e_2) \Rightarrow e_1 = e_2.$

---

[2] We omit a formal definition of *component* as its boundaries are each given by a body and its interfaces implicitly.

The set of edges from export nodes to body nodes is defined as $G_{EB} = \{e \in G_E \mid s_G(e) \in G_{Exp} \land t_G(e) \in G_{Bod}\}$.

- *Each import node is source of exactly one network edge with its target being a body node: $\forall n \in G_{Imp} : \exists e \in G_E$ with $s_G(e) = n \land t_G(e) \in G_{Bod}$ and $\forall e_1, e_2 \in G_E : s_G(e_1) \in G_{Imp} \land s_G(e_2) \in G_{Imp} \land s_G(e_1) = s_G(e_2) \Rightarrow e_1 = e_2$.*
  The set of edges from import nodes to body nodes is defined as $G_{IB} = \{e \in G_E \mid s_G(e) \in G_{Imp} \land t_G(e) \in G_{Bod}\}$.
  In addition, each import node is source of exactly one network edge targeting an export node: $\forall n \in G_{Imp} : \exists e \in G_E$ with $s_G(e) = n \land t_G(e) \in G_{Exp}$ and $\forall e_1, e_2 \in G_E : s_G(e_1) \in G_{Imp} \land s_G(e_2) \in G_{Imp} \land s_G(e_1) = s_G(e_2) \Rightarrow e_1 = e_2$.
  The set of edges from import nodes to export nodes is defined as $G_{IE} = \{e \in G_E \mid s_G(e) \in G_{Imp} \land t_G(e) \in G_{Exp}\}$. Please note, that export nodes can be referred to by an arbitrary number of import nodes.

Example 3. An example composite network graph is shown in Fig. 4. Rounded rectangles represent network nodes with meaningful letters (B: Body, E:Export, I:Import) and arrows represent network edges. Dashed edges illustrate network edges between bodies and their interfaces, dotted edges occur between interfaces. Surrounding circles group nodes of the same component for clarity. The component on the left-hand side consists of a body node and three export interface nodes while the component on the opposite side consists of one export node and two import nodes. As shown within the component on the right-hand side of Fig. 4, relations between interfaces may not only occur between different components but between the same as well. In the following, we will see that this makes sense for type graphs only.

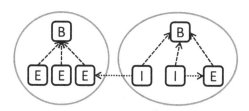

Fig. 4. Composite network graph

Definition 9 (Composite IC-Graph). *Given a composite network graph G, a composite IC-graph $\hat{G}$ over G is defined as $\hat{G} = (G, \mathcal{G}(G), \mathcal{M}(G))$ with*

- *$\mathcal{G}(G)$ being a set of IC-graphs (cf. Def. 4) with each graph refining a network node in G_N: $\mathcal{G}(G) = \{\hat{G}(n) \mid \hat{G}(n)$ is an IC-graph and $n \in G_N\}$,*
- *$\mathcal{M}(G)$ being a set of IC-graph morphisms (cf. Def. 5), each refining a network edge in G_E: $\mathcal{M}(G) = \{\hat{G}(e): \hat{G}(i) \to \hat{G}(j) \mid \hat{G}(e)$ is an IC-graph morphism and $e \in G_E$ with $s(e) = i$ and $t(e) = j\}$,*
- *$\forall e \in G_{EB} \cup G_{IB} : \hat{G}(e)$ is total, and*
- *$\forall n \in G_{Bod} : \hat{G}(n)$ is rooted.*

Please note, that we allow all refinements of network edges between import and export interfaces to be partial morphisms. If these morphisms are really partial, i.e. not total, they may indicate some sort of inconsistency of components with each other, e.g. an import interface is not fully served by an export interface.

Example 4. According to the network graph in Fig. 4, a composite IC-graph is shown in Fig. 5. Each refined network node is equipped with a meaningful name in the upper left corner. The department management component (Dep) is shown on the left while the project management component (Prj) is shown on the right. Body graphs, *DepBody* and *PrjBody*, are depicted on the top of Fig. 5 and interfaces are arranged below. Again, dashed arrows illustrate mappings between interfaces and their bodies while dotted arrows are mappings between interfaces.

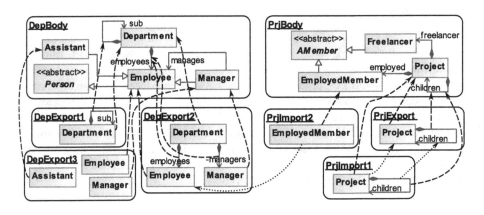

Fig. 5. Composite IC-graph over the network graph in Fig. 4

This scenario describes how two software components may be connected. On the one hand, there is a department software component providing interfaces to gather informations about published department structures, employed persons and assignments of persons to certain departments. On the other hand, a project software component may reveal its project structure and even may require other projects to be associated at the same time. Furthermore, it may depend on (exported) employees of a department to work within a project.

As required, each body graph is *rooted*. Interface graphs do not have to be rooted as they provide only elements from their corresponding rooted body graphs e.g. export *DepExport3* provides no root. Furthermore, interfaces do not need to be mapped into their corresponding body graph in one-to-one manner e.g. export graph *DepExport2* provides a structure which is different to the corresponding part in its body. This export graph is valid though since containment edge managers could be easily mapped to body containment edge employees. This technique allows to hide structural dependencies and complexity by providing simple and convenient interfaces at the same time. Both components are connected by import *PrjImport2* and export *DepExport2*. Moreover, the project management component is connected with itself by *PrjExport* and *PrjImport1*. The impact of both connections is examined in detail in Example 5.

Definition 10 (Composite IC-Graph Morphism). *Given two composite graphs $\hat{G}$ and $\hat{H}$ with network graphs G and H, a composite IC-graph morphism, short* composite IC-morphism, *written $\hat{f}: \hat{G} \to \hat{H}$, is a pair $\hat{f} = (f, m)$ where:*

- *$f: G \to H$ is a total graph morphism and*
- *m is a family of IC-morphisms (cf. Def. 5) $\{\hat{f}(n) \mid n \in G_N\}$ such that:*

  - *for all nodes i in G_N: $\hat{f}_N(i): \hat{G}(i) \to \hat{H}(f_N(i))$ is an IC-morphism and*
  - *for all edges $e: i \to j$ in G_E we have $\hat{H}(f_E(e)) \circ \hat{f}_N(i) \leq \hat{f}_N(j) \circ \hat{G}(e)$ (see the illustration in Fig. 6).*

If f and $\hat{f}(i)$ for all $i \in G_N$ are total graph morphisms, $\hat{f}$ is also called total.

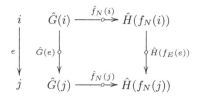

Fig. 6. Illustration of composite IC-graph morphisms

Definition 11 (Typed Composite Graph with Inheritance and Containment). *A composite IC-graph $\hat{G}(G)$ with network graph G is typed over composite IC-graph $\hat{TG}(TG)$ with network graph TG, if there is a composite IC-graph morphism ctype: $\hat{G} \to \hat{TG}$, called* composite typing IC-morphism *which is a pair $\hat{ctype} = (ctype, m)$ where:*

- *ctype: $G \to TG$ is a total graph morphism mapping body, export, and import nodes accordingly to each other:*
  - *$ctype(G_{Bod}) \subseteq TG_{Bod}$,*
  - *$ctype(G_{Exp}) \subseteq TG_{Exp}$,*
  - *$ctype(G_{Imp}) \subseteq TG_{Imp}$, and*
- *m is a family of typing IC-morphisms (cf. Def. 6) $\{\hat{ctype}(n) \mid n \in G_N\}$.*

Example 5. Considering Fig. 5 as composite type graph, Fig. 7 shows a valid example typed composite IC-graph. Each graph is equipped again with a meaningful name according to its typing body or interface graph. On the left, a department management component instance is given with one body and two export interfaces. Each interface is typed over a different export type graph. On the right, two project management component instances are given. In order to distinguish both, suffixes ".1" and ".2" are added to related component parts. Mappings between edges are neglected again.

Export *DepExport2* shows that managers, employees, and departments can be exported. Furthermore, it shows that they do not have to be exported, since some instances are missing. Moreover, although Martin is contained in a department (see *DepBody*), this detail does not have to be exported either as shown. In export *DepExport1*, the department structure is exposed. Again, departments can be exported but do not have to.

On the right-hand side, import *PrjImport2.1* shows a morphism to export *DepExport2*. However, this morphism is partial, since one EmployedMember instance in import interface *PrjImport2.1* is not served by an appropriate exported instance node. This can be interpreted as inconsistency between components. Furthermore, consider import and export graphs *PrjImport1.1* and *PrjExport.2*. They illustrate how component instances which are typed over the same component type can relate to each other. Please note the related mapping on type level in the bottom right of Fig. 5 between interfaces of the same component.

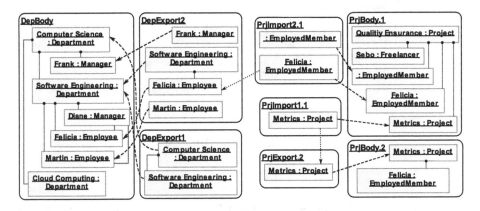

Fig. 7. Typed composite IC-graph

Proposition 3. *Composite IC-graphs and composite IC-graph morphisms form a category, called* COMPICGRAPH.

Proof idea. *Based on distributed structures presented in [8], we define* COMPIC-GRAPH *as category of distributed structures and morphisms over category* ICGRAPH *(see [7] for the complete proof).*

As natural extension of typed composite IC-graphs, we consider typed composite IC-graph morphisms.

Definition 12 (Typed Composite IC-Graph Morphism). *Given two composite IC-graphs $\hat{G}$ and $\hat{H}$ typed over a third composite IC-graph $\hat{TG}$ by $ct\hat{y}pe_G$ and $ct\hat{y}pe_H$. A typed composite IC-graph morphism is a composite IC-graph morphism $\hat{f} \colon \hat{G} \to \hat{H}$ such that $ct\hat{y}pe_H \circ \hat{f} = ct\hat{y}pe_G$.*

Note that COMPTICGRAPH, *the category of typed composite IC-graphs and morphisms is defined as slice category (*COMPICGRAPH/ $\hat{TG}$*).*

4   Towards Transformation of Composite Graphs

After having defined the category of composite IC-graphs and morphisms on the basis of distributed structures, we head towards composite IC-graph transformations. In [8], distributed transformations are defined as double pushouts within the category of distributed structures. However, they are not constructed component-wise which can force unspecified component synchronizations. In [9], additional application conditions are required for distributed rule applications such that component-wise pushout construction is possible. However, this result is given for simple graphs only.

In [8], it is shown that category DISC of distributed structures is complete and co-complete if the underlying category C is so. Since we know that category GRAPH^P of graphs with partial morphisms is co-complete (see [10]) and typed co-complete categories are also co-complete (see [5]), we can follow that category TGRAPH^P of typed graphs with partial morphisms is co-complete and thus, category DISTGRAPH^P of composite typed graphs with composite morphisms is co-complete.[3] However, it is rather straightforward to see that category ICGRAPH does not have pushouts in general. The objects of ICGRAPH are typed graphs as in TGRAPH^P, but additional conditions have to be satisfied. To understand the problem, we consider the following situation: The containment constraint "at most one container" can easily lead to pushout objects not fulfilling this constraint although the original graphs are IC-graphs.

Please note that we try to construct pushouts as in the underlying category and check additional conditions afterwards. For typed graphs with inheritance we have shown that this construction is always possible (see [5]), but not for graphs with containment structures as discussed above. An obvious solution to this problem is to restrict the kind of double pushouts. In [6], we restrict the rules (and therefore the kind of pushouts) to be containment-consistent and show that containment-consistent rule applications to IC-graphs lead to IC-graphs again. It is promising to lift this result to composite IC-graph rules and their application where we have to elaborate a notion of containment-consistency for composite graphs. In addition there may be special application conditions ensuring component-wise construction of composite transformations.

Based on the classification in [1], we distinguish four kinds of composite transformations: (1) internal transformations which run in bodies only and do not change any interface, (2) component transformations which are concerned with manipulations of single components, i.e. body transformations with interface adaptations, (3) synchronized transformations consisting of several parallel component transformations, and (4) network reconfigurations to change the component structure together with adaptations at already existing components. It is up to future work to define consistent composite transformations in general and to characterize these special kinds as well as their interrelations.

5   Related Work

In the following, we compare distribution concepts for EMF models and graphs.

5.1   EMF Models with Remote References

EMF already offers the possibility to spread models over a number of so-called resources e.g. files (compare [2]). The question is how this conventional distribution approach compares with the approach presented in this paper.

---

[3] Please note that this argumentation can be easily lifted to attributed graphs.

EMF models can be distributed by exploiting remote references. Such a remote reference is an ordinary association whose source is an object node in the local model while its target is an object node in a remote model. Each object node can be referred to by remote references. Thus no information hiding takes place here. The only restriction is given by the typing of such a reference i.e. the referred object type is restricted but any reference type may be utilized as remote reference. In accordance to distributed graphs, EMF models with remote references can be considered as body graphs with interfaces having been defined implicitly. Target nodes of remote references induce the implicit import while all model elements are exported implicitly. The imported element is identical to the corresponding exported element.[4] Moreover, an imported model element can be utilized to reveal connected elements of the remote model. Consequently, several physically distributed EMF models can be considered as one single model logically.

Our approach differs from the one in [2] in a number of aspects. In particular, our approach can be compared to component-oriented distribution of models. Possibilities to distribute and to interconnect graphs or model graphs are predefined on the type level. The types of components and object structures that can be exported and imported are declared. Furthermore at the instance level, only those object nodes are visible which are part of an export interface, and vice versa only those object nodes can be accessed which are explicitly imported. This explicit structuring of model graphs is not trivial and deserves careful considerations by the engineers. As a side effect, it deserves a structured design process followed by the engineers. In contrast to EMF models with remote references we consider import elements to be some kind of delegate objects. In particular, they do not need to have the same type as the targeted one and are especially not the same objects on instance level. Rather they shall define features e.g. containments and references (and attributes which can be added to the formal setting) which are mapped to corresponding features of targeted elements. Accessing such an imported feature leads to a delegation to the corresponding exported feature. Delegate objects have further advantages: In addition to importing (or delegating) features, they may have a number of own features. They may inherit or be inherited, they may contain or be contained, they may define own local references, etc. As a consequence, a composite graph is not considered as a single huge model but as a composition of separate component graphs. Thus, EMF-typical constraints such as acyclic containments, do not have to be checked on the large composed model, but can be checked on each of its components locally.

5.2 Graph-Oriented Approaches

In [11], an approach to distributed graphs and graph transformations is presented which allows the distribution of graph parts, but does not support explicit interfaces. In this sense, the distribution concepts of that approach and of EMF models with remote references are quite similar. In [12], Mezei et.al. present distributed model transformations based on graph transformation concepts. Model transformations are not distributed logically, but in the sense that they are performed in a distributed way in order to increase

---

[4] In fact, proxy objects represent remotely targeted nodes until they are accessed the first time. In that case a proxy is resolved and replaced by the original node.

efficiency. This means that transformations are distributed automatically. Again, interfaces are handled implicitly. View-oriented modeling has already been specified by distributed graph transformation in [13] using the same formal basis as composite graphs. However, the structuring of graphs is more elaborated in this paper, since we consider components with explicit interfaces. In [14], Heidenreich et al. propose an approach to extend a meta-model inversely in order to compose its instance models. This merging is done along predefined interface-like structures. In contrast, we present a component concept for models which consist of separate components with possibly different meta-models sharing explicitly revealed information only.

6   Conclusion and Future Work

Typed graphs with inheritance and containment structures are well suited to describe the essentials of EMF models. Building up on basic definitions in [6] we allow partial graph morphisms and show that IC-graphs and graph morphisms form a category. Furthermore, we developed a component concept based on distributed graphs as presented in [1] and originally introduced in [9]. Using the category of IC-graphs and morphisms as underlying category, we show that composite IC-graphs and morphisms also form a category. As pointed out in Section 4, the presented definitions and basic results can lead to a formal definition of consistent composite IC-graph transformations. Following this line, we head towards a well-defined component concept for EMF models including composite EMF model transformations.

References

1. Jurack, S., Taentzer, G.: Towards Composite Model Transformations Using Distributed Graph Transformation Concepts. In: Schürr, A., Selic, B. (eds.) MODELS 2009. LNCS, vol. 5795, pp. 226–240. Springer, Heidelberg (2009)
2. EMF: Eclipse Modeling Framework (2010), http://www.eclipse.com/emf
3. Steinberg, D., Budinsky, F., Patenostro, M., Merks, E.: EMF: Eclipse Modeling Framework, 2nd edn. Addison Wesley, Reading (2008)
4. OMG: Meta Object Facilities 2.0 Specification (2010),
   http://www.omg.org/spec/MOF/2.0/
5. Ehrig, H., Ehrig, K., Prange, U., Taentzer, G.: Fundamentals of Algebraic Graph Transformation. In: Monographs in Theoretical Computer Science. An EATCS Series. Springer, Heidelberg (2006)
6. Biermann, E., Ermel, C., Taentzer, G.: Precise Semantics of EMF Model Transformations by Graph Transformation. In: Czarnecki, K., Ober, I., Bruel, J.-M., Uhl, A., Völter, M. (eds.) MODELS 2008. LNCS, vol. 5301, pp. 53–67. Springer, Heidelberg (2008)
7. Jurack, S., Taentzer, G.: A Component Concept for Typed Graphs with Inheritance and Containment Structures: Long Version. Technical report, Philipps-Universität Marburg (2010), http://www.uni-marburg.de/fb12/forschung/berichte/berichteinformtk/pdfbi/bi2010-1.pdf
8. Ehrig, H., Orejas, F., Prange, U.: Categorical Foundations of Distributed Graph Transformation. In: Corradini, A., Ehrig, H., Montanari, U., Ribeiro, L., Rozenberg, G. (eds.) ICGT 2006. LNCS, vol. 4178, pp. 215–229. Springer, Heidelberg (2006)

 9. Taentzer, G.: Parallel and Distributed Graph Transformation: Formal Description and Application to Communication Based Systems. PhD thesis, Technical University of Berlin (1996)
10. Ehrig, H., Heckel, R., Korff, M., Löwe, M., Ribeiro, L., Wagner, A., Corradini, A.: Algebraic Approaches to Graph Transformation - Part II: Single Pushout Approach and Comparison with Double Pushout Approach. In: Rozenberg, G. (ed.) Handbook of Graph Grammars and Computing by Graph Transformations. Foundations, vol. 1, pp. 247–312. World Scientific, Singapore (1997)
11. Ranger, U., Lüstraeten, M.: Search Trees for Distributed Graph Transformation Systems. In: Electronic Communication of the EASST, vol. 4 (2006)
12. Mezei, G., Juhasz, S., Levendovsky, T.: A distribution technique for graph rewriting and model transformation systems. In: Burkhart, H. (ed.) Proc. of the IASTED Int. Conference on Parallel and Distributed Computing Networks. IASTED/ACTA Press (2007)
13. Goedicke, M., Meyer, T., Taentzer, G.: ViewPoint-oriented Software Development by Distributed Graph Transformation: Towards a Basis for Living with Inconsistencies. In: Proc. 4th IEEE Int. Symposium on Requirements Engineering (RE 1999), University of Limerick, Ireland, June 7-11. IEEE Computer Society, Los Alamitos (1999), ISBN 0-7695-0188-5
14. Heidenreich, F., Henriksson, J., Johannes, J., Zschaler, S.: On Language-Independent Model Modularisation. In: T. Aspect-Oriented Software Development VI, pp. 39–82 (2009)

Combining Termination Criteria by Isolating Deletion

Dénes Bisztray and Reiko Heckel

Department of Automation and Applied Informatics,
Budapest University of Technology and Economics
bisztray@aut.bme.hu
Department of Computer Science,
University of Leicester
reiko@mcs.le.ac.uk

Abstract. The functional behaviour of a graph transformation system is a crucial property in several application domains including model transformations and visual language engineering. Termination is one of the ingredients of functional behaviour and thus equally important. However, the termination of graph transformation systems is generally undecidable. Hence, most of the published termination criteria focus on specific classes of graph transformations. Unfortunately graph transformations with lots of production rules usually do not fit into one of these classes. It would be advantageous if different sets of the production rules in the graph transformation system could be verified using different criteria. This paper addresses this problem by providing structural conditions on the rules enabling such combination of termination criteria.

Keywords: Termination, Graph Transformations, Model Transformations.

1 Introduction

Termination is a fundamental property of graph transformation systems implementing functions over sets of graphs, such as in the case of model transformations. A graph transformation system is *terminating* if all its transformation sequences are finite. Proving that a system has this property for all graphs in the input set is a difficult task, undecidable in general [Plu95].

The work presented here was inspired by the *Activity Diagram to CSP* transformation published in [VAB+08, DB08]. In [EEdL+05], the rules of a graph transformation system are sorted into layers. These layers are either *deletion* or *nondeletion* layers. The conditions for forming the *deletion layers* express that the last creation of a node of a certain type should precede the first deletion of a node with the same type. The *nondeletion layer* conditions ensure that if an element of a specific type occurs in the LHS of a rule, then all elements of the same type were already created in previous layers [EEdL+05]. However, the transformation builds a CSP abstract syntax tree from the Activity Diagram: the various node types in CSP (e.g *Process* or *Event*) are created by almost every

H. Ehrig et al. (Eds.): ICGT 2010, LNCS 6372, pp. 203–217, 2010.

rule. Hence, they cannot be sorted into creation layers (i.e. almost all rules would be in one big layer) and thus the criteria introduced in [EEdL+05] cannot be applied. Although the termination criteria from [LPE07] is reasonably generic, one needs to investigate every combination of the rules pairwise. The transformation contains 18 rules with around 8-14 elements contained in the LHS rule graphs and 10-18 elements in the RHS rule graphs. Checking all the possible combinations by hand without making errors is unlikely.

A general termination criterion used as a technique in manual proofs is the following. A graph transformation system (TG, P) consisting of a type graph TG and a set of typed graph productions P is terminating if its transformations are monotonic with respect to a well-founded partial order $\succ$ over the instances of TG [Bog95, BKPPT05]. This ordering reflects the state of transformation, i.e. $a \succ b$ if b is closer to completion. To formalise this notion of closeness to completion, a metric $\mathcal{M}$ can be defined that assigns a natural number to every instance of TG. Then, $\mathcal{M}(a) > \mathcal{M}(b)$ implies $a \succ b$. Unfortunately, this technique is hard to apply in practice since defining the partial order or suitable metric is a task which requires much ingenuity and experience.

Recent research has focused on the transformation of this general criterion into a more applicable form. By now there are criteria of termination for specific classes of graph transformation systems [EEdL+05, LPE07].

When transformations are complex, it is less likely that they fit any single one of these classes. Indeed, it would be advantageous if they could be combined by proving termination of subsets of rules according to specific criteria and combining the resulting terminating subsets into a provably terminating global system. This is the approach of the present paper. The idea is based on the observation that in complex transformations, there are distinct rule sets working on different parts of the host graph. While these rules are applied nondeterministically, i.e., the execution of rules from different sets is often interleaved, their effect is related. Thus, termination of such rule set is independent of that of other rule sets provided that the sets are suitably isolated from one another. This notation of isolation is based on the absence and acyclicity of certain dependency relations between rules.

We approach the problem in two steps. First, termination of transformations consisting of only non-deleting rules with self-disabling negative conditions is shown. Where existing approaches are using layered rule sets [EEdL+05], we analyse a given unstructured set for *produce-enable* dependencies, i.e., sequential dependencies between rules that could lead to the creation of a match for one rule by another through creation of elements. If this relation is acyclic, the system terminates because self-disabling rules can only be applied once at each match and only finitely many matches can be created.

At the next stage, the result is extended to GTSs with general (possibly deleting) rules that preserve the start graph. Assuming that the effect of the deletion can be isolated into rule groups, we extend the dependency relation to self-contained rule groups. Termination follows if the rule groups are terminating by themselves and the dependency relation is acyclic. The termination of the rule groups themselves can be established using arbitrary termination criteria.

The outline of the paper is as follows. Section 2 presents the required basic definitions. In Section 3 the termination criterion for non-deleting graph transformation systems is established. In Section 4 the criterion is extended for deleting transformations. Section 5 presents an application of the approach to a complex transformation of UML activity diagram into CSP. We conclude the paper with Section 6.

2   Basic Definitions

In this section we collect some fundamental definitions from existing literature. We use the double-pushout approach to typed attributed graph transformation with negative application conditions according to [EEPT06]. In the DPO approach, a graph K is used. K is the common interface of L and R, i.e. their intersection. Hence, a rule is given by a span $p : L \leftarrow K \rightarrow R$.

Definition 1. (Graph Production [CMR+97]). *A* (typed) graph production $p = (L \overset{l}{\leftarrow} K \overset{r}{\rightarrow} R)$ *consists of (typed) graphs* L, K, R, *called the left-hand side, gluing graph (or interface graph) and the right-hand side respectively, and two injective (typed) graph morphisms* l *and* r.

A graph production is nondeleting if the morphism $l : K \rightarrow L$ *is the identity.*

Definition 2. (Graph Transformation [CMR+97]). *Given a (typed) graph production* $p = (L \overset{l}{\leftarrow} K \overset{r}{\rightarrow} R)$ *and a (typed) graph* G *with a (typed) graph morphism* $m : L \rightarrow G$, *called the match, a direct (typed) graph transformation* $G \overset{p,m}{\Rightarrow} H$ *from* G *to a (typed) graph* H *is given by the following double-pushout (DPO) diagram, where (1) and (2) are pushouts in the category* **Graphs** *(or* **Graphs**$_{TG}$ *respectively):*

$$
\begin{array}{ccccc}
L & \xleftarrow{\ l\ } & K & \xrightarrow{\ r\ } & R \\
\downarrow{\scriptstyle m} & (1) & \downarrow{\scriptstyle k} & (2) & \downarrow{\scriptstyle n} \\
G & \xleftarrow{\ f\ } & D & \xrightarrow{\ g\ } & H
\end{array}
$$

A sequence $G_0 \Rightarrow G_1 \Rightarrow ... \Rightarrow G_n$ *of direct (typed) graph transformations is called a (typed) graph transformation and is denoted by* $G_0 \overset{*}{\Rightarrow} G_n$.

Definition 3. (Negative Application Condition [EEPT06]). *A* negative application condition *or* $NAC(n)$ *on* L *is an arbitrary morphism* $n : L \rightarrow N$. *A morphism* $g : L \rightarrow G$ *satisfies* $NAC(n)$ *on* L *i.e.* $g \models NAC(n)$ *if and only if does not exists and injective* $q : N \rightarrow G$ *such that* $q \circ n = g$.

$$
\begin{array}{ccc}
L & \xrightarrow{\ n\ } & N \\
\downarrow{\scriptstyle m} & \overset{x}{\diagdown} & \\
G & \diagup{\scriptstyle q} &
\end{array}
$$

A set of NACs on L is denoted by $NAC_L = \{NAC(n_i)|i \in I\}$. A morphism $g : L \to G$ satisfies NAC_L if and only if g satisfies all single NACs on L i.e. $g \models NAC(n_i)\forall i \in I$.

Definition 4. (GT System, Graph Grammar [CMR+97]). A typed graph transformation system $GTS = (TG, P)$ consists of a type graph TG and a set of typed graph productions P.

We may use the abbreviation GT system for typed graph transformation system.

A fundamental notion in the context of termination is that of *essential match*. It deals with the possible application of a production to essentially the same match into different graphs in a sequence.

Definition 5. (Tracking Morphism and Essential Match [EEdL+05])
Given a (typed) graph transformation system with injective matches and start graph G, a nondeleting production $p : (L \xleftarrow{l} K \xrightarrow{r} R)$ with an injective morphism $r : L \to R$ and injective match $m : L \to G$ leading to a direct transformation $G \overset{p,m}{\Rightarrow} H$ via (p, m) defined by the pushout (1) of r and m. The morphism $d : G \to H$ is called tracking morphism of $G \overset{p,m}{\Rightarrow} H$:

$$
\begin{array}{ccc}
L & \xrightarrow{r} & R \\
\downarrow{m} & (1) & \downarrow{m^*} \\
G & \xrightarrow{d} & H
\end{array}
\qquad
\begin{array}{ccc}
 & & L \\
 & \overset{m_0}{\swarrow} & \downarrow{m_1} \\
G_0 & \xrightarrow{d_1} & H_1
\end{array}
$$

Since both r and m are injective, the pushout properties of (1) imply that also d and m^* are injective.

Given a transformation $G_0 \overset{*}{\Rightarrow} H_1$, i.e. a sequence of direct transformations with an induced injective tracking morphism $d_1 : G_0 \to H_1$, a match $m_1 : L \to H_1$ of L in H_1 has an essential match $m_0 : L \to G_0$ of L in G_0 if we have $d_1 \circ m_0 = m_1$.

Note that because of d_1 is injective, there is at most one essential match m_0 for m_1. The notions of *essential match* and *tracking morphism* are relevant for nondeleting rules only, because a deleting rule consumes elements from the match and thus cannot be applied on the same match again. A non-deleting rule is *self-disabling* if it has a NAC that prohibits the existence of the pattern that the rule creates.

Definition 6. (Self-Disabling Production). Given a nondeleting production $p : (L \xleftarrow{l} K \xrightarrow{r} R)$ with an injective morphism $r : L \to R$ and NAC $n : L \to N$. The negative application condition $NAC(n)$ is self-disabling if there is an injective $n' : N \to R$ such that $n' \circ n = r$. Production p is self-disabling if it has a self-disabling NAC.

The following lemma establishes that a *self-disabling* production cannot be applied on the same match again, and extends it to graph transformations that consists of *self-disabling* rules only.

Lemma 1. (Essential Match Applicability [EEdL+05]). *In every transformation starting from G_0 of a nondeleting (typed) graph transformation system $GG = (TG, P)$ with injective matches and self-disabling productions, each production $p \in P$ with $r : L \to R$ can be applied at most once with the same essential match $m_0 : L \to G_0$ where $m_0 \models NAC(n)$.*

3   Termination of Nondeleting Transformations

In this section we present a termination criterion for nondeleting graph transformations. We define a precedence on production rules based on the *produce-enable* sequential dependency. The transitive closure of the precedence relation gives us the possible application order of production rules. If the transitive closure of the precedence relation is irreflexive, the GTS is terminating.

Figure 1 shows two simple rules responsible for creating CSP *process assignments*. Node PA represents a process assignment with two P nodes representing processes. The symbol n is an attribute of E and P nodes; a and b denote different constant values matched in the host graph.

$Rule_A$ shown in Figure 1(a) transforms an *activity edge* to an empty *process declaration* by creating nodes PA,P with edge $\xrightarrow{a}$ for every E node identified by the n attribute. $Rule_B$ creates a *process definition* from an *action* by connecting an edge and node $\xrightarrow{p} P$ to a PA node for every interconnected E and A node instance as shown in Figure 1(b).

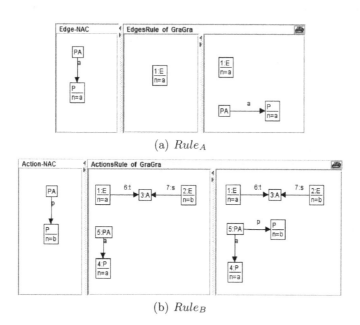

(a) $Rule_A$

(b) $Rule_B$

Fig. 1. $R_{sample} = \{Rule_A, Rule_B\}$

The criteria introduced in [EEdL⁺05] cannot be applied. The rules can be sorted into one or two layers. If the rules are in one layer, this only layer does not satisfy the deletion layer conditions for trivial reasons: $Rule_A$ does not decrease the number of graph items since it is nondeleting. When sorted into two layers ($Rule_A$ in layer 1 and $Rule_B$ in layer 2), layer 2 does not satisfy the nondeletion layer conditions: elements of types P and PA can be found in the LHS of $Rule_B$, but all elements of the same type were not already created in this or previous layer.

In spite of missing layers, a *de-facto* rule application precedence can be observed: $Rule_B$ is only applied once $Rule_A$ created the necessary $PA \xrightarrow{a} P$ nodes to a particular E node. Rules are applied nondeterministically, but still, on a certain match $Rule_B$ follows $Rule_A$. This notion of *precedence* can be formalised as a *produce-enable* dependency: production rules p_1, p_2 are *produce-enable* dependent if p_1 produces some objects that enable the application of p_2.

There are other types of sequential dependencies, besides *produce-enable*. In *delete - forbid* dependency, one rule application deletes a graph object which is in the match of another rule application. The *change - use attribute* dependency means that one rule application changes attributes being in the match of another rule application. In the *produce - forbid* dependency, one rule application generates graph objects in a way that a graph structure would occur which is prohibited by a NAC of another rule application [AGG07]. Since both *delete - forbid* and *change - use attribute* dependencies involve deletion (i.e. attribute change involves deletion of attribute edges from graph to data nodes), they would not occur in nondeleting transformations. Even though it is possible to have *produce - forbid* dependent rule pairs in the presence of NACs, they will not create new matches. As we are interested in connections where new matches may be created, only the *produce-enable* dependency remains.

Definition 7. (Produce-Enable Dependency). *Given a graph transformation system $GTS = (TG, P)$, rules p_1, p_2 with NACs NAC_{p_1}, NAC_{p_2} are in a produce-enable dependency, denoted by $p_1 \xrightarrow{pe} p_2$ if there exist two direct graph transformations $t_1 : G \xRightarrow{p_1, m_1} H_1$, $t_2 : H_1 \xRightarrow{p_2, m_2} H_2$ such that*

1. $\nexists h_{21} : L_2 \to D_1 : e_1 \circ h_{21} = m_2$.
2. $\exists h_{12} : R_1 \to D_2 : d_2 \circ h_{12} = m_1'$.

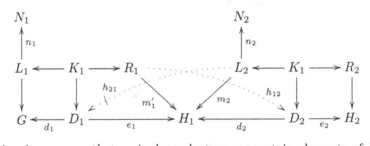

Condition 1 expresses that p_2 is dependent on p_1: certain elements of L_2 that are not gluing items in p_1 and not present in the host graph beforehand. Condition 2 ensures that the items created by p_1 are not deleted by p_2 (i.e. they are gluing items). Condition 2 is always satisfied for non-deleting rules—however, the definition is required for general rules as detailed in Section 4.

The *produce-enable* relation is a *binary relation* on production rules. Since two different rules may be mutually produce-enable dependent on each other ($p_1 \overset{pe}{\to} p_2$ and $p_2 \overset{pe}{\to} p_1$ for $p_1 \neq p_2$), it not in general *symmetric* or *antisymmetric*. If the transitive closure $\overset{pe*}{\to}$ of $\overset{pe}{\to}$ is irreflexive, no production rule is *produce-enable* dependent directly or indirectly on itself. Hence rules would not be able to produce corresponding matches for each other indefinitely, resulting in infinite rule application sequence. Hence the system must be terminating.

Theorem 1. (Termination of Nondeleting GTS). *Given a graph transformation system $GTS = (TG, P)$ such that all rules are nondeleting and have self-disabling NACs. If the start graph G_0 and the set of rules P are finite and the transitive closure of the* produce-enable *dependency is irreflexive, the GTS is terminating.*

Proof. Termination is proven by contradiction, hence we assume the existence of an infinite rule application sequence σ_{inf}.

Since P contains m rules, and σ_{inf} is an infinite application sequence, by the Pigeonhole principle, all rule applications cannot be distinct. Thus, there is set of rules $\{r_i, ..., r_j\} \in P$ that are applied infinitely many times, where $0 < i, j \leq m$. (We do not assume in general that all rules are in one cycle.)

For each direct derivation $G_i \overset{r_i}{\Longrightarrow} G_{i+1}$ with injective matches there is an injective morphism $d_i : G_i \to G_{i+1}$ because r_i is nondeleting. Each match $m_{i+1} : L_i \to G_{i+1}$ has an *essential match* $m_i : L_i \to G_i$ with $d_i \circ m_i = m_{i+1}$. From Lemma 1 we conclude that we have at most one application of the rule r_i on the essential match m_i.

Thus the rules in the set $\{r_i, ...r_j\}$ create matches for each other. This means that there is at least one cycle where r_i creates a match for r_{i+1}, then r_{i+1} creates a match for r_{i+2}, etc, and finally r_j creates a match for r_i. This means that $r_i \overset{pe}{\to} r_{i+1} \overset{pe}{\to} ... \overset{pe}{\to} r_j \overset{pe}{\to} r_i$. However, this means that the transitive closure relation is not irreflexive, i.e. $r_i \overset{pe*}{\to} r_i$, which contradicts our assumptions. Thus, an infinite sequence of rule applications cannot exist.

4   Termination with GTS Components

Unfortunately Theorem 1 does not carry over to GTSs with general rules. Although deleting rules cannot be applied twice on the same match, a rule may delete elements contained in the NAC of another rule, thus enabling it again on the same essential match and resulting in a possibly endless cycle. Thus we extend the criterion presented in Section 3 to GTSs with deleting rules. We assume that the start graph is kept intact and thus the effect of the deleting rules can be isolated into subsets of rules, called components. Then, we show that if the termination of all individual components can be proven, the entire GTS is terminating. The main advantage of this approach is that the criteria used to prove the termination of a component can be chosen freely.

Definition 8. (Protective Transformation). *A graph transformation system GTS = (TG, P) with NACs is* protective *if its rules never delete an element of any given start graph, that is, for any TG-typed start graph G_0, all transformations $G_0 \Longrightarrow^* H$ preserves G_0.*

It is easy to mistake a *protective* transformation to be *nondeleting*. The difference is that while in *nondeleting* transformations no deletion is allowed at all (i.e. all rules are nondeleting), in *protective* transformations the rules may delete objects created by the transformation.

This condition does not present a serious limitation for exogenous transformations, such as *PIM to PSM* transformations and others, generating one model from another one while keeping the source models intact. Also, all one-way transformations generated from bidirectional rules in triple graph grammars [Sch94] satisfy this criterion.

For a motivating example, we add two rules to R_{sample}: $Rule_{D_1}$ and $Rule_{D_2}$ as shown in Figure 2. Both *NACs* are connected to the LHSs of the respective rules. The rules are converting a *1-to-n* junction to a binary tree bottom up. First, an arbitrary *E*-node is matched creating the lowest element of the tree. Then, the tree is built by adding the elements one-by-one. However, $Rule_{D_2}$ is deleting: as shown in Figure 2(b), the *p* edge from the *4:PA* node to the *6:C* node is deleted. Hence Theorem 1 does not apply directly to R_{sample} because it contains deleting rules. The strategy to deal with the deleting rules is to isolate them.

R_{sample} can be sorted into subsets responsible for transforming certain elements. For instance, $Rule_{D_1}$ and $Rule_{D_2}$ will not interfere with the elements matched by $Rule_B$. This means that such a component is self-contained: deletion is contained within that group; one rule deletes only the product of other rules in the same group. Two rules will be in one such component if they are either *produce-delete* dependent, or *delete-enable* dependent.

Definition 9. (Delete-Enable Dependency). *Given a graph transformation system GTS = (TG, P), rules p_1, p_2 with NAC_{p_1}, NAC_{p_2} are in a* delete-enable dependency, *denoted by $p_1 \overset{de}{\rightarrow} p_2$ if there exist two direct transformations t_1 : $G \overset{p_1, m_1}{\Rightarrow} H_1$, $t_2 : H_1 \overset{p_2, m_2}{\Rightarrow} H_2$ such that for all NAC $n_2 \in NAC_{p_2}$ there exists an injective $e : N_2 \rightarrow G$, but no injective $e' : N_2 \rightarrow D_1$ with $d_1 \circ e' = e$.*

That means, p_1 deletes elements that are part of an occurrence of the negative pattern N_2 prohibiting the application of rule p_2 at match $e \circ n_2$.

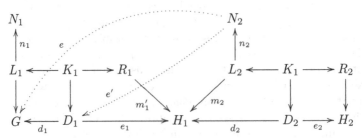

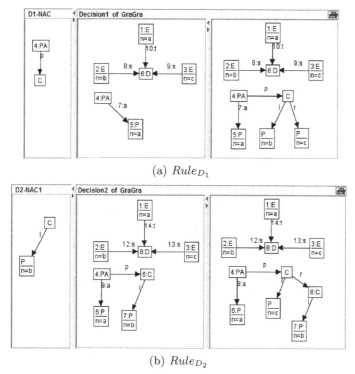

(a) $Rule_{D_1}$

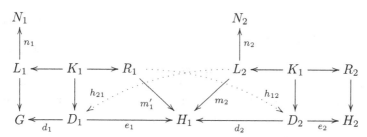

(b) $Rule_{D_2}$

Fig. 2. Rules $Rule_{D_1}$ and $Rule_{D_2}$

Definition 10. (Produce-Delete Dependency). *Given a graph transformation system $GTS = (TG, P)$, rules p_1, p_2 are in a produce-delete dependency, denoted by $p_1 \overset{pd}{\to} p_2$ if there exist two direct graph transformations $t_1 : G \overset{p_1, m_1}{\Rightarrow} H_1$, $t_2 : H_1 \overset{p_2, m_2}{\Rightarrow} H_2$ with neither $h_{12} : R_1 \to D_2$ such that $d_2 \circ h_{12} = m_1'$ nor $h_{21} : L_2 \to D_1$ such that $e_1 \circ h_{21} = m_2$.*

$$
\begin{array}{ccc}
N_1 & & N_2 \\
\uparrow n_1 & & \uparrow n_2 \\
L_1 \longleftarrow K_1 \longrightarrow R_1 \cdots \cdots \cdots \cdots L_2 \longleftarrow K_1 \longrightarrow R_2 \\
\downarrow \quad\quad \downarrow \quad h_{21} \qu\quad m_1' \searrow \quad m_2 \swarrow \quad h_{12} \quad \downarrow \qu\quad \downarrow \\
G \longleftarrow_{d_1} D_1 \longrightarrow_{e_1} H_1 \longleftarrow_{d_2} D_2 \longrightarrow_{e_2} H_2
\end{array}
$$

That means, p_2 deletes elements that were created by p_1 and enabled the application of p_2.

Definition 11. (GTS Component). *Given a typed graph transformation system $GTS = (TG, P)$, a subset $P_i \subset P$ is a component (of GTS) if, when $p \in P_i$ is produce-delete dependent or delete-enable dependent on $q \in P$, then $q \in P_i$.*

From a different perspective, this means that deletion only happens within the subgraph that was created by the rules of the component. If p is nondeleting and there is no $q \in P$ such that q is sequentially dependent, but not *produce-enable* dependent on p, then it forms its own component, i.e. $P_i = \{p\}$.

Let us show the formation of components in our example. Consider the rules $Rule_A$, $Rule_B$, $Rule_{D_1}$ and $Rule_{D_2}$ in R_{sample}. The dependency matrix of these rules is shown in the table below:

	$Rule_A$	$Rule_B$	$Rule_{D_1}$	$Rule_{D_2}$
$Rule_A$	no	$\overset{pe}{\rightarrow}$	$\overset{pe}{\rightarrow}$	no
$Rule_B$	no	no	no	no
$Rule_{D_1}$	no	no	no	$\overset{pd}{\rightarrow}$
$Rule_{D_2}$	no	no	no	$\overset{pd}{\rightarrow}$

As the P node created by $Rule_A$ may trigger an application of $Rule_B$ or $Rule_{D_1}$, $Rule_B$ and $Rule_{D_1}$ are *produce-enable* dependent on $Rule_A$. Since they are not *delete-enable* or *produce-delete* dependent on each other, according to Definition 11 there is no reason to put them in the same component. Thus they will be in different components. The dependency between $Rule_{D_1}$ and $Rule_{D_2}$ is more interesting: they seem to be *produce-enable* dependent, as $Rule_{D_2}$ matches the $C \rightarrow P$ nodes created by $Rule_{D_1}$. However, according to Condition 2 in Definition 7 the connected C and P nodes created by $Rule_{D_1}$ should be gluing items in $Rule_{D_2}$. Apparently they are not, because the p edge between nodes $4{:}PA$ and $6{:}C$ is deleted. This is exactly why Condition 2 is necessary: the deleting rule has to be grouped with the rules it deletes from. Thus we have three components:

$$\{Rule_A\}, \{Rule_B\}, \{Rule_{D_1}, Rule_{D_2}\}$$

It is important to see that components are not layers according to [EEdL+05]. The most apparent difference is the rule application mechanism. Layers follow each other sequentially: the transition to layer $n + 1$ occurs after all rules are applied to all possible matches in layer n. Rule application in a graph transformation system with components is 'normal': rules are applied nondeterministically. The application of rules between different components can hence interleave. For instance, after an application of $Rule_A$ and $Rule_{D_1}$ there is no restriction to $Rule_{D_2}$. We can apply $Rule_A$ or $Rule_{D_1}$ again (on a different match obviously). The second difference between layers and components is their meaning. Layers are creation and deletion layers dedicated to rules creating or deleting a specific type. Components on the other hand cover rules that are sequentially dependent on each other and thus work on the same subgraphs.

Definition 12. (Non-Interfering Rule System). *Given a graph transformation system $GTS = (TG, P)$. The subsets of rules $R_1, R_2, \ldots R_n \subset P$ form a non-interfering rule system if they are pairwise disjoint, their union equals P and each R_i is a component of GTS.*

It is important to see the difference between a simple *rule set*, a *rule component*, and a *non-interfering rule system*. Given a graph transformation system $GTS =$

(TG, P), a simple *rule set* R is trivially an arbitrary set of rules $R \subseteq P$. However, if we have several rule sets R_i that are pairwise disjoint and their union gives P it is *not* a non-interfering rule system. A non-interfering rule system consists of such rule sets that are *rule components* as well, i.e. *produce-delete* and *delete-enable* rules grouped together. Thus a *rule-system* can be interfering in two ways: (i) the rule sets are not components, or (ii) the components are not pairwise disjoint or their union does not equal P.

In order to prove termination, we lift the definition of *produce-enable dependency* (Def. 7) to incorporate components.

Definition 13. *(PE Dependency for Components).* *Components P and R are produce-enable dependent, if there exist rules $r \in R$ and $p \in P$ such that $r \xrightarrow{pe} p$.*

In the following we extend Theorem 1 to GTS components. The idea is that arbitrary termination criteria can be used to prove the termination of the individual components. Once their termination is ascertained, we use the transitive closure of the precedence relation to show that they would not produce corresponding matches for each other infinitely.

In order to establish Theorem 2 it is important to show that if p enables the application of q, they are either in one component, or they are *produce-enable* dependent.

Lemma 2. *(Enabling Rules).* *Given a graph transformation system $GTS = (TG, P)$ with rules $p, q \in P$. If p creates a match for q, then either $p \xrightarrow{pe} q$ or there exists a component R_i of GTS such that $p, q \in R_i$.*

Proof. A match can be created by p for q if q is either *produce-enable*, *produce-delete*, or *delete-enable* dependent on p. Any other dependencies do not create new matches for q, but prevent the application of p after q. Thus, the lemma follows from Definition 11.

Theorem 2. *(Termination of GTSs with With Non-Interfering Rule System).* *Given a protective graph transformation system $GTS = (TG, P)$ with non-interfering rule system $P = R_1, R_2, \ldots R_n$ such that all rules have a self-disabling NAC (Def. 6). If the start graph G_0 is finite, the transitive closure relation of the produce-enable dependency on components is irreflexive and the components R_i terminate individually, then GTS is terminating.*

Proof. Termination is proven by contradiction, hence we assume the existence of an infinite rule application sequence σ_{inf}. By the termination of each component σ_{inf} can be decomposed into a sequence of finite sequences $\sigma_1 \sigma_2 \ldots$ with each σ_i being a maximal sequence of transformations using rules from a single component only.

Since the sum of rule applications of $\sigma_1 \sigma_2 \ldots$ is finite, and σ_{inf} is an infinite application sequence, by the Pigeonhole principle, all rule applications cannot be distinct. Thus, there is a set of $m \leq n$ rules $\{r_i, \ldots, r_j\} \in P$ that are applied infinitely many times, where $0 < i, j \leq m$. (We do not assume in general that all rules are in one cycle.) The rules $\{r_i, \ldots, r_j\} \in P$ cannot be in one component as their termination was assumed.

According to Lemma 2, a rule $r_k \in R_k$, $(k \in [n])$ creates matches for $r_l \in R_l$ $(l \in [n])$ only if they are *produce-enable* dependent or belong to the same component.

Thus, the individual rules $\{r_i, \ldots, r_j\} \in P$ are nondeleting and form components themselves. For each direct derivation $G_i \overset{r_i}{\Longrightarrow} G_{i+1}$ with injective matches there is an injective morphism $d_i : G_i \to G_{i+1}$ because r_i is nondeleting. Each match $m_{i+1} : L_i \to G_{i+1}$ has an *essential match* $m_i : L_i \to G_i$ with $d_i \circ m_i = m_{i+1}$. From Lemma 1 we conclude that we have at most one application of rule r_i on essential match m_i.

Thus the rules in set $\{r_i, \ldots, r_j\}$ create matches for each other, i.e., there is at least one cycle, where $r_i \overset{pe}{\to} r_{i+1} \overset{pe}{\to} \ldots \overset{pe}{\to} r_j \overset{pe}{\to} r_i$. However, this means that the transitive closure is not irreflexive, i.e. $r_i \overset{pe*}{\to} r_i$, which contradicts our assumptions. Thus, an infinite rule application sequence does not exist.

5   Application in Practice

In order to apply the termination criteria established in Theorems 1 and 2, three tasks need to be accomplished:

1. Determination of the sequential dependencies between the rules.
2. The definition of the *rule components*.
3. Search for the directed cycles within the sequential dependencies.

At the moment, only the first task is automated. The Attributed Graph Grammar System (AGG) [AGG07] provides verification facilities for graph transformations. Besides critical pair analysis, AGG can generate all dependencies of rule applications. It is important to note that AGG finds all possible dependency types between rules, so we need to classify and select *produce-enable*, *produce-delete* and *delete-enable* dependencies for our purposes.

We used the theoretical results to prove the termination of the *Activity Diagram to CSP* transformation [VAB+08, DB08]. The transformation which provides a denotational *semantic mapping* for UML Activity Diagrams [OMG06] into Communicating Sequential Processes (CSP) [Hoa85] is denoted by GTS_{smc} with type graph TG_{root} and set of rules P_{smc} (listed in Figure 4). The individual rule design of the semantic mapping was inspired by triple graph grammars, although TGGs were never used for implementation. The creation of target elements, in combination with negative application conditions on the target model, allows us to retain the input model and restrict ourselves to protective rules, preserving the input model. These two properties are important for Theorem 1. As mentioned in Section 1 existing termination criteria were insufficient for our case.

The basic elements of CSP are processes. A *process* is the behaviour pattern of an object with an alphabet of a limited set of events. Processes are defined using recursive process equations with guarded expressions. The syntax of the process equations is the following.

$$P ::= event \rightarrow P \mid P \,\square\, Q \mid P \parallel Q \mid P \setminus a \mid SKIP \mid STOP$$

The prefix $a \rightarrow P$ performs action a and then behaves like P. The process $P \,\square\, Q$ represents external choice between processes P and Q. The process $P \parallel Q$ behaves as P and Q engaged in a lock-step synchronisation. Hiding $P \setminus a$ behaves like P except that all occurrences of event a are hidden. $SKIP$ represents successful termination, $STOP$ is a deadlock [Hoa85].

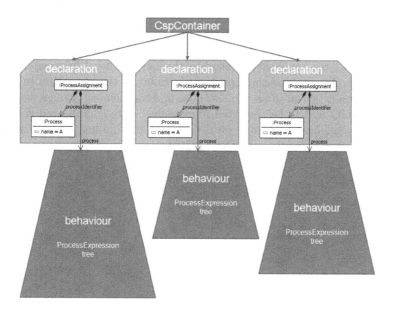

Fig. 3. The Structure of the CSP graph

Figure 3 shows a generic outline of the abstract syntax tree of CSP: the root node is a *Csp-Container* instance connected to process declarations. A *process declaration* is a *process assignment* that is identified by a *process* instance contained via the *processAssignment* aggregation. Further depth is given to the tree by the connected behaviour. We call these connected behaviour-trees *process expression subtrees*, *PE-trees* for short. The transformation builds the abstract syntax tree of CSP expressions top-down: the elements closer to the root are created first, then the branches. The process declarations are created first from control flow edges of the activity diagram. Then, the empty declarations are completed with behaviour in terms of *PE-trees*. One PE-subtree corresponds to one activity element, i.e. decision node, forknode, action. The rule-design reflects this: P_{smc} is sorted into named subsets $P_i \subset P_{smc}$, each responsible for transforming a certain element of the activity diagram. It is important to observe that one rule group does not interfere with the *PE-tree* of another group. For instance, the rules responsible for building the subtree associated with a *decision node* will not modify or create a subtree that describes the behaviour of a *join node*.

The behaviour transformation follows a delocated design. First, all the edges are transformed to the corresponding process declarations by the *BhEdge* rule which is similar to $Rule_A$ in Figure 1(a). Then, the various nodes fill the empty process definitions. The *BhAction* rule, similar to $Rule_B$ in Figure 1(b), transforms an *action node* to a prefix in CSP. *BhInit* creates the root process, *BhFinal* fills the declaration with a $SKIP$ process. $Rule_{D_1}$ and $Rule_{D_2}$ introduced in Figure 2 are the skeletons for the *BhDecision1,BhDecision2*, *BhFork1,BhFork2* and *BhForkNoJoin1,BhForkNoJoin2* rule groups. While the the D nodes are *decision nodes* and the C nodes are *choices* in the *decision* group, in the *fork* groups they are *fork nodes* and *parallel compositions* respectively. The *BhJoin* synchronises and the *BhMerge* merges the multiple flows back to one process.

Using AGG, we managed to generate the dependencies between the rules used to specify the semantic mapping. Unfortunately this operation takes long hours and cannot be ran in one go. Because of Java memory management characteristics, AGG had to be restarted repeatedly before complicated rule pairs. With the dependencies at hand, we determined the minimal *rule components* and *PE-dependency graph* as shown in Figure 4.

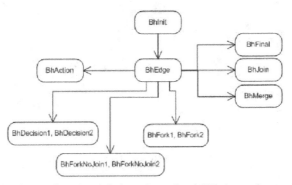

Fig. 4. Dependency graph based on the AGG dependency check

According to Figure 4, there are no directed cycles in the *PE-dependency* graph. Furthermore, the rules form a non-interfering rule system and all rules have self-disabling NACs. Thus, after showing the termination of the three rule components using [LPE07], we concluded that the *UML to CSP* transformation system, i.e. $GTS_{smc} = (P_{smc}, TG_{root})$ with a finite start graph G_0 and injective matches is *terminating*.

6   Conclusions

In this paper we addressed the termination problem for graph transformations. A generic termination criterion was established for nondeleting graph transformations. To enable the combination of various termination criteria, structural conditions were provided on the rules of a graph transformations system. To apply the theoretical results in practice, AGG was employed to check minimal dependencies in the rules of graph transformation systems.

Future work includes to investigate the possibilities of designing rule sets that are terminating in the first place, i.e. turning the termination criteria into a design method.

References

[AGG07]   AGG - Attributed Graph Grammar System Environment (2007),
           `http://tfs.cs.tu-berlin.de/agg`
[BKPPT05] Bottoni, P., Koch, M., Parisi-Presicce, F., Taentzer, G.: Termination of
           high-level replacement units with application to model transformation.
           Electronic Notes Theoretical Computer Science 127(4), 71–86 (2005)
[Bog95]   Bognar, M.: A survey of abstract rewriting. Master's thesis, VU University Amsterdam (1995)
[CMR+97]   Corradini, A., Montanari, U., Rossi, F., Ehrig, H., Heckel, R., Löwe, M.:
           Algebraic approaches to graph transformation - part i: Basic concepts
           and double pushout approach. In: Handbook of Graph Grammars, pp.
           163–246 (1997)
[DB08]   Ehrig, H., Bisztray, D., Heckel, R.: Verification of architectural refactoring rules. Technical report, Department of Computer Science, University
           of Leicester (2008),
           `http://www.cs.le.ac.uk/people/dab24/refactoring-techrep.pdf`
[EEdL+05] Ehrig, H., Ehrig, K., de Lara, J., Taentzer, G., Varró, D., Varró-Gyapay,
           S.: Termination criteria for model transformation. In: Cerioli, M. (ed.)
           FASE 2005. LNCS, vol. 3442, pp. 49–63. Springer, Heidelberg (2005)
[EEPT06]   Ehrig, H., Ehrig, K., Prange, U., Taentzer, G.: Fundamentals of Algebraic
           Graph Transformation (Monographs in Theoretical Computer Science).
           An EATCS Series. Springer, New York (2006)
[Hoa85]   Antony, C., Hoare, R.: Communicating Sequential Processes. Prentice
           Hall International Series in Computer Science. Prentice Hall, Englewood
           Cliffs (April 1985)
[LPE07]   Levendovszky, T., Prange, U., Ehrig, H.: Termination criteria for dpo
           transformations with injective matches. Electron. Notes Theor. Comput.
           Sci. 175(4), 87–100 (2007)
[OMG06]   OMG. Unified Modeling Language, version 2.1.1 (2006),
           `http://www.omg.org/technology/documents/formal/uml.htm`
[Plu95]   Plump, D.: On termination of graph rewriting. In: Graph-Theoretic Concepts in Computer Science, pp. 88–100 (1995)
[Sch94]   Schürr, A.: Specification of graph translators with triple graph grammars.
           In: Tinhofer (ed.) WG 1994, vol. 903, pp. 151–163. Springer, Heidelberg
           (1994)
[VAB+08]   Varró, D., Asztalos, M., Bisztray, D., Boronat, A., Dang, D.-H., Geiß,
           R., Greenyer, J., Gorp, P., Kniemeyer, O., Narayanan, A., Rencis, E.,
           Weinell, E.: Transformation of uml models to csp: A case study for graph
           transformation tools, pp. 540–565 (2008)

Graph Rewriting in Span-Categories

Michael Löwe

Fachhochschule für die Wirtschaft Hannover
michael.loewe@fhdw.de

Abstract. There are three variations of algebraic graph rewriting, the double-pushout, the single-pushout, and the sesqui-pushout approach. In this paper, we show that all three approaches can be considered special cases of a general rewriting framework in suitable categories of spans over a graph-like base category. From this new view point, it is possible to provide a general and unifying theory for all approaches. We demonstrate this fact by the investigation of general parallel independence. Besides this, the new and more general framework offers completely new ways of rewriting: Using spans as matches, for example, provides a simple mechanism for universal quantification. The general theory, however, applies to these new types of rewriting as well.

1 Introduction

All three main variations of algebraic graph rewriting[1] use spans, i. e. pairs $(L \xleftarrow{l} K \xrightarrow{r} R)$ of morphisms, as transformation *rules* and morphisms $m : L \to G$ as *matches* in a graph G. They differ in the format of the rules and the construction of the direct derivation for the application of a rule at a match.

In the double-pushout approach DPO [8,4,6], both rule morphisms l and r are monomorphisms. The application $(l^* : D \rightarrowtail G, r^* : D \rightarrowtail H)$ of a rule at a match m is given by the pushout complement (l^*, m') of (l, m), if it exists, and the pushout (r^*, m^*) of (r, m').

In the single-pushout approach SPO [14,7], only l is monomorphic. The application (l^*, r^*) of a rule at a match m is the pushout $((l^*, r^*), (m_i, m^*))$ of (l, r) and (id, m) as *partial* morphisms. If the match is *conflict-free*, i. e. it does not identify items in the image of l with items outside the image of l, m_i is the identity, (l^*, m') can be constructed as the final pullback complement of (m, l), and (r^*, m^*) as the pushout of (r, m').

In the sesqui-pushout approach SqPO [3], all spans are rules. The application (l^*, r^*) of a rule at a match m is given by the final pullback complement (l^*, m') of (l, m), if it exists, and the pushout (r^*, m^*) of (r, m').

Hence, all algebraic approaches are very similar as far as the syntax is concerned. We show in this paper, that they also coincide semantically. Under very

---

[1] We do not consider double-pullback rewriting as in [11], since rewriting is kind of indeterministic here. Also the pullback complements in [1] are not unique up to isomorphism, providing again some impractical indeterminism.

H. Ehrig et al. (Eds.): ICGT 2010, LNCS 6372, pp. 218–233, 2010.
© Springer-Verlag Berlin Heidelberg 2010

mild restrictions, all three constructions for the direct derivation turn out to be variants of a general construction in the category of spans over the underlying graph-like base category. This new perspective provides a unifying approach to the theory of all algebraic rewriting mechanisms. And using spans as matches extends the expressiveness of graph rewriting considerably.

The paper is organized as follows: Section 2 introduces the category of abstract spans $Span(C)$ over a category C with finite limits and colimits. Section 2.1 provides the central gluing construction for a rewrite step in $Span(C)$. This construction uses "Final-Pullback-Complements (FPC)". We also provide the general results about FPCs that are needed in the following. Section 2.2 investigates the category of spans over graph-like structures. FPCs are characterized and additional properties of FPCs in these special categories are provided. Section 3 defines a general framework for algebraic graph rewriting based on the gluing construction provided in section 2. Here, we show that all DPO-, SqPO- and almost all SPO-rewriting can be considered a special case of this new framework. The reformulation of the parallel independence properties shows that a unifying theory is possible. Section 4 demonstrates the new possibilities that are offered by the framework: We introduce a new algebraic graph rewriting approach which uses spans as matches and integrates universal quantification into the matching process. Parallel independence for this new approach turns out to be a special case of the theory developed in section 3 as well.

2   Span Categories and the Gluing Construction

In this section, we investigate a pushout-like construction in categories of abstract spans over a suitable base category. Our construction uses the ideas of [13,16], especially [16]. However, we use *abstract* spans and *do* not restrict the left- and right-hand side of the span to some subcategories a priori. We construct the final triples of [16] explicitly as a pullback and two final pullback complements. On the one hand, we lose the pushout property in the category of spans and some modelling power, especially wrt. the single-pushout approach. On the other hand, the construction becomes much simpler and easier to comprehend in concrete categories and the required composition property that is needed for parallel independence can be restated.

Definition 1. (Span Category) Let C be a category with all finite limits. A *span* over C is a pair (g, h) of morphisms with the same domain. Spans $(g, h), (g', h')$ are *equivalent* if there is an isomorphism i with $g' \circ i = g$ and $h' \circ i = h$. The class of all spans equivalent to a given one is called *abstract span*. For two spans $(g, h), (m, n)$ such that the codomain of n and g coincide, the *composition* is defined by $(g, h) \circ (m, n) = (m \circ g', h \circ n')$, where (g', n') is the pullback of (g, n). The equivalence on spans is a congruence wrt. composition and composition is associative.[2] Thus, composition is an associative operation on abstract spans. The *identity span* for an object $A \in C$ is the abstract span represented by

---

[2] Due to composition and decomposition properties of pullbacks.

(id_A, id_A). It is left and right neutral wrt. composition. Therefore, abstract spans with the given composition and identities constitute a category $Span(C)$.

2.1 Gluing of Abstract Spans

The category of abstract spans $Span(C)$ does not have all limits and colimits even if C has. Therefore we cannot not use span pushouts as a direct transformation in a rewriting system. Instead we provide a "pushout-like"[3] construction for rewriting in span categories using (special) pullbacks and pushouts of C.

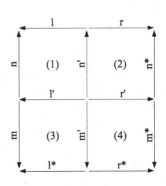

Fig. 1. Gluing

Definition 2. (Gluing of Abstract Spans) Let $Span(C)$ be the category of abstract spans over some category C. A commutative diagram as in figure 1, i. e. $(l \circ n', m^* \circ r') = (n^*, m^*) \circ (l, r) = (l^*, r^*) \circ (n, m) = (n \circ l', r^* \circ m')$ is called a *gluing diagram* if (1) (l', n') is pullback of (l, n), (2) (n^*, r') is final pullback complement of (r, n'), (3) (l^*, m') is final pullback complement of (m, l'), r' and m' are jointly monic, and (4) (r^*, m^*) is pushout of (r', m').[4]

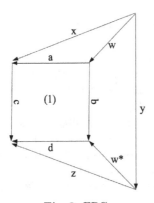

Fig. 2. FPC

The notion of gluing uses the concept of "Final-Pullback-Complement (FPC)".

Definition 3. (Final Pullback Complement) A pair of morphisms (d, b) is the *final pullback complement* of a pair (c, a), if (a, b) is the pullback of (c, d) and for each collection of morphisms (x, y, z, w), where (i) (x, y) is the pullback of (c, z) and (ii) $a \circ w = x$, there is a unique morphism w^* with $d \circ w^* = z$ and $b \circ w = w^* \circ y$ (compare figure 2).

Due to their universal property, final pullback complements are unique up to isomorphism, if they exist.

Corollary 4. (Unique Gluing) If $((n_1^*, m_1^*) : B \to D_1, (l_1^*, r_1^*) : C \to D_1)$ and $((n_2^*, m_2^*) : B \to D_2, (l_2^*, r_2^*) : C \to D_2)$ are two gluings of the same pair of abstract spans $((l, r) : A \to B, (n, m) : A \to C)$ in $Span(C)$, the objects D_1 and D_2 are isomorphic in C.

Proof. Direct consequence of the uniqueness up to isomorphism of pullbacks, final pullback complements and pushouts in C.

---

[3] The notion "pushout-like" means that the rewrite diagrams can be composed and (under some conditions) decomposed as pushouts can.

[4] In this case $((n^*, m^*), (l^*, r^*))$ is called the *gluing* of $((l, r), (n, m))$.

Proposition 5. (Composition/Decomposition of FPCs) Final pullback complements possess the following properties:

1. (Horizontal Composition) If (d, b) is FPC of (c, a) and (g, f) is FPC of (b, e), then $(d \circ g, f)$ is FPC of $(c, a \circ e)$.
2. (Vertical Composition) If (c, a) is FPC of (d, b) and (b, e) of (g, f), then $(c, a \circ e)$ is FPC of $(d \circ g, f)$.
3. (Vertical Decomposition) If (c, h) is FPC of $(d \circ g, f)$ and d is monomorphic, then the decomposition into two pullbacks provides two final pullback complements.

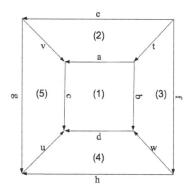

Fig. 3. Commutative Cube

Proposition 6. (FPCs are Stable under Pullbacks) Consider figure 3. If (1)-(5) is a commutative diagram, (1) is final pullback complement and (3), (4) as well as (5) are pullbacks, then (h, f) is the final pullback complement of (g, e).

Definition 7. Consider a commutative diagram in a category C as in figure 3.

1. *Pushouts are stable under pullbacks* in C if (g, h) is pushout of (e, f) whenever (1) is pushout and (2) - (5) are pullbacks and
2. *FPCs are stable under pullbacks and pushouts* in C if (h, f) is FPC of (g, e), whenever (d, b) is FPC of (c, a), (5) is pushout with jointly monic morphisms w and h and (2) as well as (4) are pullbacks.

Theorem 8. (Composition of Gluings) If pushouts are stable under pullbacks and final pullback complements are stable under pullbacks and pushouts in a category C then gluings in $Span(C)$ are compositional. I. e. if $((l^*, r^*), (n^*, m^*))$ is gluing of $((l, r), (n, m))$ and $((p^*, q^*), (\overline{n}, \overline{m}))$ of $((p, q), (n^*, m^*))$, then $((p^*, q^*) \circ (l^*, r^*), (\overline{n}, \overline{m}))$ is the gluing of $((p, q) \circ (l, r), (n, m))$.

Proof. Consider figure 4. Let (x, y) be the pullback of (r, p) such that $(l \circ x, q \circ y) = (p, q) \circ (l, r)$ and (x^*, y^*) be the pullback of (r^*, p^*) such that $(l^* \circ x^*, q^* \circ y^*) = (p^*, q^*) \circ (l^*, r^*)$. Construct (x', y') as the pullback of (r', p') and v and w as the

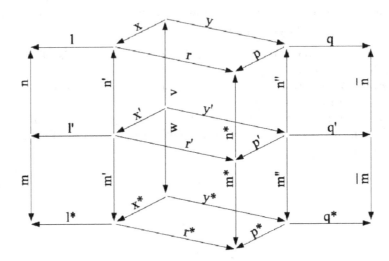

Fig. 4. Compositionality of Gluing

unique morphisms such that the whole diagram commutes. Since (p', n'') and (x', y') are pullbacks and $n^* \circ r' = r \circ n'$ and $n'' \circ y' = y \circ v$ the pair $(v \circ y, x')$ is pullback of $(p, r \circ n')$. Since (x, y) is pullback, (v, x') is pullback. Similar arguments provide that (v, y') and (x', w) are pullbacks. By compositionality of pullbacks we obtain that $(q' \circ y', v)$ is pullback of $(\overline{n}, q \circ y)$ and $(l' \circ x', w)$ is pullback of $(m, l^* \circ x^*)$. Thus the diagram commutes in $Span(C)$. Again by compositionality of pullbacks $(l \circ x, v)$ is pullback of $(n, l' \circ x')$. By proposition 6 (n'', y') is FPC of (y, v). And by proposition 5 $(\overline{n}, q' \circ y')$ is FPC of $(q \circ y, v)$. Since FPCs are stable under pushouts and pullbacks (x^*, w) is FPC of (m', x'). Proposition 5 guarantees that $(l^* \circ x^*, w)$ is FPC of $(m, l' \circ x')$. Stability of pushouts under pullbacks states that $(m'', y*)$ is pushout of (y', w). Pushouts compose. Hence $(\overline{m}, q^* \circ y^*)$ is pushout of $(w, q' \circ y')$. Let $y' \circ h = y' \circ k$ and $w \circ h = w \circ k$. Then $m' \circ x' \circ h = m' \circ x' \circ k$ and $r' \circ x' \circ h = r' \circ x' \circ k$. Since (r', m') are jointly monic $x' \circ h = x' \circ k$. Since (w, x') is pullback $h = k$. Thus (w, y') are jointly monic. Since jointly monic diagrams compose $(q' \circ y', w)$ is a pair of jointly monomorphic morphisms.

2.2   Gluing in the Category of Graphs

Final pullback complements (FPC), an essential part of the gluing construction in definition 2 do not exist for all given situations even in the category of sets and mappings. Therefore, this section investigates under which conditions FPCs exist and can be constructed in the categories of graphs. Then we show that FPCs in graphs are stable under pullbacks and pushouts (definition 7(2)). This guarantees that gluings are compositional in the category of graphs.

Definition 9. (Category of Graphs) A *graph* $G = (V; E; s, t : E \to V)$ consists of a set of vertices V, a set of edges E, and two total mappings s and t assigning the source and target vertex to every edge. A *graph morphism* $f : G \to H = (f_V : V_G \to V_H, f_E : E_G :\to E_H)$ is a pair of total mappings such that $f_V \circ s^G = s^H \circ f_E$ and $f_V \circ t^G = t^H \circ f_E$. All graphs and graph morphisms with componentwise composition and identities constitute the category $\mathcal{G}$.

Limits and colimits exist in $\mathcal{G}$ for all suitable situations and are constructed componentwise as limits resp. colimits in the category of sets for the vertices and edges. $\mathcal{G}$ is a topos, the pullback functor has a right adjoint, and pushouts are stable under pullbacks (definition 7(1)), compare [9,15].[5]
    The effective construction of final pullback complements for a pair of morphisms $(c : B \to C, a : A \to B)$ in $\mathcal{G}$ is given in two steps. First we consider the case where c is a monomorphism (construction 11).

Proposition 10. (FPCs in Graphs for Monomorphisms) The category $\mathcal{G}$ has a final pullback complement for $(c : B \to C, a : A \to B)$ if c is monomorphism.

Proposition 10 is a direct consequence of the fact that the pullback functor in $\mathcal{G}$ has a right adjoint. A concrete construction of the complement is given below.

Construction 11. (FPC Construction 1) If two graph morphisms $(c : B \rightarrowtail C, a : A \to B)$ are given and c is monomorphism, construct the final pullback complement $(d : D \to C, b : A \rightarrowtail D)$ as follows:

1. On vertices:
   $V_D = V_A + (V_C - c_V(V_B))$,
   $b_V(v) = v$ for all $v \in V_A$, and
   $$d_V(v) = \begin{cases} c_V(a_V(v)) & v \in V_A \\ v & otherwise \end{cases}.$$

2. On edges:
   $E_D = E_A + \{(s, e, t) : e \in E_C - c_E(E_B), s \in d_V^{-1}(s^C(e)), t \in d_V^{-1}(t^C(e))\}$,
   $$s^D(e) = \begin{cases} s^A(e) & e \in E_A \\ s & e = (s, e, t) \end{cases} \text{ and } t^D(e) = \begin{cases} t^A(e) & e \in E_A \\ t & e = (s, e, t) \end{cases}$$
   $b_E(e) = e$ for all $e \in E_A$, and
   $$d_E(e) = \begin{cases} c_E(a_E(e)) & e \in E_A \\ e' & e = (s, e', t) \text{ for } e' \in E_C - c_E(E_B) \end{cases}$$

For a proof that this construction yields a final pullback complement see [3]. From construction 11 we are able to extract the first necessary condition for a final pullback complement.

Condition 12. (FPC Condition 1) If $(d : D \to C, b : A \to D)$ is the final pullback complement of $(c : B \to C, a : A \to B)$ then

1. for all nodes $x \in C$ which are not in the image of c there is exactly one node $y \in D$ with $d(y) = x$ and

---

[5] Note that the following constructions can also be generalized to any slice category $\mathcal{G} \uparrow T$ of graphs 'under' T and T being a type graph.

2. for all edges $x \in C$ which are not in the image of c and each pair of nodes (z_1, z_2) with $d(z_1) = s^C(x)$ and $d(z_2) = t^C(x)$ there is exactly one edge $y \in D$ with $d(y) = x$ and $z_1 = s^D(y)$ and $z_1 = t^D(y)$.

FPCs do not always exist for arbitrary morphisms c, compare lemma 13.

Lemma 13. (Existence of FPCs in $\mathcal{G}$) The pair $(a : A \to B, c : B \to C)$ has a final pullback complement, only if (1) there is $(d : D \to C, b : A \to D)$ such that (a, b) is pullback of (c, d) and (2) for any other $(d' : D' \to C, b' : A \to D')$ such that (a, b') is pullback of (c, d') the congruence generated by b on A coincides with the congruence generated by b' on A, i. e. kernel(b) = kernel(b').

Proof. (1) is obvious. For (2) suppose $(d : D \to C, b : A \to D)$ is FPC of $(a : A \to B, c : B \to C)$. Then for any pullback (a, b') of (c, d') there is a morphism w with $w \circ b' = b$ and $d \circ w = d'$. This means that (id,b') is pullback of (b, w). Therefore w is monomorphic on the image of b' and kernel(b) = kernel(b').

The condition of lemma 13 translates to the following constructive condition for a pair of morphisms to possess a final pullback complement.

Condition 14. (FPC Condition 2) Two graph morphisms $(c : B \to C, a : A \to B)$ are called *semi-pullback* if the following conditions are satisfied:

- On nodes: if $v \neq w$ and $c_V(v) = c_V(w)$, then $|a_V^{-1}(v)| = |a_V^{-1}(w)| \leq 1$.
- On edges: if $e \neq f$ and $c_E(e) = c_E(f)$
  1. and $s^B(e) = s^B(f)$ and $t^B(e) = t^B(f)$, then for each pair $v, w \in V_A$ with $a_V(v) = s^B(e)$ and $a_V(w) = t^B(e)$, $|\{e' : e' \in a_E^{-1}(e), s^A(e') = v, t^A(e') = w\}| = |\{f' : f' \in a_E^{-1}(f), s^A(f') = v, t^A(f') = w\}| \leq 1$.
  2. and $s^B(e) = s^B(f)$ and $t^B(e) \neq t^B(f)$, then for each vertex $v \in V_A$ with $a_V(v) = s^B(e)$, $|\{e' : e' \in a_E^{-1}(e), s^A(e') = v\}| = |\{f' : f' \in a_E^{-1}(f), s^A(f') = v,\}| \leq 1$,
  3. and $s^B(e) \neq s^B(f)$ and $t^B(e) = t^B(f)$, then for each vertex $v \in V_A$ with $a_V(v) = t^B(e)$, $|\{e' : e' \in a_E^{-1}(e), t^A(e') = v\}| = |\{f' : f' \in a_E^{-1}(f), t^A(f') = v,\}| \leq 1$,
  4. and $s^B(e) \neq s^B(f)$ and $t^B(e) \neq t^B(f)$, then $|a_E^{-1}(e)| = |a_E^{-1}(f)| \leq 1$.

Construction 15. (FPC Construction 2) Let $(c : B \twoheadrightarrow C, a : A \to B)$ be a semi-pullback and let c be an epimorphism. Construct the completing pair $(b : A \to D, d : D \to C)$ as follows: $D = C|_\equiv$ where $\equiv$ is generated by $\sim =$ $\{(o_1, o_2) : a(o_1) \neq a(o_2), c(a(o_1)) = c(a(o_2))\}$, b is the natural homomorphism, and d is the unique morphism making the diagram commute.

Proposition 16. The pair $(c : B \twoheadrightarrow C, a : A \to B)$ with epimorphism c has a final pullback complement $(d : D \to C, b : A \twoheadrightarrow D)$ if and only if (c, a) is a semi-pullback. The FPC can be constructed by construction 15.

The results achieved in this section can be summarized as follows: The pair (d, b) exists as a FPC for (c, a) if and only if (c, a) is semi-pullback. It can be constructed by an decomposition of c in an epimorphism e and a monomorphism m, i. e. $c = m \circ e$, using construction 15 for (e, a) providing (b_1, x) and construction 11 for (m, x) providing (d, b_2), such that $b = b_2 \circ b_1$. Moreover the constructions 11 and 15 provide an effective criterion for FPCs in $\mathcal{G}$.

Corollary 17. (Criterion for FPC) A pullback $(a : A \to B, b : A \to D)$ of $(c : B \to C, d : D \to C)$ is a FPC for (a, c) if it satisfies conditions 12 and 14.

Proposition 18. FPCs in $\mathcal{G}$ are stable under pullbacks and pushouts.

Proof. Let a commutative cube as in figure 3 be given in $\mathcal{G}$, where (d, b) is FPC of (c, a), (2) and (4) are pullbacks, and (5) is a pushout with jointly monic morphisms g and v. We check the criterion of corollary 17. Let $x \neq y$ and $g(x) = g(y)$. Then $v(x) \neq v(y)$ since (g, v) is jointly monic and $c(v(x)) = c(v(y))$ since (5) is commutative. (1) is FPC and (c, a) is semi-pullback. If x and y are nodes, $v(x)$ and $v(y)$ have either both exactly one preimge wrt. a or no preimage. Since (2) is pullback, the same property holds for x, y wrt. e. If x and y are edges (c, a) satisfies (2) - (4) of condition 14. Since (2) is pullback, properties (2) - (4) of condition 14 are true for x and y. Thus, (g, e) is a semi-pullback.

Let x be outside the image of h. Then $u(x)$ is not in the image of c and has no other preimage besides x, since (5) is pushout. If x is a node $u(x)$ has exactly one preimage wrt. d due to (1) being FPC. Since (4) is pullback, x has exactly one preimage wrt. h. If x is an edge, $u(x)$ has a preimage z wrt. d with $s(x) = n_1$ and $t(x) = n_2$ for every pair of of nodes n_1, n_2 with $d(n_1) = s(x)$ and $d(n_2) = t(x)$. Since (4) is pullback the same property carries over to x and its preimages wrt. h. ∎

Corollary 19. Gluings are compositional in $\mathcal{G}$.

As a direct consequence of the constructions 11 and 15 we obtain:

Corollary 20. (FPC and Pushout Complements)

1. If (d, b) is pushout complement of (c, a) and a is monomorphic, then (d, b) is final pullback complement of (c, a).
2. If (d, b) is final pullback complement of (c, a) and a and c are monomorphic, then (d, b) is pushout complement of (c, a).

3 A Framework for Algebraic Graph Rewriting

In this section, we start designing a general framework for algebraic graph rewriting in the category $\mathcal{G}$ of graphs based on the gluing construction of definition 2. We show that the double-pushout approach, the single-pushout approach at conflict-free matches, and the sesqui pushout approach with its two variants, monomorphic rules and conflict-free matches and monomorphic matches with arbitrary morphisms in the rule's left-hand side, are instances of this framework. We study general parallel independence and show that the results for all four algebraic approaches are specialisations of a general result.

Definition 21. (Algebraic Graph Rewriting System) An *algebraic graph rewriting system* $GRS = (T, P = (P_t)_{t \in T})$ consists of a class of abstract spans T, called transformation rules, and a class of matching abstract spans P_t for each $t \in T$, such that $p = (n, m) \in P_{t=(l,r)}$ implies codomain(l) = codomain(n) and the gluing of (t, p), as a notion of direct derivation with t at p, exists, compare definition 2 and figure 1.

Example 22. (Monic Double-Pushout Rewriting) The double-pushout approach with monic rule morphisms is defined by the algebraic graph rewriting system $GRS_{DPO} = (T_{DPO}, P^{DPO})$: T_{DPO} contains all abstract spans where both components are monomorphic and $p = (n, m) \in P_t^{DPO}$ for a rule $t = (l, r)$, if and only if n is the identity and the pushout complement of m and l exists.

To see that every double-pushout transformation in $\mathcal{G}$ is a gluing in the category of $Span(\mathcal{G})$, add to the two pushouts (1) and (2) (compare figure 5) the rectangles (3) and (4) using a copy of the rule and identities between the original and the copy. (3) and (4) are pullbacks and it is easy to check that (4) is final pullback complement. By proposition 20, (1) is final pullback complement, since l is monomorphic. The pair (r, m') is jointly monic since r is monomorphic. This standard completion also shows that rewriting in examples 23 - 24 below is a gluing construction.

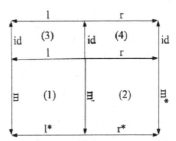

Fig. 5. DPO as Gluing

Example 23. (Sesqui-Pushout Rewriting) The two variants of sesqui-pushout rewriting provide the systems $GRS_{SePO1} = (T_{SePO1}, P^{SePO1})$ and $GRS_{SePO2} = (T_{SePO2}, P^{SePO2})$. $T_{SePO1} = T_{DPO}$ and $p = (n, m) \in P_t^{SePO1}$, if n is the identity. The rules in T_{SePO2} are all spans with a monomorphic right-hand side[6] Matches $(n, m) \in P_t^{SePO2}$ require $n = id$ and monomorphic m.

General single-pushout rewriting with arbitrary rules and matches goes slightly beyond the framework presented here. The existence of the final pullback complement requires conflict-free matches.[7] And the property of jointly monic morphisms in the pushout part of the gluing additionally restricts the matches.

Example 24. (Conflict-free Single-Pushout Rewriting) Single-pushout rewriting at conflict-free matches leads to the system $GRS_{SPO} = (T_{SPO}, P^{SPO})$ where rules are spans with monomorphic left hand sides and matches $p = (n, m) \in P_{(l,r)}$ satisfy (again) that n is the identity.[8]

---

[6] For arbitrary right-hand sides, see next section.

[7] Note that most of the theory for single-pushout rewriting requires conflict-free matches!

[8] Note that $p = (n, m) \in P_{(l,r)}^{SPO}$ requires in $\mathcal{G}$ that $m(x) = m(y) \implies (x = y \lor x, y \notin$ image(l) $\lor (x = l(x'), y = l(y'), r(x') \neq r(y'))$.

The examples show that the notion of algebraic graph rewriting system covers almost all well-known algebraic approaches to graph transformation. Since it generalises all existing approaches also theoretical results can be generalised. A first impression of the possible general theory is given by the following investigation of parallel independence.

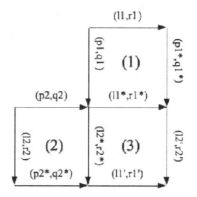

Fig. 6. Parallel Independence

Definition 25. (Parallel Indepedence in GRS) Two transformations (l_1^*, r_1^*) : $G \rightsquigarrow H$, and (l_2^*, r_2^*) : $G \rightsquigarrow K$ with rule $t_1 = (l_1, r_1)$ at match (p_1, q_1) and with $t_2 = (l_2, r_2)$ at (p_2, q_2) are *parallel independent* in a graph rewriting system, if (1) $(l_1^*, r_1^*) \circ (p_2, q_2) \in P_{t_2}$ and $(l_2^*, r_2^*) \circ (p_1, q_1) \in P_{t_1}$ and (2) the gluing of (l_1^*, r_1^*) and (l_2^*, r_2^*) exists. Parallel independence in a graph rewriting system is called *syntactical* if condition (1) implies condition (2).

Theorem 26. (Parallel Independence in GRS) If two transformations $G \rightsquigarrow H$ with rule t_1 and $G \rightsquigarrow K$ with rule t_2 are parallel independent, then there are transformations $H \rightsquigarrow X$ with rule t_2 and $K \rightsquigarrow X$ with rule t_1.

Proof. Parallel independence guarantees that t_2 is applicable after t_1 and vice versa at the induced matches. Consider figure 6. Parallel independence guarantees that the the gluing of (l_1^*, r_1^*) and (l_2^*, r_2^*) indicated by (3) exists. Since gluings compose (2)+(3) is the gluing that represents the transformation with t_2 after t_1 and (1)+(3) is the gluing, which represents the transformation of t_1 after t_2. Thus, t_1 after t_2 produces the same graph[9] as t_2 after t_1, q.e.d. Even the two traces $(l_1', r_1') \circ (l_2^*, r_2^*)$ and $(l_2', r_2') \circ (l_1^*, r_1^*)$ coincide, since (3) commutes.

That the notion of parallel independence is a generalisation of the well-known notions in the several algebraic approaches is demonstrated by the following propositions.

Proposition 27. (DPO Independence) Two transformations $G \rightsquigarrow H$ with rule t_1 and $G \rightsquigarrow K$ with rule t_2 are parallel independent in the double-pushout approach, if and only if the are parallel independent in GRS_{DPO}[10]. The two transformations $H \rightsquigarrow X$ with rule t_2 and $K \rightsquigarrow X$ with rule t_1, that exist by theorem 26, are double-pushout transformations. Parallel independence in GRS_{DPO} is syntactical.

Proof. Independence in the double-pushout approach requires two morphisms p and q, such that $l_2^* \circ p = m_1$ and $l_1^* \circ q = m_2$, figure 7. Since $a \circ b = c$ and a monic,

---

[9] Up to isomorphism.
[10] Compare example 22!

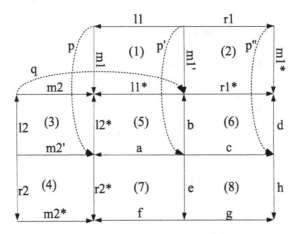

Fig. 7. DPO Independence

if and only if (b, id) is pullback of (a, c), we obtain $(l_1^*, r_1^*) \circ (id, m_2) = (id, r_1^* \circ q)$ and $(l_2^*, r_2^*) \circ (id, m_1) = (id, r_2^* \circ p)$ and, vice versa, if $(l_1^*, r_1^*) \circ (id, m_2) = (id, x)$ and $(l_2^*, r_2^*) \circ (id, m_1) = (id, y)$, $x = r_1^* \circ q$, and $y = r_2^* \circ p$. Since r_1^* and r_2^* are monic, (6) and (7) can be constructed as FPCs providing monic c resp. e. Therefore c and e are jointly monic in the pushout (8). Thus the pullback (5) together with (6) - (8) is the gluing of (l_1^*, r_1^*) and (l_2^*, r_2^*).

We show, that the combination of the diagrams (1), (5), and (7) as well as (2), (6), and (8) are pushouts. Since (5) is pullback and $l_1^* \circ m_1' = l_2^* \circ p \circ l_1$, there is p' with $b \circ p' = m_1'$ and $a \circ p' = p \circ l_1$. (l_1, p') is pullback since (1) and (5) are. Stability of pushouts under pullbacks provides that it is also a pushout, because (1) is. (7) is FPC and by proposition 20 also pushout. Thus (7) together with $(a, p', l_1 p)$ is pushout. (2) is pushout and r_1 is monomorphic. In all topoi, (2) is also pullback. Since (6) is FPC and $b \circ p' = m_1'$, there is p'' with $d \circ p'' = m_1^*$, $c \circ p' = p'' \circ r_1$, and (c, p', r_1, p'') is pullback. Stability of pushouts under pullbacks provides that (c, p'') is the pushout of (r_1, p'). (8) is pushout by construction and, therefore (8) together with (c, p', r_1, p'') is pushout.

With this proof technique we can show a new result for sesqui-pushout rewriting.

Proposition 28. (Parallel Independence in GRS_{SePO2}) Two transformations $(l_1^*, r_1^*) : G \rightsquigarrow H$ with rule $t_1 = (l_1, r_1)$ at match m_1 and $(l_2^*, r_2^*) : G \rightsquigarrow K$ with rule $t_2 = (l_2, r_2)$ at match m_2 are parallel independent in GRS_{SePO2} if only if there are morphisms p, q such that (p, id) is pullback of (l_2^*, m_1) and (q, id) is pullback of (l_1^*, m_2). The two transformations $H \rightsquigarrow X$ with rule t_2 and $K \rightsquigarrow X$ with rule t_1, that exist by theorem 26, are sesqui-pushout transformations. Parallel independence in GRS_{SePO2} is syntactical.

Proof. (Sketch) Both proofs follow the schema of the proof in proposition 27. Instead of the stability of pushouts under pullbacks for (1) in figure 7, we have to use stability of final pullback complements, i. e. proposition 6.

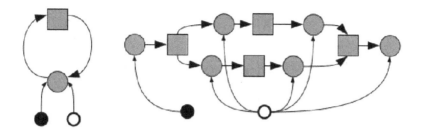

Fig. 8. Petri-Net Type Graph and a Sample Net

Parallel independence in the single-pushout approach also turns out to be syntactical and to coincide with the given notion in GRS_{SPO} for the subset of conflict-free redices that are considered here.

4   Algebraic Graph Rewriting with Context

As we have seen above, almost all algebraic graph rewriting can be modeled by the gluing construction of definition 2. But gluing of spans goes beyond the traditional approaches. Gluings as direct derivations allow to use spans as matches, compare definition 21. In this section we show, from a more intutive point of view, the potentials of spans as matches.

We use the well-known example of event switching in Petri-nets. The left part of figure 8 depicts the type graph for simple condition event nets. Events are drawn as grey boxes, conditions are depicted by grey circles, and the state of a condition is indicated by a connection to a black circle (true-token) or a white circle (false-token). Every concrete net satisfies the invariant that there is exactly one such connection to each condition. The token themselves are singletons, i. e. in every net there is exactly one black node and exactly one white node. The right part of figure 8 shows a sample net in which exactly one event is enabled.

We want to model the switching behavior of events by a single graph transformation rule since the switching step is always the same. For this purpose we need some kind of universal quantification, since we do not know at modelling time how many preconditions and postconditions a concrete event may have.

But not all objects in the rule's left-hand side can be universally quantified because we want to model the switching of an individual event. Therefore we must distinguish at least two parts in the left-hand sides of the rules: (1) the existentially quantified part E the objects in which are matched by *exactly one* object in the host graph and (2) the universally quantified part U the objects in which are matched with *all* fitting objects in the host graph.

The universally quantified part U shall be the *whole context* of E. An object o_1 is in the context of another object o_2 if there is a path from o_1 to o_2 navigating the edges of the graph in both directions. Given this definition, every event or condition in a connected Petri-net is in the context of any other event or

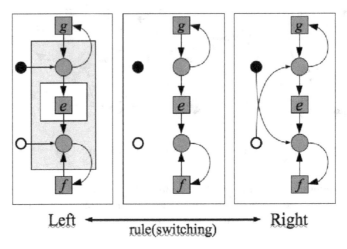

Left $\longleftarrow$ rule(switching) $\longrightarrow$ Right

Fig. 9. Net Switching

condition. This would mean that each match will map the whole host graph to the rule's left-hand side. This is clearly unsatifactory because all transformations become global transformations in this way. Therefore we distinguish a third part in the rule's left-hand side, the context L. If an object c of the host graph gets mapped to such an object the context of c is not required to be part of the match.

Having these ideas in mind, figure 9 shows a rule for general switching of an event e in condition event nets. The left-hand side is structured into three parts:

1. The existentially quantified part E is indicated by the inner white rectangle and contains a single event e.
2. The universally quantified part U is indicated by the grey area. It contains E and the pre- and postconditions of e.
3. The context L contains U and coincides with the complete left-hand side.

The span removes the token-connections from the pre- and postconditions of e and inserts new connections on the right-hand side of the rule. Intuitively, this is switching in condition event nets.

In order to formalise this intuition, a matching span $(n : X \rightarrow L, m : X \rightarrowtail G)$ shall satisfy the following conditions, if the rule's left-hand side L is structured by two monomorphisms $e : E \rightarrowtail U, u : U \rightarrowtail L$:

1. The morphism m determines a subgraph of the host graph G, i. e. m is monomorphism.
2. All objects in the existentially quantified part E of L possess *exactly one* preimage under n, i. e. there is a monomorphism $m_e : E \rightarrowtail X$ such that (id_E, m_e) is the pullback of $(u \circ e, n)$.
3. If n maps node o' to some node in the universally quantified part U of L, i. e. $n(o') = u(o)$, all edges having o' as target or source are in X.

Note that condition (3) provides the universal quantification and local negative application conditions in the sense of [10]. For example, the rule of figure 9 is not applicable to an event with one precondition having a false-token or with one postcondition having a true-token. In these cases there is no chance to map the whole context of this condition to the rule's left-hand side. Hence there is no match. Note also that the final pullback complement construction with the rule's right-hand side (compare construction 11) is reponsible for the insertion of an unknown number of new connection, one for each pre- and postcondition. These ideas lead to the following definition of contextual graph rewriting:

Definition 29. (Contextual Graph Rewriting System) A *contextual rewriting rule* (e, u, l, r) is a span $(l : K \to L, r : K \to R)$ together with two monomorphisms $(u : U \rightarrowtail L, e : E \rightarrowtail U)$ structuring the left-hand side. An abstract span $p = (L \xleftarrow{n} X \xrightarrowtail{m} G)$ is a *contextual match* for a rule (e, u, l, r) if (1) m is monomorphic, (2) there is a monomorphism $m_e : E \to X$ such that (id_E, m_e) is the pullback of $(u \circ e, n)$, and (3) X is the smallest subgraph of G that contains (3a) the unique part of the match, i. e. $m_e(E)$, and (3b) the *complete context* of the unique part of the match, i. e. for all nodes $o' \in X$ with $n(o') = u(o)$ all adjacent edges e in G [i. e. edges such that $s^G(e) = m(o')$ or $t^G(e) = m(o')$] have a preimage under m.

The general result about parallel independence holds also for this system.

Corollary 30. (Parallel Independence of Contextual Rewriting) Given two contextual transformations $(l_1^*, r_1^*) : G \rightsquigarrow H$, and $(l_2^*, r_2^*) : G \rightsquigarrow K$ with rule t_1 at match p_1 and with t_2 at p_2 that are *parallel independent*, i. e. $(l_1^*, r_1^*) \circ p_2$ is contextual match for t_2 and $(l_2^*, r_2^*) \circ p_1$ is contextual match for t_1, then there are transformations $H \rightsquigarrow X$ with rule t_2 and $K \rightsquigarrow X$ with rule t_1.

Note that parallel independence one to one corresponds to independence in condition event nets if switching is modeled by the rule in figure 9.

5   Conclusion

In this paper, we introduced a new way of graph rewriting using a gluing construction of abstract spans. It turned out that most of the algebraic approaches to graph rewriting, including the double-pushout approach, a great body of the single-pushout approach, and the variants of the sesqui-pushout approach introduced in [3], are special cases of the new framework.

Since the new approach allows to use spans as matches, it offers completely new possibilities to define rewriting systems. In the last section, we demonstrated that, for example, universal quantification over the context of a match can be incorporated into the rewriting. The introduced contextual rewriting is a first attempt to exploit the new features. It is a task of future research to adjust the rewriting wrt. practical needs and to compare the new framework to the existing practical approaches, for example [5,18].

There is a good chance to transfer the known theory of the algebraic approaches, mainly of the double pushout approach [6], to this more general framework. We demonstrated it by an investigation of general parallel independence. Parallel independence means that the gluing of the two direct derivations exists. The proof for the central theorem 26 that, in an independent situation, every sequence of rule application provides the same result is as easy as in the single-pushout approach[14]. It is an interesting question for future research if also results wrt. sequential independence, concurrency, amalgamation, etc. carry over. A prerequisite for these results will be suitable decomposition properties of the gluing construction comparable to the decomposition of pushouts.

Another research topic is the comparison of the gluing construction of spans to other formalisms that provide universal quantification like for example [12].

References

1. Bauderon, M., Jacquet, H.: Pullback as a generic graph rewriting mechanism. Applied Categorical Structures 9(1), 65–82 (2001)
2. Corradini, A., Ehrig, H., Montanari, U., Ribeiro, L., Rozenberg, G. (eds.): Proceedings of Graph Transformations, Third International Conference, ICGT 2006, Natal, Rio Grande do Norte, Brazil. LNCS, vol. 4178. Springer, Heidelberg (2006)
3. Corradini, A., Heindel, T., Hermann, F., König, B.: Sesqui-pushout rewriting. In: Corradini, et al. (eds.) [2], pp. 30–45
4. Corradini, A., Montanari, U., Rossi, F., Ehrig, H., Heckel, R., Löwe, M.: Algebraic approaches to graph transformation - part i: Basic concepts and double pushout approach. In: Rozenberg (ed.) [17], pp. 163–246
5. Drewes, F., Hoffmann, B., Janssens, D., Minas, M., Van Eetvelde, N.: Adaptive star grammars. In: Corradini, et al. (eds.) [2], pp. 77–91
6. Ehrig, H., Ehrig, K., Prange, U., Taentzer, G.: Fundamentals of Algebraic Graph Transformation. Springer, Heidelberg (2006)
7. Ehrig, H., Heckel, R., Korff, M., Löwe, M., Ribeiro, L., Wagner, A., Corradini, A.: Algebraic approaches to graph transformation - part ii: Single pushout approach and comparison with double pushout approach. In: Rozenberg (ed.) [17], pp. 247–312
8. Ehrig, H., Pfender, M., Schneider, H.J.: Graph-grammars: An algebraic approach. In: FOCS, pp. 167–180. IEEE, Los Alamitos (1973)
9. Goldblatt, R.: Topoi. Dover Publications, Mineola (1984)
10. Habel, A., Heckel, R., Taentzer, G.: Graph grammars with negative application conditions. Fundam. Inform. 26(3/4), 287–313 (1996)
11. Heckel, R., Ehrig, H., Wolter, U., Corradini, A.: Double-pullback transitions and coalgebraic loose semantics for graph transformation systems. Applied Categorical Structures 9(1), 83–110 (2001)
12. Kahl, W.: A relation-algebraic approach to graph structure transformation. Habil. Thesis 2002-03, Fakultät für Informatik, Univ. der Bundeswehr München (2001)
13. Kennaway, R.: Graph rewriting in some categories of partial morphisms. In: Ehrig, H., Kreowski, H.-J., Rozenberg, G. (eds.) Graph Grammars 1990. LNCS, vol. 532, pp. 490–504. Springer, Heidelberg (1990)
14. Löwe, M.: Algebraic approach to single-pushout graph transformation. Theor. Comput. Sci. 109(1&2), 181–224 (1993)

15. McLarty, C.: Elementary Categories, Elementary Toposes. Oxford Science Publications/Clarendon Press, Oxford (1992)
16. Monserrat, M., Rossello, F., Torrens, J., Valiente, G.: Single pushout rewriting in categories of spans i: The general setting. Technical Report LSI-97-23-R, Department de Llenguatges i Sistemes Informàtics, Universitat Politècnica de Catalunya (1997)
17. Rozenberg, G. (ed.): Handbook of Graph Grammars and Computing by Graph Transformations. Foundations, vol. 1. World Scientific, Singapore (1997)
18. Schürr, A., Klar, F.: 15 years of triple graph grammars. In: Ehrig, H., Heckel, R., Rozenberg, G., Taentzer, G. (eds.) ICGT 2008. LNCS, vol. 5214, pp. 411–425. Springer, Heidelberg (2008)

Finitary $\mathcal{M}$-Adhesive Categories

Benjamin Braatz[1], Hartmut Ehrig[2], Karsten Gabriel[2], and Ulrike Golas[2]

[1] Université du Luxembourg
benjamin.braatz@uni.lu
[2] Technische Universität Berlin, Germany
{ehrig,kgabriel,ugolas}@cs.tu-berlin.de

Abstract. Finitary $\mathcal{M}$-adhesive categories are $\mathcal{M}$-adhesive categories with finite objects only, where the notion $\mathcal{M}$-adhesive category is short for weak adhesive high-level replacement (HLR) category. We call an object finite if it has a finite number of $\mathcal{M}$-subobjects. In this paper, we show that in finitary $\mathcal{M}$-adhesive categories we do not only have all the well-known properties of $\mathcal{M}$-adhesive categories, but also all the additional HLR-requirements which are needed to prove the classical results for $\mathcal{M}$-adhesive systems. These results are the Local Church-Rosser, Parallelism, Concurrency, Embedding, Extension, and Local Confluence Theorems, where the latter is based on critical pairs. More precisely, we are able to show that finitary $\mathcal{M}$-adhesive categories have a unique $\mathcal{E}$-$\mathcal{M}$ factorization and initial pushouts, and the existence of an $\mathcal{M}$-initial object implies in addition finite coproducts and a unique $\mathcal{E}'$-$\mathcal{M}'$ pair factorization. Moreover, we can show that the finitary restriction of each $\mathcal{M}$-adhesive category is a finitary $\mathcal{M}$-adhesive category and finitariness is preserved under functor and comma category constructions based on $\mathcal{M}$-adhesive categories. This means that all the classical results are also valid for corresponding finitary $\mathcal{M}$-adhesive systems like several kinds of finitary graph and Petri net transformation systems. Finally, we discuss how some of the results can be extended to non-$\mathcal{M}$-adhesive categories.

1 Introduction

The concepts of adhesive [1] and (weak) adhesive high-level-replacement (HLR) [2] categories have been a break-through for the double pushout approach (DPO) of algebraic graph transformations [3]. Almost all main results in the DPO-approach have been formulated and proven in these categorical frameworks and instantiated to a large variety of HLR systems, including different kinds of graph and Petri net transformation systems. These main results include the Local Church-Rosser, Parallelism, and Concurrency Theorems, the Embedding and Extension Theorem, completeness of critical pairs, and the Local Confluence Theorem.

However, in addition to the well-known properties of adhesive and (weak) adhesive HLR categories $(\mathbf{C}, \mathcal{M})$, also the following additional HLR-requirements have been needed in [2] to prove these main results: finite coproducts compatible with $\mathcal{M}$, $\mathcal{E}'$-$\mathcal{M}'$ pair factorization usually based on suitable $\mathcal{E}$-$\mathcal{M}$ factorization

H. Ehrig et al. (Eds.): ICGT 2010, LNCS 6372, pp. 234–249, 2010.

of morphisms, and initial pushouts. It is an open question up to now under which conditions these additional HLR-requirements are valid in order to avoid an explicit verification for each instantiation of an adhesive or (weak) adhesive HLR category. In [4], this has been investigated for comma and functor category constructions of weak adhesive HLR categories, but the results hold only under strong preconditions. In this paper, we close this gap showing that these additional properties are valid in finitary $\mathcal{M}$-adhesive categories. We use the notion "$\mathcal{M}$-adhesive category" as short hand for "weak adhesive HLR category" in the sense of [2]. Moreover, an object A in an $\mathcal{M}$-adhesive category is called finite, if A has (up to isomorphism) only a finite number of $\mathcal{M}$-subobjects, i.e., only finite many $\mathcal{M}$-morphisms $m: A' \to A$ up to isomorphism. The category $\mathbf{C}$ is called finitary, if it has only finite objects. Note, that the notion "finitary" depends on the class $\mathcal{M}$ of monomorphisms and "$\mathbf{C}$ being finitary" must not be confused with "$\mathbf{C}$ being finite" in the sense of a finite number of objects and morphisms. In the standard cases of **Sets** and **Graphs** where $\mathcal{M}$ is the class of all monomorphisms, finite objects are exactly finite sets and finite graphs, respectively.

Although in most application areas for the theory of graph transformations only finite graphs are considered, the theory has been developed for general graphs, including also infinite graphs, and it is implicitly assumed that the results can be restricted to finite graphs and to attributed graphs with finite graph part, while the data algebra may be infinite. Obviously, not only **Sets** and **Graphs** are adhesive categories but also the full subcategories **Sets**$_{\mathrm{fin}}$ of finite sets and **Graphs**$_{\mathrm{fin}}$ of finite graphs. But to our knowledge it is an open question whether for each adhesive category $\mathbf{C}$ also the restriction $\mathbf{C}_{\mathrm{fin}}$ to finite objects is again an adhesive category. As far as we know this is true, if the inclusion functor $I: \mathbf{C}_{\mathrm{fin}} \to \mathbf{C}$ preserves monomorphisms, but we are not aware of any adhesive category, where this property fails, or whether this can be shown in general. In this paper, we consider $\mathcal{M}$-adhesive categories $(\mathbf{C},\mathcal{M})$ with restriction to finite objects $(\mathbf{C}_{\mathrm{fin}},\mathcal{M}_{\mathrm{fin}})$, where $\mathcal{M}_{\mathrm{fin}}$ is the restriction of $\mathcal{M}$ to morphisms between finite objects. In this case, the inclusion functor $I: \mathbf{C}_{\mathrm{fin}} \to \mathbf{C}$ preserves $\mathcal{M}$-morphisms, such that finite objects in $\mathbf{C}_{\mathrm{fin}}$ w.r.t. $\mathcal{M}_{\mathrm{fin}}$ are exactly the finite objects in $\mathbf{C}$ w.r.t. $\mathcal{M}$. More generally, we are able to show that the finitary restriction $(\mathbf{C}_{\mathrm{fin}},\mathcal{M}_{\mathrm{fin}})$ of any $\mathcal{M}$-adhesive category $(\mathbf{C},\mathcal{M})$ is a finitary $\mathcal{M}$-adhesive category. Moreover, finitariness is preserved under functor and comma category constructions based on $\mathcal{M}$-adhesive categories.

In Section 2, we introduce basic notions of finitary $\mathcal{M}$-adhesive categories including finite coproducts compatible with $\mathcal{M}$, $\mathcal{M}$-initial objects, finite objects, and finite intersections, which are essential for the theory of finitary $\mathcal{M}$-adhesive categories. The first main result, showing that the additional HLR-requirements mentioned above are valid for finitary $\mathcal{M}$-adhesive categories, is presented in Section 3. In Section 4, we show as second main result that the finitary restriction of an $\mathcal{M}$-adhesive category is a finitary $\mathcal{M}$-adhesive category such that the results of Section 3 are applicable. In Section 5, we show that functorial constructions, including functor and comma categories, applied to finitary $\mathcal{M}$-adhesive categories are again finitary $\mathcal{M}$-adhesive categories under suitable conditions. In Section 6, we analyze how some of the results in Section 3 can be shown in a weaker form for (finitary) non-$\mathcal{M}$-adhesive categories, like the category of simple graphs with all monomorphisms $\mathcal{M}$. Especially, we consider the construction of weak initial pushouts which are the basis for the gluing condition in order to construct (unique) minimal pushout complements in such categories, while initial pushouts are the basis for the construction of (unique) pushout complements in (finitary) $\mathcal{M}$-adhesive categories. In the conclusion, we summarize the main results and discuss open problems for future research. The full proofs for all propositions can be found in the unabridged technical report [5].

2 Basic Notions of Finitary $\mathcal{M}$-Adhesive Categories

Adhesive categories have been introduced by Lack and Sobociński in [1] and generalized to (weak) adhesive HLR categories in [6, 2] as a categorical framework for various kinds of graph and net transformation systems.

An $\mathcal{M}$-*adhesive category* $(\mathbf{C}, \mathcal{M})$, called weak adhesive HLR category in [2], consists of a category $\mathbf{C}$ and a class $\mathcal{M}$ of monomorphisms in $\mathbf{C}$, which is closed under isomorphisms, composition, and decomposition ($g \circ f \in \mathcal{M}$ and $g \in \mathcal{M}$ implies $f \in \mathcal{M}$), such that $\mathbf{C}$ has pushouts and pullbacks along $\mathcal{M}$-morphisms, $\mathcal{M}$-morphisms are closed under pushouts and pullbacks, and pushouts along $\mathcal{M}$-morphisms are weak van Kampen (VK) squares.

A *weak VK square* is a pushout as at the bottom of the cube in the adjacent figure with $m \in \mathcal{M}$, which satisfies the *weak VK property*, i.e., for any commutative cube, where the back faces are pullbacks and ($f \in \mathcal{M}$ or $b, c, d \in \mathcal{M}$), the following statement holds: The top face is a pushout if and only if the front 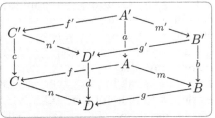 faces are pullbacks. In contrast, the (non-weak) VK property does not assume ($f \in \mathcal{M}$ or $b, c, d \in \mathcal{M}$).

Well-known examples of $\mathcal{M}$-adhesive categories are the categories $(\mathbf{Sets}, \mathcal{M})$ of sets, $(\mathbf{Graphs}, \mathcal{M})$ of graphs, $(\mathbf{Graphs}_{TG}, \mathcal{M})$ of typed graphs, $(\mathbf{ElemNets}, \mathcal{M})$ of elementary Petri nets, $(\mathbf{PTNets}, \mathcal{M})$ of place/transition nets, where for all these categories $\mathcal{M}$ is the class of all monomorphisms, and $(\mathbf{AGraphs}_{ATG}, \mathcal{M})$ of typed attributed graphs, where $\mathcal{M}$ is the class of all injective typed attributed graph morphisms with isomorphic data type component (see [2]).

The compatibility of the morphism class $\mathcal{M}$ with (finite) coproducts was required for the construction of parallel rules in [2], but in fact finite coproducts (if they exist) are always compatible with $\mathcal{M}$ in $\mathcal{M}$-adhesive categories.

Proposition 1 (Finite Coproducts Compatible with $\mathcal{M}$). *For each $\mathcal{M}$-adhesive category $(\mathbf{C}, \mathcal{M})$ with finite coproducts, finite coproducts are compatible with $\mathcal{M}$, i.e., $f_i \in \mathcal{M}$ for $i = 1, \ldots, n$ implies that $f_1 + \cdots + f_n \in \mathcal{M}$.*

Proof (Idea). The proof constructs a coproduct $f + g \colon A + B \to A' + B'$ of morphisms as the composition of $f + \mathrm{id}_B$ and $\mathrm{id}_{A'} + g$ resulting from pushouts. Since $\mathcal{M}$ is closed under pushouts and composition this implies $f + g \in \mathcal{M}$. □

For the construction of coproducts, it often makes sense to use pushouts over $\mathcal{M}$-initial objects in the following sense.

Definition 1 ($\mathcal{M}$-Initial Object). *An initial object I in $(\mathbf{C}, \mathcal{M})$ is called $\mathcal{M}$-initial if for each object $A \in \mathbf{C}$ the unique morphism $i_A \colon I \to A$ is in $\mathcal{M}$.*

Note that if $(\mathbf{C}, \mathcal{M})$ has an $\mathcal{M}$-initial object then all initial objects are $\mathcal{M}$-initial due to $\mathcal{M}$ being closed under isomorphisms and composition.

In the $\mathcal{M}$-adhesive categories $(\mathbf{Sets}, \mathcal{M})$, $(\mathbf{Graphs}, \mathcal{M})$, $(\mathbf{Graphs}_{TG}, \mathcal{M})$, $(\mathbf{ElemNets}, \mathcal{M})$, and $(\mathbf{PTNets}, \mathcal{M})$ we have $\mathcal{M}$-initial objects defined by the empty set, empty graphs, and empty nets, respectively. But in $(\mathbf{AGraphs}_{ATG}, \mathcal{M})$, there is no $\mathcal{M}$-initial object. The initial attributed graph $(\varnothing, T_{DSIG})$ with term algebra T_{DSIG} of the data type signature $DSIG$ is not $\mathcal{M}$-initial because the data type part of the unique morphism $(\varnothing, T_{DSIG}) \to (G, D)$ is, in general, not an isomorphism.

The existence of an $\mathcal{M}$-initial object implies that we have finite coproducts.

Proposition 2 (Existence of Finite Coproducts). *For each $\mathcal{M}$-adhesive category $(\mathbf{C}, \mathcal{M})$ with $\mathcal{M}$-initial object, $(\mathbf{C}, \mathcal{M})$ has finite coproducts, where the injections into coproducts are in $\mathcal{M}$.*

Proof (Idea). Coproducts can be constructed as pushouts under the $\mathcal{M}$-initial object. Since the morphisms from the $\mathcal{M}$-initial object are in $\mathcal{M}$ and $\mathcal{M}$ is closed under pushouts, the injections into the resulting coproduct are also in $\mathcal{M}$. □

Note that an $\mathcal{M}$-adhesive category may still have coproducts even if it does not have an $\mathcal{M}$-initial object. The $\mathcal{M}$-adhesive category $(\mathbf{AGraphs}_{ATG}, \mathcal{M})$, e.g., has finite coproducts as shown in [2].

Now, we are going to consider finite objects in $\mathcal{M}$-adhesive categories. Intuitively, we are interested in those objects where the graph or net part is finite. This can be expressed in a general $\mathcal{M}$-adhesive category by the fact that we have only a finite number of $\mathcal{M}$-subobjects. An $\mathcal{M}$-*subobject* of an object A is an isomorphism class of

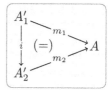

$\mathcal{M}$-morphisms $m \colon A' \to A$, where $\mathcal{M}$-morphisms $m_1 \colon A'_1 \to A$ and $m_2 \colon A'_2 \to A$ belong to the same $\mathcal{M}$-subobject of A if there is an isomorphism $i \colon A'_1 \xrightarrow{\sim} A'_2$ with $m_1 = m_2 \circ i$.

Definition 2 (Finite Object and Finitary $\mathcal{M}$-Adhesive Category). *An object A in an $\mathcal{M}$-adhesive category $(\mathbf{C}, \mathcal{M})$ is called* finite *if A has finitely many $\mathcal{M}$-subobjects.*

An $\mathcal{M}$-adhesive category $(\mathbf{C}, \mathcal{M})$ is called finitary, *if each object $A \in \mathbf{C}$ is finite.*

In $(\mathbf{Sets}, \mathcal{M})$, the finite objects are the finite sets. Graphs in $(\mathbf{Graphs}, \mathcal{M})$ and $(\mathbf{Graphs}_{TG}, \mathcal{M})$ are finite if the node and edge sets have finite cardinality, while

TG itself may be infinite. Petri nets in (**ElemNets**, $\mathcal{M}$) and (**PTNets**, $\mathcal{M}$) are finite if the number of places and transitions is finite. A typed attributed graph $AG = ((G, D), t)$ in (**AGraphs**$_{ATG}$, $\mathcal{M}$) with typing $t\colon (G, D) \to ATG$ is finite if the graph part of G, i. e., all vertex and edge sets except the set V_D of data vertices generated from D, is finite, while the attributed type graph ATG or the data type part D may be infinite, because $\mathcal{M}$-morphisms are isomorphisms on the data type part.

In the following, we will use finite $\mathcal{M}$-intersections in various constructions. Finite $\mathcal{M}$-intersections are a generalization of pullbacks to an arbitrary, but finite number of $\mathcal{M}$-subobjects and, thus, a special case of limits.

Definition 3 (Finite $\mathcal{M}$-Intersection). *Given an $\mathcal{M}$-adhesive category* (**C**, $\mathcal{M}$) *and morphisms* $m_i\colon A_i \to B \in \mathcal{M}$ *($i \in \mathcal{I}$ for finite $\mathcal{I}$) with the same codomain object B, a finite $\mathcal{M}$-intersection of m_i ($i \in \mathcal{I}$) is an object A with morphisms $n_i\colon A \to A_i$ ($i \in \mathcal{I}$), such that $m_i \circ n_i = m_j \circ n_j$ ($i, j \in \mathcal{I}$) and for each other object A' and morphisms $n_i'\colon A' \to A_i$*

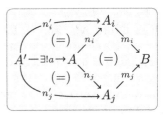

($i \in \mathcal{I}$) with $m_i \circ n_i' = m_j \circ n_j'$ ($i, j \in \mathcal{I}$) there is a unique morphism $a\colon A' \to A$ with $n_i \circ a = n_i'$ ($i \in \mathcal{I}$).

Note that finite $\mathcal{M}$-intersections can be constructed by iterated pullbacks and, hence, always exist in $\mathcal{M}$-adhesive categories. Moreover, since pullbacks preserve $\mathcal{M}$-morphisms, the morphisms n_i are also in $\mathcal{M}$.

3   Additional HLR-Requirements for Finitary $\mathcal{M}$-adhesive Categories

In order to prove the main classical results for $\mathcal{M}$-adhesive systems based on $\mathcal{M}$-adhesive categories additional HLR-requirements have been used in [2]. For the Parallelism Theorem, binary coproducts compatible with $\mathcal{M}$ are required in order to construct parallel rules. Initial pushouts are used in order to define the gluing condition and to show that consistency in the Embedding Theorem is not only sufficient, but also necessary. In connection with the Concurrency Theorem and for completeness of critical pairs, an $\mathcal{E}'$-$\mathcal{M}'$ pair factorization is used such that the class $\mathcal{M}'$ satisfies the $\mathcal{M}$-$\mathcal{M}'$ pushout-pullback decomposition property. Moreover, a standard construction for $\mathcal{E}'$-$\mathcal{M}'$ pair factorization uses an $\mathcal{E}$-$\mathcal{M}$ factorization of morphisms in **C**, where $\mathcal{E}'$ is constructed from $\mathcal{E}$ using binary coproducts.

As far as we know, these additional HLR-requirements cannot be concluded from the axioms of $\mathcal{M}$-adhesive categories, at least we do not know proofs for non-trivial classes $\mathcal{E}$, $\mathcal{E}'$, $\mathcal{M}$, and $\mathcal{M}'$. However, in the case of finitary $\mathcal{M}$-adhesive categories (**C**, $\mathcal{M}$) we are able to show that these additional HLR-requirements are valid for suitable classes $\mathcal{E}$ and $\mathcal{E}'$, and $\mathcal{M}' = \mathcal{M}$. Note that for $\mathcal{M}' = \mathcal{M}$, the $\mathcal{M}$-$\mathcal{M}'$ pushout-pullback decomposition property is the $\mathcal{M}$ pushout-pullback decomposition property which is valid already in general $\mathcal{M}$-adhesive categories.

The reason for the existence of an $\mathcal{E}$-$\mathcal{M}$ factorization of morphisms in finitary $\mathcal{M}$-adhesive categories is the fact that we only need finite intersections of $\mathcal{M}$-subobjects and not infinite intersections as would be required in general $\mathcal{M}$-adhesive categories. Moreover, we fix the choice of the class $\mathcal{E}$ to extremal morphisms w.r.t. $\mathcal{M}$.

The dependencies are shown in Fig. 1, where the additional assumptions of finitariness and $\mathcal{M}$-initial objects are shown in the top row, the HLR-requirements shown in this paper in the center and the classical theorems in the bottom row.

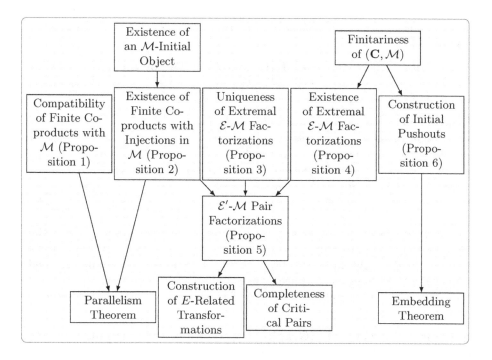

Fig. 1. Dependency graph

Definition 4 (Extremal $\mathcal{E}$-$\mathcal{M}$ Factorization). *Given an $\mathcal{M}$-adhesive category $(\mathbf{C}, \mathcal{M})$, the class $\mathcal{E}$ of all extremal morphisms w.r.t. $\mathcal{M}$ is defined by $\mathcal{E} := \{e \text{ in } \mathbf{C} \mid \text{for all } m, f \text{ in } \mathbf{C} \text{ with } m \circ f = e\colon m \in \mathcal{M} \text{ implies } m \text{ isomorphism}\}$. For a morphism $f\colon A \to B$ in $\mathbf{C}$ an extremal $\mathcal{E}$-$\mathcal{M}$ factorization of f is given by an object $\bar{B}$ and morphisms $e\colon A \to \bar{B} \in \mathcal{E}$ and $m\colon \bar{B} \to B \in \mathcal{M}$, such that $m \circ e = f$.*

Remark 1. Although in several example categories the class $\mathcal{E}$ consists of all epimorphisms, we will show below that the class $\mathcal{E}$ of extremal morphisms w.r.t. $\mathcal{M}$ is not necessarily a class of epimorphisms. But if we require $\mathcal{M}$ to be the class of all monomorphisms and e and f in the definition of $\mathcal{E}$ in Definition 4 to be epimorphisms then $\mathcal{E}$ is the class of all extremal epimorphisms in the sense of [7].

Proposition 3 (Uniqueness of Extremal $\mathcal{E}$-$\mathcal{M}$ Factorizations). *Given an $\mathcal{M}$-adhesive category $(\mathbf{C}, \mathcal{M})$, then extremal $\mathcal{E}$-$\mathcal{M}$ factorizations are unique up to isomorphism.*

Proof (Idea). For two extremal $\mathcal{E}$-$\mathcal{M}$ factorizations of a morphism $f \colon A \to B$, we construct a pullback over the two morphisms in $\mathcal{M}$ with the resulting morphisms also in $\mathcal{M}$. The universal property of the pullback induces a unique morphism from A into the pullback object which, together with the pullback morphisms, factors the two morphisms in $\mathcal{E}$. Since these are extremal, the pullback morphisms are isomorphisms and the two extremal $\mathcal{E}$-$\mathcal{M}$ factorizations are isomorphic.   □

Proposition 4 (Existence of Extremal $\mathcal{E}$-$\mathcal{M}$ Factorizations). *Given a finitary $\mathcal{M}$-adhesive category $(\mathbf{C}, \mathcal{M})$, then we can construct an extremal $\mathcal{E}$-$\mathcal{M}$ factorization $m \circ e = f$ for each morphism $f \colon A \to B$ in $\mathbf{C}$.*

Construction: $m \colon \bar{B} \to B$ is constructed as $\mathcal{M}$-intersection of all $\mathcal{M}$-subobjects $m_i \colon B_i \to B$ for which there exists $e_i \colon A \to B_i$ with $f = m_i \circ e_i$, leading to a suitable finite index set $\mathcal{I}$, and $e \colon A \to \bar{B}$ is the induced unique morphism with $\bar{m}_i \circ e = e_i$ for all $i \in \mathcal{I}$.

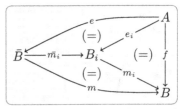

Proof (Idea). The construction is always possible, since there is at least the trivial $\mathcal{M}$-subobject $m_i = \mathrm{id}_B$ with $e_i = f$ and at most finitely many $\mathcal{M}$-subobjects. It results in $\bar{m}_i \in \mathcal{M}$ and $m \in \mathcal{M}$, because $\mathcal{M}$ is closed under pullbacks and composition. The induced morphism e is in $\mathcal{E}$, since each factorization $e = m' \circ e'$ leads to another subobject $m_i = m \circ m'$ of B with $e_i = e'$ and the intersection induces an inverse of m'.   □

In the categories $(\mathbf{Sets}, \mathcal{M})$, $(\mathbf{Graphs}, \mathcal{M})$, $(\mathbf{Graphs_{TG}}, \mathcal{M})$, $(\mathbf{ElemNets}, \mathcal{M})$, and $(\mathbf{PTNets}, \mathcal{M})$, the extremal $\mathcal{E}$-$\mathcal{M}$ factorization $f = m \circ e$ for $f \colon A \to B$ with finite A and B is nothing else but the well-known epi-mono factorization of morphisms, which also works for infinite objects A and B, because these categories have not only finite but also general intersections. For $(\mathbf{AGraphs_{ATG}}, \mathcal{M})$, the extremal $\mathcal{E}$-$\mathcal{M}$ factorization of $(f_G, f_D) \colon (G, D) \to (G', D')$ with finite (or also infinite) G and G' is given by $(f_G, f_D) = (m_G, m_D) \circ (e_G, e_D)$ with $(e_G, e_D) \colon (G, D) \to (\bar{G}, \bar{D})$ and $(m_G, m_D) \colon (\bar{G}, \bar{D}) \to (G', D')$, where e_G is an epimorphism, m_G a monomorphism and m_D an isomorphism. In general, e_D and, hence, also (e_G, e_D) is not an epimorphism, since m_D is an isomorphism and, therefore, e_D has to be essentially the same as f_D. This means that the class $\mathcal{E}$, which depends on $\mathcal{M}$, is not necessarily a class of epimorphisms.

Given an $\mathcal{E}$-$\mathcal{M}'$ factorization and binary coproducts, we are able to construct an $\mathcal{E}'$-$\mathcal{M}'$ pair factorization in a standard way (see [2]), where we will consider the special case $\mathcal{M}' = \mathcal{M}$. First we recall $\mathcal{E}'$-$\mathcal{M}'$ pair factorizations.

Definition 5 ($\mathcal{E}'$-$\mathcal{M}'$ Pair Factorization). *Given a morphism class $\mathcal{M}'$ and a class $\mathcal{E}'$ of morphism pairs with common codomain in a category $\mathbf{C}$, then $\mathbf{C}$ has an $\mathcal{E}'$-$\mathcal{M}'$ pair factorization if for each pair of morphisms $f_A \colon A \to D$, $f_B \colon B \to D$ there is, unique up to isomorphism, an object C and mor-*

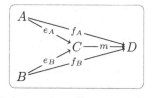

phisms $e_A \colon A \to C$, $e_B \colon B \to C$, and $m \colon C \to D$ with $(e_A, e_B) \in \mathcal{E}'$, $m \in \mathcal{M}'$, $m \circ e_A = f_A$ and $m \circ e_B = f_B$.

Proposition 5 (Construction of $\mathcal{E}'$-$\mathcal{M}'$ Pair Factorization). *Given a category $\mathbf{C}$ with an $\mathcal{E}$-$\mathcal{M}'$ factorization and binary coproducts, then $\mathbf{C}$ has also an $\mathcal{E}'$-$\mathcal{M}'$ pair factorization for the class $\mathcal{E}' = \{(e_A \colon A \to C, e_B \colon B \to C) \mid e_A, e_B \in \mathbf{C}$ with induced $e \colon A + B \to C \in \mathcal{E}\}$.*

Proof (Idea). For morphisms $f_A \colon A \to D$ and $f_B \colon B \to D$, we use the $\mathcal{E}$-$\mathcal{M}'$ factorization $f = m \circ e$ of the induced morphism $f \colon A + B \to D$ and obtain e_A and e_B by composing the respective coproduct inclusions with e. □

Remark 2. With the previous facts, we have extremal $\mathcal{E}$-$\mathcal{M}$ factorizations and corresponding $\mathcal{E}'$-$\mathcal{M}$ pair factorizations for all finitary $\mathcal{M}$-adhesive categories with $\mathcal{M}$-initial objects and these factorizations are unique up to isomorphism.

Finally, let us consider the construction of initial pushouts in finitary $\mathcal{M}$-adhesive categories. Similar to the extremal $\mathcal{E}$-$\mathcal{M}$ factorization, we are able to construct initial pushouts by finite $\mathcal{M}$-intersections of $\mathcal{M}$-subobjects in finitary $\mathcal{M}$-adhesive categories, but not in general ones. First we recall the definition.

Definition 6 (Initial Pushout). *A pushout (1) over a morphism $m \colon L \to G$ with $b, c \in \mathcal{M}$ in an $\mathcal{M}$-adhesive category $(\mathbf{C}, \mathcal{M})$ is called* initial *if the following condition holds: for all*

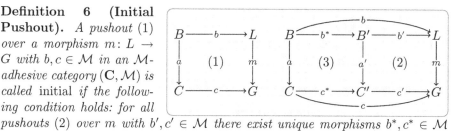

pushouts (2) over m with $b', c' \in \mathcal{M}$ there exist unique morphisms $b^*, c^* \in \mathcal{M}$ such that $b' \circ b^* = b$, $c' \circ c^* = c$, and (3) is a pushout.

Remark 3. As shown in [2], the initial pushout allows to define a gluing condition, which is necessary and sufficient for the construction of pushout complements. Given $m \colon L \to G$ with initial pushout (1) and $l \colon K \to L \in \mathcal{M}$, which can be considered as the left-hand side of a rule, the gluing condition is satisfied if there exists

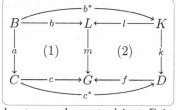

$b^* \colon B \to K$ with $l \circ b^* = b$. In this case, the pushout complement object D in (2) is constructed as pushout object of a and b^*.

Proposition 6 (Initial Pushouts in Finitary $\mathcal{M}$-Adhesive Categories). *Each finitary $\mathcal{M}$-adhesive category has initial pushouts.*

Construction: Given $m\colon L$
$\to G$, *we consider all those*
$\mathcal{M}$*-subobjects* $b_i\colon B_i \to L$
of L *and* $c_i\colon C_i \to G$ *of* G
such that there is a pushout
(P_i) *over* m*. Since* L *and* G
are finite this leads to a fi-

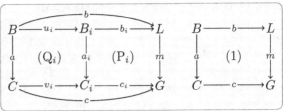

nite index set $\mathcal{I}$ *for all* (P_i) *with* $i \in \mathcal{I}$*. Now construct* $b\colon B \to L$ *as the finite*
$\mathcal{M}$*-intersection of* $(b_i)_{i\in\mathcal{I}}$ *and* $c\colon C \to G$ *as the finite* $\mathcal{M}$*-intersection of* $(c_i)_{i\in\mathcal{I}}$*.*
Then there is a unique $a\colon B \to C$ *such that* (Q_i) *commutes for all* $i \in \mathcal{I}$ *and the*
outer diagram (1) *is the initial pushout over* m*.*

Proof (Idea). We have to show that (1) is a pushout. This is done by constructing
the finite $\mathcal{M}$-intersections B and C by iterated pullbacks. In each step, the weak
VK property is used to show that the pushouts are pulled back and composition
of pushouts then leads to the diagonal square also being a pushout. The pushout
(1) is also initial, since for each comparison pushout (1′) there is an $i_0 \in \mathcal{I}$ for
which (P_{i_0}) is isomorphic to (1′) and the initiality property is given by the
corresponding pushout (Q_{i_0}). $\square$

The following theorem summarizes that the additional HLR-requirements mentioned above are valid for all finitary $\mathcal{M}$-adhesive categories.

**Theorem 1 (Additional HLR-Requirements in Finitary $\mathcal{M}$-adhesive
Categories).** *Given a finitary $\mathcal{M}$-adhesive category* $(\mathbf{C}, \mathcal{M})$*, the following additional HLR-requirements are valid:*

1. $(\mathbf{C}, \mathcal{M})$ *has initial pushouts.*
2. $(\mathbf{C}, \mathcal{M})$ *has a unique extremal $\mathcal{E}$-$\mathcal{M}$ factorization, where $\mathcal{E}$ is the class of all
   extremal morphisms w. r. t. $\mathcal{M}$.*

If $(\mathbf{C}, \mathcal{M})$ *has an $\mathcal{M}$-initial object, we also have that:*

3. $(\mathbf{C}, \mathcal{M})$ *has finite coproducts compatible with $\mathcal{M}$.*
4. $(\mathbf{C}, \mathcal{M})$ *has a unique $\mathcal{E}'$-$\mathcal{M}'$ pair factorization for the classes $\mathcal{M}' = \mathcal{M}$ and
   $\mathcal{E}'$ induced by $\mathcal{E}$.*

Proof. Requirement 1 follows from Proposition 6, Requirement 2 from Propositions 3 and 4, Requirement 3 from Propositions 2 and 1, and, finally, Requirement
4 from Proposition 5. $\square$

4   Finitary Restriction of $\mathcal{M}$-Adhesive Categories

In order to construct $\mathcal{M}$-adhesive categories it is important to know that (**Sets**,
$\mathcal{M}$) is an $\mathcal{M}$-adhesive category, and that $\mathcal{M}$-adhesive categories are closed under
product, slice, coslice, functor, and comma category constructions, provided that
suitable conditions are satisfied (see [2]). This allows to show that (**Graphs**, $\mathcal{M}$),
(**Graphs$_{\mathbf{TG}}$**, $\mathcal{M}$), (**ElemNets**, $\mathcal{M}$), and (**PTNets**, $\mathcal{M}$) are also $\mathcal{M}$-adhesive

categories. However, it is more difficult to show similar results for the additional HLR-requirements considered in Section 3, especially there are only weak results concerning the existence and construction of initial pushouts [4].

We have already shown that these additional HLR-requirements are valid in finitary $\mathcal{M}$-adhesive categories under weak assumptions. It remains to show how to construct finitary $\mathcal{M}$-adhesive categories. In the main result of this section, we show that for any $\mathcal{M}$-adhesive category $(\mathbf{C}, \mathcal{M})$ the restriction $(\mathbf{C}_{\mathrm{fin}}, \mathcal{M}_{\mathrm{fin}})$ to finite objects is a finitary $\mathcal{M}$-adhesive category, where $\mathcal{M}_{\mathrm{fin}}$ is the corresponding restriction of $\mathcal{M}$. Moreover, we know how to construct pushouts and pullbacks in $\mathbf{C}_{\mathrm{fin}}$ along $\mathcal{M}_{\mathrm{fin}}$-morphisms, because the inclusion functor $I_{\mathrm{fin}} \colon \mathbf{C}_{\mathrm{fin}} \to \mathbf{C}$ creates and preserves pushouts and pullbacks along $\mathcal{M}_{\mathrm{fin}}$ and $\mathcal{M}$, respectively.

Definition 7 (Finitary Restriction of $\mathcal{M}$-adhesive Category). *Given an $\mathcal{M}$-adhesive category $(\mathbf{C}, \mathcal{M})$ the restriction to all finite objects of $(\mathbf{C}, \mathcal{M})$ defines the full subcategory $\mathbf{C}_{\mathrm{fin}}$ of $\mathbf{C}$, and $(\mathbf{C}_{\mathrm{fin}}, \mathcal{M}_{\mathrm{fin}})$ with $\mathcal{M}_{\mathrm{fin}} = \mathcal{M} \cap \mathbf{C}_{\mathrm{fin}}$ is called finitary restriction of $(\mathbf{C}, \mathcal{M})$.*

Remark 4. Note, that an object A in $\mathbf{C}$ is finite in $(\mathbf{C}, \mathcal{M})$ if and only if A is finite in $(\mathbf{C}_{\mathrm{fin}}, \mathcal{M}_{\mathrm{fin}})$. If $\mathcal{M}$ is the class of all monomorphisms in $\mathbf{C}$ then $\mathcal{M}_{\mathrm{fin}}$ is not necessarily the class of all monomorphisms in $\mathbf{C}_{\mathrm{fin}}$. This means that for an adhesive category $\mathbf{C}$, which is based on the class of all monomorphisms, there may be monomorphisms in $\mathbf{C}_{\mathrm{fin}}$ which are not monomorphisms in $\mathbf{C}$, such that it is not clear whether the finite objects in $\mathbf{C}$ and $\mathbf{C}_{\mathrm{fin}}$ are the same. This problem is avoided for $\mathcal{M}$-adhesive categories, where finitariness depends on $\mathcal{M}$.

In order to prove that with $(\mathbf{C}, \mathcal{M})$ also $(\mathbf{C}_{\mathrm{fin}}, \mathcal{M}_{\mathrm{fin}})$ is an $\mathcal{M}$-adhesive category, we have to analyze the construction and preservation of pushouts and pullbacks in $(\mathbf{C}, \mathcal{M})$ and $(\mathbf{C}_{\mathrm{fin}}, \mathcal{M}_{\mathrm{fin}})$. This corresponds to the following creation and preservation properties of the inclusion functor $I_{\mathrm{fin}} \colon \mathbf{C}_{\mathrm{fin}} \to \mathbf{C}$.

Definition 8 (Creation and Preservation of Pushout and Pullback). *Given an $\mathcal{M}$-adhesive category $(\mathbf{C}, \mathcal{M})$, the inclusion functor $I_{\mathrm{fin}} \colon \mathbf{C}_{\mathrm{fin}} \to \mathbf{C}$ creates pushouts along $\mathcal{M}$ if for each pair of morphisms f, h in $\mathbf{C}_{\mathrm{fin}}$ with $f \in \mathcal{M}_{\mathrm{fin}}$ and pushout (1) in $\mathbf{C}$ we have already $D \in \mathbf{C}_{\mathrm{fin}}$ such that (1) is a pushout in $\mathbf{C}_{\mathrm{fin}}$ along $\mathcal{M}_{\mathrm{fin}}$.*

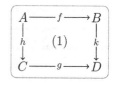

Similarly, I_{fin} creates pullbacks along $\mathcal{M}$ if for each pullback (1) in $\mathbf{C}$ with $g \in \mathcal{M}_{\mathrm{fin}}$ and $B, C, D \in \mathbf{C}_{\mathrm{fin}}$ also $A \in \mathbf{C}_{\mathrm{fin}}$ such that (1) is a pullback in $\mathbf{C}_{\mathrm{fin}}$ along $\mathcal{M}_{\mathrm{fin}}$.
I_{fin} preserves pushouts (pullbacks) along $\mathcal{M}_{\mathrm{fin}}$ if each pushout (pullback) (1) in $\mathbf{C}_{\mathrm{fin}}$ with $f \in \mathcal{M}_{\mathrm{fin}}$ ($g \in \mathcal{M}_{\mathrm{fin}}$) is also a pushout (pullback) in $\mathbf{C}$ with $f \in \mathcal{M}$ ($g \in \mathcal{M}$).

Proposition 7 (Creation and Preservation of Pushout and Pullback). *Given an $\mathcal{M}$-adhesive category $(\mathbf{C}, \mathcal{M})$ the inclusion functor $I_{\mathrm{fin}} \colon \mathbf{C}_{\mathrm{fin}} \to \mathbf{C}$ creates pushouts and pullbacks along $\mathcal{M}$ and preserves pushouts and pullbacks along $\mathcal{M}_{\mathrm{fin}}$.*

Proof (Idea). 1. I_{fin} creates pullbacks along $\mathcal{M}$, because $\mathcal{M}$ is closed under pullbacks and composition and, therefore, each $\mathcal{M}$-subobject of A in Definition 8 is also an $\mathcal{M}$-subobject of B. Hence, B being finite implies that A is finite.

2. I_{fin} creates pushouts along $\mathcal{M}$, because $\mathcal{M}$ is closed under pushouts and, moreover, we can show (using the weak VK property) that each $\mathcal{M}$-subobject of D in Definition 8 corresponds up to isomorphism to a pair of $\mathcal{M}$-subobjects of B and C obtained by pullback constructions. Hence, B and C being finite implies that D is finite.

3. I_{fin} preserves pushouts along $\mathcal{M}_{\text{fin}}$, because given pushout (1) in $\mathbf{C}_{\text{fin}}$ with $f \in \mathcal{M}_{\text{fin}}$ also $f \in \mathcal{M}$. Since I_{fin} creates pushouts along $\mathcal{M}$ by Item 2, the pushout (1') of $f \in \mathcal{M}$ and h in $\mathbf{C}$ is also a pushout in $\mathbf{C}_{\text{fin}}$. By uniqueness of pushouts this means that (1) and (1') are isomorphic and hence (1) is also a pushout in $\mathbf{C}$.

4. Similarly, we can show that I_{fin} preserves pullbacks along $\mathcal{M}_{\text{fin}}$ using the fact that I_{fin} creates pullbacks along $\mathcal{M}$ as shown in Item 1.   □

Now we are able to show the second main result.

Theorem 2. *The finitary restriction* $(\mathbf{C}_{\text{fin}}, \mathcal{M}_{\text{fin}})$ *of any $\mathcal{M}$-adhesive category* $(\mathbf{C}, \mathcal{M})$ *is a finitary $\mathcal{M}$-adhesive category.*

Proof. According to Remark 4, an object A in $\mathbf{C}$ is finite in $(\mathbf{C}, \mathcal{M})$ if and only if it is finite in $(\mathbf{C}_{\text{fin}}, \mathcal{M}_{\text{fin}})$. Hence, all objects in $(\mathbf{C}_{\text{fin}}, \mathcal{M}_{\text{fin}})$ are finite.

Moreover, $\mathcal{M}_{\text{fin}}$ is closed under isomorphisms, composition, and decomposition, because this is valid for $\mathcal{M}$. $(\mathbf{C}_{\text{fin}}, \mathcal{M}_{\text{fin}})$ has pushouts along $\mathcal{M}_{\text{fin}}$ because $(\mathbf{C}, \mathcal{M})$ has pushouts along $\mathcal{M}$ and I_{fin} creates pushouts along $\mathcal{M}$ by Proposition 7. This implies also that $\mathcal{M}_{\text{fin}}$ is preserved by pushouts along $\mathcal{M}_{\text{fin}}$ in $\mathbf{C}_{\text{fin}}$. Similarly, $(\mathbf{C}_{\text{fin}}, \mathcal{M}_{\text{fin}})$ has pullbacks along $\mathcal{M}_{\text{fin}}$ and $\mathcal{M}_{\text{fin}}$ is preserved by pullbacks along $\mathcal{M}_{\text{fin}}$ in $\mathbf{C}_{\text{fin}}$.

Finally, the weak VK property of $(\mathbf{C}, \mathcal{M})$ implies that of $(\mathbf{C}_{\text{fin}}, \mathcal{M}_{\text{fin}})$ using that I_{fin} preserves pushouts and pullbacks along $\mathcal{M}_{\text{fin}}$ and creates pushouts and pullbacks along $\mathcal{M}$.   □

A direct consequence of Theorem 2 is the fact that finitary restrictions of (**Sets**, $\mathcal{M}$), (**Graphs**, $\mathcal{M}$), (**Graphs**$_{\mathbf{TG}}$, $\mathcal{M}$), (**ElemNets**, $\mathcal{M}$), (**PTNets**, $\mathcal{M}$), and (**AGraphs**$_{\mathbf{ATG}}$, $\mathcal{M}$) are all finitary $\mathcal{M}$-adhesive categories satisfying not only the axioms of $\mathcal{M}$-adhesive categories, but also the additional HLR-requirements stated in Theorem 1, where, however, the existence of finite coproducts in (**AGraphs**$_{\mathbf{ATG}}$, $\mathcal{M}$) is valid (see [2]), but cannot be concluded from the existence of $\mathcal{M}$-initial objects as required in Items 3 and 4 of Theorem 1. Note, that I is an $\mathcal{M}_{\text{fin}}$-initial object in $(\mathbf{C}_{\text{fin}}, \mathcal{M}_{\text{fin}})$ if I is an $\mathcal{M}$-initial object in $(\mathbf{C}, \mathcal{M})$.

Remark 5. From Theorem 1 and Theorem 2 we can conclude that the main results for the DPO approach, like the Local Church-Rosser, Parallelism, Concurrency, Embedding, Extension, and Local Confluence Theorems, are valid in all finitary restrictions of $\mathcal{M}$-adhesive categories. This includes also corresponding results with nested application conditions [8], because shifts along morphisms and rules preserve finiteness of the objects occuring in the application conditions.

5   Functorial Constructions of Finitary $\mathcal{M}$-Adhesive Categories

Similar to general $\mathcal{M}$-adhesive categories, also finitary $\mathcal{M}$-adhesive categories are closed under product, slice, coslice, functor, and comma categories under suitable conditions [2]. It suffices to show this for functor and comma categories, because all others are special cases.

Proposition 8 (Finitary Functor Categories). *Given a finitary $\mathcal{M}$-adhesive category $(\mathbf{C}, \mathcal{M})$ and a category $\mathbf{X}$ with a finite class of objects then also the functor category $(\mathbf{Funct}(\mathbf{X}, \mathbf{C}), \mathcal{M}_F)$ is a finitary $\mathcal{M}$-adhesive category, where $\mathcal{M}_F$ is the class of all $\mathcal{M}$-functor transformations $t \colon F' \to F$, i. e., $t(X) \colon F'(X) \to F(X) \in \mathcal{M}$ for all objects X in $\mathbf{X}$.*

Proof (Idea). Functor categories over $\mathcal{M}$-adhesive categories are $\mathcal{M}$-adhesive due to Theorem 4.15.3 in [2]. It remains to show that each $F \colon \mathbf{X} \to \mathbf{C}$ is finite. Since $\mathbf{X}$ has a finite class of objects and for each X in $\mathbf{X}$ we have only finitely many $t(X) \colon F'(X) \to F(X) \in \mathcal{M}$, we also have only finitely many $t \colon F' \to F$ with all morphisms in $\mathcal{M}$ up to isomorphism. ☐

Remark 6. For infinite (discrete) $\mathbf{X}$ we have $\mathbf{Funct}(\mathbf{X}, \mathbf{C}) \cong \prod_{i \in \mathbb{N}} \mathbf{C}$. With $\mathbf{C} = \mathbf{Sets}_{\mathrm{fin}}$ the object $(2_i)_{i \in \mathbb{N}}$ with $2_i = \{1, 2\}$ has an infinite number of subobjects $(1_i)_{i \in \mathbb{N}}$ of $(2_i)_{i \in \mathbb{N}}$ with $1_i = \{1\}$, because in each component $i \in \mathbb{N}$ we have two choices of injective functions $f_{1/2} \colon \{1\} \to \{1, 2\}$. Hence $\mathbf{Funct}(\mathbf{X}, \mathbf{C})$ is not finitary, because $(2_i)_{i \in \mathbb{N}}$ in $\prod_{i \in \mathbb{N}} \mathbf{C}$ is not finite.

Proposition 9 (Finitary Comma Categories). *Given finitary $\mathcal{M}$-adhesive categories $(\mathbf{A}, \mathcal{M}_1)$ and $(\mathbf{B}, \mathcal{M}_2)$ and functors $F \colon \mathbf{A} \to \mathbf{C}$ and $G \colon \mathbf{B} \to \mathbf{C}$, where F preserves pushouts along $\mathcal{M}_1$ and G preserves pullbacks along $\mathcal{M}_2$, then the comma category $\mathbf{ComCat}(F, G; \mathcal{I})$ with $\mathcal{M} = (\mathcal{M}_1 \times \mathcal{M}_2) \cap \mathbf{ComCat}(F, G; \mathcal{I})$ is a finitary $\mathcal{M}$-adhesive category.*

Proof (Idea). Comma categories under $\mathcal{M}$-adhesive categories are $\mathcal{M}$-adhesive due to Theorem 4.15.4 in [2]. It remains to show that each (A, B, op) in $\mathbf{ComCat}$ $(F, G; \mathcal{I})$ is finite. Since $(\mathbf{A}, \mathcal{M}_1)$ and $(\mathbf{B}, \mathcal{M}_2)$ are finitary we have only finitely many choices for $m \colon A' \to A$ and $n \colon B' \to B$ in a subobject (A', B', op'). Moreover, we have at most one choice for each op'_k with $k \in \mathcal{I}$, since $G(n)$ is a monomorphism by G preserving pullbacks along $\mathcal{M}_2$. ☐

Remark 7. Note that $\mathcal{I}$ in $\mathbf{ComCat}(F, G; \mathcal{I})$ is not required to be finite.

6   Extension to Non-$\mathcal{M}$-Adhesive Categories

There are some relevant categories in computer science which are not $\mathcal{M}$-adhesive for non-trivial choices of $\mathcal{M}$. The categories $\mathbf{SGraphs}$ of simple graphs (i. e., there is at most one edge between each pair of vertices) and $\mathbf{RDFGraphs}$ of Resource Description Framework graphs [9, 10] are, e. g., only $\mathcal{M}$-adhesive if $\mathcal{M}$

is chosen to be bijective on edges which is not satisfactory, since $\mathcal{M}$ is the class used for transformation rules and these should be able to add and delete edges. This difference between multi and simple graphs is due to the fact that colimits implicitly identify equivalent edges for simple graphs and similar structures and, hence, behave radically differently than in the case of multi graphs.

Similar behaviour can be expected for a wide variety of categories in which the objects contain some kind of relational structure. Since relational structures are omnipresent in computer science – in databases, non-deterministic automata, logical structures – the study of transformations for these categories is also highly relevant.

Moreover, pushout complements are not even unique in these categories leading to double-pushout transformations being non-deterministic even for determined rule and match. We can, however, canonically choose a minimal pushout complement (MPOC), which is the approach taken in [9, 10]. This leads to a new variant of the double-pushout transformation framework which is applicable to such categories of relational structures.

Therefore, it is interesting to explore to what extent the results on finitary $\mathcal{M}$-adhesive categories presented in this paper are also valid in such non-$\mathcal{M}$-adhesive categories in order to transfer as much as possible of the extensive theoretical results from $\mathcal{M}$-adhesive categories to the MPOC framework and possibly also other approaches.

Definition 9 ($\mathcal{M}$-Category). *A category $\mathbf{C}$ together with a class $\mathcal{M}$ of monomorphisms is called $\mathcal{M}$-category $(\mathbf{C}, \mathcal{M})$ if $\mathcal{M}$ is closed under composition, decomposition and isomorphisms, pushouts and pullbacks along $\mathcal{M}$ exist, and $\mathcal{M}$ is closed under pushouts and pullbacks.*

An object A in $(\mathbf{C}, \mathcal{M})$ is called finite *if the number of $\mathcal{M}$-subobjects of A is finite and the $\mathcal{M}$-category is called* finitary *if each object A in $(\mathbf{C}, \mathcal{M})$ is finite.*

Propositions 1–5 regarding coproducts and factorizations are already valid for (finitary) $\mathcal{M}$-categories, where we need in addition $\mathcal{M}$-initial objects for Propositions 2 and 5. Moreover, Propositions 8 and 9 remain valid for finitary $\mathcal{M}$-categories, but this problem is open for the creation of pushouts in Proposition 7 and, hence, also for Theorem 2.

By contrast, initial pushouts as they are defined in Definition 6 and constructed in Proposition 6 do not, in general, exist in finitary $\mathcal{M}$-categories. The problem is that the squares between the initial pushout and the comparison pushouts have to be pushouts themselves. Therefore, we define a weaker variant of initial pushouts, which does not require these squares to be pushouts but just to be commutative.

Definition 10 (Weak Initial Pushout). *Given an $\mathcal{M}$-category $(\mathbf{C}, \mathcal{M})$, a pushout (1) as in Definition 6 over a morphism $m\colon L \to G$ with $b, c \in \mathcal{M}$ is called* weak initial *if for all pushouts (2) over m with $b', c' \in \mathcal{M}$ there exist unique morphisms $b^*, c^* \in \mathcal{M}$, such that $b' \circ b^* = b$, $c' \circ c^* = c$, and (3) commutes.*

Remark 8. Observe that in $\mathcal{M}$-adhesive categories each weak initial pushout is already an initial pushout, since the initial pushout can be decomposed by

$\mathcal{M}$-pushout-pullback-decomposition which holds in $\mathcal{M}$-adhesive categories, because the comparison pushout is also a pullback in $\mathcal{M}$-adhesive, but not in general $\mathcal{M}$-categories.

Now, we show the existence and construction of weak initial pushouts for finitary $\mathcal{M}$-categories, provided that $\mathcal{M}$-pushouts are closed under pullbacks in the following sense.

Definition 11 (Closure of $\mathcal{M}$-Pushouts under Pullbacks). *Given an $\mathcal{M}$-category $(\mathbf{C}, \mathcal{M})$, we say that $\mathcal{M}$-pushouts are closed under pullbacks if for each morphism $m \colon L \to G$ and commutative diagram with pushouts over m in the right squares, pullbacks in the top and bottom and $b_1, b_2 \in \mathcal{M}$ (and, hence,*

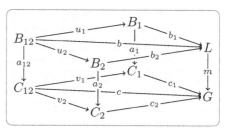

$c_1, c_2, u_1, u_2, v_1, v_2 \in \mathcal{M}$) *it follows that the diagonal square is a pushout.*

Proposition 10 (Existence of Weak Initial Pushouts). *Finitary $\mathcal{M}$-categories have weak initial pushouts, provided that $\mathcal{M}$-pushouts are closed under pullbacks.*

Proof (Idea). Similarly to Proposition 6, we obtain the weak initial pushout by finite $\mathcal{M}$-intersections B and C constructed by iterated pullbacks. Now, the closure of $\mathcal{M}$-pushouts under pullbacks is used to directly show that the diagonal square is a pushout in each iteration.   □

Note that the required closure of $\mathcal{M}$-pushouts under pullbacks already holds in $\mathcal{M}$-adhesive categories. Moreover, the closure holds in the categories **SGraphs** and **RDFGraphs**, allowing us to construct weak initial pushouts in these categories.

Remark 9. Similar to Remark 3, weak initial pushouts allow to define a gluing condition, which in this case is necessary and sufficient for the existence and uniqueness of minimal pushout complements (see [10]).

7   Conclusion

We have introduced finite objects in weak adhesive HLR categories [2], called $\mathcal{M}$-adhesive categories for simplicity in this paper. This leads to finitary $\mathcal{M}$-adhesive categories, like the category $\mathbf{Sets}_{\text{fin}}$ of finite sets and $\mathbf{Graphs}_{\text{fin}}$ of finite graphs with class $\mathcal{M}$ of all monomorphisms. In order to prove the main results like the Local Church-Rosser, Parallelism, Concurrency, Embedding, Extension, and Local Confluence Theorems not only the well-known properties of $\mathcal{M}$-adhesive categories are needed in [2], but also some additional HLR-requirements, especially initial pushouts, which are important to define the gluing condition and pushout complements, but often tedious to be constructed explicitly. In this paper, we have shown that for finitary $\mathcal{M}$-adhesive categories initial pushouts can be constructed by finite $\mathcal{M}$-intersections. Moreover, also the other additional HLR-requirements are valid in finitary $\mathcal{M}$-adhesive categories and, hence, the main results are valid for all $\mathcal{M}$-adhesive systems with finite objects, which are especially important for most of the application domains.

In order to construct finitary $\mathcal{M}$-adhesive categories we can either restrict $\mathcal{M}$-adhesive categories to all finite objects or apply suitable functor and comma category constructions (known already for general $\mathcal{M}$-adhesive categories [2]).

Finally, we have extended some of the results to non-$\mathcal{M}$-adhesive categories, like the category of simple graphs. Although adhesive categories [1] are special cases of $\mathcal{M}$-adhesive categories for the class $\mathcal{M}$ of all monomorphisms we have to be careful in specializing the results to finitary adhesive categories. While an object is finite in an $\mathcal{M}$-adhesive category $\mathbf{C}$ if and only if it is finite in the finitary restriction $\mathbf{C}_{\mathrm{fin}}$ (with $\mathcal{M}_{\mathrm{fin}} = \mathcal{M} \cap \mathbf{C}_{\mathrm{fin}}$) this is only valid in adhesive categories if the inclusion functor $I\colon \mathbf{C}_{\mathrm{fin}} \to \mathbf{C}$ preserves monomorphisms. It is to our knowledge an open problem for which kind of adhesive categories this condition is valid. Concerning categories $\mathbf{C}$ with a class $\mathcal{M}$ of monomorphisms, called $\mathcal{M}$-categories, it is open whether there are non-$\mathcal{M}$-adhesive categories such that the finitary restriction $(\mathbf{C}_{\mathrm{fin}}, \mathcal{M}_{\mathrm{fin}})$ is a finitary $\mathcal{M}$-adhesive category. For non-$\mathcal{M}$-adhesive categories it would be interesting to find a variant of the Van-Kampen-property which is still valid and allows to prove at least weak versions of the main results known for $\mathcal{M}$-adhesive systems. The closure of $\mathcal{M}$-pushouts under pullbacks is a first step in this direction, because it allows to construct weak initial pushouts for finitary $\mathcal{M}$-categories.

It remains open to compare our notion of finite objects in $\mathcal{M}$-categories with similar notions in category theory [11, 7] and to investigate other examples of $\mathcal{M}$-categories. Moreover, the relationships to work on (finite) subobject lattices in adhesive categories in [12, 13] are a valuable line of further research.

References

[1] Lack, S., Sobociński, P.: Adhesive Categories. In: Walukiewicz, I. (ed.) FOSSACS 2004. LNCS, vol. 2987, pp. 273–288. Springer, Heidelberg (2004)

[2] Ehrig, H., Ehrig, K., Prange, U., Taentzer, G.: Fundamentals of Algebraic Graph Transformation. EATCS Monographs. Springer, Heidelberg (2006)

[3] Rozenberg, G. (ed.): Handbook of Graph Grammars and Computing by Graph Transformation. Foundations, vol. 1. World Scientific, Singapore (1997)

[4] Prange, U., Ehrig, H., Lambers, L.: Construction and Properties of Adhesive and Weak Adhesive High-Level Replacement Categories. Applied Categorical Structures 16(3), 365–388 (2008)

[5] Gabriel, K., Braatz, B., Ehrig, H., Golas, U.: Finitary $\mathcal{M}$-adhesive categories. Unabridged version. Technical report, Technische Universität Berlin, Fakultät IV – Elektrotechnik und Informatik (to appear 2010)

[6] Ehrig, H., Padberg, J., Prange, U., Habel, A.: Adhesive High-Level Replacement Systems: A New Categorical Framework for Graph Transformation. Fundamenta Informaticae 74(1), 1–29 (2006)

[7] Adámek, J., Herrlich, H., Strecker, G.: Abstract and Concrete Categories. Wiley, Chichester (1990)

[8] Habel, A., Pennemann, K.H.: Correctness of High-Level Transformation Systems Relative to Nested Conditions. MSCS 19(2), 245–296 (2009)

[9] Braatz, B., Brandt, C.: Graph transformations for the Resource Description Framework. In: Ermel, C., Heckel, R., de Lara, J. (eds.) Proc. GT-VMT 2008. Electronic Communications of the EASST, vol. 10 (2008)

[10] Braatz, B.: Formal Modelling and Application of Graph Transformations in the Resource Description Framework. Dissertation, Technische Universität Berlin (2009)

[11] MacLane, S.: Categories for the Working Mathematician. Graduate Texts in Mathematics, vol. 5. Springer, Heidelberg (1971)

[12] Baldan, P., Bonchi, F., Heindel, T., König, B.: Irreducible objects and lattice homomorphisms in adhesive categories. In: Pfalzgraf, J. (ed.) Proc. of ACCAT 2008, Workshop on Applied and Computational Category Theory (2008)

[13] Baldan, P., Bonchi, F., Corradini, A., Heindel, T., König, B.: A lattice-theoretical perspective on adhesive categories (submitted for publication 2010)

Hereditary Pushouts Reconsidered

Tobias Heindel*

Formal Methods and Tools Group, EWI-INF, University of Twente
PO Box 217, 7500 AE, Enschede, The Netherlands
tobias.heindel@uni-due.de

Abstract. The introduction of adhesive categories revived interest in the study of properties of pushouts with respect to pullbacks, which started over thirty years ago in the category of graphs. Adhesive categories provide a single property of pushouts that suffices to derive lemmas that are essential for central theorems of double pushout rewriting such as the local Church-Rosser Theorem.

The present paper shows that the same lemmas already hold for pushouts that are hereditary, i.e. those pushouts that remain pushouts when they are embedded into the associated category of partial maps. Hereditary pushouts – a twenty year old concept – induce a generalization of adhesive categories, which will be dubbed partial map adhesive. An application relevant category that does not fit the framework of adhesive categories and its variations in the literature will serve as an illustrating example of a partial map adhesive category.

1 Introduction

Adhesive categories [15] provide an established theoretical foundation for double pushout rewriting [7]. Motivated by the fact that there are high-level-replacement (HLR) systems [5] that are neither adhesive (nor quasiadhesive [16]), several variants of adhesive categories have been proposed (see e.g. [6, 8, 11]). However, the resulting frameworks have lost (some of) the theoretical elegance of adhesive categories. For instance, the main requirement on pushouts in weak adhesive HLR categories [8] distinguishes between two different types of pushouts while adhesive categories have a single, uniform condition; moreover, the definition of weak adhesive HLR categories additionally *assumes* that monos (of a suitable class) are stable under pushout whereas in adhesive categories we can *prove* that monos are pushout-stable.

The goal of the present paper is to combine the theoretical elegance of adhesive categories with the generality of its variants. To achieve this goal, we will argue that a category is suitable for double pushout rewriting if it has (enough) pushouts that are hereditary [14], i.e. it suffices if pushouts in the category of rewriting induce pushouts in the associated category of partial maps.

To illustrate the concept of a hereditary pushout informally, we appeal to geometric intuition and assume that the objects in our category of rewriting can

---

* This work was partially supported by the EU Artemis project CHARTER.

H. Ehrig et al. (Eds.): ICGT 2010, LNCS 6372, pp. 250–265, 2010.

be faithfully represented by (a drawing on) some "ideal fabric". Moreover, we suppose that every *morphism* between structures can be realized as a "smooth" deformation of the "fabric".

An arbitrary morphism f between structures is depicted as an arrow as shown on the left in Figure 1. For example f could be a function between two sets, or an algebra homomorphism, or any continuous map between topological spaces. To keep the analogy with these concrete examples, the *image of f* in its codomain is marked as a darker area.

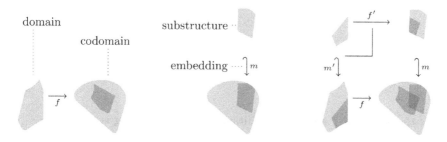

Fig. 1. Illustration of a morphism, an embedding, and a (codomain) restriction

As mentioned before, hereditary pushouts in a category are those pushouts that induce pushouts in the associated category of *partial maps*. A partial map consist of a usual morphism that is defined on a *domain of definition* that is embedded into the actual domain of the partial map. An embedding is depicted in the middle of Figure 1; an example illustration of a partial map is the pair $\langle m', f' \rangle$ on the right in Figure 1. Here, the partial map $\langle m', f' \rangle$ is obtained by restricting the morphism f to the embedding m, which yields the *inverse image m' (under f)* of the embedding m and the *(codomain) restriction f'*; the latter constructions are used to compose partial maps.

A pushout or a gluing of structures [4] is constructed from a *span* of morphisms, i.e. we start with a pair of morphisms f and g with a common domain. An example of such a span is illustrated in Figure 2(a): the morphisms f

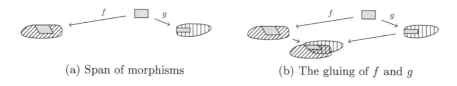

(a) Span of morphisms   (b) The gluing of f and g

Fig. 2. Gluing of structures

and g deform a rectangle in two different ways and embed it into the respective codomains. Now, to obtain the gluing of f and g, we combine the deformations of the rectangle and adjoin the remainder of the codomains of f and g as illustrated in Figure 2(b).

Finally, we are ready to sketch what it means that a gluing is hereditary. Consider two embeddings m_1 and m_2 of structures into the respective codomains of f and g that have the same inverse image in the common domain of f and g, namely the upper half of the rectangle in the left one of the below diagrams.

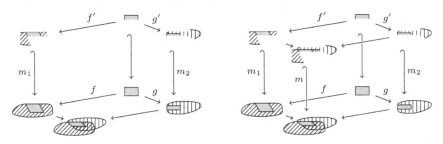

Now, if the gluing of f and g is hereditary then the following is true.

1. For any embedding m into the gluing of f and g such that the embeddings m_1 and m_2 arise as inverse images of m, the domain of the embedding m can be obtained as the gluing of the restricted morphisms f' and g'.

2. Conversely, the gluing of the restrictions f' and g' induces an embedding m into the gluing of f and g that has m_1 and m_2 as inverse images.

First, we establish the fact that the sketched property of gluings (which is a variant of the defining property of *Van Kampen squares* in the definition of adhesive categories [16]) actually characterizes hereditary pushouts [14].

Further, we shall see that in the usual categories of graphs *all* pushouts are hereditary. In fact, we wonder if this could be "the" explanation for the "good behaviour" of double pushout rewriting in the usual category of graphs. The reason is that we can prove the majority of the results that have been established for adhesive categories in [16] also in categories with (enough) hereditary pushouts. Finally, inspired by [2], we present an example of a category with "enough" hereditary pushouts that is neither an adhesive category nor a weak adhesive HLR category.

In view of these results, we have decided to call a category *partial map adhesive* if pushouts along monos (of a suitable class) exist and are hereditary.

Structure of the paper. We first review categories of partial maps and the definition of adhesive categories in Section 2. Then we define partial map adhesive categories in Section 3, show that they are a generalization of adhesive categories, and give examples of partial map adhesive categories; the main objective is to justify that it is appropriate to define yet another variation of adhesive categories. In Section 4, we provide the details of the so-called HLR conditions [18] that partial map adhesive categories share with adhesive categories. Finally, Section 5 gives a summary of the results and concludes with the conjecture that we have actually achieved our goal to unify adhesive categories and their variants.

2   Preliminaries

We review partial map categories and recall the definition of adhesive categories. We assume familiarity with basic concepts of category theory as introduced in [19]; the standard reference is [17]. We fix a category $\mathbb{C}$ and all objects and morphisms are assumed to belong to $\mathbb{C}$.

We take the following definition of partial functions as starting point.

Definition 1 (Partial function). *Let A and B be two sets. A partial function from A to B, denoted by $\varphi\colon A \to B$, is a pair $\varphi = \langle A', f \rangle$ where $A' \subseteq A$ is the domain of definition, and $f\colon A' \to B$ is a (total) function.*

To generalize partial functions to partial maps, the inclusion $A' \subseteq A$ is replaced by a mono $m\colon A' \rightarrowtail A$.

Definition 2 (Spans and partial maps). *A span is a pair of morphisms with a common domain, denoted by $A \xleftarrow{m} X \xrightarrow{f} B$; such a span is a partial map span if m is a mono. A partial map from A to B, denoted by $\varphi\colon A \to B$, is an isomorphism class of a partial map span, i.e.*

$$\varphi = \langle m, f \rangle = \left\{ A \xleftarrowtail{n} Y \xrightarrow{g} B \; \middle| \; \begin{array}{l} \textit{For some isomorphism } i\colon Y \xrightarrow{\cong} X \\ \begin{array}{c} {}_{m} \nearrow {}^{X} \searrow {}^{f} \\ A \Longleftarrow i\uparrow \Longrightarrow B \quad \textit{commutes.} \\ {}_{n} \searrow {}_{Y} \nearrow {}_{g} \end{array} \end{array} \right\}$$

for some representative partial map span $A \xleftarrowtail{m} X \xrightarrow{f} B$.

Often, the mono m of a partial map $\langle m, f \rangle$ is required to belong to a class $\mathcal{M}$ that contains "well-behaved" monos; to denote members of such a class $\mathcal{M}$, we shall use a special arrow $m\colon X \hookrightarrow A$. To obtain (sub-)categories of $\mathcal{M}$-*partial maps*, such a class of monos has to be *admissible* (see [21, Proposition 1.1]).

Definition 3 (Admissible classes of monos). *Let $\mathcal{M}$ be a class of monos in $\mathbb{C}$; each element $m \in \mathcal{M}$ is called an $\mathcal{M}$-morphism. The class $\mathcal{M}$ is stable (under pullback) if for each pair of morphisms $B \xrightarrow{f} A \xleftarrow{m} C$ with $m \in \mathcal{M}$ and each pullback $B \xleftarrow{m'} D \xrightarrow{f'} C$ of $B \xrightarrow{f} A \xleftarrow{m} C$, the mono m' belongs to $\mathcal{M}$.*

The class $\mathcal{M}$ of monos is admissible, *if*

1. *the category $\mathbb{C}$ has pullbacks along $\mathcal{M}$-morphisms;*
2. *the class $\mathcal{M}$ is stable under pullback;*
3. *the class $\mathcal{M}$ contains all identities;*
4. *the class $\mathcal{M}$ is closed under composition, i.e. for any two $\mathcal{M}$-morphisms $m\colon A \hookrightarrow B$ and $n\colon B \hookrightarrow C$, their composition $n \circ m\colon A \hookrightarrow C$ is an $\mathcal{M}$-morphism.*

Typical admissible classes are $\mathcal{M}ono$, the class of all monos, and $\mathcal{R}eg$, the class of all regular monos, i.e. those monos that arise as equalizers. Whenever $\mathcal{M}$ is an admissible class of monos, we can define a category of partial maps.

Definition 4 (M-categories and partial map categories). *An M-category is a pair $\langle \mathbf{C}, \mathcal{M} \rangle$ where $\mathbf{C}$ is a category and $\mathcal{M}$ is an admissible class of monos.*

In such an M-category $\langle \mathbf{C}, \mathcal{M} \rangle$, the category of M-partial maps, denoted by $\mathrm{Par}_{\mathcal{M}}(\mathbf{C})$, is defined as follows: it has the same objects as the category $\mathbf{C}$, and the morphisms between two objects $A, B \in \mathrm{Par}_{\mathcal{M}}(\mathbf{C})$ are the elements of

$$\mathrm{Par}_{\mathcal{M}}(\mathbf{C})(A, B) = \left\{ \langle m, f \rangle \colon A \rightharpoonup B \,\middle|\, A \stackrel{m}{\longleftarrow} X \stackrel{f}{\longrightarrow} B \ \& \ m \in \mathcal{M} \right\},$$

which contains all M-partial maps from A to B.

The identity on an object A is $\langle \mathrm{id}_A, \mathrm{id}_A \rangle$, and the composition of two partial maps $\langle m, f \rangle \colon A \rightharpoonup B$ and $\langle k, h \rangle \colon B \rightharpoonup C$ is $\langle k, h \rangle \circ \langle m, f \rangle = \langle m \circ p, h \circ q \rangle$ where $X \stackrel{p}{\leftarrowtail} U \stackrel{q}{\to} Z$ is some arbitrary pullback of $X \stackrel{f}{\to} B \stackrel{k}{\leftarrowtail} Z$.

The category of partial functions $\mathbf{Pfn}$ is (isomorphic to) $\mathrm{Par}_{\mathrm{Mono}}(\mathbf{Set})$ where $\mathbf{Set}$ is the usual category of sets and functions and Mono is the collection of all monos. The fact that each total function can be seen as a particular partial one has a natural counterpart for arbitrary M-categories. The obvious inclusion of an M-category $\langle \mathbf{C}, \mathcal{M} \rangle$ into $\mathrm{Par}_{\mathcal{M}}(\mathbf{C})$ is via the *graphing*[1] functor.

Definition 5 (Graphing functor). *Let $\langle \mathbf{C}, \mathcal{M} \rangle$ be an M-category. The graphing functor, denoted $\Gamma \colon \mathbf{C} \to \mathrm{Par}_{\mathcal{M}}(\mathbf{C})$, acts as the identity on objects, i.e. $\Gamma A = A$ for all $A \in \mathbf{C}$, and it maps each morphism $f \colon A \to B$ in $\mathbf{C}$ to $\Gamma f = \langle \mathrm{id}, f \rangle \colon A \rightharpoonup B$.*

We finally recall the definition of an adhesive category [16], which is phrased in terms of *Van Kampen squares*.

Definition 6 (Adhesive category). *A category $\mathbf{C}$ is adhesive if*

- *it has pullbacks,*
- *it has pushouts along monos, i.e. a pushout of a span $B \leftarrow f - A \rightarrowtail m \to C$ exists if m is a mono, and*
- *pushouts along monos yield Van Kampen (VK) squares, i.e. if $B -n \to D \leftarrow g - C$ is the pushout of $B \leftarrow f - A \rightarrowtail m \to C$*

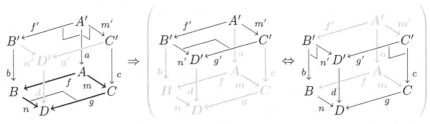

then in each commutative cube over this square as shown above on the left, which has pullbacks as back faces, the top face is a pushout if and only if the front faces are pullbacks (as illustrated above on the right).

---

[1] This terminology is taken from [9] and is based on the concept of a graph of a function as used in analysis.

3   Why Yet Another Variant of Adhesive Categories?

First, we claim that neither adhesive categories nor any of its variations that have been proposed in the literature so far [6, 8] impose only a *single* requirement on pushouts (along M-morphisms in some M-category) that

1. can be easily verified for common example categories;
2. holds for a large variety of structures that are intuitively "graph-like" or "set-like";
3. implies that double pushout rewriting is sufficiently well-behaved by virtue of the relevant properties of pushouts along monos in adhesive categories that have been discussed in the seminal paper [16].

Second, we shall propose a generalization of adhesive categories – partial map adhesive ones – and argue that it enjoys all of the above properties.

The single requirement for pushouts (along M-morphisms) that we shall require is that they should be hereditary in the sense of [14, Theorem 3.4]. In Section 3.2, we illustrate for two examples of "graph-like" categories that the proof of the fact that *all* pushouts are hereditary is relatively easy (if we compare them with the proofs of [13]). In Section 3.3, we give an example of a "set-like" category that is partial map adhesive but neither adhesive nor (weak) HLR adhesive [6, 8]; this example category provides evidence for the generality of partial map adhesive categories. Finally, in Section 4, we shall argue that partial map adhesive categories are (almost) as good as adhesive categories as a framework for double pushout rewriting.

3.1   Hereditary Pushouts and Partial Map Adhesive Categories

Partial map adhesive categories can be succinctly described as M-categories that have *hereditary* pushouts along M-morphisms, i.e. pushouts along M-morphisms exists and remain pushouts when they are considered in the "larger universe" of partial maps (see the illustration to the right). The formal definition is as follows.

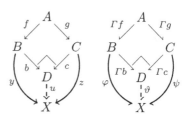

Definition 7 (Hereditary pushouts & partial map adhesive categories).
Let $\langle \mathbf{C}, \mathcal{M} \rangle$ *be an M-category. Let* $B \leftarrow f - A - g \rightarrow C$ *be a span; a hereditary pushout of the span* $B \leftarrow f - A - g \rightarrow C$ *[14, Theorem 3.4] is a cospan* $B - b \rightarrow D \leftarrow c - C$ *such that* $B - \Gamma b \rightarrow D \leftarrow \Gamma c - C$ *is a pushout of* $B \leftarrow \Gamma f - A - \Gamma g \rightarrow C$ *in* $\mathrm{Par}_{\mathcal{M}}(\mathbf{C})$ *(as illustrated above) where* $\Gamma \colon \mathbf{C} \to \mathrm{Par}_{\mathcal{M}}(\mathbf{C})$ *is the graphing functor (see Definition 5).*
*   An M-category* $\langle \mathbf{C}, \mathcal{M} \rangle$ *is M-partial map adhesive if*

– *it has pushouts along M-morphisms, and*
– *pushouts along M-morphisms are hereditary.*

It might come as a surprise that Van Kampen squares (or variations of them) do not occur in this definition. However, as we show next, we would obtain an equivalent definition if we replace the requirement 'are hereditary' by 'are *partial Van Kampen squares*'.

Definition 8 (Partial Van Kampen square). *A pushout square $B \stackrel{A}{\underset{D}{\leftrightarrows}} C$ is a partial Van Kampen (PVK) square if for each commutative cube on top of the pushout square as shown in Figure 3 on the left, which has pullback squares as back faces such that both b and c are M-morphisms, its top face is a pushout square if and only if the front faces are pullback squares and the morphism d is an M-morphism (as illustrated in Figure 3 on the right).*

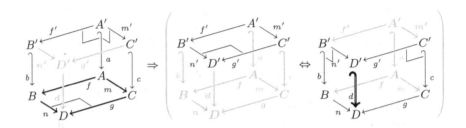

Fig. 3. Partial Van Kampen square property

Roughly, PVK squares combine features of Van Kampen squares and the so-called "weak" Van Kampen squares of [8]. Under natural assumptions, hereditary pushouts and PVK squares are the same; this can be made precise as follows.

Theorem 9 (Hereditary pushout characterization). *Let $\langle \mathbf{C}, \mathcal{M} \rangle$ be an M-category such that $\mathbf{C}$ has pushouts (along M-morphisms).*

Then pushouts (along M-morphisms) are hereditary if and only if they are partial Van Kampen squares.

Proof. This theorem is a direct consequence of Theorem B.4 of [11]. The crucial observation is that in the left cube diagram in Figure 3, the partial maps $\langle c, g' \rangle \colon C \to D'$, $\langle a, g' \circ m' \rangle \colon A \to D'$ and $\langle b, n' \rangle \colon B \to D'$ form a cocone of the span $B \leftharpoonup\!\ulcorner\! f \!- A -\!\ulcorner\! m \!\rightharpoonup C$ in Par($\mathbf{C}$); moreover, $\langle d, \mathrm{id} \rangle \colon D \to D'$ is the mediating partial map from the pushout $B - \ulcorner\! n \!\rightharpoonup D \leftharpoonup\!\ulcorner\! g \!- C$ in the rightmost cube diagram.   □

Remark 10. The characterization of partial map adhesive categories in terms of PVK squares yields the usual closure properties (see [16, Proposition 3.5] and [8, Theorem 1]).

Proposition 11. *Adhesive categories are Mono-partial map adhesive.*

Proof. The desired is a direct consequence of the main theorem of [12].   □

The definition of partial map adhesive categories in terms of hereditary pushouts naturally arises from the work [12], which shows that adhesive categories are exactly those categories with pullbacks and pushouts along monos in which pushouts along monos "survive" in the "larger universe" of *spans*. For partial map adhesive categories we use the "universe" of partial maps, which lies in between the "universe" of the usual (total) morphisms and the "universe" of spans. The first benefit of this change of "universe" is a useful proof technique that often can be applied to verify that a category is partial map adhesive.

3.2   Partial Map Adhesivity via Partial Map Classifiers

We have claimed that the most common example categories are easily shown to be partial map adhesive. To substantiate our claim, we consider simple graphs and "undirected" multigraphs as example categories: the former are the archetypal graphs in Computer Science and the latter capture the main features of the category of Petri nets that is described in [20, Definition 5]. To show that they form partial map adhesive categories, we provide concrete constructions of partial map classifiers, although for the case of simple graphs their existence is a standard fact of topos theory [1, Definition 28.7]; moreover, for the case of "undirected" multigraphs, partial map adhesivity follows from (an adapted version of) Theorem 1 of [8].

We shall use a proof technique that is based on a generalization of the well-known fact that each partial function $\langle A', f \rangle \colon A \rightharpoonup B$ corresponds to a total function $f_\perp \colon A \to B \uplus \{\perp\}$. More precisely, we have the inclusion $\eta_X \colon X \rightarrowtail X \uplus \{\perp\}$ for any set X, and each partial function $\langle A', f \rangle \colon A \rightharpoonup B$ corresponds to a unique total function $f_\perp \colon A \to B \uplus \{\perp\}$ that yields the equation $\langle A', f \rangle = \langle B, \mathrm{id} \rangle \circ \langle A, f_\perp \rangle$ in $\mathbb{Pfn}$. This situation naturally lifts to arbitrary $\mathcal{M}$-categories.

Definition 12 ($\mathcal{M}$-partial map classifiers). *Let $\langle \mathbb{C}, \mathcal{M} \rangle$ be an $\mathcal{M}$-category.*

$$
\begin{array}{ccc}
X \overset{m}{\hookrightarrow} A & X \overset{m}{\hookrightarrow} A & X \overset{m}{\dashrightarrow} A \\
\downarrow{\scriptstyle f} \quad {\scriptstyle \langle m, f\rangle} & \downarrow{\scriptstyle f} \qquad \Rightarrow & {\scriptstyle f}\downarrow \quad \lrcorner \quad \downarrow{\scriptstyle f'} \\
B \underset{\eta_B}{\hookrightarrow} \mathcal{L}B & B \underset{\eta_B}{\hookrightarrow} \mathcal{L}B & B \underset{\eta_B}{\hookrightarrow} \mathcal{L}B
\end{array}
$$

Then $\mathbb{C}$ has $\mathcal{M}$-partial map classifiers if for every object $B \in \mathbb{C}$, there is an $\mathcal{M}$-morphism $\eta_B \colon B \hookrightarrow \mathcal{L}B$ such that for each partial map span $B \leftarrow f - X \hookleftarrow m \to A$ with $m \in \mathcal{M}$ there is a unique morphism $f' \colon A \to \mathcal{L}B$ such that $B \hookleftarrow \eta_B \to \mathcal{L}B \leftarrow f' - A$ is the pullback of $B \hookleftarrow \eta_B \to \mathcal{L}B \leftarrow f' - A$.

That an $\mathcal{M}$-category has $\mathcal{M}$-partial map classifiers means exactly that the graphing functor has a right adjoint $\mathcal{L}$, which (up to natural isomorphism) is determined by the following property: for each partial map $\langle m, f \rangle \colon A \rightharpoonup B$ there exists a unique morphism $f' \colon A \to \mathcal{L}B$ such that the equation $\langle m, f \rangle = \langle \eta_B, \mathrm{id} \rangle \circ \Gamma f'$ holds in $\mathrm{Par}(\mathbb{C})$. Hence Γ is a left adjoint and thus preserves *all* colimits; in particular, we have proved the following lemma.

Lemma 13. *Every $\mathcal{M}$-category that has $\mathcal{M}$-partial map classifiers and pushouts along $\mathcal{M}$-morphisms is partial map adhesive.*

This lemma immediately yields the following examples of partial map adhesive categories.

Example 14 (Simple graphs). The category of simple graphs, denoted sGraph, has pairs $\langle V, E \rangle$ as objects where V is a set of *vertices* and $E \subseteq V \times V$ is a binary relation and its elements $(u, v) \in E$ are called *edges*. A morphism is a relation preserving function between the node sets, i.e. a morphism from a graph $\langle V, E \rangle$ to another one $\langle V', E' \rangle$ is a function $f \colon V \to V'$ such that $\big(f(u), f(v)\big) \in E'$ holds for all edges $(u, v) \in E$.

The category sGraph is not (quasi-)adhesive [13] although it is a quasi-topos and thus it has $\mathcal{R}eg$-partial map classifiers [1, Definition 28.7]. The regular monos are the *edge reflecting* ones, i.e. a mono $f \colon \langle V, E \rangle \to \langle V', E' \rangle$ is regular iff the implication $\big(f(u), f(v)\big) \in E' \Rightarrow (u, v) \in E$ holds for all edges $(u, v) \in E$. The concrete construction of $\mathcal{R}eg$-partial map classifiers in sGraph is as follows.

Lemma 15 (Partial map classifiers in sGraph). *Let $G = \langle V, E \rangle$ be a simple graph. The partial map classifier for G is (isomorphic to) $\mathcal{L}G = \langle V', E' \rangle$ where*

$$V' = V \uplus \{\bot\} \qquad E' = E \uplus (V \times \{\bot\}) \uplus (\{\bot\} \times V) \uplus \{(\bot, \bot)\}.$$

Proof. A $\mathcal{R}eg$-partial map from a graph $H = \langle U, D \rangle$ into G is essentially a subgraph $H' = \langle U', D' \rangle \subseteq H$ such that the inclusion $U' \subseteq U$ reflects edges together with a partial function $\langle U', f \rangle \colon U \to V$. Now, the corresponding (total) morphism $f' \colon H \to \mathcal{L}G$ must map a node $u \in U$ either to $f(u)$ (if $u \in U'$) or to $\bot$. It remains to verify that f' preserves edges. For $(v, w) \in D$ the only interesting case is the one in which both v and w belong to U'. However $(v, w) \in D'$ follows from regularity of the inclusion $U' \subseteq U$, and thus $(f'(v), f'(w)) \in E'$ holds. $\square$

Our second example of a partial map adhesive category has "undirected" multigraphs as objects. It is inspired by the category of Petri nets that has been used in [20] as an example of a weak adhesive HLR category.

Example 16. An *undirected multigraph* is a triple $\langle V, E, \mathrm{cnct} \rangle$ where V is the set of *nodes*, E is the set of *edges*, and $\mathrm{cnct} \colon E \to V^{\oplus}$ is the *connection function*, which connects each edge to a multiset of nodes, i.e. $\langle V^{\oplus}, \varnothing, \oplus \rangle$ is the free commutative monoid over V. A morphism from $\langle V, E, \mathrm{cnct} \rangle$ to $\langle V', E', \mathrm{cnct}' \rangle$ is a pair of functions $f = \langle f_V \colon V \to V', f_E \colon E \to E' \rangle$ such that the equation $\mathrm{cnct}' \circ f_E = f_V^{\oplus} \circ \mathrm{cnct}$ holds where $f_V^{\oplus}$ is the extension of f_V to $V^{\oplus}$.

Up to isomorphism, the $\mathcal{M}ono$-partial map classifier for an undirected multigraph $G = \langle V, E, \mathrm{cnct} \rangle$ is $\mathcal{L}G = \langle V', E', \mathrm{cnct}' \rangle$ where

$$V' = V \uplus \{\bot\}$$

$$E' = E \uplus \bigcup_{n \in \mathbb{N}_0} \left\{ v_1 \oplus \cdots \oplus v_n \;\middle|\; \{v_1, \ldots, v_n\} \subseteq (V \uplus \{\bot\}) \right\}$$

$$\mathrm{cnct}'(e') = \begin{cases} \mathrm{cnct}(e) & \text{if } e' \in E \\ e' & \text{otherwise} \end{cases}.$$

It is straightforward to verify that this construction yields $\mathcal{M}ono$-partial map classifiers.

Further examples of partial map adhesive categories with partial map classifiers include the category of topological spaces (see [1, Example 28.2(4)]). However, not all partial map adhesive categories have partial map classifiers as it will be the case in our main example.

3.3   A New Member in the Family

We give an example of a partial map adhesive category that is not adhesive and does not fit any of the frameworks in the literature that have been inspired by adhesive categories. The common point of all these frameworks is the requirement that pushouts of pairs of $\mathcal{M}$-morphisms must be (proper) Van Kampen squares (where in the case of adhesive categories $\mathcal{M} = \mathcal{M}ono$). This property will not hold in our example category, which is based on the list graphs of [2]. The counterexample (in Figure 4 below) involves a central feature of list graph morphisms that allows to map a so-called list node to a list of "actual" nodes, i.e. an element of the free monoid $\langle Y_V^*, \varepsilon, \rangle$ if Y_V is the set of nodes. For simplicity, we consider only the (full sub-)category of *list sets*.

Example 17 (List sets). A *list set* is a pair $X = (X_V, X_L)$ of disjoint sets X_V and X_L such that moreover $X_L \cap X_V^* = \varnothing$ holds; the set X_V contains *nodes* (ranged over by $u, v, w, \ldots$) and X_L is the set of *list variables* or just *lists* (ranged over by $\tilde{u}, \tilde{v}, \tilde{w}, \ldots$) and the *carrier* of X is $|X| = X_V \uplus X_L$.

A *(list set) morphism* between list sets X and Y is just a pair of functions $f = (f_V, f_L) \colon X \to Y$ of the following type:

$$f_V \colon X_V \to Y_V \quad \text{and} \quad f_L \colon X_L \to (Y_L \uplus Y_V^*).$$

The *carrier function* $|f| \colon |X| \to |Y|^*$ of f is defined by

$$|f|(x) = \begin{cases} f_V(x) & \text{if } x \in X_V \\ f_L(x) & \text{if } x \in X_L \end{cases},$$

and f is an *inclusion* if $|f|(x) = x$ holds for all $x \in |X|$ and $f_L(\tilde{w}) \in Y_L$ holds for all $\tilde{w} \in X_L$.

The identity $\mathrm{id}_X \colon X \to X$ on a list graph X is *the* inclusion of X into X and the composition $f \circ g \colon X \to Z$ of two morphisms $f \colon X \to Y$ and $g \colon Y \to Z$ is defined by

$$(g \circ f)_V = f_V \circ g_V \quad \text{and} \quad (g \circ f)_L(\tilde{w}) = \begin{cases} g_L \circ f_L(\tilde{w}) & \text{if } f_L(\tilde{w}) \in Y_L \\ g_V^*(f_L(\tilde{w})) & \text{if } f_L(\tilde{w}) \in Y_V^* \end{cases}$$

where g_V^* is the extension of g_V to X_V^*. The resulting category is denoted by $\mathbb{l}\mathsf{Set}$.

List sets can be considered "set-like" insofar as $\textsf{ISet}$ contains both $\textsf{Set}$ and $\textsf{Pfn}$ as full subcategories. The category $\textsf{Set}$ is embedded by mapping each set M to $(M, \varnothing)$ and each function $f : M \to N$ to $(f, \varnothing)$. Similarly, to embed $\textsf{Pfn}$, we map each set M to $(\varnothing, M)$ and each partial function $\langle M', f \rangle : M \to N$ to $(\varnothing, f_\varepsilon)$ where $f_\varepsilon(m) = f(m)$ for all $m \in M'$ and $f_\varepsilon(m) = \varepsilon$ otherwise.

As example of a "new" morphism that actually uses the feature of list variables, take the morphism from $(\varnothing, \{\tilde{w}\})$ to $(\{u, v\}, \varnothing)$ that maps the list variable $\tilde{w}$ to the list uv. This morphism is in fact both a mono and an epi but not an isomorphism; hence $\textsf{ISet}$ is not adhesive (by [16, Lemma 4.9]).

Lemma 18. *The category of list sets is not adhesive.*

To show that $\textsf{ISet}$ is $\mathcal{R}eg$-partial map adhesive, we use the fact that the regular monos are essentially the inclusions.

Proposition 19. *The category of list sets is a $\mathcal{R}eg$-category.*

Proof (idea). As inclusions are closed under composition and regular monos are pullback stable in any category, it is enough to show that pullbacks along inclusions exist.

Given an inclusion $j : X \hookrightarrow Y$ and a morphism $f : A \to Y$, the pullback object B has the carrier

$$|B| = \left\{ a \in |A| \;\middle|\; |f|(a) \in |X| \cup X_L^* \right\}.$$

It is straightforward to verify that the inclusion $i : B \hookrightarrow A$ and the restriction of f to B form a pullback, i.e. the pullback of $X \hookrightarrow{j} Y \leftarrow{f} A$ is $X \leftarrow{f'} B \hookrightarrow{i} A$ where $|f'|(b) = |f|(b)$ for all $b \in |B|$. □

Contrary to what had been originally claimed in [2], the category of list sets is not a (weak) adhesive HLR category w.r.t. $\mathcal{R}eg$ [8]. The counterexample is the pushout in Figure 4(a), which is "inherited" from $\textsf{Set}$. To see that it is not a Van

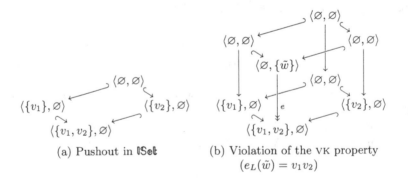

(a) Pushout in $\textsf{ISet}$

(b) Violation of the VK property
$(e_L(\tilde{w}) = v_1 v_2)$

Fig. 4. A pushout of a pair of regular monos that is not a Van Kampen square

Kampen square, consider the cube in Figure 4(b): all of its faces are pullback squares except for the top face, which is also not a pushout. However, $\mathbb{I}\mathbf{Set}$ has pushouts along regular monos and they yield PVK squares.

Theorem 20. *The $\mathcal{R}eg$-category $\langle \mathbb{I}\mathbf{Set}, \mathcal{R}eg \rangle$ is partial map adhesive.*

Proof (sketch). As regular monos are essentially co-product injections in $\mathbb{I}\mathbf{Set}$, it is enough to construct pushouts for spans of the form $U + A \leftarrow_{\iota_1} A -f\rightarrow B$, and the pushout of such a span is given by $U + A -(\mathrm{id}+f)\rightarrow U + B \leftarrow_{\iota_1} B$. Finally, one directly verifies the PVK property (as $\mathbb{I}\mathbf{Set}$ does not have $\mathcal{R}eg$-partial map classifiers). $\qquad\Box$

In the remainder of the paper we shall see that partial map adhesive categories and adhesive categories enjoy the same set of HLR conditions [18]; thus, partial map adhesive categories are a viable framework for double pushout rewriting.

4   Properties of Partial Van Kampen Squares

We finally show that *all* HLR properties of Van Kampen squares and adhesive categories that have been discussed in [16] also hold for PVK squares and partial map adhesive categories. We also comment on the role of each property in the context of (single and) double pushout rewriting [3] or give pointers to the literature. Wherever it will be appropriate, we speak about partial Van Kampen squares in general, as there are many categories in which all pushouts are hereditary. For the remainder of this section we fix an $\mathcal{M}$-category $\langle \mathbf{C}, \mathcal{M} \rangle$.

The first property of partial Van Kampen squares that we discuss is an extended version of the Pushout-Pullback Lemma from [4], which also takes into account Lemma 2.3 of [16]. It will directly imply that in any partial map adhesive category, double pushout rewriting in $\mathbf{C}$ is single pushout rewriting in $\mathrm{Par}(\mathbf{C})$ (see [3]). Our *extended* Pushout-Pullback Lemma is illustrated in Figure 5: Given

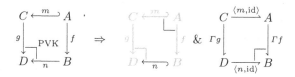

Fig. 5. The extended Pushout-Pullback Lemma for PVK squares

a pushout $C -g\rightarrow D \leftarrow n- B$ of a span $C \leftarrow m\rightharpoonup A -f\rightarrow B$ with an $\mathcal{M}$-morphism m that yields a PVK square, the morphism n is again an $\mathcal{M}$-morphism, the square is a pullback and induces a "reversed" pushout in $\mathrm{Par}(\mathbf{C})$.

Lemma 21 (Pushout-Pullback Lemma). *Let $C \leftarrow m\rightharpoonup A -f\rightarrow B$ be a $\mathbf{C}$-span with $m \in \mathcal{M}$; let $C -g\rightarrow D \leftarrow n- B$ be a hereditary pushout of $C \leftarrow m\rightharpoonup A -f\rightarrow B$.*
Then the morphism n belongs to the class $\mathcal{M}$, $C \leftarrow m\rightharpoonup A -f\rightarrow B$ is a pullback of $C -g\rightarrow D \leftarrow n\rightharpoonup B$, and $D -\langle n, \mathrm{id}\rangle\rightarrow B \leftarrow \Gamma f- A$ is a pushout of $D \leftarrow \Gamma g- C -\langle m, \mathrm{id}\rangle\rightarrow A$ in $\mathrm{Par}(\mathbf{C})$.

Proof (sketch). The first part of the proof is essentially the same as the one of Lemma 2.3 in [16].

For the second part we first proceed as for Lemma 2.8 in [16] and verify that for each $\mathcal{M}$-morphism $d\colon D' \hookrightarrow D$ we have $\mathbf{C}{\downarrow}C(g^*d, m) \cong \mathbf{C}{\downarrow}D(d, n)$ for any pullback $D' \leftarrow g'- X \hookrightarrow g^*d\to C$ of $D' \hookrightarrow d\to D \leftarrow g- C$. We then use the latter property to verify that $D -\langle n,\mathrm{id}\rangle\to B \leftarrow \ulcorner f-A$ is a pushout of the span $D -\ulcorner g- C -\langle m,\mathrm{id}\rangle\to A$ in $\mathrm{Par}(\mathbf{C})$.   □

Corollary 22 (Uniqueness of pushout complements). *Let $\mathbf{C}$ be a partial map adhesive category, let $C \leftarrow m\rightharpoonup A -f\to B$ be a $\mathbf{C}$-span with $m \in \mathcal{M}$, and let $C -g\to D \leftarrow n- B$ be its pushout.*

For any composable pair $D \leftarrow n'- B' \leftarrow f'- A$ such that $C -g\to D \leftarrow n'- B'$ is a pushout of $C \leftarrow m\rightharpoonup A -f'\to B'$, there is a unique isomorphism $i\colon B \to B'$ such that both $n = n' \circ i$ and $i \circ f = f'$ hold.

Corollary 23 (Double pushout is single pushout). *If in the left double pushout diagram below*

the morphism l is an $\mathcal{M}$-morphism and the two pushouts are partial Van Kampen squares (in $\mathbf{C}$), then the square on the right is a pushout in $\mathrm{Par}(\mathbf{C})$.

Remark 24. If in Corollary 23, the morphism $g\colon L \to A$ is assumed to be an $\mathcal{M}$-morphism, i.e. if we require matches to be $\mathcal{M}$-morphisms, then partial map adhesive categories are exactly those categories in which double pushout rewriting (in $\mathbf{C}$) is single pushout rewriting (in $\mathrm{Par}(\mathbf{C})$). We do not know whether a similar statement is true for weak adhesive HLR categories since we do not know whether arbitrary pushouts along $\mathcal{M}$-morphisms yield PVK squares (see also [11, Proposition 8.1.3]).

The second crucial lemma that holds in partial map adhesive categories is the so-called Pushout Pullback Decomposition property. It implies parallel and sequential commutativity, which are central theorems of double pushout rewriting (see e.g. [10]). In fact, the next lemma subsumes Lemmas 4.6 and 4.7 of [16].

Lemma 25 (Pushout Pullback Decomposition). *Let $B -h\to D \leftarrow g- C$ be a hereditary pushout of $B \leftarrow f- A -k\to C$, and let $h\colon B \to D$ factor as $m \circ \ell$ for a morphism $\ell\colon B \to M$ and an $\mathcal{M}$-morphism $m\colon M \hookrightarrow D$ as in the left one of the below diagrams.*

$$
\begin{array}{ccc}
A \xrightarrow{k} C & A \overset{k}{\underset{j}{\cdots\!\to}} N \underset{n}{\rightharpoonup} C & A \xrightarrow{j} N \underset{n}{\rightharpoonup} C \\
f\downarrow \quad \downarrow g \;\&\; f\downarrow \; i\downarrow \; \downarrow g \;\Rightarrow\; f\downarrow \; i\downarrow \; \downarrow g \\
B \xrightarrow{\ell} M \xrightarrow{m} D & B \xrightarrow{\ell} M \xrightarrow{m} D & B \xrightarrow{\ell} M \xrightarrow{m} D \\
\underset{h}{\smile} & \underset{h}{\smile} &
\end{array}
$$

Further let $M \xleftarrow{i} N \xhookrightarrow{n} C$ be the pullback of $M \xhookrightarrow{m} D \xleftarrow{g} C$, and let $j: A \to N$ be the unique morphism satisfying $i \circ j = \ell \circ f$ and $n \circ j = k$ as shown above. Then the cospan $B \xrightarrow{\ell} M \xleftarrow{i} N$ is a pushout of $B \xleftarrow{f} A \xrightarrow{j} N$.

Proof. The proof of this lemma is essentially the same as the one given for Lemmas 4.6 and 4.7 in [16]. □

There remain three properties to be established for PVK squares: the Special Pullback-Pushout Property, the so-called Twisted-Triple-Pushout Condition, and the Cube-Pushout-Pullback Lemma. All three properties are best understood in the context of high-level-replacement systems [5].

Lemma 26 (Special Pullback-Pushout Property). *Let $\langle \mathbf{C}, \mathcal{M} \rangle$ be a partial map adhesive category. Then the Special Pullback-Pushout Property holds in $\mathbf{C}$.*

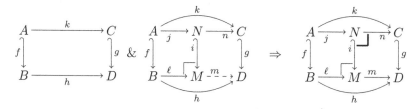

Proof. The proof is omitted due to space constraints. □

The Cube Pushout Pullback Lemma is a direct consequence of the characterization of partial Van Kampen squares given in Theorem 9.

Lemma 27 (Cube Pushout Pullback Lemma). *For any cube diagram in which all but the vertical morphisms are $\mathcal{M}$-morphisms as shown below*

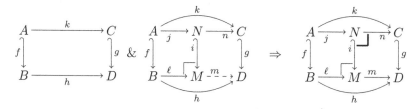

where the front faces are two partial Van Kampen squares and the top face is a pullback, the following is true: the bottom face is a pullback if and only if the back faces are pushouts.

Proof. The proof is an adaption of the proof of Lemma 8.4 of [16]. □

Finally, the Twisted-Triple-Pushout condition holds if the category $\mathbf{C}$ has all pushouts and these pushouts yield $\mathcal{M}$-partial Van Kampen squares.

Lemma 28 (Twisted-Triple-Pushout Lemma). *Let $A -f\to B -p\to E$ be a pair of morphisms, let $A \hookrightarrow^{m}\to C$ be an $\mathcal{M}$-morphism, let $C -g\to D -q\to F$ be a pair of morphisms, and let $E -k\to F$ be a morphism such that $E -k\to F \leftarrow q\circ g- C$ is a PVK pushout of $E \leftarrow p\circ f- A \hookrightarrow^{m}\to C$.*

$$
\begin{array}{ccccc}
A \stackrel{m}{\hookrightarrow} C \stackrel{c}{\leftarrow} X & & A \stackrel{m}{\hookrightarrow} C \stackrel{c}{\leftarrow} X & & A \stackrel{m}{\hookrightarrow} C \\
{\scriptstyle f}\downarrow \quad {\scriptstyle g}\downarrow \quad {\scriptstyle z}\downarrow & & {\scriptstyle f}\downarrow \quad {\scriptstyle g}\downarrow \quad {\scriptstyle z}\downarrow & \Rightarrow & {\scriptstyle f}\downarrow \quad {\scriptstyle g}\downarrow \\
B \quad D \leftarrow_{v} Z & \& & B \xrightarrow{n} D \quad Z & & B \xrightarrow{n} D \\
{\scriptstyle p}\downarrow \quad {\scriptstyle q}\uparrow & & {\scriptstyle p}\downarrow \quad {\scriptstyle q}\downarrow \quad {\scriptstyle \mathrm{id}_Z}\downarrow & & \\
E \xrightarrow{k} F & & E \xrightarrow{k} F \stackrel{\leftarrow}{q} D \stackrel{\leftarrow}{v} Z & &
\end{array}
$$

Further let $C \leftarrow c- X -z\to Z$ be a span and let $D \leftarrow v- Z$ be an arrow such that $C -g\to D \leftarrow v- Z$ is a PVK pushout of $C \leftarrow c- X -z\to Z$ and also $C \leftarrow c- X -z\to Z$ is a pullback of $C -q\circ g\to F \leftarrow q\circ v- Z$. Finally let $B -n\to D$ be an arrow such that the equation $g \circ m = n \circ f$ holds and the span $E \leftarrow p- B -n\to D$ is a pullback of $E -k\to F \leftarrow q- D$. Then $B -n\to D \leftarrow g- C$ is a pushout of $B \leftarrow f- A \hookrightarrow^{m}\to C$.

Proof. The proof for Lemma 8.5 of [16] can be easily adapted.   □

5   Conclusion

We have defined partial map adhesive categories, which generalize adhesive categories and provide a framework for double pushout rewriting because (pushouts in) partial map adhesive categories satisfy a sufficient set of HLR conditions. We have illustrated that common example categories are easily verified to be partial map adhesive by proving the existence of partial map classifiers. As evidence for the generality of the new framework, we have provided an example of a partial map adhesive category that is not a weak adhesive HLR category [8].

Motivated by preliminary results [11, Proposition 8.1.3], we conclude with the conjecture that all weak adhesive HLR categories are also partial map adhesive and that indeed hereditary pushouts are a fundamental, common concept of adhesive categories and its variants.

Acknowledgments. I would like to thank Arend Rensink and Maarten de Mol for inspiring discussions about list graphs and the presentation of the results of this paper; moreover I would like to thank the referees for their constructive criticism.

References

1. Adámek, J., Herrlich, H., Strecker, G.E.: Abstract and concrete categories: the joy of cats. Wiley, Chichester (1990)
2. de Mol, M., Rensink, A.: On a graph formalism for ordered edges. In: International Workshop on Graph Transformation and Visual Modeling Techniques, GT-VMT (to appear 2010)

3. Ehrig, H., Heckel, R., Korff, M., Löwe, M., Ribeiro, L., Wagner, A., Corradini, A.: Algebraic approaches to graph transformation – part ii: Single pushout approach and comparison with double pushout approach. In: Rozenberg, G. (ed.) Handbook of Graph Grammars, pp. 247–312. World Scientific, Singapore (1997)
4. Ehrig, H., Kreowski, H.-J.: Pushout-properties: An analysis of gluing constructions for graphs. Mathematische Nachrichten 91, 135–149 (1979)
5. Ehrig, H., Löwe, M.: Categorical principles, techniques and results for high-level-replacement systems in computer science. Applied Categorical Structures 1(1), 21–50 (1993)
6. Ehrig, H., Padberg, J., Prange, U., Habel, A.: Adhesive high-level replacement systems: A new categorical framework for graph transformation. Fundamenta Informaticae 74(1), 1–29 (2006)
7. Ehrig, H., Pfender, M., Schneider, H.J.: Graph-grammars: An algebraic approach. In: 14th Annual Symposium on Switching and Automata Theory, Institute of Electrical and Electronics Engineers, pp. 167–180 (1973)
8. Ehrig, H., Prange, U.: Weak adhesive high-level replacement categories and systems: A unifying framework for graph and Petri net transformations. In: Futatsugi, K., Jouannaud, J.-P., Meseguer, J. (eds.) Algebra, Meaning, and Computation. LNCS, vol. 4060, pp. 235–251. Springer, Heidelberg (2006)
9. Freyd, P.J., Scedrov, A.: Categories Allegories. North-Holland, Amsterdam (1990)
10. Habel, A., Müller, J., Plump, D.: Double-pushout graph transformation revisited. Mathematical Structures in Computer Science 11(5), 637–688 (2001)
11. Heindel, T.: A Category Theoretical Approach to the Concurrent Semantics of Rewriting – Adhesive Categories and Related Concepts. PhD thesis, Universität Duisburg-Essen (2009)
12. Heindel, T., Sobocinski, P.: Van Kampen colimits as bicolimits in span. In: Kurz, A., Lenisa, M., Tarlecki, A. (eds.) CALCO 2009. LNCS, vol. 5728, pp. 335–349. Springer, Heidelberg (2009)
13. Johnstone, P.T., Lack, S., Sobociński, P.: Quasitoposes, quasiadhesive categories and artin glueing. In: Mossakowski, T., Montanari, U., Haveraaen, M. (eds.) CALCO 2007. LNCS, vol. 4624, pp. 312–326. Springer, Heidelberg (2007)
14. Kennaway, R.: Graph rewriting in some categories of partial morphisms. In: Ehrig, H., Kreowski, H.-J., Rozenberg, G. (eds.) Graph-Grammars and Their Application to Computer Science. LNCS, vol. 532, pp. 490–504. Springer, Heidelberg (1990)
15. Lack, S., Sobociński, P.: Adhesive categories. In: Walukiewicz, I. (ed.) FOSSACS 2004. LNCS, vol. 2987, pp. 273–288. Springer, Heidelberg (2004)
16. Lack, S., Sobociński, P.: Adhesive and quasiadhesive categories. Theoretical Informatics and Applications 39(2), 511–546 (2005)
17. Lane, S.M.: Categories for the Working Mathematician. Graduate Texts in Mathematics, vol. 5. Springer, Heidelberg (1998)
18. Padberg, J.: Survey of high-level replacement systems. Technical report, Technische Universität Berlin (1993)
19. Pierce, B.C.: Basic Category Theory for Computer Scientists. MIT Press, Cambridge (1991)
20. Prange, U.: Algebraic High-Level Nets as Weak Adhesive HLR Categories. Electronic Communications of the EASST 2, 1–13 (2007)
21. Robinson, E., Rosolini, G.: Categories of partial maps. Information and Computation 79(2), 95–130 (1988)

Graph Transformation for Domain-Specific Discrete Event Time Simulation

Juan de Lara[1], Esther Guerra[2], Artur Boronat[3],
Reiko Heckel[3], and Paolo Torrini[3]

[1] Universidad Autónoma de Madrid (Spain)
Juan.deLara@uam.es
[2] Universidad Carlos III de Madrid (Spain)
eguerra@inf.uc3m.es
[3] University of Leicester (UK)
{aboronat,reiko,pt95}@mcs.le.ac.uk

Abstract. Graph transformation is being increasingly used to express the semantics of domain specific visual languages since its graphical nature makes rules intuitive. However, many application domains require an explicit handling of time in order to represent accurately the behaviour of the real system and to obtain useful simulation metrics.

Inspired by the vast knowledge and experience accumulated by the discrete event simulation community, we propose a novel way of adding explicit time to graph transformation rules. In particular, we take the event scheduling discrete simulation world view and incorporate to the rules the ability of scheduling the occurrence of other rules in the future. Hence, our work combines standard, efficient techniques for discrete event simulation (based on the handling of a future event set) and the intuitive, visual nature of graph transformation. Moreover, we show how our formalism can be used to give semantics to other timed approaches.

1 Introduction

Graph Transformation [5] (GT) is becoming increasingly popular as a means to express and analyse the behaviour of systems. For example, it has been extensively used to describe the operational semantics of Domain Specific Visual Languages (DSVLs) in areas such as reliable messaging in SOA [7], web services [12], gaming [16] and manufacturing [3]. The success of GT is partly because rules are intuitive and allow the designer to use the concrete syntax of the DSVLs.

When used to specify the DSVL semantics, the rules define a simulator, and their execution accounts for the state change of the system. This is enough for languages with a discrete, *untimed* semantics, where the time elapsed between two state changes is not important. However, for its use in simulation applications or real-time systems, where system behaviour depends on explicit timing (e.g. time-outs in network protocols) and performance metrics are essential, a mechanism is needed to model how time progresses during the GT execution.

Computer simulation [6] is the activity of performing virtual experiments on the computer (instead of in the real world) by representing real systems by means

H. Ehrig et al. (Eds.): ICGT 2010, LNCS 6372, pp. 266–281, 2010.

of computational models. Simulation is intrinsically multi-disciplinary, and is at the core of research areas as diverse as real-time systems, ecology, economy and physics. Hence, users of simulations are frequently domain experts (but not necessarily computer scientists) who are hardly proficient in programming languages but on the domain-specific notations used in their scientific domain.

Discrete-event simulation (DES) [2,6] studies systems where time is modelled in a continuous way ($\mathbb{R}$), but in which there is only a finite number of events or state changes in a finite time interval. Many languages, systems and tools have been proposed over the years in the DES domain [6]. However, these require specialized knowledge that domain experts usually lack, or consist of libraries for programming languages like Java. Therefore simulationists would strongly benefit from a domain-specific, graphical language to describe simulations.

In this paper, we propose such a language by incorporating an explicit handling of time into the GT formalism. In this way, based on the *event scheduling* approach to simulation [14], we allow rules to program the occurrence of other rules in the future. For this purpose, our approach makes use of two concepts: (i) explicit rule invocation (and cancellation) with parameter passing between invocations and (ii) time scheduling of those matches. This improves efficiency in two ways: rule execution is guided by the parameter passing, and the global time is increased to the time of the next occurring event (instead of doing small increments). Our goal is to provide the simplest possible time handling primitive, on top of which other more advanced constructs can be added. We show that *scheduling* is one such primitive mechanism, and demonstrate its use to model (stochastic) delays, timers, durations and periodic activities.

Paper organization. §2 gives an overview to DES and *event scheduling*. §3 introduces the use of (untimed) GT to describe the semantics of DSVLs. Next, §4 extends GT with rule invocations and parameter passing, called *flow grammars*. These are extended with time scheduling in §5. §6 discusses how to model other timed approaches with ours. §7 covers related research and §8 concludes.

2   Discrete Event Simulation: World Views

Discrete-event systems can be modelled using different styles, or *world-views* [2,6]. Each world-view focuses on a particular aspect of the system: events, activities or processes. An *event* is an instantaneous change in an object's state. An *object activity* is the state of an object during a time interval, between two events. A *process* is a succession of object states defining its simulation life-cycle. Therefore, there are three different approaches to describe discrete time models [2]: *Event Scheduling* (ES) focussing on events, *Activity Scanning* (AS) focussing on activities, and *Process Interaction* (PI) focussing on object processes.

ES languages offer primitives to describe events, their effect on the current state, and the scheduling of future events. Time is managed efficiently by simply advancing the simulation time to the time of the next event to occur. AS languages focus on describing the conditions enabling the start of activities. They are less efficient because, lacking the concept of event to signal state changes, they have to advance the time using a small discrete increment. To increase efficiency, the *three-phase approach* combines ES and AS so that the start of new activities is only checked after handling an event. Finally, PI provides constructs to describe the life-cycle of the active entities of the system.

Among the three approaches, ES is the most primitive, as events delimit the start and end of activities, and a flow of activities makes up a process. Hence, we concentrate on the ES approach, and in particular on the Event Graphs notation [14], an example of which is shown to the left of Fig. 1. The event graph models a simple communication network, where a node sends messages periodically to a receiver node through a channel with limited capacity.

Nodes in the event graph represent events, and there are two special ones: the start and the end of the simulation, identified with a tick and a double circle respectively. The state is represented with variables (ch being the load of the channel and w the number of messages waiting). Below event nodes, a sequence of variable expressions describes state changes. Arrows between events represent schedulings. For instance, the arrow from the event *start* to *end* means that, once *start* happens, an occurrence of *end* will happen after t_f time units. If no time is indicated (like in the arrow from *start* to *create*) then the target event is scheduled to occur immediately. Arrows can be decorated with a condition that is evaluated after processing the source event, and that must be *true* in order to schedule the target event at the indicated time. For example, the arrow from *create* to *send* means that after creating a message, this will be sent only if there is some message waiting and the channel has enough capacity. Finally, although not shown in the example, event graphs can also contain *event-cancelling* edges, represented as dashed arrows. These edges indicate the deletion of all events of the target type scheduled after the indicated time units, if the condition (if given) holds at the time the source event is processed [14].

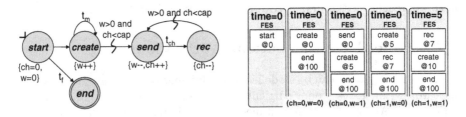

Fig. 1. An event graph model (left). An execution of the model (right).

DES simulators use a future event set (FES) which contains the events scheduled to occur in the future. The simulation proceeds by taking the event with earliest occurrence time and executing its specification as given by the event graph (i.e. modifying the system state and scheduling new events). Many algorithms and data structures exist to handle the FES efficiently [6].

The right of Fig. 1 shows some execution steps of the model, using as parameters $t_m = 5$, $t_f = 100$, $cap = 1$, $t_{ch} = 7$. Each state transition consumes the earliest event in the set, updates the current time to the time of this event, modifies the variables, and schedules new events according to the model. The simulation continues until processing the *end* event.

The execution of event graphs is efficient, but its modelling sometimes lacks intuitivity. This is so because event graphs are not domain-specific and force the use of scattered variables for expressing state changes instead of full-fledged models and DSVLs. Next we show how GT provides such intuitive formalism, but lacks time handling capabilities, which we subsequently add in §5.

3   Rule-Based, Domain-Specific, Untimed Simulation

In this section we give an overview of the use of GT to describe the semantics of DSVLs. The syntax of DSVLs is usually defined through a meta-model, or type graph, which contains the node and edge types that models can use. For example, Fig. 2 shows on the left an example meta-model describing a DSVL in the domain of communication networks and protocols. In this language, a network is made of nodes which exchange messages through channels. Messages can be either requests or replies. There are two special kinds of nodes: initiators, whose attribute *isInit* is *true*, and terminal, whose attribute *isFinal* is *true*.

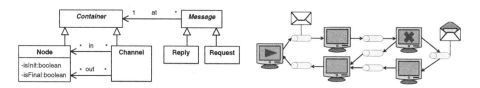

Fig. 2. The DSVL meta-model (left). A model (right).

The right of Fig. 2 shows an example model in concrete syntax, with an initiator node to the left (marked with a "play" icon), and a terminal node marked with a cross to the right. Requests are shown as closed envelopes (like the one to the left) and replies as open envelopes (like the one to the right). Channels are depicted as pipes.

We are using this DSVL to describe the dynamics of a simple protocol where the messages are propagated through the network at random. When a request reaches a terminal node, this node sends back a reply which traverses the network randomly until it reaches the initiator. Since channels can lose messages, the initiator sends a new request periodically. We also model changes in the topology, so that nodes are connected and disconnected from channels.

We define this protocol with DPO GT rules [5], in which rules have the form $p : \langle L \xleftarrow{l} K \xrightarrow{r} R, NAC = \{n_i : L \rightarrow N_i\}_{i \in I}\rangle$. NAC is a set of Negative Application Conditions. In this paper, we sometimes depict rules using just their LHS and RHS, and use the concrete syntax of the DSVL.

A morphism $m: L \rightarrow G$ is a valid match for rule p in the host graph G, written $m \models_G p$, if m satisfies the gluing conditions [5], and if $\forall n_i: L \rightarrow N_i \in NAC$ $\nexists n: N_i \rightarrow G$ with $n \circ n_i = m$. If $m \models_G p$, we can perform a direct derivation.

Fig. 3 shows some GT rules of the simulator. Messages are generated from initiators by rule *init*. Nodes can *send* and *receive* messages. As in [5], objects in rules can be matched to objects in the host graph with a more concrete type. For example, rule *send* contains an object m of type message (depicted as a dotted envelope) which can be matched to both requests and replies. The rule does not apply if the message is a request and the node is terminal (first NAC), or if the message is a reply and the node is initial (second NAC). The first case is handled by rule *reply*, which processes the request and generates a reply, whereas the second case is handled by rule *end*, which removes the reply from the net. Rules *createConnection* and *deleteConnection* model the creation and deletion of connections from nodes to channels. The former only applies to nodes without output channels. Finally, rule *lose* simulates the loss of messages.

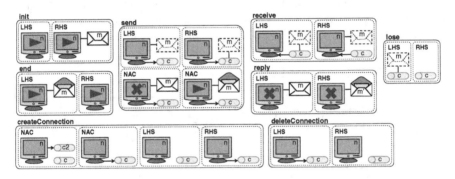

Fig. 3. Some rules of the DSVL simulator

For the purpose of simulation, the standard approach to GT has two drawbacks. First, even though these rules capture the untimed semantics of the language, they cannot represent time-outs, delays or be used to obtain metrics about the system performance. For example, we would like to set a time-out in the initiator so that it sends requests each 50 time units, and also to model transmission delays and the average rate at which channels lose messages.

Second, rules represent events which signal the start or end of activities of the entities in the system. Thus, the focus on active entities requires an explicit model for event processing (a *process*) which identifies the context in which events are executed in order to pass part of this context to subsequent events. This would result in more efficient simulations. Moreover, different processes may interact, e.g. we would like to prevent the deletion of connections if they are being used to send a message. Hence, next we extend GT with these two features.

4   Flow Graph Grammars

An important need in modelling DES is the ability to describe the order in which events should be executed and their context of execution. For example, a message has to be sent before it is received. Even though these conditions can be encoded in the LHS and NACs of the rules, it is sometimes simpler to resort to rule invocations, as well as more efficient to provide a data dependency between rules so that the context is passed as a parameter. These dependencies are conceptually the edges of the event graph (cf. Fig. 1), where for the moment we are not taking into account the time scheduling.

This feature is already present in tools like Fujaba, GReAT and VMTS[1], but here we give a novel formalization in terms of DPO, and include event cancelling edges, also new in the GT literature. This formalization will be used in the next section to incorporate a time scheduling distribution function to rule invocations. By separating rule invocation from time scheduling we show how to extend existing tools to handle time.

Hence we start by defining *flow grammars* as a set of productions P with two sets I and C of invocation and cancelling edges between productions. Each edge defines in addition a parameter passing from the source to the target rule. For technical reasons, we define an auxiliary empty rule $\bot = \langle \emptyset \leftarrow \emptyset \rightarrow \emptyset \rangle$, which is used to invoke the initial rules of the flow.

Def. 1 (Flow grammar). *A flow grammar $FG = \langle P \cup \{\bot\}, end, I, C, G_0 \rangle$ is made of a set $P \cup \{\bot\}$ of rules; a set $end \subseteq P$ of final rules; a set of invocation edges $I = \{(p_i, R_i \leftarrow M_{ik} \rightarrow L_k, p_k)\}$, where $p_i, p_k \in P \cup \{\bot\}$, R_i is p_i's RHS and L_k is p_k's LHS; a set $C = \{(p_j, R_j \leftarrow M_{jl} \rightarrow L_l, p_l)\}$ of cancelling edges; and an initial graph G_0.*

Given a rule $p_i \in P \cup \{\bot\}$, we use the notation $I(p_i) = \{s = (p_i, R_i \leftarrow M_{ik} \rightarrow L_k, p_k) \mid s \in I\}$ and $C(p_i) = \{s = (p_i, R_i \leftarrow M_{ik} \rightarrow L_k, p_k) \mid s \in C\}$.

Remark. The structure $R_i \leftarrow M_{ij} \rightarrow L_j$ of invocation and cancelling edges is used to pass the context of execution from R_i to L_j. M_{ij} identifies the elements of R_i and L_j that have to be matched in the same elements of the host graph. If M_{ij} is empty, there is no data dependency, but still rule invocation.

Example. Fig. 4 shows to the left the definition of an invocation edge which passes the node and linked message from rule *init*'s RHS to rule *send*'s LHS. The typing of the message in the $M_{init,send}$ component is abstract, as the typing of the message in L_{send} is abstract too. The figure shows in the center a visual representation of a flow grammar built using the rules of Fig. 3 plus the rule *channelCheck* shown to the right. We use a notation similar to that of event graphs, where each node represents a rule, but in the edges we depict the parameters passed between rules (i.e. the M_{ij}), as these are more informative. For example, the invocation edge depicted to the left is represented as a directed

---

[1] http://www.fujaba.de, http://www.isis.vanderbilt.edu/tools/GReAT,
     http://avalon.aut.bme.hu/~tihamer/research/vmts/

edge decorated with $M_{init,send}$ in this visual representation. The rules marked with a tick are the initial rules, which receive an invocation from rule $\bot$. We take the convention of not showing the rule $\bot$ and its invocation edges $I(\bot)$. This event graph representation reveals three processes, given by the connected components. Hence there are three active entities: messages (their process starts in *init* and ends in *end*), connections and channels. The latter makes each channel to lose a message periodically. Processes can interact explicitly by means of cancelling edges. For example, if a connection is used by rule *send*, then we cancel its programmed (i.e. invoked) deletions so that the network connectivity is optimized to the most used connections.

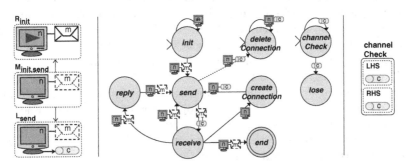

Fig. 4. Parameter passing (left). Flow grammar (center). Additional rule (right).

In order to define the semantics of a flow grammar, first we need to define the system state. This is made of the host graph, plus a set of events storing rule invocations (i.e. elements in I) together with the match in which the rules should be applied (i.e. matches of the invoked rule's LHS).

Def. 2 (Event and state). *Given a flow grammar FG, an* event *is a tuple $e = \langle m \colon L_j \to G, s \rangle$, with $s = (p_i, R_i \leftarrow M_{ij} \to L_j, p_j) \in I$ and $m \models_G p_j$. We write $m(e) = m$, $s(e) = s$, $p(e) = p_j$ to refer to e's match, edge and invoked rule.*

A state *$S = \langle G, E \rangle$ is a tuple made of a graph G, and a set E of events such that $\forall e \in E, m(e) \models_G p(e)$.*

The execution of a flow grammar starts from the matches of the initial rules (those invoked from $\bot$). These matches are converted into events to populate the event set E_0 of the initial system state.

Def. 3 (Initial state). *Given a flow grammar FG, the* initial system state *init (FG) is given by $S_0 = \langle G_0, E_0 \rangle$, where G_0 is the initial graph of FG, and $E_0 = \{(m \colon L_i \to G_0, s = (\bot, \emptyset \leftarrow \emptyset \to L_i, p_i)) | s \in I(\bot) \text{ and } m \models_{G_0} p_i\}$.*

Example. The initial system state for the flow grammar of Fig. 4, taking as initial graph the one shown to the right of Fig. 2, contains one event e whose production $p(e)$ is *init*, 6 events due to matches of *deleteConnection*, and 6 events due to matches of *channelCheck*. Thus, there is an event for each match of the initial rules. These events are the initial starting points for the autonomous execution of each of the three processes in the grammar.

A direct derivation of a flow grammar from a state $\langle G, E \rangle$ consists of taking one event $e \in E$ (if more than one exist, one is taken at random), performing a standard DPO direct derivation using the match $m(e)$, and then calculating the new set of enabled events E'. This set E' contains the old matches in E that were not destroyed by the application of $p(e)$ (set OLD in the definition), and incorporates at most one event for each rule invoked from $p(e)$ (set NEW). Moreover, E' excludes e from the system state, as well as the events cancelled by the cancelling edges $C(p(e))$ (set $CANC$). Please note that the events in E whose match is destroyed by the application of $p(e)$ are not transferred into E'.

Def. 4 (Derivation). *Given a flow grammar $FG = \langle P \cup \{\bot\}, end, I, C, G_0 \rangle$, and a state $S = \langle G, E \rangle$, a direct derivation $S = \langle G, E \rangle \overset{e}{\Longrightarrow} S' = \langle H, E' \rangle$ due to the event $e = \langle m_i, (p_s, R_s \leftarrow M_{si} \rightarrow L_i, p_i) \rangle \in E$ is performed as follows:*

- *H is obtained by a standard DPO direct derivation $G \overset{m_i, p_i}{\Longrightarrow} H$, as the left of Fig. 5 shows, where $p(e) = p_i$ and $m(e) = m_i \colon L_i \rightarrow G$,*
- *$E' = NEW \cup (OLD \setminus CANC)$, where:*
  - *$NEW = \{(m_k \colon L_k \rightarrow H, s) \mid s = (p_i, R_i \leftarrow M_{ik} \rightarrow L_k, p_k) \in I, \nexists (m'_k, s) \neq (m_k, s) \in NEW, (1) \text{ commutes in Fig. 5 and } m_k \models_H p_k\}$,*
  - *$OLD = \{(h \circ m'_j, s_j = (p_k, R_k \leftarrow M_{kj} \rightarrow L_j, p_j)) \mid e_j = (m_j \colon L_j \rightarrow G, s_j) \in E, e_j \neq e, \exists m'_j \colon L_j \rightarrow D \text{ with } d \circ m'_j = m_j \text{ (see left of Fig. 5)} \text{ and } h \circ m'_j \models_H p_j\}$,*
  - *$CANC = \{(m_c \colon L_c \rightarrow H, s'_c) \in OLD \mid s_c = (p_i, R_i \leftarrow M_{ic} \rightarrow L_c, p_c) \in C, p(s'_c) = p_c \text{ and } (2) \text{ commutes to the right of Fig. 5}\}$.*

A derivation $S_0 \Rightarrow^* S_n$ is a sequence of zero or more direct derivations.

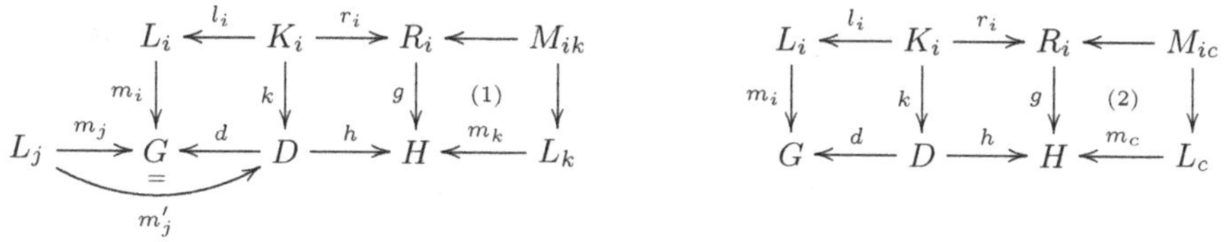

$$
\begin{array}{ccccccc}
L_i & \overset{l_i}{\longleftarrow} & K_i & \overset{r_i}{\longrightarrow} & R_i & \longleftarrow & M_{ik} \\
{\scriptstyle m_i}\downarrow & & {\scriptstyle k}\downarrow & & {\scriptstyle g}\downarrow \;(1) & & \downarrow \\
L_j \overset{m_j}{\longrightarrow} G & \overset{d}{\longleftarrow} & D & \overset{h}{\longrightarrow} & H & \overset{m_k}{\longleftarrow} & L_k
\end{array}
\qquad
\begin{array}{ccccccc}
L_i & \overset{l_i}{\longleftarrow} & K_i & \overset{r_i}{\longrightarrow} & R_i & \longleftarrow & M_{ic} \\
{\scriptstyle m_i}\downarrow & & {\scriptstyle k}\downarrow & & {\scriptstyle g}\downarrow \;(2) & & \downarrow \\
G & \overset{d}{\longleftarrow} & D & \overset{h}{\longrightarrow} & H & \overset{m_c}{\longleftarrow} & L_c
\end{array}
$$

Fig. 5. NEW and OLD events (left). CANC events (right).

Remarks. The condition $\nexists (m'_k, s) \neq (m_k, s) \in NEW$ ensures that at most one match of each invoked rule is added to the set of new enabled matches NEW. Hence, if more than one match exists, one is chosen non deterministically.

The set NEW contains at most one event for each rule invoked from e. Such events are demanded to contain a valid match of the LHS of the invoked rule. If no match is found for a certain invoked rule, then no event is generated for it. In this way, the LHS and NACs of the invoked rules are conditions for programming the rules, although in contrast to traditional event graphs, we do not visually show these conditions in the edges (cf. Fig. 4). The set OLD contains the existing events in the system state whose matches are preserved by the rule execution. In fact, all matches that are right-parallel independent with the execution of e are preserved. Finally, the set $CANC$ contains those events in OLD which are cancelled due to the execution of e. It can be noted that cancellation only affects to events in OLD (pre-existing events), so that if an event e both invokes and cancels the same kind of events, invocation prevails.

Example. Fig. 6 shows an example of derivation. The initial system state is given by the graph G and the events to the left (the actual matches in the events are given by equality of identifiers in L_i and G). Applying the match for *send* in the upper left gives as a result graph H. The set of events is updated as shown to the right of the figure: (i) the applied event is removed, (ii) a new event *receive* is added due to the invocation edge coming out from *send* in the flow grammar, and (iii) the old event *deleteConnection* for the match given by objects n and c is removed due to the cancelling edge. Note how the cancelling edge only removes one of the *deleteConnection* events in the system state, namely the one that contains the node and channel involved in the execution of *send*, which are passed as parameters (cf. Fig. 4). As this shows, cancellation edges cannot always be modelled easily with NACs.

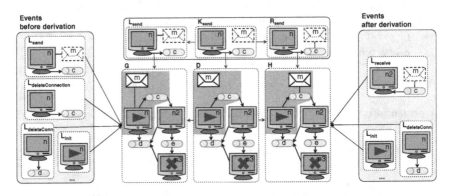

Fig. 6. Example of derivation

Parallelism. A direct derivation adds to *NEW* *at most one* match from each invoked rule. However, for certain applications (e.g. to model broadcasting in networks), it is interesting to introduce *all enabled matches* instead. In that case we would just have to remove the condition $\nexists(m'_k, s) \neq (m_k, s) \in NEW$ in Def. 4. This feature is related to the degree of parallelism of the system, called *server semantics* in timed Petri nets [10]. The *single server semantics* assumes that the system can process one invocation at a time, which corresponds to the original Def. 4. The *infinite server* semantics takes into account all enabled matches. The k-server semantics limits the parallelism to at most k matches. These semantics can be included in our model by adding a function $par: I \to \mathbb{N} \cup \{*\}$ ("*" for unbounded). We visually annotate the invocation edges in the event graph by placing the value of this function near the arrow end. If no annotation is used in the arrow end, then we assume it is the default value 1.

Next we define the semantics of a flow grammar as the set of all derivations whose last direct derivation was performed by a final rule. We use a set of traces (instead of a set of reachable graphs) in order to take performance metrics.

Def. 5 (Flow grammar semantics). *Given a flow grammar FG, its semantics is defined as* $SEM(FG) = \{init(FG) \Rightarrow^* S_n \overset{e}{\Rightarrow} S_{n+1}|p(e) \in end\}$.

5   Time Scheduling

A flow grammar describes the structure of an event graph, but still lacks the ability to handle time explicitly. That is, we need to introduce an implicit notion of simulation time, and to decorate the edges of the event graph with explicit time values. To this purpose, we extend our grammars with scheduling functions, associating edges with relative time values, or more generally with probability density functions $p(t)$. These distributions give the relative likelihood $p(t)$ of the target rule to be scheduled at relative time t. In this way, we can model either specific times (e.g., 4 using a degenerate distribution δ_4), as well as discrete and continuous distributions, like the uniform, normal and exponential negative.

Def. 6 (Scheduling grammar). *A scheduling grammar* $SG = \langle FG, t_I, t_C \rangle$ *is made of a flow grammar FG, a time scheduling function* $t_I : I \to \mathbb{R} \to [0, 1]$, *and a time cancelling function* $t_C : C \to \mathbb{R} \to [0, 1]$.

Remark. Given $s \in I$, $t_I(s)$ maps s to a probability density function $t_I(s) : \mathbb{R} \to [0, 1]$, which assigns each time value $x \in \mathbb{R}$ a probability $t_I(s)(x)$. Hence at a particular derivation step, we make use of a random variable X_i with density $t_I(s)$, which in the case of scheduling edges can be interpreted as the waiting time before the corresponding rule is applied, and therefore added to the simulation time gives the absolute time the rule application is scheduled for.

Example. The left of Fig. 7 shows part of the example flow grammar annotated with time. For instance, when *send* happens, an event *receive* is scheduled with uniform probability between 5 and 7 units of time later. Rule *init* is scheduled to happen periodically each 50 units of time. The cancelling edge and the invocation edge from *init* to *send* have no timing annotation, so 0 is assumed. *channelCheck* schedules itself at times given by a normal distribution, and then deletes one message, if there is any. This periodic behaviour is similar to a timer.

Now we define the semantics of a timed grammar. For this purpose, we extend states and events presented in Def. 2 with a concrete absolute occurrence time. The time of events should be greater or equal than the current simulation time. The occurrence time of an event is produced when it gets scheduled.

Def. 7 (Timed event and timed state). *Given a scheduling grammar SG, a timed event is a tuple* $e = \langle m : L \to G, p, t \rangle$, *where* $\langle m : L \to G, p \rangle$ *is an event according to Def. 2 and* $t \in \mathbb{R}$. *We write* $t(e) = t$, *to refer to e's scheduled time.*

A timed state $S = \langle G, FES, t \rangle$ *is made of a state* $\langle G, FES \rangle$ *and the current simulation time* $t \geq 0$, *where* $\forall e \in FES, t(e) \geq t \land m(e) \models_G p(e)$.

Remark. We use *FES* (future event set) instead of E to remark the similarity of this concept with that of discrete-event systems.

The initial state of a scheduling grammar is a state $S_0 = \langle G_0, FES_0, 0 \rangle$, where G_0 is the grammar initial graph, zero is the simulation start time, and FES_0 contains one event for each valid match of the initial rules. These are scheduled to occur at an absolute time given by a set of variables X_i that follow the density function assigned to the scheduling edges from $\perp$. For brevity we do not include the formal definition, straightforward from Def. 3.

A timed derivation step is performed according to Def. 4, but we select the event with lowest time (if several, one is taken at random), and we update the current simulation time to the time of this selected event. In addition, when we schedule a new event, we choose an absolute time equal to the actual time plus a random variable with the probability distribution of the scheduled edge $e \in I$. Finally, given a cancelling edge $c \in C$, we cancel all events that have a greater occurrence time than the current time plus a random variable that follows the probability distribution $t_C(c)$. In the interest of brevity, we avoid duplicating Def. 4, only indicating how the times for events and states are calculated.

Def. 8 (Timed derivation). *Given a scheduling grammar* $SG = \langle FG, t_I, t_C \rangle$, *and a timed state* $S = \langle G, FES, t \rangle$, *a direct timed derivation or state change* $S = \langle G, FES, t \rangle \stackrel{e}{\Longrightarrow} S' = \langle H, FES', t' \rangle$ *due to the event* $e = \langle m_i, (p_s, R_s \leftarrow M_{si} \rightarrow L_i, p_i), t' \rangle \in FES$ *can be performed iff* $\nexists e' \in FES$ *with* $t(e') < t(e)$. *The resulting state* S' *is calculated as in the untimed case (see Def. 4), while the time of events and the set* $CANC$ *are calculated as follows:*

- $\forall e_i \in NEW, t(e_i) = t' + X_i$, *s.t.* X_i *is a random variable with density* $t_I(s(e_i))$.
- $\forall e_i \in OLD$, *its occurrence time* $t(e_i)$ *remains unchanged (so that* e_i *"ages").*
- $CANC = \{(m_c \colon L_c \rightarrow H, s'_c, t'_c) \in OLD \mid s_c = (p_i, R_i \leftarrow M_{ic} \rightarrow L_c, p_c) \in C$, $p(s'_c) = p_c$, *(2) commutes to the right of Fig. 5,* $t'_c \geq t' + X_c$, *with* X_c *being a random variable with density* $t_C(s_c)\}$.

Remark. Two conditions are needed for cancelling an event: its match should commute as square (2) in Fig. 5 indicates, and the absolute time of the cancelled event should be greater or equal than the current time plus the relative time the cancelling edge indicates (through a probability distribution). Usually, the relative time t_C of cancelling edges is zero.

Example. The right of Fig. 7 shows a timed derivation like the one in Fig. 6 but considering time. Before applying the timed derivation, the simulation time is 40 and there are scheduled the following events: *send* at time 50, *deleteConnection* at two different matches at time 70, and *init* at time 100. Applying the first scheduled rule, which is *send*, updates the system state as follows: (i) the host graph is modified by the derivation of the DPO rule (not shown, it is performed as depicted in Fig. 6), (ii) the simulation time advances to 50 (as this was the scheduled time for the event), (iii) a new event *receive* is scheduled at time $50 + 6 = 56$, and (iv) one of the *deleteConnection* events is cancelled.

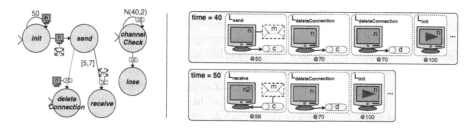

Fig. 7. Scheduling of events (left). Update of events in timed derivation example (right).

The language of a scheduling grammar is similar to that of a flow grammar, but each state is decorated with its absolute time. This is useful to take metrics, as demonstrated next.

5.1   Metrics

One of the objectives of simulation is to obtain metrics on the system behaviour. We can take metrics in three ways. The first one is just observing the occurrence time of events. In particular, the time of the final event in our example tells us the time taken for the initiator node to get a response. The second is by counting the occurrences of events of different types. In our case we can, e.g., count the number of lost messages. The third way involves defining domain-specific metrics. For this purpose we define graph constraints [5] and use them to check the states in which the constraints start to be satisfied or are no longer satisfied, so that we obtain the different time intervals in which the constraint holds. For example, the figure to the right shows the definition of a constraint that is satisfied whenever a channel has at least one message. Then, for all channels (matches of P), we check the states in which we find a message (matches of Q). This allows measuring the utilization time for each channel.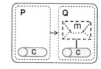

Other more advanced metrics can be taken by *counting* matches. For instance, in our example, we not only check that a match for Q exists, but we count how many of them are in order to measure the utilization level of the channels. Even though theoretically the metrics are defined on derivations of the language, for practical purposes these metrics are taken while the simulation is running, to avoid storing all intermediate states.

6   Modelling Higher-Level Timed Primitives

Now we show that our formalism is low-level and general enough to give semantics to other timing schemes and primitives [1,4,9,13,17].

Three phase approach. One of the features of standard GT is that, when the host graph changes, new matches for the rules of the grammar can be created and then "discovered" by the pattern matching algorithm. In our approach, matches for a certain rule are only sought if the rule is explicitly scheduled.

Inspired by the *three phase approach* [6], we can combine scheduling and activity scanning by extending the definition of scheduling grammar with an additional set $act \subseteq P$. The rules in act represent the start of activities, so that whenever we execute a rule in $P \setminus act$, in addition to scheduling events, we seek *all* matches from rules in act and schedule them for immediate execution. This does not increase the expressive power of our original formalism, but is a shortcut notation that can be modelled by just adding explicit schedulings from all rules in $P \setminus act$ to each rule in act (we schedule all matches, as the "*" indicates), at relative time 0, with empty M_{ij}, as shown to the left of Fig. 8.

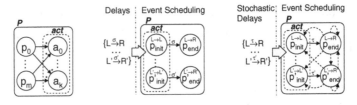

Fig. 8. Activities (left). Rules with delays (center). Stochastic delays (right).

Delays. Delays are used in [4,17] to extend GT with time. Once a valid match for a rule is found, the execution of the rule at such match is delayed by a time σ (an interval in [4] and other distributions in [17]). We write these rules as $p = \langle L \xrightarrow{\sigma} R \rangle$. Our events can be used to give semantics to delays. Delayed rules can be seen as activities that do not modify the system state when they start but only when they finish after a delay of σ. Hence, we split a delayed rule p in two, p_{init} and p_{end}, with the former scheduling the latter after σ. p_{init} is the identity rule $L \to L$, p_{end} is the original rule, and the dependency passes L from p_{init} to p_{end}. The scheme is shown in the center of Fig. 8.

In the semantics of [4], new matches are sought whenever a delayed rule is executed. Its infinite server semantics corresponds to our three-phase approach, where the set act of initial conditions for starting the activities (which in this case are the delayed rules) is given by the events p_{init}. To model the single server semantics of [4], we need to ensure at most one activity of the same type executing on the same set of objects, hence each p_{end} would have a cancelling self-loop with the context of execution as parameter.

Stochastic delays. In [9], GT rules are extended with stochastic delays given by a negative exponential distribution. A rule with stochastic delay $p = \langle L \xrightarrow{\tau} R \rangle$ has similar semantics to a delayed rule, but the difference concerns the *memory* policy when it is executed. After executing a rule, the remaining time of scheduled events has to be *restarted* and *resampled* again. We can model this by using cancelling edges. In particular, we split a stochastic rule in two as before, and in addition, we add cancelling edges from the event p_{end} to each rule p^k in the original stochastic grammar (see the right of Fig. 8). This is so as, at each derivation, we have to "forget the past" stored in the FES.

Activities, duration and conflicts. As seen before, activities are represented by an initial event, a final event, and a duration. However, as a difference from delays, activities may have an observable behaviour when started, and hence p_{init} does not need to be the identity rule. Activities can be interruptible or not. In the first case, the behaviour corresponds with the semantics of our formalism. The behaviour of non-interruptible activities is more complex to model, because an initiated activity *has to be completed*. This means that one cannot schedule the start of new activities if such activities would destroy the match of the final event of running activities. This behaviour can be modelled using FES policies. In this way, a new event at a match $m: L \to G$ cannot be scheduled to occur at absolute time t, if $\exists e \in FES$, where e is the end of some activity, with $t(e) \geq t$, and where m and $m(e)$ are in conflict (executing the rule at m breaks $m(e)$).

Timers. Several approaches associate timers to model elements [1,13]. Timers get an initial value t_o that is decremented as time progresses. When they expire, an action represented by a rule *act* is executed. As the rule *channelCheck* in our example shows, we can model timers by an identity rule identifying the element the timer should be added to, which schedules the rule *act* after t_o time units.

Periodic activities. These are activities that are repeated periodically. In our case, the final event of an activity schedules the initial event of the activity, passing certain elements in the match, like the *init* rule does.

7   Related Work

There are three ways of adding time to GT rules: (i) embedding the time in the host graph (time as data); (ii) incorporating it into the GT formalism (time as control); and (iii) embedding GT into some other simulation formalism.

In the first approach, [8] proposes using time stamps to mark the elements of the host graph. GT rules are standard untimed rules, but two conditions are demanded concerning the manipulation of local clocks: monotonicity (time should progress) and uniformity (time should progress at equal rates locally). In [15], the authors develop a timed approach with the purpose of animating the execution of GT rules. Conceptually, their rules are classified as internal or external events (the latter may be triggered by users), but the timing information is represented in the model, as additional attributes for the different elements. In [3], the author encodes the list of scheduled events in the host graph, and the events that have to be executed are modelled as edges pointing to the different graph elements. In our view, these approaches pollute the model (and the simulation formalism) with timing elements for control purposes.

In the second approach, [4] adapts concepts from timed Petri nets, so that rules are assigned a range, and rule executions are delayed with uniform probability in such range. The work of [9] takes concepts from stochastic Petri nets, so that rules are assigned a delay given by a negative exponential distribution. An important difference is that, while time is assigned to rules in [4,9,17], we assign it to schedulings. Hence, while they interpret rules as activities with unobservable initiation, we interpret rules as events, making our approach able to model all of them in a unified way. In [17], events are related to equivalence classes of matches modulo renaming, and time can follow a general distribution. Our approach, based on parameter passing and scheduling, is more efficient as we do not need to compute the equivalence classes at each derivation step.

Other approaches based on rewriting logic follow a similar purpose. In [1] elements in models can be assigned timed constructs like clocks or timers. The work of [13] provides a variety of high-level timed primitives, like periodic activities. Rules can manipulate the FES, mixing both control and data. In our case, a neat separation between control and data is achieved through the use of scheduling and cancelling relations between events.

With respect to the third approach, in [16], GT rules are embedded into the DEVS simulation formalism. Rule concurrency issues are difficult to handle and have to be solved in an ad-hoc way, whereas we use cancelling edges and the theory of GT to eliminate scheduled matches that are no longer valid.

Finally, our work also relates to the models of computations proposed by the embedded systems and systems-on-chip communities [11]. However, whereas we follow the discrete-time model of computation, our approach is not based on modules (processes) and communication channels between these. Instead, our behavioural specifications are decoupled from the actual model where they are executed, allowing its dynamic change.

8   Conclusions and Future Work

Inspired by the *Event Scheduling* world view of discrete-event simulation, we have presented a new way to incorporate time into GT. We model events as rule matches, which may explicitly schedule and cancel the occurrence of other events in the future, and may pass information (partial matches) between such event occurrences for efficiency purposes. We have presented the approach in two steps. *Flow grammars* organize rule flows into processes with parameter passing, formalizing a mechanism that is present in several tools such as Fujaba, VMTS or GReAT. *Scheduling grammars* are built on top of flow grammars adding time in a modular way so that other (untimed) approaches can be extended in a similar way. We have shown that the approach is general enough to model other timing approaches to GT. Finally, the visual nature of GT makes the approach suitable in application domains where simulation is used.

In the future, we will implement tool support for the approach. We also plan to work on analysis methods, both taken from Event Graphs theory [14] and from GT theory, in particular the analysis of rule independence.

Acknowledgements. Work partially sponsored by the Spanish Ministry of Science and Innovation, under project "METEORIC" (TIN2008-02081) and mobility grants JC2009-00015 and PR2009-0019, as well as by the R&D programme of the Community of Madrid, project "e-Madrid" (S2009/TIC-1650). We are grateful to the anonymous reviewers, which helped in improving the paper.

References

1. Boronat, A., Ölveczky, P.C.: Formal real-time model transformations in MOMENT2. In: Rosenblum, D.S., Taentzer, G. (eds.) Fundamental Approaches to Software Engineering. LNCS, vol. 6013, pp. 29–43. Springer, Heidelberg (2010)
2. Cassandras, C.G., Lafortune, S.: Introduction to Discrete Event Systems, 2nd edn. Springer, Heidelberg (2008)
3. de Lara, J.: Meta-modelling and graph transformation for the simulation of systems. Bulletin of the EATCS 81, 180–194 (2003)
4. de Lara, J., Vangheluwe, H.: Automating the transformation-based analysis of visual languages. Formal Aspects of Computing 22(3-4), 297–326 (2010)
5. Ehrig, H., Ehrig, K., Prange, U., Taentzer, G.: Fundamentals of Algebraic Graph Transformation. Springer, Heidelberg (2006)
6. Fishman, G.S.: Discrete-Event Simulation: Modeling, Programming, and Analysis. Springer, Heidelberg (2001)
7. Gönczy, L., Kovács, M., Varró, D.: Modeling and verification of reliable messaging by graph transformation systems. ENTCS 175(4), 37–50 (2007)
8. Gyapay, S., Varró, D., Heckel, R.: Graph transformation with time. Fundam. Inform. 58(1), 1–22 (2003)
9. Heckel, R., Lajios, G., Menge, S.: Stochastic graph transformation systems. Fundam. Inform. 74(1), 63–84 (2006)
10. Marsan, M.A., Balbo, G., Conte, G., Donatelli, S., Franceschinis, G.: Modelling with Generalized Stochastic Petri Nets. John Wiley & Sons, Chichester (1995)
11. Mathaikutty, D.A., Patel, H.D., Shukla, S.K., Jantsch, A.: SML-Sys: a functional framework with multiple models of computation for modeling heterogeneous system. Des. Autom. Embed. Syst. 12, 1–30 (2008)
12. Naeem, M., Heckel, R., Orejas, F., Hermann, F.: Incremental service composition based on partial matching of visual contracts. In: Rosenblum, D.S., Taentzer, G. (eds.) Fundamental Approaches to Software Engineering. LNCS, vol. 6013, pp. 123–138. Springer, Heidelberg (2010)
13. Rivera, J.E., Durán, F., Vallecillo, A.: A graphical approach for modeling time-dependent behavior of DSLs. In: VL/HCC 2009, pp. 51–55. IEEE, Los Alamitos (2009)
14. Schruben, L.: Simulation modeling with event graphs. Commun. ACM 26(11), 957–963 (1983)
15. Strobl, T., Minas, M.: Specifying and generating editing environments for interactive animated visual models. In: GT-VMT 2010 (2010)
16. Syriani, E., Vangheluwe, H.: Programmed graph rewriting with DEVS. In: Schürr, A., Nagl, M., Zündorf, A. (eds.) AGTIVE 2007. LNCS, vol. 5088, pp. 136–151. Springer, Heidelberg (2008)
17. Torrini, P., Heckel, R., Rath, I.: Stochastic graph transformation with regions. In: GT-VMT 2010 (2010)

Counterpart Semantics
for a Second-Order μ-Calculus*

Fabio Gadducci[1], Alberto Lluch Lafuente[2], and Andrea Vandin[2]

[1] Department of Computer Science, University of Pisa, Italy
[2] IMT Institute for Advanced Studies Lucca, Italy

Abstract. We propose a novel approach to the semantics of quantified μ-calculi, considering models where states are algebras; the evolution relation is given by a counterpart relation (a family of partial homomorphisms), allowing for the creation, deletion, and merging of components; and formulas are interpreted over sets of state assignments (families of substitutions, associating formula variables to state components). Our proposal avoids the limitations of existing approaches, usually enforcing restrictions of the evolution relation: the resulting semantics is a streamlined and intuitively appealing one, yet it is general enough to cover most of the alternative proposals we are aware of.

Keywords: Quantified μ-calculi, counterpart semantics, graph transformation.

1 Introduction

Any assessment on the usability of a visual specification formalism should rely on the existence of languages for expressing properties, as well as on the availability of tools for their verification. As far as graph transformation systems are concerned, after the seminal work of Courcelle [8], suitable variants/fragments of graph logics have been proposed and their connection with topological properties of graphs thoroughly investigated [9].

The need to reason about the possible transformations in a graph topology led more recently to the idea of combining temporal and graph logics. Before that, many authors studied decidability and complexity of temporal first-order logics, developed for reasoning about the evolution of individual components within a software system. Unfortunately, such logics are in general not decidable (see e.g. [12,16] and the references therein). As a consequence, many efforts have been invested on defining logics (or identifying fragments) that sacrifice expressiveness in favour of efficient computability, thus providing verification tools where logics become effective specification mechanisms (see § 6).

Recent approaches [1] propose variants of quantified μ-calculi, a combination of the fix-point and modal operators of temporal logics with monadic second-order logic for graphs [8]. Albeit less expressive than full second-order proposals,

---

* Supported by the MIUR Project SisteR.

H. Ehrig et al. (Eds.): ICGT 2010, LNCS 6372, pp. 282–297, 2010.

since the class of admissible predicates is restricted to first-order equality, these logics fit at the right level of abstraction for graph transformation systems: if state systems are graphs, and state components are thus graph items, one is not only interested in the topological structure of each reachable graph alone, but on its evolution as well. As a concrete example, consider graphs that represent the communication topology of a distributed leader election algorithm on a ring network that proceeds by iteratively discarding processes (edges) for the leadership. On the one hand, one would like to claim that eventually a leader will be found (the only self-closed remaining edge). This can be achieved with a formula like $\mu Z.[\exists edge.source(edge) = target(edge) \vee \Diamond Z]$. Note that the formula is fundamentally propositional in its temporal dimension: it asserts that at some reachable graph there will be a self-closed edge (the leader). Instead, we might be interested in expressing the eventual existence a process (an edge) which will be become the leader (self-closed). This is obtained moving to a purely first-order μ-calculus formula such as $\mu Z.\exists edge.[source(edge) = target(edge) \vee \Diamond Z]$.

The situation concerning the semantical models for such logics is less clearly cut. While it is obvious that a closed formula should be valued as the set of states where it holds, consider instead the open formula $source(edge) = target(edge) \vee \Diamond Z$: once the value of $edge$ is chosen in the current state, how is such value passed to the states denoted by the fix-point variable Z? The issue is denoted in the quantified temporal logic literature as the *trans-world identity* problem (see [15] as well as [2] for a survey of the related philosophical issues). From a practitioner point of view, a typical solution follows the so-called "Kripke semantics" approach: roughly, a set of universal (graph) items is chosen, and its elements are used to form each state. Another solution exploits *counterpart relations*, i.e. (partial) functions among states, explicitly relating elements of different states. The first solution is the most widely adopted, and it underlines most of the proposals we are aware of (as they are briefly surveyed in § 6 of the paper): it is e.g. implicitly used also in the approach discussed in [1] (admittedly the most similar in the chosen syntax/semantics to the one that we are going to introduce), since the counterpart relations used there for modelling the association of items belonging to different graphs are actually partial inclusions.

However, Kripke-like solutions do not fit too well with the possibility that items might be merged or that the evolution relation might form cycles: if the value of an open formula is a set of states, how to account for an item of a state that is first deleted and then added again? The latter problem is often solved by restricting the class of admissible evolution relations among states: this may force to reformulate the state transition relation modelling the system evolution, though such solutions tend to hamper the intuitive meaning of the logic.

In this paper we introduce a novel, purely counterpart-like semantics for quantified μ-calculi. We instantiate our proposal by considering a simple second-order syntax, and considering models where states are algebras and the evolution relation is given by a family of partial homomorphisms. Most importantly, open formulas are interpreted over sets of pairs (σ, w), for w a state and σ an assignment over w (that is, a substitution associating formula variables to components of the state w): the resulting model thus faithfully represents also the presence of cycles in the evolution relation. Our proposal avoids the limitations of existing approaches, since it dispenses with the reformulation of the transition relation: the resulting semantics is a streamlined and intuitively appealing one, yet it is general enough to cover most of the alternatives we are aware of.

Synopsis. The opening § 2 presents our counterpart model, roughly based on [15], yet considering suitable algebras as states. Then, § 3 presents the syntax of our logic, a second-order μ-calculus that is reminiscent of the one proposed in [1]. Finally, § 4 presents the core contribution of the paper: the semantics for our logic, based on sets of assignments. The proposal is put at work with a series of simple examples in § 5. And while § 6 discusses related works, focusing on those logics applied to the verification of visual formalisms, the closing § 7 concludes the paper and outlines future research avenues.

2   Counterpart Model

In this section we define the class of models over which our logic is interpreted. We begin recalling the definition of multi-sorted algebras and their homomorphisms, which lies at the basis of the structure of our worlds.

Definition 1 (Multi-sorted algebra). *A (multi-sorted) signature Σ is a pair (S_Σ, F_Σ) composed by a set of sorts $S_\Sigma = \{\tau_1, \cdots, \tau_m\}$ and by a set of function symbols $F_\Sigma = \{f_\Sigma : \tau_1 \times \ldots \times \tau_n \to \tau \mid \tau_i, \tau \in S_\Sigma\}$ typed over S_Σ. A (multi-sorted) algebra $\boldsymbol{A}$ with signature Σ (a Σ-algebra) is a pair (A, F_Σ^A) such that*

- *the carrier $A = \{A_\tau \mid \tau \in S_\Sigma\}$ is a set of elements typed over S_Σ;*
- *$F_\Sigma^A = \{f_\Sigma^A : A_{\tau_1} \times \ldots \times A_{\tau_n} \to A_\tau \mid f_\Sigma : \tau_1 \times \ldots \times \tau_n \to \tau \in F_\Sigma\}$ is a family of functions on A typed over S_Σ.*

Given two Σ-algebras $\boldsymbol{A}$ and $\boldsymbol{B}$, a (partial) homomorphism ϱ is a family of possibly partial functions $\{\varrho_\tau \mid \tau \in S_\Sigma\}$ typed over S_Σ, such that for each typed function symbol $f_\Sigma : \tau_1 \times \ldots \times \tau_n \to \tau \in F_\Sigma$ and list of elements $a_1, \ldots, a_n$, if each function ϱ_{τ_i} is defined for the element a_i of type τ_i, then ϱ_τ is defined for the element $f_\Sigma^A(a_1, \ldots, a_n)$ of type τ and moreover the elements $\varrho_\tau(f_\Sigma^A(a_1, \ldots, a_n))$ and $f_\Sigma^B(\varrho_{\tau_1}(a_1), \ldots, \varrho_{\tau_n}(a_n))$ coincide.

Each typed function symbol $f_\Sigma \in F_\Sigma$ corresponds to a function f_Σ^A in F_Σ^A: the functions in F_Σ^A are called fundamental operations of A. Note that our homomorphisms can be partial, possibly decreasing the domain of definition of a function and thus modelling the removal of world elements.

Example 1 (Graph algebra). As a running example we adopt a very simple unary algebra, the one for ordinary directed graphs. More precisely, the signature for directed graphs is (S_{graph}, F_{graph}). The set S_{graph} of sorts is composed by the sort of nodes τ_N and the sort of edges τ_E, while the set F_{graph} is composed by the function symbols $s : \tau_E \to \tau_N$ and $t : \tau_E \to \tau_N$ which determine respectively the

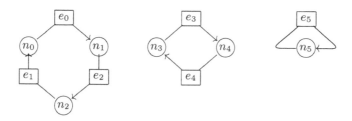

Fig. 1. Three graphs: $\mathbf{G}_0$ (left), $\mathbf{G}_1$ (middle) and $\mathbf{G}_2$ (right)

source and the target node of an edge. In Fig. 1 we find the visual representations for three graphs: $\mathbf{G}_0, \mathbf{G}_1, \mathbf{G}_2$. The first of these graph algebras is given by $\mathbf{G}_0 = (N_0 \uplus E_0, \{s^{\mathbf{G}_0}, t^{\mathbf{G}_0}\})$, where $N_0 = \{n_0, n_1, n_2\}$, $E_0 = \{e_0, e_1, e_2\}$, $s^{\mathbf{G}_0} = \{e_0 \mapsto n_0, e_1 \mapsto n_2, e_2 \mapsto n_1\}$ and $t^{\mathbf{G}_0} = \{e_0 \mapsto n_1, e_1 \mapsto n_0, e_2 \mapsto n_2\}$. Each graph can be understood as the state of the communication topology in a distributed algorithm: edges represent processes, source and target functions denote the ports of the processes and nodes are the communication channels.

We are interested in open terms, i.e. terms with variables. For this purpose we consider signatures Σ_X obtained by extending a multi-sorted signature Σ with a denumerable set X of variables typed over S_Σ: we let X_τ denote the τ-typed subset of variables and with x_τ or $x : \tau$ a variable with sort τ. Similarly, we let ϵ_τ or $\epsilon : \tau$ indicate a τ-sorted term.

Definition 2 (Terms). *Let Σ be a signature, let X be a (denumerable) set of individual variables typed over S_Σ, and let Σ_X denote the signature obtained extending Σ with X. The (multi-sorted) set $T(\Sigma_X)$ of terms obtained from Σ_X is the smallest set such that*

$$\frac{}{X \subseteq T(\Sigma_X)} \qquad \frac{\epsilon_i : \tau_i \in T(\Sigma_X),\ f : \tau_1 \times \ldots \times \tau_n \to \tau \in F_\Sigma}{f(\epsilon_1, \ldots, \epsilon_n) : \tau \ \in \ T(\Sigma_X)}$$

We omit the sort when it is clear from the context or when it is not necessary. Moreover, for an algebra $\mathbf{A}$, we let $\epsilon^{\mathbf{A}}$ denote the function associated to a term ϵ. We remark that the derived signature Σ_X does not allow to denote an individual element of the carrier directly, but only indirectly via constant symbols or variables and the well known concept of *variable assignment* σ.

Example 2 (Terms). Consider the algebra of $\mathbf{G}_0$ and let $\{x_N, x_E\} \subseteq X$ be typed variables. Then, x_N, x_E, $s(x_E)$, and $t(x_E)$ are valid terms, while n_2 and e_1 are not. Intuitively, terms are supposed to denote either a node or an edge of the graph, but they are undefined until evaluated with respect to a chosen variable assignment. For instance, given the assignment $\sigma = \{x_E \mapsto e_1\}$ process (edge) e_1 and its *source* and *target* channels (nodes) n_2, n_0 are respectively denoted by $\sigma(x_E)$, $\sigma(s(x_E))$, and $\sigma(t(x_E))$.

We can now introduce the notion of *counterpart model*.

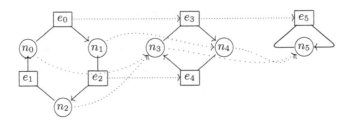

Fig. 2. A counterpart model with three sequential worlds (w_0, w_1, w_2)

Definition 3 (Counterpart model). *Let Σ be a signature, X a denumerable set of variables typed over S_Σ, and $\mathcal{A}$ the set of algebras over Σ_X. A counterpart model M is a tuple $(W, \rightsquigarrow, d, C)$ such that*

- *W is a non-empty set of worlds of the model;*
- *$\rightsquigarrow \subseteq (W \times W)$ is a binary relation, called* accessibility *relation over W;*
- *$d : W \to \mathcal{A}$ is a function assigning an algebra $A_{d(w)}$ to each world;*
- *$C : \rightsquigarrow \to (\mathcal{A} \rightharpoonup \mathcal{A})$ is a function assigning to every pair of worlds $(w, w') \in \rightsquigarrow$ a homomorphism (its* counterpart function $C_{w,w'}$*) from $A_{d(w)}$ to $A_{d(w')}$.*

Intuitively, the counterpart relations allow for creation, deletion, renaming and merging of elements in a type-respecting way, while forbidding duplication, i.e. it is not possible to associate an element of $A_{d(w)}$ to more than one element of $A_{d(w')}$. In the following we shall also use counterpart functions for sets of elements (with the obvious meaning of lifting the functions to sets).

Example 3 (Counterpart model). The example of Fig. 2 illustrates a model made of three worlds, namely w_0, w_1, and w_2, that are mapped into the graph algebras $\mathbf{G}_0$, $\mathbf{G}_1$, and $\mathbf{G}_2$ of Fig. 1, respectively. The transition relation is a simple sequence $w_0 \rightsquigarrow w_1 \rightsquigarrow w_2$ which can be understood as a sequential execution of the distributed leader election algorithm. The counterpart relations (drawn with dotted lines) reflect the fact that at each transition one process (edge) is discarded and its source and target channels (nodes) are merged: e_1 at the first transition and e_4 at the second one.

3   Syntax

Before presenting the syntax of our logic, we introduce the notion of second-order variables $\chi \in \mathcal{X}$. Intuitively, a variable of the second order χ_τ with sort $\tau \in S_\Sigma$ is evaluated in a set of elements of sort τ. An assignment σ associates the variables of first- (second-) order to the elements (set of elements, respectively) of the algebra $A_{d(w)}$ underlying a world w. Hence, fix-point variables $Z \in \mathcal{Z}$ range over the set of pairs (σ_w, w) relative to a counterpart model M, where w is a world of M, and σ_w an assignment for w.

Definition 4 (Quantified modal formulas). *Let Σ be a signature, $\mathcal{Z}$ a set of fix-point variables, and X, $\mathcal{X}$ (denumerable) sets of first- and second-order* *variables typed over S_Σ, respectively. The set $\mathcal{F}_\Sigma$ of formulas of our logic is inductively generated by the following rules*

$$\psi ::= \ tt \mid \epsilon : \tau \in_\tau \chi_\tau \mid \neg\psi \mid \psi \vee \psi \mid \exists x_\tau.\psi \mid \exists \chi_\tau.\psi \mid \Diamond\psi \mid Z \mid \mu Z.\psi$$

where $\epsilon : \tau$ is a term over Σ_X of type τ, $\in_\tau$ is a family of membership predicates typed over S_Σ indicating that (the evaluation of) a term with sort τ belongs to (the evaluation of) a second-order variable with the same sort τ, and μ denotes the least fixed point operator.

Whenever clear from the context, in the following subscripts and types will usually be omitted. We shall also use the derived symbols $\wedge$, $\rightarrow$, $\leftrightarrow$, $\forall$, as well as the modal operator $\Box$, defined as usual as $\Box\psi \equiv \neg\Diamond\neg\psi$. Moreover, as it is standard, we restrict to *monotonic* formulas, i.e., such that each fix-point variable Z occurs under the scope of an even number of negations. This is a sufficient condition for the fixed points to be well-defined.

Note that the logic is simple, yet reasonably expressive. For instance, binary equivalence can also be defined as a derived operator, namely, $\epsilon_1 : \tau =_\tau \epsilon_2 : \tau$ is defined as $\forall \chi_\tau. (\epsilon_1 : \tau \in_\tau \chi_\tau \leftrightarrow \epsilon_2 : \tau \in_\tau \chi_\tau)$.

Example 4 (Formula). Consider again the graph signature and our running example. The following are examples of formulas expressing different liveness properties of the distributed leader election algorithm: $\psi_1 \equiv \mu Z.(\exists x.s(x) = t(x)) \vee \Diamond Z$ *(eventually there will be a leader)* and $\psi_2 \equiv \mu Z.\exists x.(s(x) = t(x) \vee \Diamond Z)$ *(there is a process that eventually will become the leader)*. Intuitively, ψ_2 is satisfied by those worlds w that can reach a world where a leader (self-closed edge) is present. Instead, ψ_1 is satisfied by those worlds w which contain a process (edge) that is a leader (i.e. its *source* and *target* ports coincide) or if this process (edge) will become a leader (self-closed) in a world reachable after some finite number of steps. Formula ψ_2 has thus quite a different meaning than ψ_1: in ψ_1 the sub-formula $\Diamond Z$ is inside the scope of the existential quantifier which fixes the element associated to x (the potential leader) in the source world to keep track of its evolution.

We now introduce the notion of context that is used for decorating terms and formulas with relevant variable-related information. For the sake of simplicity, in the rest of the paper we fix a signature Σ and denumerable sets $X, \mathcal{X}, \mathcal{Z}$ of first-order, second-order, and fix-point variables, respectively.

Definition 5 (First-order context). *A first-order context Γ over X is a subset of X. We write Γ, x to indicate $\Gamma \cup \{x\}$ and $\Gamma \setminus x$ to indicate $\Gamma \setminus \{x\}$.*

We indicate with $\mathcal{C}_1$ the set of all the first-order contexts. Now, we define how terms are decorated with such contexts.

Definition 6 (Term-in-context). *A term-in-context takes the form $\epsilon : \tau[\Gamma]$ where ϵ is a term of type τ over Σ_X, and Γ is a first-order context over X. The set of well-formed terms-in-context $T^C(\Sigma_X)$ over $T(\Sigma_X)$ is defined as*

$$\frac{x \in X_\tau, \Gamma \in \mathcal{C}_1}{x[\Gamma, x] \ \in \ T^C(\Sigma_X)} \qquad \frac{\epsilon_i : \tau_i[\Gamma] \in T^C(\Sigma_X), \ f_\Sigma : \tau_1 \times \ldots \tau_n \to \tau \in F_\Sigma}{f_\Sigma(\epsilon_1, \ldots, \epsilon_n) : \tau[\Gamma] \ \in \ T^C(\Sigma_X)}$$

Example 5 (Term-in-context). Instantiating Σ with the graph signature of our running example, and considering the variables $x, y \in X_{\tau_E}$, then $s(x)[\{x\}]$ and $s(x)[\{x, y\}]$ are terms-in-context, while $s(x)[\{y\}]$ is not.

Since our logic allows for second-order quantification, we have to extend the concept of context to the second order.

Definition 7 (Second-order context). *A second-order context Δ over $\mathcal{X}$ is a subset of $\mathcal{X}$. The operations of addition and removal of a second-order variable with respect to a second-order context are defined as for the first-order case.*

As before, with $\mathcal{C}_2$ we indicate the set of all the second-order contexts. We can finally define how formulas are to be decorated with information about the variables involved. Their use is twofold: on the one-side, they allow for a smooth definition of the semantics, as it is going to be shown in § 4. Furthermore, even if it is not going to be further investigated here, contexts are needed to guarantee the normality of the logic and, in particular, the so-called K-scheme (see the remarks in the concluding section, as well as [2,22] for further details).

Definition 8 (Context of a formula). *We define the context of a formula as a pair $[\Gamma; \Delta]$ where Γ is a first-order context, and Δ is a second-order context.*

Intuitively, the context of a formula has to contain at least the free variables that are actually appearing in the formula. Hence, a *formula-in-context* is a formula decorated with such a context, recursively defined as follows. In the definition we omit Σ from $\mathcal{F}_\Sigma$ considering it fixed, and use $\psi \in \mathcal{F}^{[\Gamma;\Delta]}$ as abbreviation for $\psi[\Gamma; \Delta] \in \mathcal{F}_\Sigma^{[\Gamma;\Delta]}$.

Definition 9 (Formula-in-context). *A formula-in-context is $\psi[\Gamma; \Delta]$ where ψ is a formula in $\mathcal{F}_\Sigma$, and $[\Gamma; \Delta]$ is a context of ψ. The set of well formed formulas-in-context $\mathcal{F}_\Sigma^{[\Gamma;\Delta]}$ over $\mathcal{F}_\Sigma$ is defined as*

$$\frac{}{tt \in \mathcal{F}^{[\Gamma;\Delta]}} \qquad \frac{\epsilon:\tau[\Gamma]\in T^{[\Gamma;\Delta]}(\Sigma_X)}{\epsilon:\tau\in_\tau\chi_\tau \in \mathcal{F}^{[\Gamma;\Delta,\chi_\tau]}} \qquad \frac{\psi\in\mathcal{F}^{[\Gamma;\Delta]}}{\neg\psi \in \mathcal{F}^{[\Gamma;\Delta]}}$$

$$\frac{\{\psi_1,\psi_2\}\subseteq\mathcal{F}^{[\Gamma;\Delta]}}{\psi_1\vee\psi_2 \in \mathcal{F}^{[\Gamma;\Delta]}} \qquad \frac{\psi\in\mathcal{F}^{[\Gamma,x_\tau;\Delta]}}{\exists x_\tau.\psi \in \mathcal{F}^{[\Gamma;\Delta]}} \qquad \frac{\psi\in\mathcal{F}^{[\Gamma;\Delta,\chi_\tau]}}{\exists\chi_\tau.\psi \in \mathcal{F}^{[\Gamma;\Delta]}}$$

$$\frac{\psi\in\mathcal{F}^{[\Gamma;\Delta]}}{\Diamond\psi \in \mathcal{F}^{[\Gamma;\Delta]}} \qquad \frac{}{Z \in \mathcal{F}^{[\Gamma;\Delta]}} \qquad \frac{\psi\in\mathcal{F}^{[\Gamma;\Delta]}}{\mu Z.\psi \in \mathcal{F}^{[\Gamma;\Delta]}}$$

where $Z \in \mathcal{Z}$, $x_\tau \in X$ and $\chi_\tau \in \mathcal{X}_\tau$.

Note that, by construction, a context cannot contain a variable quantified in the formula, so that e.g. a formula like $(\exists x_\tau.\psi) \vee (x_\tau =_\tau x_\tau)$ has no associated context. This property is ensured by the lack of a weakening axiom, replaced by the rule introducing any term in context $\epsilon : \tau[\Gamma]$ inside a membership expression. This solution is adopted since it simplifies the semantics for the quantifiers in Definition 12, but it is not restrictive, since the actual identity of a bound variable is immaterial. Moreover, our inference rules for formulas-in-context could anyhow be easily generalised in order to allow at least a context for any formula.

Example 6 (Formula-in-context). Consider the formula $\mu Z.\exists x.(s(x) = t(x)\lor\Diamond Z)$ proposed in Example 4. We see that it contains a quantified variable "x", but no free variables. The context of this formula can thus consist of any set of first-order variables not containing x, and of any set of second-order variables. It can even be the empty context $[\emptyset; \emptyset]$. A more useful example comes from the examination of the construction of the cited formula. Assume $[\{x\}; \emptyset]$ as the context of the formula $s(x) = t(x)$. In order to apply the disjunction rule, $\Diamond Z$ has to have the same context $[\{x\}; \emptyset]$. We thus obtain $(s(x) = t(x) \lor \Diamond Z)[\{x\}; \emptyset]$. Now it is possible to apply the rule relative to the first-order quantifier, removing x from the context, obtaining $\exists x.(s(x) = t(x)\lor\Diamond Z)[\emptyset; \emptyset]$. Finally, applying the least fixed point rule, we obtain the formula-in-context $\mu Z.\exists x.(s(x) = t(x) \lor \Diamond Z)[\emptyset; \emptyset]$.

4   Counterpart Semantics

In this section we present the core contribution of the paper: we introduce the semantical domain for our logic, and we provide the rules for evaluating a formula-in-context as a set of assignments on a counterpart model. Once more for the sake of simplicity, in the rest of the paper we fix a counterpart model M.

Definition 10 (Assignments). *A (variable) assignment* $\sigma = (\sigma_1, \sigma_2)$ *for a world* $w \in M$ *is a pair of partial functions typed over* S_Σ *such that* $\sigma_1 : X \rightharpoonup A_{d(w)}$ *and* $\sigma_2 : X \rightharpoonup 2^{A_{d(w)}}$.

Let Ω *denote the set of pairs* (σ, w), *for* σ *an assignment over the world* w. *A (fix-point variable) assignment is a partial function* $\rho : Z \to 2^\Omega$.

In the following, we denote by $\Omega^{[\Gamma;\Delta]}$ those pairs $((\sigma_1, \sigma_2), w)$ such that the domain of definition of σ_1 and σ_2 is Γ and Δ, respectively. Moreover, $\Omega_w \subseteq \Omega$ denotes the sub-set of assignments over a world w (i.e. those pairs whose second component is the world w), and similarly for the sub-set $\Omega_w^{[\Gamma;\Delta]} \subseteq \Omega^{[\Gamma;\Delta]}$.

Another definition regards the notions of assignment extension and restriction.

Definition 11 (Extensions and restrictions). *Let* $[\Gamma; \Delta]$ *be a context and* $x \notin \Gamma$ *a variable. Given an assignment* $\sigma = (\sigma_1, \sigma_2) \in \Omega^{[\Gamma,x;\Delta]}$, *its restriction* $\sigma \downarrow_x \in \Omega^{[\Gamma;\Delta]}$ *is the assignment* $(\sigma_1 \downarrow_x, \sigma_2)$ *obtained by restricting the domain of definition of* σ_1 *to* Γ. *Vice versa, let* $a \in A_{d(w)}$ *be an element of the world* w. *Given an assignment* $\sigma \in \Omega_w^{[\Gamma;\Delta]}$, *its extension* $\sigma[^a/_x] \in \Omega_w^{[\Gamma,x;\Delta]}$ *is the assignment* $(\sigma_1[^a/_x], \sigma_2)$ *obtained by extending the domain of definition of* σ_1 *to* Γ, x *by assigning the element* a *to the variable* x.

Given a context $[\Gamma; \Delta]$ and a variable $x \notin \Gamma$, the function $2^{\downarrow_x} : 2^{\Omega^{[\Gamma,x;\Delta]}} \to 2^{\Omega^{[\Gamma;\Delta]}}$ lifts $\downarrow_x$ to sets. Vice versa, the function $\uparrow_x : \Omega^{[\Gamma;\Delta]} \to 2^{\Omega^{[\Gamma,x;\Delta]}}$ maps each assignment $\sigma \in \Omega_w^{[\Gamma;\Delta]}$ to the set $\{\sigma[^a/_x] \mid a \in A_{d(w)}\} \subseteq \Omega_w^{[\Gamma,x;\Delta]}$, for any world $w \in M$, and $2^{\uparrow_x} : 2^{\Omega^{[\Gamma;\Delta]}} \to 2^{\Omega^{[\Gamma,x;\Delta]}}$ denotes the lifting of $\uparrow_x$ to sets.

The corresponding functions $\downarrow_\chi$, $2^{\downarrow_\chi}$, $\uparrow_\chi$, and $2^{\uparrow_\chi}$, with respect to a second-order variable $\chi \notin \Delta$, are defined in the same way.

Example 7 (Assignments). Let us consider the counterpart model of Fig. 2, and let us denote $\lambda = (\lambda_1, \lambda_2)$ the empty assignment, regardless of the world. Then, each set $\Omega^{[\emptyset;\emptyset]}_{w_i}$ simply corresponds to $\{(\lambda, w_i)\}$, and consequently $\Omega^{[\emptyset;\emptyset]}$ corresponds to $\{(\lambda, w_0), (\lambda, w_1), (\lambda, w_2)\}$. If we extend $\Omega^{[\emptyset;\emptyset]}_{w_1}$ including the first-order variable x with sort τ_E, we obtain

$$2^{\uparrow_x}(\Omega^{[\emptyset;\emptyset]}_{w_1}) = \{((\lambda_1[^{e_3}/x], \lambda_2), w_1), ((\lambda_1[^{e_4}/x]\lambda_2), w_1)\}$$

which is in turn equivalent to $\{(((\{x \mapsto e_3\}, \lambda_2), w_1), \{(((\{x \mapsto e_4\}, \lambda_2), w_1)\}$.

As a final step, we have to define when a given pair of assignments is compatible with the counterpart relations.

Definition 12 (Assignment counterpart). *Let $\sigma_w \in \Omega_w$, $\sigma_{w'} \in \Omega_{w'}$ be two assignments. We say that $\sigma_{w'}$ is a counterpart of σ_w relative to $[\Gamma; \Delta]$, denoted as $\sigma_w \overset{[\Gamma;\Delta]}{\rightsquigarrow} \sigma_{w'}$, if*

1. *for each first-order variable $x \in \Gamma$, the elements assigned to x by σ_w and by $\sigma_{w'}$ are in counterpart relation, i.e. $C_{w,w'}(\sigma_w(x)) = \sigma_{w'}(x)$*
2. *for each second-order variable χ in Δ, the sets assigned to χ by σ_w and by $\sigma_{w'}$ are in counterpart relation, i.e. $2^{C_{w,w'}}(\sigma_w(\chi)) = \sigma'_w(\chi)$*

Lifting $C_{w,w'}$ to the second-order case means that for each variable χ the set denoted by its interpretation $\sigma_w(\chi)$ is precisely mapped by the counterpart relation to the set $\sigma'_w(\chi)$.

It should be noticed that we do not accept as a valid counterpart relation one where an already assigned element is canceled. Indeed, restricting the domain of discourse to existing entities is the main characterising feature of the counterpart solutions: this aspect of our proposal is going to be illustrated by examples in § 5, while it is discussed in some detail in [2,22].

We are now ready to introduce the semantical evaluation for our logic in a model M. It associates to a formula-in-context $\psi[\Gamma; \Delta]$ a set of assignments over the worlds of M contained in $\Omega^{[\Gamma;\Delta]}$. Hence, the domain of these assignments is exactly $\langle \Gamma; \Delta \rangle$, and thus our proposal is reminiscent of the semantics of temporal formulas over sets of constraints introduced in [13].

Definition 13 (Semantics). *Let $\psi[\Gamma; \Delta]$ be a formula-in-context. The evaluation of $\psi[\Gamma; \Delta]$ in M under the assignment $\rho : \mathcal{Z} \to 2^{\Omega^{[\Gamma;\Delta]}}$ is given by the function $[\![\cdot]\!]_\rho : \mathcal{F}^{[\Gamma;\Delta]} \to \Omega^{[\Gamma;\Delta]}$ defined as*

$$[\![tt[\Gamma; \Delta]]\!]_\rho = \Omega^{[\Gamma;\Delta]}$$
$$[\![(\epsilon : \tau \in_\tau \chi_\tau)[\Gamma; \Delta]]\!]_\rho = \{(\sigma, w) \in \Omega^{[\Gamma;\Delta]} \mid \sigma(\epsilon) \in \sigma(\chi_\tau)\}$$
$$[\![\neg\psi[\Gamma; \Delta]]\!]_\rho = \Omega^{[\Gamma;\Delta]} \setminus [\![\psi[\Gamma; \Delta]]\!]_\rho$$
$$[\![\psi_1 \vee \psi_2[\Gamma; \Delta]]\!]_\rho = [\![\psi_1[\Gamma; \Delta]]\!]_\rho \cup [\![\psi_2[\Gamma; \Delta]]\!]_\rho$$
$$[\![\exists x_\tau. \psi[\Gamma; \Delta]]\!]_\rho = 2^{\downarrow_{x_\tau}}([\![\psi[\Gamma, x_\tau; \Delta]]\!]_{(2^{\uparrow_x} \circ \rho)})$$
$$[\![\exists \chi_\tau. \psi[\Gamma; \Delta]]\!]_\rho = 2^{\downarrow_{\chi_\tau}}([\![\psi[\Gamma; \Delta, \chi_\tau]]\!]_{(2^{\uparrow_x} \circ \rho)})$$
$$[\![\Diamond\psi[\Gamma; \Delta]]\!]_\rho = \{(\sigma, w) \in \Omega^{[\Gamma;\Delta]} \mid \exists (\sigma', w') \in [\![\psi[\Gamma; \Delta]]\!]_\rho \,.\, \sigma \overset{[\Gamma;\Delta]}{\rightsquigarrow} \sigma'\}$$
$$[\![Z[\Gamma; \Delta]]\!]_\rho = \rho(Z)$$
$$[\![\mu Z.\psi[\Gamma; \Delta]]\!]_\rho = lfp(\lambda Y.[\![\psi[\Gamma; \Delta]]\!]_\rho[^Y/z])$$

Note that in the evaluation of the membership predicate, $\sigma(\epsilon)$ denotes the lifting of the substitution σ_1 to the set of terms over Σ_X.

In the evaluation of the quantifiers, it is pivotal to require that the assignment ρ for fix-point variables is modified, in order to account for the extensions to the newly introduced variables: it allows for a proper sorting of $\rho(Z)$, since it must now belong to the subsets of $\Omega^{[\Gamma, x; \Delta]}$ ($\Omega^{[\Gamma; \Delta, x]}$ in the second-order case).

Finally, in the evaluation of the modal operator the "renaming" of values across worlds is ensured by requiring that the assignments σ and σ' are in counterpart relation. Thus, our semantics discards those worlds that are reachable but are not in counterpart with respect to the current context. The rationale behind this is that it makes no sense to make claims about about non-existence (see [22,2]). We shall illustrate this issue with some examples in § 5.

The semantics of our logic is well-defined. In particular, the restriction to formulas where all occurrences of fix-point variables are positive guarantees that any function $\lambda Y.[\![\psi[\Gamma; \Delta]]\!]_\rho[^Y/_Z]$ is monotonic. Therefore, by Knaster-Tarski theorem, fixed points are well-defined.

The evaluation of a closed formula, i.e. of a formula $\psi[\emptyset; \emptyset]$ with an empty context, is just a set of pairs $\{(\lambda, w)\}$, for λ the empty assignment over the world w. Hence, such an evaluation characterises a set of worlds: this ensures that our proposal properly extends the standard semantics for propositional modal logics.

Example 8 (Evaluation of a formula-in-context). We have now the means for showing the evaluation of the formula-in-context $\mu Z.\exists x.(s(x) = t(x) \vee \Diamond Z)[\emptyset; \emptyset]$ proposed in Example 6 in the model presented in Example 3. Recall that the formula states that some process (edge) will eventually become the unique leader (self-closed). According to the rule semantics we have to obtain the least fixed point of $[\![\exists x.(s(x) = t(x) \vee \Diamond Z)[\emptyset; \emptyset]]\!]_\rho$. Intuitively we expect that the formula holds for all worlds with the empty assignment, i.e. $\{(\lambda, w_0), (\lambda, w_1), (\lambda, w_2)\}$, because w_2 contains the process (edge) e_5 which is a leader (self-closed), w_1 contains process (edge) e_3 which has counterpart e_5, and w_0 has the process (edge) e_0 which has counterpart e_3. Consider the assignment $\rho = (Z, \{(\lambda, w_0), (\lambda, w_1), (\lambda, w_2)\})$.

If we apply the semantics of the existential quantifier $\rho(Z)$ becomes $\{((\{x \mapsto e_0\}, w_0), (\{x \mapsto e_1\}, w_0), (\{x \mapsto e_2\}, w_0)\} \cup \{((\{x \mapsto e_3\}, w_1), (\{x \mapsto e_4\}, w_1)\} \cup \{((\{x \mapsto e_5\}, w_2)\}$. The assignment $(\{x \mapsto e_5\}, w_2)$ clearly verifies the formula $\psi_x \equiv (s(x) = t(x))$, so (λ, w_2) verifies the entire formula-in-context. Considering (λ, w_1), neither $\{x \mapsto e_3\}$ nor $\{x \mapsto e_4\}$ verify ψ_x, but $\{x \mapsto e_3\} \overset{[x;\emptyset]}{\rightsquigarrow} \{x \mapsto e_5\}$ holds, and $(\{x \mapsto e_5\}, w_2) \in \rho(Z)$. So $\{\lambda, w_1\}$ verifies the entire formula. Lastly, considering $\{\lambda, w_0\}$, as for w_1 ψ_x is not verified in w_0, but $\{x \mapsto e_0\} \overset{[x;\emptyset]}{\rightsquigarrow} \{x \mapsto e_3\}$ holds, and $(\{x \mapsto e_3\}, w_1) \in \rho(Z)$. So also $\{\lambda, w_0\}$ verifies the entire formula. Thus $\{(\lambda, w_0), (\lambda, w_1), (\lambda, w_2)\}$ is a fix-point and indeed the smallest one. Therefore, the semantics of our formula is the set of all worlds, as we intuitively expected.

5   Examples

The aim of this section is to illustrate the use of the logic to express properties of the evolution of systems and their components. Most of the examples are drawn from the literature reviewed in § 6. In order to make the syntax lighter, we do not indicate the contexts of the formulas and the types of the variables.

Death. The creation and destruction of entities has attracted the interest of various authors (e.g. [11,23]) as a mean for reasoning about the allocation and deallocation of resources or processes. It is important to understand that our logic has no built-in mechanism for that purpose, hence it allows for various interpretations of what it means for an entity to be deleted. In our setting, an entity that has no counterpart in a certain world w' is an entity that simply does not exist in w', which is different from an entity that is deallocated (whose existence we might want to remember).

To illustrate the difference between absence and deallocation, we consider absence first and then some particular flavours of deallocation. Recall that our logic mainly regards the evolution of existing entities, thus disregarding of their absence from the system. Therefore, while predicates regarding the presence and absence of entities can be defined (e.g. **present**$(x) \equiv \exists y.x = y$, **absent**$(x) \equiv \neg$**present**$(x)$) their semantics in our logic might not be meaningful. For instance, under the scope of the next-time modality, predicates over x should be intended as "as long as x is present", so that formulas like $\Box$**present**(x) might accept assignments for x in worlds that can evolve by deleting x. The key point is that our logic should be used to reason about living entities.

Now, when one is interested in reasoning about deallocated entities, one possible solution is to introduce a particular value $\bot$ in all domains and map deallocated entities (and the functions over them) onto that value (morphisms become total). This is essentially the underlying idea of [11]. Then, a deallocated entity can be characterised with predicate **dead**$(x) \equiv x = \bot$. Several issues arise when reasoning about deallocated entities. For instance, the choice of [11] is for all predicates over deallocated entities to be false, even the trivial self equality $x = x$, which is a tautology for living entities.

Our logic is flexible enough to adopt that strategy, by e.g. redefining the abbreviation $\epsilon_1 =_\tau \epsilon_2$ for binary equality as follows $\forall \chi_\tau. \left((\epsilon_1 \in_\tau \chi_\tau \leftrightarrow \epsilon_2 \in_\tau \chi_\tau) \wedge (\bot \notin \chi_\tau) \right)$. Other choices are also possible, like the one in [1] where the identity of deallocated entities is kept but quantification restricts to living entities.

Birth. When reasoning about entity creation, it is interesting to distinguish new from old entities. Our logic has no built-in mechanism (like e.g. in [11]) for this purpose, yet one can assume that this information is provided by the model (by using *new* and *old* values and a function from entities into those values).

Still, it is possible in general to define a modal predicate to capture the creation of a new entity x as follows: $\langle \mathbf{new}(x) \rangle(\psi) \equiv \exists \chi. \forall y.y \in \chi \wedge \Diamond \exists x.x \notin \chi \wedge \psi$.

Note that the defined modality is existential and restricts to non-deleting steps only, but other choices are of course possible.

Growth. Our logic is suited for expressing properties about the growth of a system. For instance, a growth bound of 3 is stated with **at-most-2** $\equiv \forall x.\forall y.\forall z.x = y \vee y = z \vee z = x$ as in [11] and is required as an invariant with $\nu Z.\textbf{at-most-2} \wedge \Box Z$, for $\nu Z.-$ the (informally defined) operator for greatest fixed point.

More interestingly, we can express properties along entity preserving behaviours. For instance, an **all-preserved** modality focussing on system evolutions where no entity is deleted is $\langle \textbf{all-preserved} \rangle(\psi) \equiv \exists \chi.\forall y.y \in \chi \wedge \Diamond \psi$.

Similarly, an abbreviation for system steps creating at least one element but preserving the rest is $\langle \textbf{one-more} \rangle(\psi) \equiv \exists \chi.\forall y.y \in \chi \wedge \Diamond \exists x.x \notin \chi \wedge \psi$.

Finally, we can then state the *possibility* of unbounded growth with the following formula: $\nu Z.\mu Z'.\langle \textbf{one-more} \rangle(Z') \wedge \langle \textbf{all-preserved} \rangle(Z)$. Instead, the *necessity* of unbounded growth (see e.g. [11]) would require a model with explicit deallocation as suggested above in order to be able to require absence of deletions.

Life. Apart from the growth in the number of entities, our logic regards the evolution of those entities. A typical example is the mobility of objects (like the message propagation in the example of [1]). Assuming an algebra of objects and locations with a function *loc* for denoting the location of an object, we can express location change for an object x with predicate $\textbf{moves}(x, \psi) \equiv \exists y.y = loc(x) \wedge \Box(loc(x) \neq y \wedge \psi)$. Then we can express that x never remains in the same location with $\nu Z.\textbf{moves}(x, Z)$.

Along the same lines, we can define other typical individual safety and liveness properties. For instance, individual mutual exclusion (used e.g. in [23]) can be stated with formula like $\nu Z.loc(x) \neq loc(y) \wedge \Box Z$ which requires x and y never to be in the same location. Another example are individual responsiveness properties, like requiring two entities to eventually meet whenever they are in separate locations: $\nu Z.loc(x) \neq loc(y) \rightarrow (\mu Z'.loc(x) = loc(y) \vee \Diamond Z') \wedge Z$.

6   Related Works

As we mentioned in the introduction, many authors addressed decidability and complexity issues concerning quantified modal logics, and many efforts have been focused on defining logics (or identifying fragments) that sacrifice expressivity in favour of efficient computability. The aim of our work is not concerned with such aspects yet, since we are interested in defining fist a generic, natural and intuitive semantics that overcomes some of the drawbacks of traditional Kripke-like semantics and follows the spirit of counterpart semantics. Thus, this section reviews some current proposals for quantified modal logics, trying to sum up the differences with the present paper, with a specific interest towards those approaches developed for the verification of visual specification formalisms.

Logics for reasoning about knowledge change (e.g. temporal description logics) have been proposed by various authors (see e.g. [12,16]), either as first-order extensions of classical linear- and branching-time temporal logics such as LTL and CTL [16], or as extensions of the modal μ-calculus [12]. The semantics is typically given in Kripke-style with a unique domain of interpretation that does not allow for merging or renaming of elements. Decidability results are given for some fragments, e.g. the *monodic* ones, roughly consisting of equality-free formulas with a restricted number of free variables under temporal operators.

Another setting where quantified temporal logics have raised interest are graph transformation systems, where software systems exhibiting features such as component or resource allocation, deallocation, reallocation or fusion are conveniently modelled using graph morphisms. For instance, the approach of [1] aims at building a verification setting where graph transition systems are abstracted into a sort of Petri nets. The specification mechanism is a logic that mixes the modal μ-calculus with Courcelle's Monadic Second-Order Logic for graphs [8]. The graph transition systems considered are not allowed to introduce merging or renaming of graph items, and the semantics is defined over the *unravelling* of the graph transition system, i.e. a tree that represents the unfolded state space and that guarantees some additional properties such as no-reuse of item names. Another example can be found [14], where a graph logic was developed for encoding a spatial logic for the π-calculus [3] in a graph-based setting. The logic extends the μ-calculus with a node-binding modal operator, quantifiers and other ingredients along the ones in [8] to describe the graphical structure of configurations. Merging and renaming is allowed for some restricted cases only.

Another graph-based approach is described in [19], focusing on finite-state graph transition systems, and using a linear-time logic whose structural aspects are expressed with regular expressions over paths. The same author investigated first-order temporal logics for various structures. In [20] he proposes an extension of CTL with first- and (monadic) second-order quantification. The semantics is interpreted over *algebra automata*, i.e. automata enriched with an algebraic structure of states, and with a morphism-like transition relation that allows for renaming elements. The model checking problem over finite automata is shown reducible to the ordinary model checking of CTL formulas over Kripke structures, while preserving the necessary structure to exploit name symmetries. A similar approach is followed in [11] but based on LTL and including predicates to reason about allocation, deallocation and reallocation of objects. The notion of name-equipped automata allows for injective renaming, but forbids merging. The semantics of the next time modality does not discard accessible worlds where elements assigned to variables are deleted, but in this case the assignment becomes undefined so that the logic allows for expressing deallocation but equality predicates over undefined variables become false (even the simple case $x = x$). Instead, [10] is concerned with the approximation of special kinds of graphs and the verification of a similar logic for verifying pointer structures on a heap. Another logic to reason about the dynamics featured in object-oriented programming languages is *evolution logic* [23], a first-order version of LTL. The model checking approach focuses on abstract interpretation rather than symmetries.

Spatio-temporal logics form another track of formalisms for describing the evolution of process and data structures. Early works aimed at reasoning about networks of processes (e.g. the *multiprocess network logic* of [18]), and were based on extensions of classical linear- and branching-time logics with first-order quantifiers. In these works, the set of processes was considered to be fixed (i.e. no dynamic creation or deletion was considered) so that the elimination of quantifiers was possible. In the last years, spatial logics evolved and were mostly defined for algebraically presented systems. We cite among others spatial logics for process calculi like the π-calculus [3,4] and mobile ambients [7], or for data structures such as graphs [5], heaps [21], and trees [6]. The common idea in such approaches is to mix temporal modalities with spatial operators that represent the dual of the operators of the algebra, like parallel (de)composition of processes or graphs, and various forms of (name) quantification. Renaming and merging of elements is typically restricted to some special cases like α-renaming and name extrusion.

7   Conclusions and Further Works

The present paper introduces a novel semantics for a second-order μ-calculus. With respect to other approaches, including those sketched in § 6, our proposal allows for a simple definition of the semantical universe by means of counterpart models. The idea of associating to (open) formulas sets of assignments, instead of just worlds, allows for a straightforward interpretation of fixed points and for their smooth integration with the evaluation of quantifiers, which often asked for a restriction of the class of admissible models.

The starting point for our proposal was the survey on quantified modal logic proposed by Belardinelli [2], further instantiated to graph transformations in the master's thesis of the third author [22]. The present article is a revised and extended version of the latter, taking into account also fix-point operators.

We foresee a few obvious directions for further research. As a start, we would like to investigate if the correspondence results between quantified μ-calculi and Petri nets logics proposed in [1] could be lifted to our framework, and its richer family of counterpart relations. We would also like to better understand the relationship with spatial logics, along the lines of [14], possibly adopting a family of labelled counterpart relations, and the richer modal operators $\Diamond_{\langle p, Y \rangle}$, basically stating that the transition between worlds is caused by a specific rule p, that may create a chosen set Y of new elements. Another interesting point is in understanding the tradeoff between expressivity and complexity regarding the choice of information being discarded in the semantics of the modal operator. As we already discussed, we ignore those reachable worlds that are not in counterpart relation with respect to the current assignment, while other choices are possible like accepting the worlds, but making assignments undefined when the assigned element is deleted [11] or not discarding anything [1].

Also the development of adequate proof systems should be pursued. We did not further investigate the topic here, yet the use of formulas-in-context guarantees the so-called K-scheme, stating that if the formula $\Box(\psi_1 \rightarrow \psi_2)[\Gamma; \Delta]$ holds, then the $\Box$ operator may distribute, i.e. also the formula $\Box(\psi_1) \rightarrow \Box(\psi_2)[\Gamma; \Delta]$ holds. The use of contexts is pivotal here, since otherwise the axiom might not always be satisfied. Instead, its validity tells us that the resulting logic is normal, which is a property of all classical modal logics [17].

Acknowledgments. We are indebted to Giacomo Lenzi for his helpful comments and advices at the very early stages of our research.

References

1. Baldan, P., Corradini, A., König, B., Lluch Lafuente, A.: A temporal graph logic for verification of graph transformation systems. In: Fiadeiro, J.L., Schobbens, P.-Y. (eds.) WADT 2006. LNCS, vol. 4409, pp. 1–20. Springer, Heidelberg (2007)
2. Belardinelli, F.: Quantified Modal Logic and the Ontology of Physical Objects. Ph.D. thesis, Scuola Normale Superiore of Pisa (2006)
3. Caires, L.: Behavioral and spatial observations in a logic for the π-calculus. In: Walukiewicz, I. (ed.) FOSSACS 2004. LNCS, vol. 2987, pp. 72–89. Springer, Heidelberg (2004)
4. Caires, L., Cardelli, L.: A spatial logic for concurrency (part I). Information and Computation 186(2), 194–235 (2003)
5. Cardelli, L., Gardner, P., Ghelli, G.: A spatial logic for querying graphs. In: Widmayer, P., Triguero, F., Morales, R., Hennessy, M., Eidenbenz, S., Conejo, R. (eds.) ICALP 2002. LNCS, vol. 2380, pp. 597–610. Springer, Heidelberg (2002)
6. Cardelli, L., Gardner, P., Ghelli, G.: Manipulating trees with hidden labels. In: Gordon, A.D. (ed.) FOSSACS 2003. LNCS, vol. 2620, pp. 216–232. Springer, Heidelberg (2003)
7. Cardelli, L., Ghelli, G.: TQL: a query language for semistructured data based on the ambient logic. Mathematical Structures in Computer Science 14(3), 285–327 (2004)
8. Courcelle, B.: The expression of graph properties and graph transformations in monadic second-order logic. In: Rozenberg, G. (ed.) Handbook of Graph Grammars and Computing by Graph Transformation, pp. 313–400. World Scientific, Singapore (1997)
9. Dawar, A., Gardner, P., Ghelli, G.: Expressiveness and complexity of graph logic. Information and Compututation 205(3), 263–310 (2007)
10. Distefano, D., Katoen, J.P., Rensink, A.: Who is pointing when to whom? In: Lodaya, K., Mahajan, M. (eds.) FSTTCS 2004. LNCS, vol. 3328, pp. 250–262. Springer, Heidelberg (2004)
11. Distefano, D., Rensink, A., Katoen, J.P.: Model checking birth and death. In: Baeza-Yates, R.A., Montanari, U., Santoro, N. (eds.) IFIP International Conference on Theoretical Computer Science (TCS 2002). IFIP Conference Proceedings, vol. 223, pp. 435–447. Kluwer, Dordrecht (2002)
12. Franconi, E., Toman, D.: Fixpoint extensions of temporal description logics. In: Calvanese, D., Giacomo, G.D., Franconi, E. (eds.) 16th International Workshop on Description Logics (DL 2003). CEUR Workshop Proceedings, vol. 81 (2003), CEUR-WS.org
13. Gadducci, F., Heckel, R., Koch, M.: A fully abstract model for graph-interpreted temporal logic. In: Ehrig, H., Engels, G., Kreowski, H.-J., Rozenberg, G. (eds.) TAGT 1998. LNCS, vol. 1764, pp. 310–322. Springer, Heidelberg (2000)

14. Gadducci, F., Lluch Lafuente, A.: Graphical encoding of a spatial logic for the π-calculus. In: Mossakowski, T., Montanari, U., Haveraaen, M. (eds.) CALCO 2007. LNCS, vol. 4624, pp. 209–225. Springer, Heidelberg (2007)
15. Hazen, A.: Counterpart-theoretic semantics for modal logic. The Journal of Philosophy 76(6), 319–338 (2004)
16. Hodkinson, I., Wolter, F., Zakharyaschev, M.: Monodic fragments of first-order temporal logics: 2000-2001 a.d. In: Nieuwenhuis, R., Voronkov, A. (eds.) LPAR 2001. LNCS (LNAI), vol. 2250, pp. 1–23. Springer, Heidelberg (2001)
17. Huth, M., Ryan, M.: Logic in Computer Science: Modelling and Reasoning about Systems. Cambridge University Press, Cambridge (2004)
18. Reif, J., Sistla, A.P.: A multiprocess network logic with temporal and spatial modalities. International Journal of Computer and System Sciences 30(1), 41–53 (1985)
19. Rensink, A.: Towards model checking graph grammars. In: Leuschel, M., Gruner, S., Lo Presti, S. (eds.) 3rd Workshop on Automated Verification of Critical Systems (AvoCS 2003). University of Southampton Technical Reports, vol. DSSE–TR–2003–2, pp. 150–160. University of Southampton (2003)
20. Rensink, A.: Model checking quantified computation tree logic. In: Baier, C., Hermanns, H. (eds.) CONCUR 2006. LNCS, vol. 4137, pp. 110–125. Springer, Heidelberg (2006)
21. Reynolds, J.: Separation logic: A logic for shared mutable data structures. In: 17th IEEE Symposium on Logic in Computer Science (LICS 2002), pp. 55–74. IEEE Computer Society, Los Alamitos (2002)
22. Vandin, A.: Algebraic models for a second-order modal logic. Master's thesis, University of Pisa (2009), http://www.di.unipi.it/~vandin/thesis.pdf
23. Yahav, E., Reps, T.W., Sagiv, S., Wilhelm, R.: Verifying temporal heap properties specified via evolution logic. Logic Journal of the IGPL 14(5), 755–783 (2006)

Declarative Mesh Subdivision Using Topological Rewriting in **MGS**

Antoine Spicher[1], Olivier Michel[1], and Jean-Louis Giavitto[2]

[1] Université Paris-Est Créteil, LACL, 61 rue du Général de Gaulle,
F-94010 Créteil, France
antoine.spicher@u-pec.fr, olivier.michel@u-pec.fr

[2] CNRS - Université d'Évry, Laboratoire IBISC,
523 place des terrasses de l'Agora, F-91000 Évry, France
giavitto@ibisc.univ-evry.fr

Abstract. Mesh subdivision algorithms are usually specified informally using graphical schemes defining local mesh refinements. These algorithms are then implemented efficiently in an imperative framework. The implementation is cumbersome and implies some tricky indices management. Smith et al. (2004) asks the question of the declarative programming of such algorithms in an index-free way. In this paper, we positively answer this question by presenting a rewriting framework where mesh refinements are described by simple rules. This framework is based on a notion of topological chain rewriting. Topological chains generalize the notion of labeled graph to higher dimensional objects. This framework has been implemented in the domain specific language **MGS**. The same generic approach has been used to implement Loop as well as Butterfly, Catmull-Clark and Kobbelt subdivision schemes.

1 Introduction

The definition and generation of smooth curves or surfaces specified from a finite and small set of control points is a fundamental problem in geometrical modeling. A possible approach is based on the concept of *mesh subdivision* which consists in iterating the replacement of coarse parts of a mesh by finer ones. Introduced by Chaikin in 1974 [1], subdivision algorithms for curves and surfaces are now a major geometric modeling technique.

Many polygon mesh algorithms operate in a local manner and are intuitively described by local graphical schemes. However, their expression and implementation rely on the use of global arrays of points with the induced indexed notation. This obscures the essence of these algorithms and makes their specification unnecessarily complex, especially for the inevitable reindexations caused by the mesh modifications.

This issue has motivated several attempts to develop implementations of such algorithms that are simultaneously *declarative* (close to the mathematical formulation), *intensional* (abstracting meshes from their implementations) and *coordinate-free* (no explicit index manipulation). These works have succeeded

H. Ehrig et al. (Eds.): ICGT 2010, LNCS 6372, pp. 298–313, 2010.

completely in the 1D case [2] and only partially for the 2D case [3]: "The problem of providing a declarative, grammar-like method for specifying subdivision
algorithms remains open. Such specification, if possible, may provide the ulti-
mately concise and clear specification of these algorithms".

In [4,5], we have developed a topological collections rewriting formalism for
simulation purposes. Topological collections extend the notion of labeled graphs
considering higher dimensional cells (w.r.t. points and edges) such as surfaces,
volumes, etc. This framework has been validated in the modeling and simulation
of morphogenesis [6,7] and self-assembly [8] as well as diagrammatic reasoning
involving arbitrary high dimensional spaces [9].

In this paper, we show that the topological collections and the topological
rewriting framework suit well the declarative, intensional and coordinate-free
specification of mesh subdivision algorithms. In the next section, we present the
notion of topological collection and we give a formal definition of topological
rewriting. In Section 3 we introduce MGS, an experimental programming language that provides an implementation of the previous concepts. MGS is used
in Section 4 to specify the Loop's algorithm, a typical subdivision scheme. The
other classical algorithms are illustrated and their complete implementations
are given in [10]. We conclude by outlining some links with more traditional
approaches in graph rewriting as well as some related and future work.

2   Topological Rewriting

2.1   Topological Collections

A *topological collection* is a weakening of the notion of *topological chain*. The
latter is developed in algebraic topology and corresponds to a labeled *cellular
complex*. An (abstract) cellular complex is a formal construction that builds a
space in a combinatorial way through more simple objects called *topological cells*.
Each cell abstractly represents a part of the whole space. The structure of the
whole space, corresponding to the partition into topological cells, is considered
through the *incidence relationships*, relating two "neighbor" cells in the partition.
A topological chain is a function from a cellular complex to a set of labels
equipped with some algebraic structure [11]. We can forget some of the technical
machinery for topological collection.

Definition 1 (Abstract Cellular Complex). *Let $(S_n)_{n \in \mathbb{N}}$ be a family of disjoint sets of symbols. An element of S_n is called an* abstract topological cell *of
dimension n, or simply an n-cell. If $\sigma \in S_n$, n is called the dimension of σ and
is denoted by $dim(\sigma)$. We write S for the set of cells $\bigcup_n S_n$.*

An abstract cellular complex $\mathcal{K}$ *on S is a partially ordered subset of S, that is
a couple $(S, \preceq)$ such that $S \subset S$ and $\preceq$ is a partial order over S (i.e., a reflexive, transitive and antisymmetric binary relation on S) satisfying the following
condition: $\forall \sigma, \tau \in S \quad \sigma \preceq \tau \Rightarrow dim(\sigma) \leq dim(\tau)$. The relation $\preceq$ is called the*
incidence relationships *of the complex $\mathcal{K}$. A complex $\mathcal{K}$ is of finite dimension if
the integer $N = \max\{dim(\sigma) \mid \sigma \in S\}$ is defined. In such a case, N is called the
dimension of $\mathcal{K}$.*

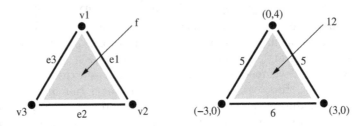

Fig. 1. On the left, an example of cellular complex: it is composed of three 0-cells (v_1, v_2, v_3), of three 1-cells (e_1, e_2, e_3) and of a single 2-cells (f). The boundary of f is constituted of its incident cells v_1, v_2, v_3, e_1, e_2 and e_3. The three edges are the faces of f, and therefore f is a common coface of e_1, e_2 and e_3. On the right, data are associated with topological cells: positions with vertices, lengths with edges and area with f.

An n-cell represents an elementary piece of space of dimension n. In particular, 0-cells are vertices, 1-cells are edges, 2-cells are surfaces, 3-cells are volumes, etc. Thus, graphs are examples of abstract cellular complexes of dimension 1. All set operations are extended to abstract cellular complexes. In particular, we will denote the empty complex $\emptyset$ and use the union $\mathcal{K}_1 \cup \mathcal{K}_2 = (S_1 \cup S_2, \prec_1 \cup \prec_2)$ and the difference of two abstract cellular complexes $\mathcal{K}_1 - \mathcal{K}_2 = (S_1 - S_2, \prec_1 /_{S_1 - S_2})$ where $\prec_1 /_{S_1 - S_2}$ represents the restriction of the relation $\prec_1$ to the elements of $S_1 - S_2$.

Notions of neighborhoods can be defined from the incidence relationships.

Definition 2 (Neighborhoods). *Let $\mathcal{K}$ be a cellular complex and let σ be a cell of $\mathcal{K}$. A cell τ of $\mathcal{K}$ is called* face *of σ iff $(\tau \prec \sigma)$ and $dim(\tau) = dim(\sigma) - 1$. The cell σ is called a* coface *of τ. This relation is denoted by $\tau < \sigma$.*

Let n and p be two integers. Two n-cells of $\mathcal{K}$, σ_1 and σ_2, are p-neighbors if there exists a p-cell in $\mathcal{K}$ that is commonly incident to σ_1 and σ_2.

The notion of incidence is close to the notion of *boundary*. For example, if two vertices v_1 and v_2 are the faces of an edge (in other words, v_1 and v_2 are 1-neighbors), they constitute the boundary of that edge. Left of Figure 1 shows an example of cellular complex. We do not detail further these notions. The interested reader should refer to [11,4].

A topological collection is a labeled cellular complex.

Definition 3 (Topological Collection). *Let S be a set of topological cells, $\mathcal{K} = (S, \preceq)$ be a complex of S, and V be an arbitrary set of values. A topological collection c over $\mathcal{K}$ with values in V is a partial function from S to V. A topological collection c is represented by a formal sum $\sum_{\sigma \in |c|} v_\sigma.\sigma$ where $v_\sigma = c(\sigma)$.*

We use the following notations: Shape(c) refers to the complex $\mathcal{K}$, $|c|$ denotes the set of cells $\sigma \in \mathcal{K}$ where $c(\sigma)$ is defined, $C_S(\mathcal{K}, V)$ denotes the set of topological collections over $\mathcal{K}$ in V, and $C_S(V)$ denotes the set of topological collections with values in V.

The indices in notations $C_S(\mathcal{K}, V)$ and $C_S(V)$ refer to the set of topological cells. By convention, when we write a collection c as a sum $c = v_1.\sigma_1 + \cdots + v_p.\sigma_p$, we insist that all σ_i are distinct. Right of Figure 1 gives an example of a topological collection $c = (0, 4).v_1 + (3, 0).v_2 + (-2, 0).v_3 + 5.e_1 + 6.e_2 + 5.e_3 + 12.f$.

2.2   Rewriting Topological Collections

Transforming topological collections using rewriting requires:

- the notion of *sub-collection* (a way to cut out a sub-part of a collection),
- the *extension* of a collection, and
- the *merge* of collections, that is a way to rebuild a collection from the elements resulting of local transformations.

Definition 4 (Sub-collection). *Let c be a collection of $C_S(\mathcal{K}, V)$. A sub-collection s of c, is an element of $C_S(\mathcal{K}, V)$ such that $|s| \subseteq |c|$ and $\forall \sigma \in |s|$, $s(\sigma) = c(\sigma)$.*

In other words, a sub-collection of a collection c is a restriction of c to a collection (with the same structure) where only a sub-part of the cells remains labeled. Note that if s is a sub-collection of c, then the collection $c - s$ is defined and is also a sub-collection of c.

Definition 5 (Extension, Collection Matching). *Let c be an arbitrary collection of $C_S(V)$ and let $\mathcal{K}$ be a complex such that $\text{Shape}(c) \subset \mathcal{K}$ (i.e. the incidence relationships of the complex underlying c are included in the incidence relationships of $\mathcal{K}$). The extension of c on $\mathcal{K}$, written $c_{|\mathcal{K}}$, is the collection c' of $C_S(\mathcal{K}, V)$ such that $c'(\sigma) = c(\sigma)$ for $\sigma \in |c|$ and c' is left undefined elsewhere.*
  We say that a collection c' matches in a collection $c \in \mathcal{K}$ if $\text{Shape}(c') \subset \mathcal{K}$ and $c'_{|\mathcal{K}}$ is a sub-collection of c.

Merging collections with similar shape naturally corresponds to the addition of their formal sum representation. To merge collections with different shapes, we first build a common abstract cellular complex on which we then perform the standard addition. Note that merge is commutative.

Definition 6 (Collections Merge). *Let $c_1 \in C_S(\mathcal{K}_1, V)$ and $c_2 \in C_S(\mathcal{K}_2, V)$ be two topological collections such that $|c_1| \cap |c_2| = \emptyset$. The merge of c_1 and c_2, denoted $c_1 \uplus c_2$, is defined by: $c_1 \uplus c_2 = c_{1|\mathcal{K}} + c_{2|\mathcal{K}}$ where $\mathcal{K} = \mathcal{K}_1 \cup \mathcal{K}_2$.*

Remark: if a collection c' matches in a collection c, then c can be written $c = c' \uplus c''$ where c'' is uniquely defined.

Definition 7 (Rewriting Relation). *Let R be a relation over $C_S(V)$. One step of rewriting generated by R is the relation $c_1 \triangleright_R c_2$ which is true either iff $c_1 = c_2$ or there exists $(l, r) \in R$ such that*

- $c_1 = c \uplus l$ *and* $c_2 = c \uplus r$, *and*
- $(\text{Shape}(r) - \text{Shape}(l)) \cap \text{Shape}(c) = \emptyset$.

In the latter case, the collection l is called the redex *of the couple $c_1 \triangleright_R c_2$.*

One step of parallel rewriting generated by R *is the relation* $c \mid \triangleright_R c'$ *which is true iff there exists a sequence* $c = c_1 \triangleright_R c_2 \triangleright_R \ldots \triangleright_R c_p = c'$ *such that the redexes l_i of the $c_i \triangleright_R c_{i+1}$ are sub-collections of c and are mutually disjoint (that is, $|l_i| \cap |l_j| = \emptyset$ for all $i \neq j$).*

In this definition, the condition $(\mathrm{Shape}(r) - \mathrm{Shape}(l)) \cap \mathrm{Shape}(c) = \emptyset$ stresses the fact that the "new" cells appearing in r w.r.t. l must really be new, even in c. Informally, the definition can be read as follows. When l matches in c_1, the application of (l, r):

1. removes the cells of l (only the elements of c_1 that do not appear in the shape of l remain with their coefficient), and
2. adds the cells of r.

In order to rebuild a collection, r may refer to some cells of $\mathrm{Shape}(c) - \mathrm{Shape}(l)$. These references correspond to the usual notion of *invariant* (or *gluing graph*) in graph rewriting [12]. If there is no matching, then the application of the rule is void and the rewriting is the identity. The parallel application of a relation R can be represented by the following diagram:

$$
\begin{array}{ccccccccc}
c_1 & = & l_1 & \uplus & \ldots & \uplus & l_n & \uplus & c \\
\downarrow{\scriptstyle |\triangleright_R} & & \downarrow{\scriptstyle \triangleright_R} & & & & \downarrow{\scriptstyle \triangleright_R} & & \downarrow{\scriptstyle \triangleright_R} \\
c_2 & = & r_1 & \uplus & \ldots & \uplus & r_n & \uplus & c
\end{array}
$$

where all the l_i are disjoint.

2.3   Transformation

The use of an explicit relation R to generate a rewriting relation is not very effective. In the following, we propose to generate a relation R from a set of rules $\alpha \to \beta$ where α is a *pattern* that matches a (sub-)collection and β is a *collection expression*. This process, called a *transformation*, relies on the use of *variables* and on the notion of *environment*.

Definition 8 (Variables, Environments and Expressions). *Let (S_n^{var}) with $n \in \mathbb{N}$, be a family of distinguished symbols. An element x of S_n^{var} is called a (topological) cell variable of dimension n. The set of all cell variables is written $S^{\mathrm{var}} = \bigcup_n S_n^{\mathrm{var}}$. Let V^{var} be a set of distinguished values called* value variables.

An environment *(resp. a* cell environment*) is a function from V^{var} (resp. S^{var}) to V (resp. S, such that the dimension of arguments and images match). The set of environments (resp. cell environments) is written Γ_V (resp. Γ_S).*

Let $\Sigma \subset (V \cup V^{\mathrm{var}})^*$ be a distinguished set of words built on V and V^{var} called expressions. *We assume that Σ is equipped by a function $\xi : \Sigma \times \Gamma_V \to V$, called the* evaluation function.

Definition 9 (Rule and Rule Occurrence). *A pattern is a topological collection of $C_{S^{\text{var}}}(V^{\text{var}})$. A collection expression is a collection of $C_{S^{\text{var}}} \cup_{S}(\Sigma)$. A rule is a couple $\alpha \to \beta$ where α is a pattern and β is a collection expression.*

Let $\alpha = a_1.x_1 + \ldots + a_p.x_p$ be a pattern and $\beta = e_1.x_1 + \ldots + e_q.x_q + e'_1.\sigma_1 + \ldots + e'_{p'}.\sigma_{p'}$, with $q \leq p$, be a collection expression. A couple of collections (l, r) is an occurrence of the rule $\alpha \to \beta$ iff it exists $\rho_V \in \Gamma_V$ and $\rho_S \in \Gamma_S$ such that:

$$l = \rho_V(a_1).\rho_S(x_1) + \ldots + \rho_V(a_p).\rho_S(x_p)$$

and

$$r = \xi(\rho_V, e_1).\rho_S(x_1) + \ldots + \xi(\rho_V, e_q).\rho_S(x_q) + \xi(\rho_V, e'_1).\sigma_1 + \ldots + \xi(\rho_V, e'_{p'}).\sigma_{p'}$$

The set of all occurrences of $\alpha \to \beta$ is written $\text{Occ}_{\alpha \to \beta}$.

We are now able to define the concept of transformation.

Definition 10 (Transformation). *Let T be a set of rules $\alpha_i \to \beta_i$. The transformation associated with T is the relation $\vert \triangleright_{R_T}$ where $R_T = \bigcup_i \text{Occ}_{\alpha_i \to \beta_i}$.*

3   The MGS Language

The previous section gives a formal description of topological collections and transformations without any explicit definition of expressions. In this section, we propose an implementation of this formalism through the definition of an experimental functional language called MGS. In particular, we focus on the definition of topological collections and on the specification of a transformation.

3.1   MGS Topological Collections

The MGS language provides the means to specify topological collections. In particular, the programmer is allowed to create new cellular complexes and to label them.

The very basic function **new_cell** creates fresh topological cells. It takes three parameters providing the dimension of the cell to be created, the sequence of its faces and the sequence of its cofaces. Associated with a recursive **letcell...in** construction (similar to a **let rec...in** in OCaml), this simple function allows to create basic complexes. Finally, the specification of a topological collection consists in associating values with cells relying on the formal sum notation given in Definition 3.

Thus, the example of Figure 1 is evaluated by the following MGS program:

```
letcell v1 = new_cell 0 ()          (e1,e3)
and       v2 = new_cell 0 ()          (e1,e2)
and       v3 = new_cell 0 ()          (e2,e3)
and       e1 = new_cell 1 (v1,v2)     (f)
and       e2 = new_cell 1 (v2,v3)     (f)
and       e3 = new_cell 1 (v1,v3)     (f)
and        f = new_cell 2 (e1,e2,e3) ()          in
     (0,4)*v1 + (3,0)*v2 + (-3,0)*v3 + + 5*e1 + 6*e2 + 5*e3 + 12*f
```

The reader is invited to pay attention that topological collections are heterogeneous, *i.e.*, any type of data can be associated with the cells within a collection (here couples of integers and integers).

3.2   MGS Patch Transformations

In Section 2, patterns and collection expressions are defined as topological collections. The difficulty of their descriptions is hidden by the implicit partial orders that they come with. As an implementation of this formalism, MGS has to make these objects as explicit as possible with a special syntax for the programmer. The objectives of the syntax is to simplify as much as possible the specification of the rewriting rules. Different rule languages have been developed. In this article, we focus on *patch transformations* (called *patches* from now on), a kind of transformation specially designed for the specification of topological surgeries.

A patch is specified as follows:

    patch P = {   *Pattern* => *Exp* ;   ...   }

The language *Pattern* has been introduced to specify the left hand side (l.h.s.) while the right hand side (r.h.s.) are basic expressions of the language where MGS operators (*e.g.*, new_cell) can be used to specify new collections.

A patch is a function having a topological collection as argument. More specifically, the expression $P(c)$ where c is an MGS collection, evaluates into a new collection c' such that $c|\triangleright_{R_P} c'$ as defined in Definition 10. Obviously, the relation R_P (which may be an infinite set) is not computed within this evaluation. In fact, a pattern matching mechanism has been developed to compute one of the possible collections c'. The interested reader should refer to [13] for an elaboration.

In the following, we detail the patch transformation language. In order to illustrate our comments, we consider the topological modification described on Figure 2: an edge named e, whose faces are called v1 and v2, is matched and replaced by two new edges e1 and e2 together with a new vertex v; the edge e1 is bound by vertices v1 and v, and e2 by v2 and v. This rule specifies the insertion of a vertex on an edge.

Fig. 2. Insertion of a vertex on an edge

Patch Patterns. Rather than specifying patterns of rules as explicit topological collections, patch patterns describe them implicitly with a list of constraints (on dimension and incidence). The topological cells of sub-collections that are matched by these patterns, have to respect the constraints.

The specification of a pattern relies on the following grammar:

> *Pattern*
>> $m ::= c \mid c\, o\, m$
>
> *Op*
>> $o ::= \varepsilon \mid < \mid >$
>
> *Clause*
>> $c ::=$ ` x:[dim=`exp_d`, faces=`exp_f`, cofaces=`exp_{cf}`, ` exp_b`]`
>>
>> $\quad\ \mid$ `~x:[dim=`exp_d`, faces=`exp_f`, cofaces=`exp_{cf}`, ` exp_b`]`

where ε represents the empty word. A pattern (*Pattern*) is a finite list of *clauses* separated by some operators (*Op*). A clause (*Clause*) corresponds to a topological cell to be matched. Each clause is characterized by some optional information that will guide the pattern matching process:

- x is a pattern variable that allows to refer to the cell matched by the clause anywhere else in the rule. A variable can be used before its definition. Nevertheless, a name refers to one and only one element. If two clauses share the same identifier, they will match the same cell (the predicates of both clauses have to hold together). Since a clause refers to an element of a collection, that is a couple (value v, cell σ), the use of the variable x in the rest of the rule might be ambiguous. We avoid such a problem by setting that x binds for the value v and the expression ^x binds for the cell σ.
- The expression exp_d associated with the field dim returns an integer constraining the dimension of the cell matched by the clause,
- Expressions exp_f and exp_{cf} respectively associated with fields faces and cofaces are evaluated in sequences of topological cells. These lists constrain the incidence relationships of the matched cell. They do not have to be exhaustive; a cell can get more (co)faces than those referred in the pattern. In these expressions, pattern variables or direct references to topological cells may be used,
- The last expression exp_b can be used to specify some arbitrary predicate the cell has to satisfy (it can be used to constrain the value associated with the cell for example),
- Two kinds of clause can be specified depending on the presence of the unary operator ~. When it is present, the clause is considered as a context for the pattern matching and the associated cell in the sub-collection is not considered as consumed by the pattern matching. If no operator ~ is specified, the topological cell is matched and consumed by the rule: no other sub-collection can refer to a not-tilded element.
  The consumption of a cell during the pattern matching ensures sub-collections to be disjoint as required by Definition 7. Operator ~ allows a same cell to be referred into distinguished sub-collections. Using notations of Definition 7, such cell belongs to Shape(l) ∩ Shape(c).

Operators *Op* correspond to syntactic sugar. The infix binary operator < (resp. >) constraints the element matched by the left operand to be a face (resp. coface) of the element matched by the right operand. Therefore, any patterns

of type a:[...] < b:[...] can be rewritten a:[..., cofaces=b] b:[...,
faces=a]. When no operator separates two clauses, there is no constraint between the matched elements.

The pattern corresponding to the l.h.s. of the rule pictures on Figure 2 is
specified by:

```
~v1 < e:[ dim = 1 ] > ~v2
```

Here, v1 and v2 are not consumed as they are used as a context for the pattern
matching of e. They are referred to in the r.h.s. but they are not rewritten.

Right Hand Side of a Rule. The r.h.s. of a rule is an MGS expression that has
to evaluate to a collection. In other words, the r.h.s. can be any MGS program. It
makes the transformation very flexible and allows program factorization. Therefore, using the MGS syntax for defining new collections given Section 3.1, the
graphical rule of Figure 2 is specified in MGS by:

```
~v1 < e:[ dim = 1 ] > ~v2 =>
  letcell v  = new_cell 0 ()        (e1,e2)
  and      e1 = new_cell 1 (^v1,v) (cofaces ^e)
  and      e2 = new_cell 1 (^v2,v) (cofaces ^e) in
     (some expression)*v
```

In the r.h.s., the three new cells are created and then a collection is defined. The
cofaces of the new defined edges e1 and e2 are the cofaces of the edge matched
by e, so that if e belongs to the boundary of a 2-cell, e1 and e2 belong to it after
the application of the patch. The new sub-collection consists of the association
of an arbitrary value with the new vertex v while the collection is left undefined
on e1 and e2.

4   Mesh Refinement Algorithms

In this section, we propose to use the rule-based programming of patches to
specify a classical but not trivial familly of algorithms in CAD: the *refinement
of meshes*.

4.1   Loop Subdivision

Subdivision algorithms generate to the limit smooth surfaces by iterating subdivisions of polygonal mesh. In [14], the descriptions of the usual subdivision
processes dedicated to the geometrical modeling and to the animation of solids
can be found. These algorithms are locally specified by *masks*. A mask is a cellular complex describing a part of a mesh centered on an element to be refined
(*e.g.*, an edge, a triangle, etc.). The refinement consists in inserting a new vertex
whose coordinates are determined by an affine combination of the other vertices
belonging to the mask. The properties of smoothness of the surface obtained to
the limit and therefore the quality of the subdivision depends on the mask. The

intended properties correspond to surfaces as smooth as possible (C^1-continuity or C^2-continuity everywhere for example) and are difficult to obtain on arbitrary meshes that exhibit irregularities on singular points. The mask is then chosen depending on the kind of the mesh (triangular, quadrangular, etc.) and the property to hold.

We are here interested in the Loop subdivision [15]. This algorithm:

- is based on *vertex insertions*: a vertex is created on each edge of the mesh within each application of the algorithm; old nodes are conserved and new nodes are linked to them by new edges;
- works on and generates *triangular meshes*;
- is *approximating*: old nodes positions are changed in the new mesh and new nodes positions are computed as a function of the old nodes positions in the old mesh (approximating subdivisions are opposed to interpolating ones for which old nodes positions are not affected).

The algorithm describes on the one hand the topological modifications and on the other hand the geometrical modifications.

Topological Modification. The mesh refinement is based on the polyhedral subdivision shown on Figure 3. A vertex is inserted on each edge and triangles are refined in 4 smaller triangles.

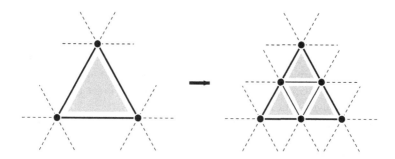

Fig. 3. Loop Algorithm: Topological Modification

Geometrical Modification. Loop's masks provide the information required to compute the positions of new vertices (as a function of the coordinates of the old ones) and the new positions of the old vertices (see Figure 4). On the left, coordinates of a new vertex v inserted on the edge whose boundary is composed of vertices v_1 and v_2 are given by:

$$\frac{3}{8}v_1 + \frac{3}{8}v_2 + \frac{1}{8}v_3 + \frac{1}{8}v_4 \tag{1}$$

On the right, new coordinates of an old vertex v is computed as a function of its 1-neighbors positions:

$$(1 - k\beta)v + \sum \beta v_i \quad \text{where } \beta = \frac{1}{k}\left(\frac{5}{8} - \left(\frac{3}{8} + \frac{1}{4}\cos\frac{2\pi}{k}\right)^2\right) \tag{2}$$

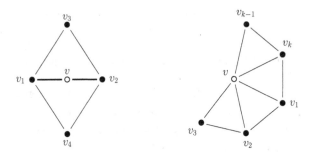

Fig. 4. Loop Algorithm: Geometrical Modification

4.2   Representation of Geometrical Objects in MGS

Meshes can be easily represented by MGS topological collections. Since they are surfaces composed of polygons, their structures are represented by a cellular complex of dimension 2. In the topological collections, only the geometrical positions of the 0-cells matter. So, we associate the constant value `'Triangle` with triangles, the constant value `'Edge` with edges, and coordinates with vertices. The coordinates of a vertex are represented by an MGS record value (that is equivalent to a C `struct`). The type of a coordinate is defined as follows:

```
record coordinate = {
     x:float, y:float, z:float, old:(coordinate|'Nil)    }
```

whose fields x, y and z encode the coordinates in the 3D space. The last field old is used to distinguish new and old vertices. For new vertices, old is set to the value `'Nil` (an MGS constant). For old vertices, old is set to the coordinates before the Loop algorithm iteration.

To simplify the description of the MGS program, we assume the existence of a function `addCoord` that sums two values of type `coordinate`, and the global constant 0 corresponding to the 3D space origin.

4.3   Loop Subdivision in MGS

Three steps are used to implement the Loop algorithm.

Update of the old vertices. In this step, we save in the field old the current coordinates of each vertex and we update the coordinates with respect to the Loop's mask (see Figure 4 on the right).

```
patch even_vertex = {
   v:[dim=0] =>
     let s = ccellsfold(addCoord, 0, v, 1) and b = β in
       { old = v, x = (1-k*b)*v.x + b*s.x, ... } * ^v
}
```

The single rule of this patch does not change the topology: one element of dimension 0 named v is matched and an elementary collection on v is computed in the r.h.s. Variables s and b correspond respectively to the sum of the coordinates of the 1-neighbors (see Definition 2) of ^v and to the β coefficient of the Loop's mask (see Equation (2)). The MGS primitive ccellsfold(f,z,σ,i) computes the sequence $(v_1, \ldots, v_n)$ of the values associated with the i-neighbors of σ and computes the value $f(v_1, (\ldots f(v_n, z) \ldots))$. Old coordinates referred by the variable v are saved in the field old.

Insertion of new vertices. The following patch is quite similar to the patch given as an example in Section 3.2.

```
patch odd_vertex = {
  ~v1 < e:[dim=1] > ~v2
  ~v1 < ~e13 > ~v3 < ~e23 > ~v2
  ~v1 < ~e14 > ~v4 < ~e24 > ~v2
  =>
    letcell v  = new_cell 0 ()        (e1,e2)
    and       e1 = new_cell 1 (^v1,v) (cofaces ^e)
    and       e2 = new_cell 1 (^v2,v) (cofaces ^e) in
      { old = 'Nil, x = vₓ,  ... }*v + 'Edge*e1 + 'Edge*e2
}
```

The pattern is extended to take into account the vertices v3 et v4 required by the Loop's mask (see Figure 4 on the right). Only the edge e is consumed to be removed and replaced. All the other clauses are used as pattern matching context. Coordinates (v_x, v_y, v_z) of the new created vertex v are computed using the field old of vertices v1, v2, v3 and v4 w.r.t. Equation (1).

Creation of the refined triangles. The previous patch leads to the transformation of the mesh triangles into hexagons. These hexagons are then removed and replaced by new smaller triangles.

```
patch subdivideFace = {
  f:[ dim = 2, faces = (^e1,^e2,^e3,^e4,^e5,^e6) ]
  ~v1 < ~e1 > ~v2:[ v2.old == 'Nil ] < ~e2 >
  ~v3 < ~e3 > ~v4:[ v4.old == 'Nil ] < ~e4 >
  ~v5 < ~e5 > ~v6:[ v6.old == 'Nil ] < ~e6 > ~v1
  =>
    letcell a1 = new_cell 1 (^v2,^v4) (f1,f4)
    and       a2 = new_cell 1 (^v4,^v6) (f2,f4)
    and       a3 = new_cell 1 (^v6,^v2) (f3,f4)
    and       f1 = new_cell 2 (a1,^e2,^e3) ()
    and       f2 = new_cell 2 (a2,^e4,^e5) ()
    and       f3 = new_cell 2 (a3,^e6,^e1) ()
    and       f4 = new_cell 2 (a1,a2,a3) () in
      'Edge*a1 + ... + 'Triangle*f4 }
```

This patch is composed of a single rule sketched in Figure 5. Note the presence of tests vi.old == 'Nil to distinguish old and new vertices.

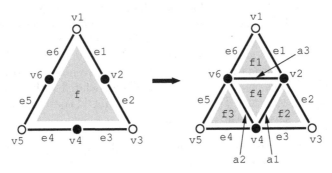

Fig. 5. Construction of the 4 refined triangles in the Loop's algorithm

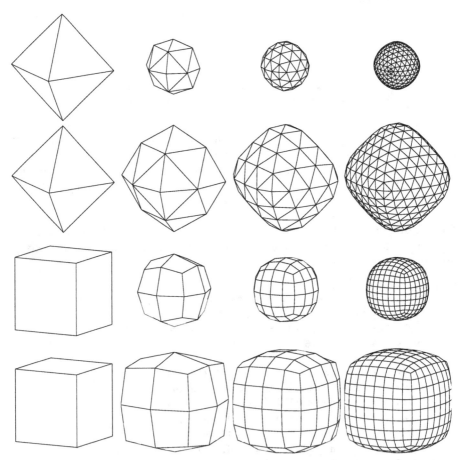

Fig. 6. Results of the application of subdivision algorithms. From top to bottom, the Loop's algorithm, the Butterfly algorithm, the Catmull-Clark's algorithm and the Kobbelt's algorithm. From the left to the right, the initial state then 3 iteration steps. These pictures have been generated by the current MGS prototype.

Figure 6 shows an example of Loop's algorithm iterations generated by the current prototype of MGS. The figure also shows outputs of three other algorithms [14] that have been implemented in the same way in MGS.

5   Conclusion

In this article, we have presented the MGS language as a positive answer to the question of a framework that allows a declarative, intensional and coordinate-free specification of mesh algorithms.

The MGS main notions, topological collections and transformations, rely on a formal definition of a topological rewriting relation. This computational model is based on the substitution of labeled cellular complexes that are an extension of graphs to higher dimensions. The expressiveness of topological collections and their transformations allows the programming of complex algorithms (in a large range of domains) in a very concise way. It has been exemplified in this paper with the specification of non trivial mesh subdivision algorithms.

Topology has already been introduced in graph transformation in different kinds of contexts [16,17,18]. Work presented in [17] is far from our purpose since the concept of topology is used to ensure structural constraints on graphs being transformed. Nevertheless, MGS and its current implementation can be used as a programming language to express and simulate the proposed model transformation rules. In [16], GRiT (Graph Rewriting in Topology), a kind of hyper-graph rewriting, is developed to take into account topological properties (from homology and homotopy) in models of parallel computing based on rewriting theory [19]. This approach has been applied in the modeling of chemical reactions, DNA computing, membrane computing, etc. MGS has also been extensively used in this context [4]. Since we focus in this paper on an application in topological modeling, our work can be compared to [18] where graph transformations have been applied to specify topological operations on *G-maps*. G-maps are a data structure used to encode cellular complexes corresponding to a specific class of topological objects called *quasi-manifolds* [20]. Roughly speaking, G-maps are graphs where vertices are named *darts* and where edges correspond to *involutions* defined between darts. Because G-maps are a specific kind of cellular complexes, they can be handled by the approach presented here. Indeed, they have been implemented in MGS (see [10] for an elaboration) as well as many of the applications proposed in [18] (especially in biology). The formal approach developped in [18] is very specific and focuses on the correct handling of involutions in G-maps construction and deletion operations.

Our contribution differs from the works mentionned above on two main points: the object to be transformed and the addressed mathematical framework. These works are all based on (hyper-)graph transformation. We propose to transform cellular complexes since the extension from graphs to complexes seems intuitive and relevant in a lot of application area [21]. Our mathematical description is not based on the usual graph morphisms and pushouts (like in [12,17,18]). Our objective was here to relate the notion of transformation to the very definition of topological collections in the context of a programming language. It requires to give explicit details on the construction of collections and how the rewriting is effectively done. Our mathematical formalization of complex transformation is inspired by [22] where graph rewriting based on a (multi-)set point of view is developed. The proposed model is close to term rewriting modulo associativity and commutativity (where the l.h.s. of a rule is removed and the r.h.s. is added) and can be applied on the formal sum notation of topological collections. This kind of approach also allows to extend results from term rewriting to topological rewriting (as we did for termination in [23]).

References

1. Chaikin, G.: An algorithm for high speed curve generation. Computer Graphics and Image Processing 3, 346–349 (1974)
2. Prusinkiewicz, P., Samavati, F.F., Smith, C., Karwowski, R.: L-system description of subdivision curves. International Journal of Shape Modeling 9(1), 41–59 (2003)
3. Smith, C., Prusinkiewicz, P., Samavati, F.: Local specification of surface subdivision algorithms. In: Pfaltz, J.L., Nagl, M., Böhlen, B. (eds.) AGTIVE 2003. LNCS, vol. 3062, pp. 313–327. Springer, Heidelberg (2004)
4. Giavitto, J.L., Michel, O.: The topological structures of membrane computing. Fundamenta Informaticae 49, 107–129 (2002)
5. Giavitto, J.L.: Invited talk: Topological collections, transformations and their application to the modeling and the simulation of dynamical systems. In: Nieuwenhuis, R. (ed.) RTA 2003. LNCS, vol. 2706, pp. 208–233. Springer, Heidelberg (2003)
6. Spicher, A., Michel, O.: Declarative modeling of a neurulation-like process. BioSystems 87, 281–288 (2006)
7. de Reuille, P.B., Bohn-Courseau, I., Ljung, K., Morin, H., Carraro, N., Godin, C., Traas, J.: Computer simulations reveal properties of the cell-cell signaling network at the shoot ape x in Arabidopsis. PNAS 103(5), 1627–1632 (2006)
8. Spicher, A., Michel, O., Giavitto, J.L.: Algorithmic self-assembly by accretion and by carving in mgs. In: Talbi, E.-G., Liardet, P., Collet, P., Lutton, E., Schoenauer, M. (eds.) EA 2005. LNCS, vol. 3871, pp. 189–200. Springer, Heidelberg (2006)
9. Valencia, E., Giavitto, J.L.: Algebraic topology for knowledge representation in analogy solving. In: ECCAI, A. (ed.) European Conference on Artificial Intelligence (ECAI 1998), Brighton, UK, Christian Rauscher, August 23-28, pp. 88–92 (1998)
10. Spicher, A.: Transformation de collections topologiques de dimension arbitraire. Application à la modélisation de systèmes dynamiques. PhD thesis, Université d'Évry (2006)
11. Munkres, J.: Elements of Algebraic Topology. Addison-Wesley, Reading (1984)
12. Ehrig, H., Pfender, M., Schneider, H.J.: Graph grammars: An algebraic approach. In: IEEE Symposium on Foundations of Computer Science, FOCS (1973)
13. Giavitto, J.L., Michel, O.: Pattern-matching and rewriting rules for group indexed data structures. In: ACM Sigplan Workshop RULE 2002, Pittsburgh, pp. 55–66. ACM, New York (2002)
14. Zorin, D.: Subdivision zoo. In: Subdivision for modeling and animation, Schröder, Peter and Zorin, Denis, pp. 65–104 (2000)
15. Loop, T.L.: Smooth subdivision surfaces based on triangle. Master's thesis, University of Utah (August 1987)

16. Liu, J.Q., Shimohara, K.: Graph rewriting in topology. i. operators and the grammar. SIG-FAI 45, 21–26 (2001)
17. Levendovszky, T., Lengyel, L., Charaf, H.: Extending the dpo approach for topological validation of metamodel-level graph rewriting rules. In: SEPADS 2005: Proceedings of the 4th WSEAS International Conference on Software Engineering, Parallel & Distributed Systems, Stevens Point, Wisconsin, USA, pp. 1–6. World Scientific and Engineering Academy and Society (WSEAS) (2005)
18. Poudret, M., Arnould, A., Comet, J.P., Gall, P.L.: Graph transformation for topology modelling. In: Ehrig, H., Heckel, R., Rozenberg, G., Taentzer, G. (eds.) ICGT 2008. LNCS, vol. 5214, pp. 147–161. Springer, Heidelberg (2008)
19. Liu, J.Q., Shimohara, K.: Graph rewriting in topology iv: Rewriting based on algebraic operators (algorithms in algebraic systems and computation theory). RIMS Kokyuroku 1268, 64–72 (2002)
20. Lienhardt, P.: Topological models for boundary representation: a comparison with n-dimensional generalized maps. Computer-Aided Design 23(1), 59–82 (1991)
21. Tonti, E.: On the mathematical structure of a large class of physicial theories. Rendiconti della Academia Nazionale dei Lincei 52(fasc. 1), 48–56 (1972); Scienze fisiche, matematiche et naturali, Serie VIII
22. Raoult, J.C., Voisin, F.: Set-theoretic graph rewriting. Technical Report RR-1665, INRIA (April 1992)
23. Giavitto, J.L., Michel, O., Spicher, A.: Spatial organization of the chemical paradigm and the specification of autonomic systems. In: Software-Intensive Systems (2008)

A Model for Distribution and Revocation of Certificates

Åsa Hagström[1] and Francesco Parisi-Presicce[2]

[1] Lawson Software
Åsa.Hagström@se.lawson.com
[2] Dipartimento di Informatica, Sapienza Universitá di Roma
parisi@di.uniroma1.it

Abstract. The distribution and revocation of public-key certificates are essential aspects of secure digital communication. As a first step towards a methodology for the development of reliable models, we present a formalism for the specification and reasoning about the distribution and revocation of public keys, based on graphs. The model is distributed in nature; each entity can issue certificates for public keys that it knows, and distribute these to other entities. Each entity has its own public key bases and can derive new certificates from this knowledge. If some of the support for the derived knowledge is revoked, then some of the derived certificates may be revoked as well. Cyclic support is avoided. Graph transformation rules are used for the management of the certificates, and we prove soundness and completeness for our model.

1 Introduction

A certificate is an assertion, made by an entity (often an "authority"), about some characteristics or privileges of another entity. We are only interested in "public-key certificates" where an asserrtion is the binding of a particular public-key with an entity, assertion signed by another entity with her private-key. In this paper we present a conceptual model to better understand the semantics of certificate distribution and revocation. The model is a graph which describes the knowledge of all entities in a system, and is distributed in nature. We **do not** assume a hierarchical model with a central Certification Authority (CA), but allow every entity to issue and distribute its own certificates to others.

The Merriam-Webster dictionary explains the act of revoking as "to annul by recalling or taking back". Thus, revocation of a certificate could be the act of a user who recalls a certificate previously passed to another user. Somehow, the revocation must cascade in the system to make sure that no information is derived from obsolete certificates.

This description of revocation seems simple enough. However, even in such a specific environment as a public-key infrastructure (PKI) – where all the information consists of certificates, each on the same form – one has to be very careful when defining *what* is being revoked. Similarly, there can be a number of reasons for a revocation. We show how different reasons can be modelled as specific actions, corresponding to the annulment of a specific piece of information.

H. Ehrig et al. (Eds.): ICGT 2010, LNCS 6372, pp. 314–329, 2010.

A stronger way to revoke a certificate is to issue its *inverse*; if there was previously a certificate (signed by A with her private key A.pr) binding B and his public key B.pu, the inverse is a certificate stating that B.pu is *not* B's public key. This annulment is time-persistent in the sense that any subsequent positive certificates for the same binding has to deal with the presence of the negative certificate. We consider any kind of annulment of information – whether by removal or by issuing the inverse – to be a form of revocation.

Our aim is to understand the meaning of revocation in the context of a PKI. The purpose is not to find an efficient implementation for revocation (in particular, we do not deal with Certificate Revocation Lists (CRL's)), but to investigate the implications and how they depend on the reason for the revocation. We also have ideas for including trust statements, but space limits us to key certificates. Many researchers use graphs to exemplify and concretize their ideas. We believe that graphs themselves constitute a powerful tool for modelling and reasoning about systems, and we have chosen to take advantage of their expressive and intuitive properties. Our formalism of choice is a graph which captures the information state of a system, and graph transformation rules which define allowed changes to the information, as well as deductions adding new knowledge.

2   Related Work

The notion of revocation can be given various interpretations. The desired result of revoking a certificate will typically depend on the reason for it. Cooper [1] divides revocation reasons into benign and malicious types, and notes that different revocation practices are needed for the two types. Both Fox and LaMacchia [2] and Gunter and Jim [3] discuss different reasons for revocation, and suitable mechanisms for each case. As for X.509, Housley et al. [4] define nine reason codes for revocation of a public-key certificate (keyCompromise, cACompromise, affiliationChanged, superseded, cessationOfOperation, certificateHold, removeFromCRL, privilegeWithdrawn, aACompromise), but do not suggest different revocation practices for the different codes. In [5], Hagström et al. define and classify eight different types of revocation schemes for an ownership-based access control system using the dimensions resilience, propagation and dominance. Some ideas from all this work will be applied in our model.

Other researchers use graphs in some form or other to visualize their ideas. When it comes to the formal treatment, however, most previous work in this area has used logic-, calculus- or language-based approaches:

Logic-based formalisms
Maurer [6] was one of the first to model a PKI using both keys and trust. A both needs to know B's public key and trust him to believe the statements he makes. Every statement is about keys or trust. Trust is given in levels; a higher level of trust in a user implies the possibility of longer chains of derived statements starting in that user. Each user's view (including all belief and trust the user has, and all recommendations made to him) is modelled separately from the

others', so it is impossible to get a global view of the system and to maintain dependencies between different users' statements.

Stubblebine and Wright [7] describe a logic for analyzing cryptographic protocols that supports the specification of time and synchronization details. Assuming that information about revoked keys cannot be immediately distributed to all parties of a system, they instead focus on policies for decisions based on information that may be revoked. There is no separate notion of trust; users are assumed to communicate honestly.

Kudo and Mathuria [8] describe timed-release public key protocols where a trusted third party binds a public-key pair to a specific future time.

Li et al. [9,10] present a logic-based knowledge representation for distributed authorization and delegation. The logic allows more general statements, and it lets users reason about other users' beliefs. The authors deal with the problem of non-monotonicity and use overriding policies to determine which statements take precedence. It seems impossible to completely remove statements and their consequences from the system.

Liu et al. [11] use a typed modal logic to specify and reason about trust in PKIs. A certification relation and a trust relation are used to specify which entities are allowed to certify other entities' keys, and which users they trust. The former is static, while the latter may change dynamically. To accept a statement, a user must find a path to that statement starting with a trusted certificate. Revocation is enforced by an overriding policy.

Calculus-based formalisms

Simmons and Meadows [12] focus on the role of trust in information integrity protocols by adding trust (delegation) to shared control schemes. The authors develop a methodology to compute the consequences of trust relations, i.e. to examine all possible concurrences of participants who may be able to exercise a protocol, and to decide whether the risks associated with those concurrences are acceptable. If not, the trust relations in the system may need to be changed, but this "revocation" is not done dynamically.

In [13], Kohlas and Maurer propose a calculus for deriving conclusions from a given user's view, which consists of evidence and inference rules that are valid in that user's world. Statements can be beliefs or recommendations about public keys or trust, commitments to statements and transfer of rights. There are no negative statements or other possibilities for revocation.

Language-based formalisms

Gunter and Jim [3] define the programming language QCM, used to define a general PKI with support for revocation through negative statements and an overriding policy. Trust is not considered.

These papers all model certificate distribution and/or revocation, much in the same way that we need for our purposes. There has also been some work in this field that uses graphs in a formal way, but that is further away from our basic problem formulation:

- Wright et al. [14] define depender graphs, where an edge from A to B indicates that A has agreed to forward revocation information to B. Graph properties are used to discuss how many dependers are needed for the scheme to be fault-tolerant.
- Aura [15] defines delegation networks to pass authorizations between users. Although the authorizations are allegedly transferred in certificates, the usage of keys and signatures has been abstracted away. Graph searching algorithms are used to find support for authorizations.
- Buldas et al. [16] use authenticated search trees to model undeniable attesters, used for long-term certificate management supporting key authenticity attestation and non-repudiation.

These modelsdeal with certificates and revocations, but graphs are only a convenient data structure not a formal tool for reasoning about the problems.

3   Background

We first define the different types of information passed in the system:

- *Certificate*: a signed statement binding a public key to an entity
- *Knowledge*: a certificate where the signer, distributor and recipient are equal
- *Quotation*: a certificate where the distributor is different from the signer

Every certificate has a number of attributes associated with it:

- *Signing key identifier*: the id of the private key used to sign a certificate
- *Signer*: the entity that owns the signing key
- *Subject*: the entity being bound to a key
- *Public key*: the public key that is bound to the subject, along with its id
- *Recipient*: the entity receiving a certificate
- *Distributor*: the entity passing/sending a certificate to the recipient
- *Value*: a certificate is either *positive* or *negative*: a positive certificate asserts the validity of the binding within it, a negative certificate denies it.

Note that not all the attributes in the previous list are actually present in the X.509 standard, but can be derived in the framework (with the exception of negative certificates). In addition, a certificate can be *active* or *inactive* depending on whether the expiration date is still to come or has past, respectively.

A certificate is either of the types O or I (outside or inside). The validity of O certificates has been established outside the system, in a secure, out-of-band procedure. I certificates are either quoted within the system or deduced from other certificates, and are required to be supported by a chain of other certificates, with an O certificate at its root. A certificate is either active or inactive, inactive when the public key has expired.

We now formalize the concepts of *support*, *collisions* and *valid state*.

Observation. All O certificates are supported.

Definition 1 (Support for deduced knowledge). *A deduced knowledge ρ with signer/recipient A is supported if and only if:*

(1) A is the recipient of a supported certificate σ (signed with $B.pr$) certifying the same binding subject – public key and with the same value as ρ, and

(2) A has positive, supported knowledge stating that he recognizes $B.pu$ as B's public key, and

(3) σ and ρ are either both active or both inactive.

Definition 2 (Support for quotations). *A quotation ρ, signed by A with $A.pr$, subject B, public key $B.pu$ and distributor C is supported if and only if C is the recipient of a certificate σ that:*

(1) certifies the same binding (B and $B.pu$) as ρ, and

(2) has the same value as ρ, and

(3) is signed with $A.pr$, and

(4) is active if ρ is active and inactive if ρ is inactive, and

(5) is supported.

In both cases, we say that σ supports ρ. In the first case, it provides a connection from B to A, and in the second from A to C.

Definition 3 (Path and Support Set)

(1) A path is a string over the set of entity names. A valid path of a certificate σ is a path representing an acyclic supported chain of certificates from a type O certificate via quotations and deductions to σ. A support set of a certificate is a set of paths

(2) The support set of a certificate σ is valid if it contains only valid paths of σ; it is complete if it contains all valid paths of σ.

Paths are denoted by lower-case Greek letters, and for support sets we use upper-case Greek letters. Also, $\Pi A = \{\pi_1 A, \pi_2 A, \ldots, \pi_i A\}$

It is possible that one entity certifies to a certain binding, another one certifies the opposite, and both communicate their assertion to the same recipient

Definition 4 (Certificate collision). *Given two certificates σ and ρ with opposite signs, both certifying the binding between B and $B.pu$, there is a certificate collision if they have the same recipient C.*

Definition 5 (Valid state). *A system is in a valid state if and only if:*

(1) All deduced knowledge is supported

(2) All quotations are supported

(3) The support set of each certificate is valid and complete

(4) There are no certificate collisions

(5) Where there is support for deduced knowledge, the deduction is made.

We now briefly review basic graph concepts and some notation.

A **graph** G over the sets of (node, edge) attributes $(V\text{-}ATT, E\text{-}ATT)$, is a tuple $G = (V, E, end, v\text{-}att, e\text{-}att)$ where V and E are sets (of vertices and of edges, respectively), $end : E \to V^2$ assigns an ordered pair of vertices to each edge, and $v\text{-}att : V \to V\text{-}ATT$ and $e\text{-}att : E \to E\text{-}ATT$ assign attributes from $V\text{-}ATT$ and $E\text{-}ATT$ to vertices and edges, respectively.

A **(partial) graph morphism** $f : G_1 \to G_2$ between graphs $G_i = (V_i, E_i, end_i, v\text{-}att_i, e\text{-}att_i), i = 1, 2$, consists of a pair of (partial) functions $f_V : V_1 \to V_2$, $f_E : E_1 \to E_2$ such that $v\text{-}att_2(f_V(v)) = v\text{-}att_1(v)$ and $e\text{-}att_2(f_E(e)) = e\text{-}att_1(e)$ (attribute preserving) and $end_2(f_E(e)) = f_V^*(end_1(e))^1$ (structure preserving)

Graph Transformation Rules can be used to construct and modify the *system state*, i.e. the graph that represents the knowledge of every user. In the context of the SPO approach, a rule is a partial function $r : L \rightharpoonup R$ where L dom r is intended to be deleted, dom$r$ is intended to be kept and used to "connect" the context to the part to be added (R codom r). The application of r to a graph G requires an **occurrence** of L in G (i.e., a total morphism $m : L \to G$, called *matching*); the result of applying r to G via m is the PO of r and g.

Rules can have **negative application conditions** (NACs) which specify the parts that must not be present for the rule to apply. A NAC for a rule $r : L \rightharpoonup R$ is a total injective morphism $c : L \to N$. A matching $m : L \to G$ satisfies the NAC $c : L \to N$ if there is no morphism $d : N \to G$ such that $d \circ c = m$ (that is, the matching cannot be extended to all of N). In our diagrams, we denote the rule $r : L \rightharpoonup R$ with NAC $c : L \to N$ by representing the left-hand side with N, with its L-part drawn solid, and the $(N - L)$-part drawn dotted.

In figure 1, the left-hand side contains the entities C and D (represented by circles), and a certificate passed from C to D (represented by a box on the edge between them). When this rule is applied to a specific graph, C, D, $A.pr$ and $B.pu$ must all be instantiated to nodes and edges present in that graph, thus matching the left-hand side. The effect of the rule is to remove the edge.

Fig. 1. Example Rule – removing a certificate

Rule expressions ([17]) are used to control the application of graph transformation rules. Here we only need expressions of the form `asLongAsPossible` R `end` which applies R until there are no more matches of the left side of R.

4   Graph Model

Our system is formalized as a graph $C\text{-}graph=(V, E, end, v\text{-}att, e\text{-}att)$ where
– V is the set of entities in the system,

---

[1] Where f_V^* is the usual extension of f_V to sequences.

- E is the set of signed statements about public keys
- the mapping $end : E \to V^2$ assigns to each edge its distributor and recipient

The mapping $v\text{-}att : V \to V\text{-}ATT$ assigns names to the nodes.

The set $E\text{-}ATT$ consists of tuples (key-id, key, type, color, value) where

- the key-id belongs to ID_{Pr} (the set of private key IDs),
- the key is an element of $ID_{Pu} \times Pu$ (the set of public key IDs and corresponding public keys)
- a type belongs to $T = \{O, I\}$ (outside or inside knowledge),
- a color belongs to $C = \{a, i\}$ (active or inactive certificate), and
- a value belongs to $S = \{+, -\}$ (positive or negative certificate).

For simplicity, the mapping $e\text{-}att$ is split into four: *cert*, *type*, *col* and *value*, where the last three associate to an edge the last three componets of the 5-tuple.

Graphically, these attributes are represented in different ways. The certificate is represented by a box on each edge, containing the signing key identifier and the identifier of the public key being certified. Edges representing O certificates are double boxes; edges representing I certificates are single boxes. Any graph transformation rule that applies to an edge of type I also applies to one of type O. Finally, inactive edges are shown with a shaded box whereas active edges have clear boxes. Rules that apply to active edges do not apply to inactive ones.

Any entity can have any number of private/public key pairs associated with it, easily distinguished with an index; the keys of entity A can be numbered $A.pu_1$, $A.pu_2$, etc. In the following rules we will leave out this index to simplify the notation, but note that in a matching of $A.pr$ and $A.pu$, the private and public keys are assumed to have the same index.

We are aware that implementations of public-key certificates contain more attributes than those mentioned. However, as noted before, our goal is to model the information flow of a PKI, not to propose a new implementation model. We focus on the distribution of keys, and keep auxiliary attributes to a minimum. We do not model time explicitly, but view expiration as a form of revocation.

In figure 2 there are two entities, A and B. The edge represents a certificate signed by A, telling herself about the binding of B and $B.pu$. The double lines marking the box implicate that the knowledge has been obtained outside the graph system (A may have been told by B in person). The support set - below the certificate box - has only one member, the path A. O certificates are always the first edges on a path, and are always marked with the signer as the path attribute. For clarity, we have included the node representing B in this figure. In subsequent figures, the signer and subject are implied when they are not identical to the distributor or recipient.

The left side of figure 1 is similar, but here the recipient is D. This diagram represents a certificate signed by A, binding B and $B.pu$. C is quoting (distributing) the certificate to D. The path attribute $\Pi = \{\pi_1, \ldots, \pi_k\}$ represents the valid and complete support set of the certificate. Note that no verification of the signature takes place until the certificate is used to derive new information.

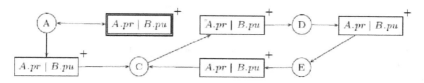

Fig. 2. Entity A has knowledge of a binding

Fig. 3. Example of a cycle

When moving from a hierarchical model with a CA that distributes every certificate, to a decentralized one where every entity can send certificates, care must be taken to avoid cycles. Consider figure 3, where no support sets are marked. From C's point of view, she has two incoming edges with A's certificate for $B.pu$, and one outgoing edge with a quotation of that certificate. However, from a global point of view, it is the edge from A that connects C to A's outside knowledge, which supports all the other certificates. If A removes the edge between himself and C, C's link to A's outside knowledge has been severed, but C is unaware of this - she still has an incoming edge from E, and this supplies support for her outgoing edge. Globally, there is a cycle involving C, D and E, but graph transformation rules act locally and are not able to capture these kinds of patterns. To handle the cycle problem, one can either not allow cycles to form or handle them at revocation time, making sure that cycles are not considered as support. We have opted for the latter approach and our solution is to include support sets below every certificate. Since the support sets hold information about all supporting paths but do not include cycles, it is possible to see when a certificate is disconnected from all its supporting paths.

The objective of this formalization is to characterize the graphs that represent states where there are no contradictions and every statement has verifiable support, possibly external. So we are looking for the following kind of graph

Definition 6 (Validity). *A C-graph is said to be* valid *if it represents a valid state (see definition 5).*

5   Model Rules

By specifying graph transformation rules, we model a system for key distribution and revocation with a C-graph. The rules are either functions that the users can call if certain conditions are met (active rules), or automatically applied deductions (deductive rules) that derive new knowledge or remove unsupported knowledge after an active rule has been applied. Note that the rules do not describe the protocol for exchanging information. For example, in rule 4, we do not know (or care) why D chooses to quote the certificate to E.

We will present the rules in four groups: first the rules that add knowledge to the graph, then three different kinds of revocation.

5.1   Adding, Quoting and Deducing Knowledge

Rules 1 and 2 are two simple administrative rules, allowing the addition and removal of entities to the system. The removal of an entity may leave edges disconnected - note that dangling edges are always automatically removed by the application of graph transformation rules.

Fig. 4. Adding an entity (rule 1); removing an entity (rule 2)

Rule 3 (called by A) lets entities add information about knowledge that they have acquired outside of the system. The rule has two negative application conditions: the top one with the shaded certificate ensures that A's public key has not been inactivated, and the bottom one keeps entities from adding duplicates.

Fig. 5. Adding knowledge (rule 3)

Quotation is described by rule 4: if D has been told by C about A's certification of $B.pu$, she can spread that information to E. Note that this rule enforces the assumption that users only can spread information that has support in another certificate, whether D has been told by another entity, or has original knowledge. The quotation of an active certificate is active, and inactive certificates have inactive quotations.

Rule 5 updates the support set of quotations when necessary. The rule adds all existing paths to Ψ, while excluding cyclic paths through the condition that there must not be a path in Ψ that is a prefix of ω, the path to be added.

Rule 6 works as follows: if Z has (active or inactive) knowledge that $X.pu$ is X's public key, and a certificate σ signed by X for $Y.pu$, Z can verify the

Fig. 6. Quotation (rule 4)

$$\exists \psi \in \Psi : \omega = \psi \phi$$

Fig. 7. Updating the support set (rule 5)

signature and will deduce knowledge about the binding of Y and $Y.pu$. The support set of the new certificate is derived from the received certificate. When new certificates are given to Z on the same form as σ, rule 7 will add that support to the support set of Z's knowledge, similarly to rule 5. In both rule 6 and 7, σ and the deduced knowledge must have the same color.

5.2   Revocation Due to Expiration

In a PKI, keys typically become inactive after a specified lifetime. To capture the mechanics of revocation due to expiration, we use inactivation.

Even when a key has expired it is necessary to store it, to be able to recover information that was encrypted while it was still valid. However, no new encryption with a given key should be allowed after its expiration. We address this issue by keeping inactivated certificates in the graph, but not allowing them to be used for new encryption (as ensured by the upper NAC of rule 3).

The graph transformation above applies to a specific public key. We use a rule expression to apply the inactivation to all occurrences of $A.pu$:

Rule 8: `asLongAsPossible` *inactivate* `end`
When calling rule 8, an instance of $A.pu$ has to be specified. The rule will then apply the inactivation to all certificates where the binding of A and $A.pu$ is made, i.e. to all possible instantiations of X, Y and Z.

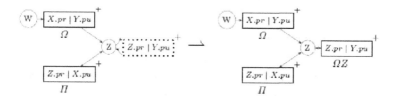

Fig. 8. Deducing new knowledge (rule 6)

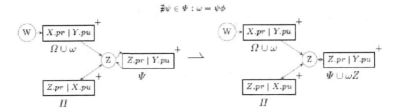

Fig. 9. Updating the support set (rule 7)

Fig. 10. Inactivation

5.3   Revocation Due to Uncertainty

We will now describe how removal of a positive assertion will affect the graph. This should be interpreted as the revoking user no longer being able to vouch for the authenticity of the certificate in question.

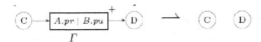

Fig. 11. Removing a certificate (rule 9)

Rule 9 lets entities remove (active or inactive) knowledge of either type (O or I). It is called by C, the distributor. Note that although an entity may remove a deduced edge, it will reappear automatically as long as the premises are true. Removing a certificate can affect the paths of quotations. Rule 10 removes path information that no longer holds. If all support is missing after its application, rule 11 detects it and removes the unsupported quotation.

The second way that revocation can propagate in the graph is by removing deduced knowledge. Rule 12 updates the support set of deduced knowledge, to keep the path information current in a similar way to rule 10. Rule 13 is the inverse of rule 6, and removes deduced knowledge that is no longer supported.

5.4   Revocation through Negative Assertions

There are many times when a revocation should be interpreted as the revoking user making a negative assertion about the authenticity of the certificate, e.g. when the certified key or the signing key has been compromised. Issuing a negative certificate is the strongest way to revoke a positive certificate; by doing so, entities can spread information on the form "$X.pu$ is *not* X's public key".

$$w \neq \epsilon$$

$$\boxed{U} \blacktriangleright \boxed{X.pr \mid Y.pu}^{+} \blacktriangleright \boxed{V} \rightarrow \boxed{X.pr \mid Y.pu}^{+} \rightarrow \boxed{W} \quad \longrightarrow \quad \boxed{V} \rightarrow \boxed{X.pr \mid Y.pu}^{+} \rightarrow \boxed{W}$$
$$\Omega \cup \omega \qquad\qquad \Psi \cup \omega V \qquad\qquad\qquad\qquad \Psi$$

Fig. 12. Updating the support set (rule 10)

$$\boxed{V} \longrightarrow \boxed{X.pr \mid Y.pu}^{+} \rightarrow \boxed{W} \quad \longrightarrow \quad \boxed{V} \qquad \boxed{W}$$
$$\emptyset$$

Fig. 13. Removal of unsupported quotations (rule 11)

Negative certificates are marked with a minus sign in the upper right corner. When a certificate collision occurs, the policy will be to give precedence to negative certificates. When a negative certificate is added where there is already a corresponding positive, the negative will be spread along any existing positive paths, and the positive certificates will be removed from the graph. Negative certificates do not expire, but may be removed and revoked if they lose support.

We will not show the simplest rules for dealing with negative certificates, as they are similar to their positive counterparts. A negative certificate can be added (rule 14), which corresponds to rule 3, but since negative certificates cannot be inactivated we do not need the upper part of the NAC of rule 3. Rules 15 and 16 handle quotation of negative certificates, just like rules 4 and 5.

Rule 17 describes how knowledge of a negative certificate is deduced. Note that the knowledge of $X.pu$ still needs to be in the form of a positive certificate, since the signature on the certificate received from W must be verified. Rule 18 updates the support set when necessary.

Rules 19 and 20 work in concert. When a negative certificate is added to a C-graph by a user A, rule 19 will propagate it along the paths of the positive certificates signed by A. Rule 20 detects the end of each such path and removes the last positive certificate on the path, thus backtracking to the beginning until no positive certificates signed by A are left.

The revocation rules 9-13 have negative counterparts: rule 21 actively removes negative certificates, rules 22 and 24 update the support set of negative quotations and deductions, resp., and rules 23 and 25 remove unsupported negative quotations and deductions, resp..

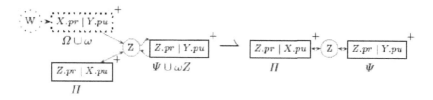

Fig. 14. Updating the support set (rule 12)

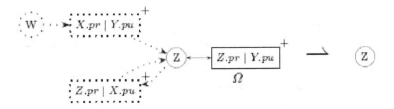

Fig. 15. Removal of deduced knowledge (rule 13)

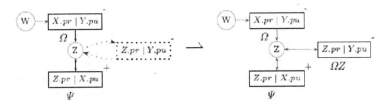

Fig. 16. Deducing negative knowledge (rule 17r)

5.5 Graph Rule Layers

When the graph is in certain states, two or more rules may affect each other's applicability (i.e. they are not parallel independent, see [18]). To avoid such situations we define the following graph rule layers:

-1. Active rules: 1, 2, 3, 4, 8, 9, 14, 15, 21
-2. Deductive adding rules: 5, 6, 7, 16
-3. Spreading negatives: 17, 18, 19, 20
-4. Deductive removal rules: 10, 11, 12, 13, 22, 23, 24, 25

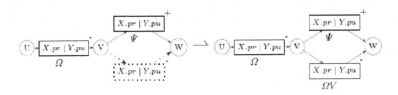

Fig. 17. Propagating negative certificates (rule 19)

Fig. 18. Backtracking (rule 20)

After a user applies a rule from layer 1, the rules of layer 2 will be applied as long as possible, then the rules from layer 3, and finally layer 4. Within each layer the rules are parallel independent (we omit the proof for lack of space).

5.6 Soundness and Completeness

We now investigate the formal properties of our graph model. In particular, we show (proofs are only sketched) that the rules characterize exactly the states where certificates have correct support and are not in contradiction (the notion of validity defined in section 3). Observe that a certificate in a C-graph generated by the rules above can only represent outside knowledge, a deduced knowledge, or a quotation, and nothing else.

Definition 7 (Stability). *A C-graph is said to be* stable *if no deductive rule from layers 2, 3 or 4 can be applied.*

In such graphs, there cannot be any contradiction or one of the rules 19 and 20 would be applicable.

Lemma 1 (Freedom from collisions). *In a stable C-graph there are no certificate collisions.*

Lemma 2 (Support sets are valid and complete). *In a stable C-graph generated by the rules in subsection 5.5, the support set of any certificate σ is valid and complete.*

Proof (Proofsketch). Consider the support set Ω of a certificate σ. If it contains a cyclic path, it must have been added (for positive certificates) by one of the rules 3, 6 or 7 and, case by case, this cannot happen by the NACs. If it does not, the proof is by induction on the length of the path, again by cases according to the only 3 kinds of certificates generated by the rules.

The following two results show that the rules, when applied according to the layers, precisely capture the notion of valid state and the intended effect of the different types of revocation.

Theorem 1 (Soundness). *Any stable C-graph generated by the rules in subsection 5.5 is valid.*

Proof (Proofsketch). If a given stable C-graph generated by the rules is not valid, at least one of the properties of definition 5 is not satisfied. If property (1) is not satisified, then there exists at least one deduced knowledge ρ that is not supported, in which case, by lemma 2, the support set of ρ must be empty. Hence, rule 11 applies, so the C-graph cannot be stable. Similarly for properties (2) and (3). A violation of property (4) is in contradisction with lemma 1 while a violation of property (5) contradicts the stability of the C-graph.

Theorem 2 (Completeness). *Any valid C-graph is stable and can be generated by the rules in subsection 5.5.*

6   Concluding Remarks

There are many different reasons why a certificate could be revoked [2, 1, 4, 3]; this is why we need different revocation schemes. Here we have described the revocation mechanisms by local operations, but it is more interesting to consider the entire chain of a revocation. Hagstrom et al. [5] describe and classify different revocation mechanisms in an access control system, investigating the consequences of a revocation for a graph as a whole. This approach can be used in our context after translating the C-graph into permissions and grants. Think of the information that is passed from one user A to another B as permissions, and the information that A has (either as knowledge or a received quotation) as a positive grant option (meaning she has the right to grant permissions to others). Revocation due to uncertainty describes revocation of the signature on the certificate, used e.g. when an entity changes affiliation. The removal of As knowledge/received quotation correspons to a weak global delete operation. The revocation propagates in the system, removing statements that are no longer supported, but only if no other statements remain that do support them.

We have presented a graphical framework for reasoning about certificates. The model allows negative certificates and resolves conflicts by giving priority to negatives. Each certificate in the system that is deduced from other certificates has support in knowledge established outside the system; cycles are prevented from being considered as support through the use of support sets. We have proved soundness and completeness of the model with respect to the definition of a valid state. Our model is global in the sense that it includes the knowledge of all entities in the system, and therefore allows the deduction of new knowledge based on other entities statements. Nevertheless, it is possible that two users have conflicting knowledge. From this perspective, the model is local. The model is decentralized, because this is the most general way to capture a PKI. Adapting the model to be hierarchical or centralized would be easy, e.g. by separating the entities into CAs and end users, and allowing only CAs to form outside knowledge. In forthcoming work we will add trust statements, not assuming that all entities are trustworthy, and define the rules to deal with them. Instead of always giving priority to negative certificates, weights can be used to decide between conflicting statements.

References

1. Cooper, D.A.: A closer look at revocation and key compromise in public key infrastructures. In: Proceedings of the 21st National Information Systems Security Conference, pp. 555–565 (1998)
2. Fox, B., LaMacchia, B.: Certificate revocation: Mechanics and meaning. In: Hirschfeld, R. (ed.) FC 1998. LNCS, vol. 1465, pp. 158–164. Springer, Heidelberg (1998)
3. Gunter, C.A., Jim, T.: Generalized certificate revocation. In: Proceedings of the 27th ACM SIGPLAN-SIGACT Symposium on Principles of Programming Languages, pp. 316–329 (2000)

4. Housley, R., Ford, W., Polk, T., Solo, D.: Internet x.509 public key infrastructure certificate and certificate revocation list (crl) profile. Technical Report RFC 3280, IETF X.509 Public Key Infrastructure Working Group, PKIX (2002)
5. Hagström, Å., Jajodia, S., Parisi-Presicce, F., Wijesekera, D.: Revocations — a classification. In: Proceedings of the 14th IEEE Computer Security Foundations Workshop, Cape Breton, Nova Scotia, Canada (2001)
6. Maurer, U.: Modelling a public-key infrastructure. In: Martella, G., Kurth, H., Montolivo, E., Bertino, E. (eds.) ESORICS 1996. LNCS, vol. 1146, pp. 325–350. Springer, Heidelberg (1996)
7. Stubblebine, S.G., Wright, R.N.: An authentication logic supporting synchronization, revocation, and recency. In: Proceedings of the 3rd ACM Conference on Computer and Communications Security, New Delhi, India, pp. 95–105 (1996)
8. Kudo, M., Mathuria, A.: An extended logic for analyzing timed-release public-key protocols. In: ISICS, pp. 183–198 (1999)
9. Li, N., Feigenbaum, J., Grosof, B.N.: A logic-based knowledge representation for authorization with delegation (extended abstract). In: Proceedings of the 12th IEEE Computer Security Foundations Workshop (1999)
10. Li, N.: Delegation Logic: A Logic-based Approach to Distributed Authorization. PhD thesis, New York University, Chapter 4: A Nonmonotonic Delegation Logic (2000)
11. Liu, C., Ozols, M., Cant, T.: An axiomatic basis for reasoning about trust in pkis. In: Varadharajan, V., Mu, Y. (eds.) ACISP 2001. LNCS, vol. 2119, pp. 274–291. Springer, Heidelberg (2001)
12. Simmons, G.J., Meadows, C.: The role of trust in information integrity protocols. Journal of Computer Security 3, 71–84 (1995)
13. Kohlas, R., Maurer, U.: Reasoning about public-key certification: On bindings between entities and public keys. In: Franklin, M.K. (ed.) FC 1999. LNCS, vol. 1648, p. 86. Springer, Heidelberg (1999)
14. Wright, R.N., Lincoln, P.D., Millen, J.K.: Efficient fault-tolerant certificate revocation. In: [19], pp. 19–24
15. Aura, T.: On the structure of delegation networks. In: Proceedings of the 11th IEEE Computer Security Foundations Workshop, Rockport, MA (1998)
16. Buldas, A., Laud, P., Lipmaa, H.: Acountable certificate management using undeniable attestations. In: [19]
17. Bottoni, P., Koch, M., Parisi-Presicce, F., Taentzer, G.: Termination of high-level replacement units with application to model transformation. ENTCS 127(4), 71–86 (2005)
18. Rozenberg, G. (ed.): Handbook of Graph Grammars and Computing by Graph Transformation. Foundations, vol. I. World Scientific, Singapore (1997)
19. Samarati, P. (ed.): Proceedings of the 7th ACM Conference on Computer and Communications Security, Athens, Greece (2000)

Local Confluence for Rules with Nested Application Conditions

Hartmut Ehrig[1], Annegret Habel[2], Leen Lambers[3],
Fernando Orejas[4], and Ulrike Golas[1]

[1] Technische Universität Berlin, Germany
{ehrig,ugolas}@cs.tu-berlin.de
[2] Carl v. Ossietzky Universität Oldenburg, Germany
habel@informatik.uni-oldenburg.de
[3] Hasso Plattner Institut, Universität Potsdam, Germany
Leen.Lambers@hpi.uni-potsdam.de
[4] Universitat Politècnica de Catalunya, Spain
orejas@lsi.upc.edu

Abstract. Local confluence is an important property in many rewriting and transformation systems. The notion of critical pairs is central for being able to verify local confluence of rewriting systems in a static way. Critical pairs are defined already in the framework of graphs and adhesive rewriting systems. These systems may hold rules with or without negative application conditions. In this paper however, we consider rules with more general application conditions — also called nested application conditions — which in the graph case are equivalent to finite first-order graph conditions. The classical critical pair notion denotes conflicting transformations in a minimal context satisfying the application conditions. This is no longer true for combinations of positive and negative application conditions — an important special case of nested ones — where we have to allow that critical pairs do not satisfy all the application conditions. This leads to a new notion of critical pairs which allows to formulate and prove a Local Confluence Theorem for the general case of rules with nested application conditions. We demonstrate this new theory on the modeling of an elevator control by a typed graph transformation system with positive and negative application conditions.

Keywords: Critical pairs, local confluence, nested application conditions.

1 Introduction

Confluence is a most important property for many kinds of rewriting systems. For instance, in the case of rewriting systems that are used as a computation model, confluence is the property that ensures the functional behaviour of a given system [1]. This is important in the case of graph transformation systems (GTSs) that are used to specify the functionality of a given software system, or to describe model transformations (see, e.g. [2]). Unfortunately, confluence of rewriting systems is undecidable. A special case is local confluence which, in

H. Ehrig et al. (Eds.): ICGT 2010, LNCS 6372, pp. 330–345, 2010.

the case of terminating rewriting systems, is equivalent to confluence. In addition, local confluence is also interesting in the sense that it shows the absence of some kinds of conflicts in the application of rules. The standard approach for proving local confluence was originally developed by Knuth and Bendix [3] for term rewriting systems (TRSs) [1]. The idea of this approach is to check the joinability (or confluence) of *critical pairs*, which represent conflicting rules in a minimal context, the minimal possible sources of non-confluence. This technique has also been applied to check local confluence for GTSs (see, e.g. [4,5,2]). However, checking local confluence for GTSs is quite more difficult than for TRSs. Actually, local confluence is undecidable for terminating GTSs, while it is decidable for terminating TRSs [5].

In standard GTSs, if we find a valid match of the left-hand side of a rule into a given graph, that rule can be applied to that graph. But we may want to limit the applicability of rules. This leads to the notion of *application conditions (AC)*, constraining the matches of a rule that are considered valid. An important case of an application condition is a *negative application condition (NAC)*, as introduced in [6], which is just a graph N that extends the left hand side of the rule. Then, that rule is applicable to a graph G if – roughly speaking – N is not present in G. However, the use of application conditions poses additional problems for constructing critical pairs. The first results on checking local confluence for GTS with ACs are quite recent and are restricted to the case of NACs [7]. Although NACs are quite useful already, they have limited expressive power: we cannot express forbidden contexts which are more complex than a graph embedding, nor can we express positive requirements on the context of applications, i.e. positive application conditions. The running example used in this paper, describing the specification of an elevator system, shows that this increase of descriptive power is needed in practical applications.

To overcome the expressive limitations of NACs, in [8] a very general kind of conditions, called *nested application conditions*, is studied in detail. In particular, in the graph case, this class of conditions has the expressive power of the first-order fragment of Courcelle's logic of graphs [9]. Following that work, in this paper we study the local confluence of transformation systems, where the rules may include arbitrary nested application conditions. However, dealing with this general kind of conditions poses new difficulties. In particular, NACs are in a sense monotonic. If a match does not satisfy a NAC then any other match embedding that one will not satisfy the NAC either [7]. This is not true for ACs in general and the reason why a different kind of critical pair is needed.

The paper is organized as follows: In Section 2, we review the problems related to checking the confluence of GTSs. In Section 3, we introduce the concepts of nested application conditions and rules. Moreover, we also introduce our running example on an elevator control, which is used along the paper to illustrate the main concepts and results. In Section 4, we define our new notion of critical pairs and state a completeness theorem (Theorem 1), saying that every pair of dependent direct derivations is an extension of a critical pair. In Section 5, we state as main result the Local Confluence Theorem (Theorem 2) for rules with ACs. A conclusion including related work is given in Section 6.

The presentation is, to some extent, informal (e.g., no proofs are included), but a technical report [10] supplies all the missing technical details. Moreover, our explanations may suggest that our results just apply to standard graph transformation systems, even if our examples use typed graphs. Actually, as can be seen in [10], our results apply to transformation systems[1] over any adhesive category [11] with an epi-$\mathcal{M}$-factorization (used in Lemma 1), a unique $\mathcal{E}$-$\mathcal{M}$ pair factorization (used in Thm 1) where $\mathcal{M}$ is the class of all monomorphisms, and with initial pushouts over $\mathcal{M}$-morphisms (used in Thm 2). The category $\mathbf{Graphs}_{TG}$ of graphs and morphisms typed over TG is adhesive and holds these properties [2]. In particular, a unique $\mathcal{E}$-$\mathcal{M}$ pair factorization is obtained when $\mathcal{E}$ is the class of pairs of jointly surjective graph morphisms and $\mathcal{M}$ is the class of injective morphisms.

2   Confluence in Graph Transformation

In this section, we review some results about the local confluence of graph transformation systems. In this sense, we assume that the reader has a reasonably good knowledge of the double-pushout approach to graph transformation. For more detail, the interested reader may consult e.g. [2].

Confluence is the property ensuring the functional behavior of transformation systems. A system is confluent if whenever a graph K can be transformed into two graphs P_1 and P_2, these graphs can be transformed into a graph K', as shown in diagram (1) below. A slightly weaker property than confluence is local confluence, represented by diagram (2) below. In this case, we just ask that P_1 and P_2 can be transformed into K' when these graphs can be obtained in exactly one transformation step from K. Confluence coincides with local confluence when the given transformation system is terminating, as shown by Newman in [12].

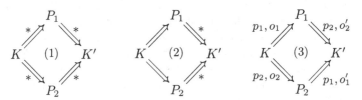

A result in the theory of Graph Transformation that ensures confluence of some derivations is the Local Church-Rosser Theorem for parallel independent transformations (see, e.g., [2]). In particular, the application of two rules p_1 and p_2 with matches o_1 and o_2 to a graph K are parallel independent, essentially, if none of these applications deletes any element which is part of the match of the

---

[1] The results can be generalized also to transformation systems over (weak) adhesive high-level replacement categories [2] like petri nets, hypergraphs, and algebraic specifications.

other rule. More precisely, if given the diagram below depicting the two double pushouts defined by these applications there exist morphisms $d_{12} \colon L_1 \to D_2$ and $d_{21} \colon L_2 \to D_1$ such that the diagram commutes. Then, the Local Church-Rosser Theorem states that in this situation we may apply these rules to P_1 and P_2 so that we obtain K' in both cases, as depicted by diagram (3) above. In general, however, not all pairs of transformations are parallel independent. It remains to analyze the parallel dependent ones.

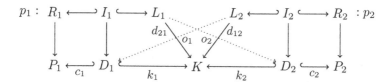

The intuition of using critical pairs to check (local) confluence is that we do not have to study all the possible cases of pairs of rule applications which are parallel dependent, but only some minimal ones which are built by gluing the left-hand side graphs of each pair of rules. A *critical pair* for the rules p_1, p_2 is a pair of parallel dependent transformations, $P_1 \Leftarrow_{p_1,o_1} K \Rightarrow_{p_2,o_2} P_2$, where o_1 and o_2 are jointly surjective. A completeness lemma, showing that every pair of parallel dependent rule applications embeds a critical pair, justifies why it is enough to consider the confluence of these cases.

As shown in [4,5], the confluence of all critical pairs is not a sufficient condition for the local confluence of a graph transformation system (as it is in the case of term rewriting). Suppose that we have two parallel dependent rule applications $G_1 \Leftarrow G \Rightarrow G_2$. By the completeness of critical pairs, there exists a critical pair $P_1 \Leftarrow K \Rightarrow P_2$ which is embedded in $G_1 \Leftarrow G \Rightarrow G_2$. Now, suppose that there are derivations from P_1 and P_2 into a common graph K', i.e. $P_1 \overset{*}{\Rightarrow} K' \overset{*}{\Leftarrow} P_2$. We could expect that these two derivations could be embedded into some derivations $G_1 \overset{*}{\Rightarrow} G' \overset{*}{\Leftarrow} G_2$ and, hence, the confluence of the critical pair would imply the confluence of $G_1 \Leftarrow G \Rightarrow G_2$. However, this is not true in general. For instance, if we try to apply in a larger context the rules in the derivations $P_1 \overset{*}{\Rightarrow} K' \overset{*}{\Leftarrow} P_2$ the gluing conditions may not hold and, as a consequence, applying these rules may be impossible. But this is not the only problem. Even if the transformations $P_1 \overset{*}{\Rightarrow} K'$ and $P_2 \overset{*}{\Rightarrow} K'$ can be embedded into transformations of $G_1 \overset{*}{\Rightarrow} H_1$ and $G_2 \overset{*}{\Rightarrow} H_2$, respectively, in general we cannot ensure that H_1 and H_2 are isomorphic. To avoid this problem, a stronger notion of confluence is needed. This notion is called *strict confluence*, but in this paper we will call it *plain strict confluence*, in the sense that it applies to transformations with *plain rules*, i.e. rules without application conditions. Essentially, a critical pair $P_1 \Leftarrow_{p_1,o_1} K \Rightarrow_{p_2,o_2} P_2$ is plain strictly confluent if there exist derivations $P_1 \overset{*}{\Rightarrow} K' \overset{*}{\Leftarrow} P_2$ such that every element in K which is preserved by the two transformations defining the critical pair is also preserved by the two derivations. In [5,2] it is shown for plain rules that plain strict confluence of all critical pairs implies the local confluence of a graph transformation system.

3   Rules with Nested Application Conditions

In this section, we introduce nested application conditions (in short, application conditions, or just ACs) and define how they can be used in graph transformation. Moreover, we introduce our running example on the modeling of an elevator control using typed graph transformation rules with application conditions.

Example 1 (*Elevator*). The type of control that we consider is meant to be used in buildings where the elevator should transport people from or to one main stop. This situation occurs, for example, in apartment buildings or multi-storey car parks. Each floor in the building is equipped with one button in order to call the elevator. The elevator car stops at a floor for which an internal stop request is given. External call requests are served by the elevator only if it is in downward mode in order to transport people to the main stop. The direction of the elevator car is not changed as long as there are remaining requests in the running direction. External call requests as well as internal stop requests are not deleted until the elevator car has arrived.

On the right of Fig. 1, a type graph *TG* for *Elevator* is depicted. This type graph expresses that an elevator car of type elevator exists, which can be on a specific floor. Moreover, the elevator can be in upward or downward mode. Floors are connected by next_up edges expressing which floor is directly above another floor. Moreover, higher_than edges express that a floor is arranged higher in the building than another floor. Each floor can hold requests of two different types. The first type is a call request expressing that an external call for the elevator car on this floor is given. The second type is a stop request expressing that a call in the elevator car is given for stopping it on this floor. On the left of Fig. 1 a graph *G*, typed over this type graph is shown, describing a four-storey building, where the elevator car is on the second floor in downward mode with a call request on the ground floor. Note that *G* contains higher_than edges from each floor which is higher than each other floor (corresponding to transitive closure of opposite edges of next_up), but they are not depicted because of visibility reasons. In Example 3, some graph transformation rules with ACs modeling the elevator

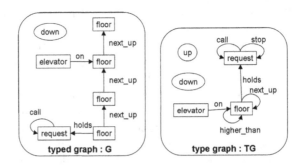

Fig. 1. A typed graph of *Elevator* together with its type graph

control are presented. We continue with the running example in Section 4 and 5, where we explain how to conclude local confluence for the parallel dependent pair of transformations with ACs as depicted in Fig. 3 by analyzing the corresponding critical pair with ACs.

Nested application conditions are defined in [8]. They generalize the corresponding notions in [6,13,14], where a negative (positive) application condition, short NAC (PAC), over a graph P, denoted $\neg\exists a$ ($\exists a$) is defined in terms of a morphism $a : P \rightarrow C$. Informally, a morphism $m : P \rightarrow G$ satisfies $\neg\exists a$ ($\exists a$) if there does not exist a morphism $q : C \rightarrow G$ extending m (if there exists q extending m). Then, an AC (also called *nested AC*) is either the special condition true or a pair of the form $\exists(a, \mathrm{ac}_C)$ or $\neg\exists(a, \mathrm{ac}_C)$, where the first case corresponds to a PAC and the second case to a NAC, and in both cases ac_C is an additional AC on C. Intuitively, a morphism $m : P \rightarrow G$ satisfies $\exists(a, \mathrm{ac}_C)$ if m satisfies a and the corresponding extension q satisfies ac_C. Moreover, ACs (and also NACs and PACs) may be combined with the usual logical connectors.

Definition 1 (application condition). An *application condition* ac_P over a graph P is inductively defined as follows: true is an application condition over P. For every morphism $a: P \rightarrow C$ and every application condition ac_C over C, $\exists(a, \mathrm{ac}_C)$ is an application condition over P. For application conditions c, c_i over P with $i \in I$ (for all index sets I), $\neg c$ and $\wedge_{i \in I} c_i$ are application conditions over P. We define inductively when a morphism *satisfies* an application condition: Every morphism satisfies true. A morphism $p: P \rightarrow G$ satisfies an application condition $\exists(a, \mathrm{ac}_C)$, denoted $p \models \exists(a, \mathrm{ac}_C)$, if there exists an injective morphism q such that $q \circ a = p$ and $q \models \mathrm{ac}_C$. A morphism $p: P \rightarrow G$ satisfies $\neg c$ if p does not satisfy c and satisfies $\wedge_{i \in I} c_i$ if it satisfies each c_i ($i \in I$).

$$\exists(\ P \xrightarrow{\ \ a\ \ } C, \blacktriangleleft\!\!\!\square\ \mathrm{ac}_C\ )$$
$$p \searrow \underset{=}{\ } \swarrow q \quad \nvDash$$
$$G$$

Note that $\exists a$ abbreviates $\exists(a, \mathrm{true})$ and $\forall(a, \mathrm{ac}_C)$ abbreviates $\neg\exists(a, \neg\mathrm{ac}_C)$.

Example 2. Examples of ACs are given below, where the first one is a standard PAC considered already in [6], while the second one is properly nested. Note that $\hookrightarrow$ denotes the inclusion.

$\exists(\underset{1}{\bigcirc}\ \underset{2}{\bigcirc} \hookrightarrow \underset{1}{\bigcirc}\!\rightarrow\!\underset{2}{\bigcirc})$	There is an edge from the image of 1 to that of 2.
$\exists(\underset{1}{\bigcirc} \hookrightarrow \underset{1}{\bigcirc}\!\rightarrow\!\underset{2}{\bigcirc},$ $\forall(\underset{1}{\bigcirc}\!\rightarrow\!\underset{2}{\bigcirc} \hookrightarrow \underset{1}{\bigcirc}\!\rightarrow\!\underset{2}{\bigcirc}\!\rightarrow\!\underset{3}{\bigcirc},$ $\exists(\underset{1}{\bigcirc}\!\rightarrow\!\underset{2}{\bigcirc}\!\rightarrow\!\underset{3}{\bigcirc} \hookrightarrow$ $\underset{1}{\bigcirc}\!\rightarrow\!\underset{2}{\bigcirc}\!\rightarrow\!\underset{3}{\circlearrowleft})))$	For the image of node 1, there exists an outgoing edge such that, for all edges outgoing from the target, the target has a loop.

ACs are used to restrict the application of graph transformation rules to a given graph. The idea is to equip the left-hand side of rules with an application

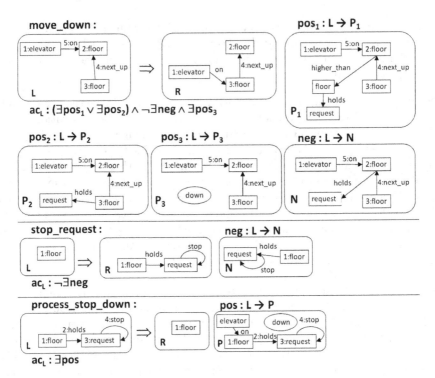

Fig. 2. Rules for *Elevator*

condition[2]. Then we can only apply a given rule to a graph G if the corresponding match morphism satisfies the AC of the rule. However, for technical reasons[3], we also introduce the application of rules *disregarding* the associated ACs.

Definition 2 (rules and transformations with ACs). A *rule* $\rho = \langle p, \mathrm{ac}_L \rangle$ consists of a *plain rule* $p = \langle L \hookleftarrow I \hookrightarrow R \rangle$ with $I \hookrightarrow L$ and $I \hookrightarrow R$ injective morphisms and an application condition ac_L over L.

$$
\begin{array}{ccccc}
\mathrm{ac}_L \blacktriangleright & L & \longleftarrow I & \longrightarrow R \\
& m \downarrow & (1) \quad \downarrow & (2) \quad \downarrow m^* \\
& G & \longleftarrow D & \longrightarrow H
\end{array}
$$

A *direct transformation* consists of two pushouts (1) and (2), called DPO, with match m and comatch m^* such that $m \models \mathrm{ac}_L$. An *AC-disregarding direct transformation* $G \Rightarrow_{\rho,m,m^*} H$ consists of DPO (1) and (2), where m does not necessarily need to satisfy ac_L.

---

[2] We could have also allowed to equip the right-hand side of rules with an additional AC, but Lemma 2 shows that this case can be reduced to rules with left ACs only.

[3] For example, a critical pair with ACs (see Def. 5) consists of transformations that do not need to satisfy the associated ACs.

Example 3 (typed graph rules of *Elevator*). We show three rules (only their left and right-hand sides[4]) modeling part of the elevator control as given in Example 1. In Fig. 2, first, we have rule move_down with combined AC on L, consisting of three PACs ($pos_i: L \to P_i, i = 1, \ldots, 3$) and a NAC (neg: $L \to N$), describing that the elevator car moves down one floor under the condition that some request is present on the next lower floor (pos_2) or some other lower floor (pos_1)[5], no request is present on the elevator floor (neg), and the elevator car is in downward mode (pos_3). As a second rule, we have stop_request, describing that an internal stop request is made on some floor under the condition that no stop request is already given for this floor. Rule process_stop_down describes that a stop request is processed for a specific floor under the condition that the elevator is on this floor and it is in downward mode.

Application conditions can be shifted over morphisms.

Lemma 1 (shift of ACs over morphisms [8,15]). There is a transformation Shift from morphisms and application conditions to application conditions such that for each application condition ac_P and for each morphisms $b: P \to P'$, Shift transforms ac_P via b into an application condition $\text{Shift}(b, ac_P)$ over P' such that for each morphism $n: P' \to H$ it holds that $n \circ b \models ac_P \Leftrightarrow n \models \text{Shift}(b, ac_P)$.

Construction. The transformation Shift is inductively defined as follows:

$P \xrightarrow{b} P'$ $\text{Shift}(b, \text{true}) = \text{true}.$

$a\downarrow \quad (1) \quad \downarrow a'$ $\text{Shift}(b, \exists(a, ac_C)) = \bigvee_{(a',b') \in \mathcal{F}} \exists(a', \text{Shift}(b', ac_C))$ if $\mathcal{F} =$

$C \xhookrightarrow{b'} C'$ $\{(a', b') \mid (a', b') \text{ jointly epimorphic, } b' \in \mathcal{M}, (1) \text{ commutes}\} \neq \emptyset$

$\triangle$ $\text{Shift}(b, \exists(a, ac_C)) = \text{false if } \mathcal{F} = \emptyset$. For Boolean formulas over

ac_C ACs, Shift is extended in the usual way.

Application conditions of a rule can be shifted from right to left and vice versa. We only describe the right to left case here, since the left to right case is symmetrical.

Lemma 2 (shift of ACs over rules [8]). There is a transformation L from rules and application conditions to application conditions such that for every application condition ac_R on R of a rule ρ, L transforms ac_R via ρ into the application condition $\text{L}(\rho, ac_R)$ on L such that we have for every direct transformation $G \Rightarrow_{\rho,m,m^*} H$ that $m \models \text{L}(\rho, ac_R) \Leftrightarrow m^* \models ac_R$.

---

[4] Thereby, the intermediate graph I and span monomorphisms can be derived as follows. Graph I consists of all nodes and edges occurring in L and R that are labeled by the same number. The span monomorphisms map nodes and edges according to the numbering. Analogously, the morphisms of the ACs consist of mappings according to the numbering of nodes and edges.

[5] Since we check that ACs are satisfied by inspecting the presence or absence of injective morphisms, we need pos_1 as well as pos_2.

Construction. The transformation L is inductively defined as follows:

$$L(\rho, \text{true}) = \text{true}$$

$L(\rho, \exists(a, \text{ac}_X)) = \exists(b, L(\rho^*, \text{ac}_X))$ if $\langle r, a \rangle$ has a pushout complement (1) and $\rho^* = \langle Y \hookleftarrow Z \hookrightarrow X \rangle$ is the derived rule by constructing the pushout (2).

$L(\rho, \exists(a, \text{ac}_X)) = \text{false}$, otherwise. For Boolean formulas over ACs, L is extended in the usual way.

4 Critical Pairs and Completeness

The use of application conditions in transformation rules complicates considerably the analysis of the confluence of a graph transformation system. For instance, two rule applications that are parallel independent if we disregard the ACs may not be confluent when considering these conditions, as we can see in Example 4 below. In this section we present the kind of critical pairs that are needed for ensuring the local confluence of a transformation system. First, we present the notion of parallel dependence for the application of rules with ACs, in order to characterize all the confluence conflicts. Then, we present a simple but weak notion of critical pair, which is shown to be complete, in the sense that every parallel dependent pair of rule applications embeds a weak critical pair. However, not every weak critical pair may be embedded in a parallel dependent pair of rule applications. To end this section, we present the adequate notion of critical pair, in the sense that they are also complete and each of them is embedded in, at least, one case of parallel dependence.

The intuition of the concept of *parallel independence*, when working with rules with ACs, is quite simple. We need not only that each rule does not delete any element which is part of the match of the other rule, but also that the resulting transformation defined by each rule application still satisfies the ACs of the other rule application. More precisely:

Definition 3 (parallel independence with ACs). A pair of direct transformations $H_1 \Leftarrow_{\rho_1, o'_1} G \Rightarrow_{\rho_2, o'_2} H_2$ with ACs is parallel independent if there exists a morphism $d'_{12} : L_1 \to D'_2$ such that $k'_2 \circ d'_{12} = o'_1$ and $c'_2 \circ d'_{12} \models \text{ac}_{L_1}$ and there exists a morphism $d'_{21} : L_2 \to D'_1$ such that $k'_1 \circ d'_{21} = o'_2$ and $c'_1 \circ d'_{21} \models \text{ac}_{L_2}$.

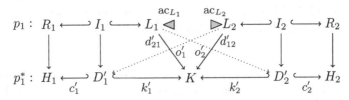

The intuition of the notion of *weak critical pair* is also quite simple. We know that all pairs of rule applications are potentially non confluent, even if they are parallel independent when disregarding the ACs. Then, we define as weak critical pairs all the minimal contexts of all pairs of AC-disregarding rule applications.

Definition 4 (weak critical pair). Given rules $\rho_1 = \langle p_1, \text{ac}_{L_1}\rangle$ and $\rho_2 = \langle p_2, \text{ac}_{L_2}\rangle$, a *weak critical pair* for $\langle\rho_1,\rho_2\rangle$ is a pair $P_1 \Leftarrow_{\rho_1,o_1} K \Rightarrow_{\rho_2,o_2} P_2$ of AC-disregarding transformations, where o_1 and o_2 are jointly surjective. Every weak critical pair induces the ACs ac_K and ac_K^* on K defined by:

$\text{ac}_K = \text{Shift}(o_1, \text{ac}_{L_1}) \wedge \text{Shift}(o_2, \text{ac}_{L_2})$, called *extension AC*, and
$\text{ac}_K^* = \neg(\text{ac}_{K,d_{12}}^* \wedge \text{ac}_{K,d_{21}}^*)$, called *conflict-inducing AC*
with $\text{ac}_{K,d_{12}}^*$ and $\text{ac}_{K,d_{21}}^*$ given as follows

$$\text{if } (\exists\, d_{12} \text{ with } k_2 \circ d_{12} = o_1) \text{ then}$$
$$\text{ac}_{K,d_{12}}^* = L(p_2^*, \text{Shift}(c_2 \circ d_{12}, \text{ac}_{L_1})) \text{ else } \text{ac}_{K,d_{12}}^* = \text{false}$$
$$\text{if } (\exists\, d_{21} \text{ with } k_1 \circ d_{21} = o_2) \text{ then}$$
$$\text{ac}_{K,d_{21}}^* = L(p_1^*, \text{Shift}(c_1 \circ d_{21}, \text{ac}_{L_2})) \text{ else } \text{ac}_{K,d_{21}}^* = \text{false}$$

where $p_1^* = \langle K \stackrel{k_1}{\hookleftarrow} D_1 \stackrel{c_1}{\hookrightarrow} P_1\rangle$ and $p_2^* = \langle K \stackrel{k_2}{\hookleftarrow} D_2 \stackrel{c_2}{\hookrightarrow} P_2\rangle$ are defined by the corresponding double pushouts.

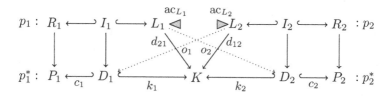

The two ACs, ac_K and ac_K^*, are used to characterize the extensions of K that may give rise to a confluence conflict. If $m : K \to G$ and $m \models \text{ac}_K$ then $m \circ o_1$ and $m \circ o_2$ are two matches of p_1 and p_2, respectively, that satisfy their associated ACs. If these two rule applications are parallel independent when disregarding the ACs then ac_K^* is precisely the condition that ensures that the two applications are parallel dependent when considering the ACs. That is, if $m \models \text{ac}_K^*$ then the two transformations $H_1 \Leftarrow_{\rho_1,m\circ o_1} G \Rightarrow_{\rho_2,m\circ o_2} H_2$ are parallel dependent.

Example 4 (weak critical pair). Consider the parallel dependent pair of direct transformations $H_1 \Leftarrow_{\rho_1,m_1} G \Rightarrow_{\rho_2,m_2} H_2$ with ACs in Fig. 3. Note that these transformations are plain parallel independent. However, they are parallel

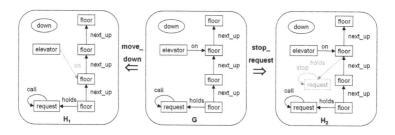

Fig. 3. Parallel dependent pair of direct transformations with ACs of *Elevator*

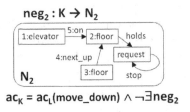

$$ac_K = ac_L(move\_down) \wedge \neg \exists neg_2$$

Fig. 4. ac_K for critical pair of *Elevator*

dependent because (see Fig. 2) rule move_down can not be applied to H_2 since rule stop_request adds a stop request to the elevator floor, which is forbidden by the AC of rule move_down. The weak critical pair $P_1 \Leftarrow K \Rightarrow P_2$ for the rules move_down and stop_request that is embedded in the above parallel dependent transformations is depicted in Figure 5 at the end of Section 5. In this case, K coincides with the left-hand side of move_down. Note that this weak critical pair consists of a pair of AC-disregarding transformations. In particular, ac_L of rule move_down is not fulfilled in K, because e.g. the PAC $\exists pos_3$ is not satisfied by $o_1 = id_L$. The extension condition ac_K of this weak critical pair is shown in Fig. 4. It is equal to the conjunction of the AC of rule move_down and $\nexists neg_2$, stemming from the NAC of rule stop_request by shifting over morphism o_2 (see Lemma 1). The conflict-inducing AC ac_K^* is a bit more tedious to compute, but it turns out to be equivalent to true for each monomorphic extension morphism, because it holds a PAC of the form $\exists id_K$. In particular, this means that any pair of transformations $H_1 \Leftarrow G \Rightarrow H_2$, embedding that critical pair, are parallel dependent since the corresponding extension would trivially satisfy ac_K^*.

As a consequence of the definition of ac_K and ac_K^*, it can be proven that an extension $m : K \to G$ satisfies ac_K and ac_K^* if and only if the transformations $H_1 \Leftarrow_{\rho_1, m \circ o_1} G \Rightarrow_{\rho_2, m \circ o_2} H_2$ are parallel dependent. This means that weak critical pairs without an extension m satisfying ac_K and ac_K^* are useless for checking local confluence. Our notion of critical pair makes use of this fact. In particular, a critical pair is a weak critical pair such that there is at least one extension satisfying ac_K and ac_K^*.

Definition 5 (critical pair). Given rules $\rho_1 = \langle p_1, ac_{L_1} \rangle$ and $\rho_2 = \langle p_2, ac_{L_2} \rangle$, a *critical pair* for $\langle \rho_1, \rho_2 \rangle$ is a weak critical pair $P_1 \Leftarrow_{\rho_1, o_1} K \Rightarrow_{\rho_2, o_2} P_2$ with induced ACs ac_K and ac_K^* on K, if there exists an injective morphism $m : K \to G$ such that $m \models ac_K \wedge ac_K^*$ and $m_i = m \circ o_i$, for $i = 1, 2$, satisfy the gluing conditions.

Note that this new critical pair notion is different from the one for rules with NACs [7]. For example, here critical pairs are AC-disregarding. One may wonder if, in the case where all the ACs are NACs, our current critical pairs coincide with the notion defined in [7]. The answer is no, although they are in some sense equivalent. Each critical pair with NACs, according to the previous notion, corresponds to a critical pair with NACs, according to the new notion. In [7] produce-forbid critical pairs, in addition to an overlap of the left-hand sides of the rules, may also contain a part of the corresponding NACs. In the new notion, this additional part would be included in ac_K^*.

As said above it can be proven that critical pairs are complete. Moreover, the converse property also holds, in the sense that every critical pair can be extended to a pair of parallel dependent rule applications (see Lemma 5 in [10]).

Theorem 1 (completeness of critical pairs with ACs [10]). For each pair of parallel dependent direct transformations $H_1 \Leftarrow_{\rho_1,m_1} G \Rightarrow_{\rho_2,m_2} H_2$ with ACs, there is a critical pair $P_1 \Leftarrow_{\rho_1,o_1} K \Rightarrow_{\rho_2,o_2} P_2$ with induced ac_K and ac_K^* and an "extension diagram" with an injective morphism m and $m \models ac_K \wedge ac_K^*$.

$$
\begin{array}{ccc}
P_1 \xleftarrow{\rho_1,o_1} & K \xrightarrow{\rho_2,o_2} & P_2 \\
\downarrow & \downarrow{m} & \downarrow \\
H_1 \xleftarrow[\rho_1,m_1]{} & G \xrightarrow[\rho_2,m_2]{} & H_2
\end{array}
$$

Moreover, for each critical pair $P_1 \Leftarrow_{\rho_1,o_1} K \Rightarrow_{\rho_2,o_2} P_2$ for (ρ_1,ρ_2) there is a parallel dependent pair $H_1 \Leftarrow_{\rho_1,m_1} G \Rightarrow_{\rho_2,m_2} H_2$ and injective morphism $m: K \to G$ such that $m \models ac_K \wedge ac_K^*$ leading to the above extension diagram.

Example 5 (critical pair). Because of Theorem 1, the weak critical pair described in Example 4 is a critical pair. In particular, the parallel dependent transformations depicted in Fig. 3 satisfy ac_K and ac_K^*.

Example 6 (critical pair with non-trivial ac_K^*). As noticed already in Example 4, the conflict-inducing condition ac_K^* is equivalent to true for each monomorphic extension morphism of the critical pair, leading to conflicting transformations for each extension of the critical pair. Now we illustrate by a toy example that, in general, ac_K^* may also be of a non-trivial kind. Consider the rule $r_1 : \bullet \leftarrow \bullet \to \bullet\bullet$, producing a node, and the rule $r_2 : \bullet \leftarrow \bullet \to \bullet\bullet$ with NAC $\nexists neg : \bullet \to \bullet\bullet\bullet$, producing a node only if two other nodes do not exist already. In this case, $ac_K = \nexists neg : \bullet \to \bullet\bullet\bullet$ and $ac_K^* = (\exists pos_1 : \bullet \to \bullet\bullet) \vee (\exists pos_2 : \bullet \to \bullet\bullet\bullet)$ with $K = \bullet$. The extension condition ac_K expresses that each extension morphism $m : K \to G$ leads to a pair of valid direct transformations via r_1 and r_2, whenever G does not contain two additional nodes to the one in K. The conflict-inducing condition ac_K^* expresses that each extension morphism $m : K \to G$ into a graph G leads to a pair of conflicting transformations via r_1 and r_2, whenever G contains one additional node to the one in K. The graph $G = \bullet\bullet$, holding one additional node to K, demonstrates that the weak critical pair $P_1 \Leftarrow_{r_1} K \Rightarrow_{r_2} P_2$ is indeed a critical pair.

5 Local Confluence

In this section, we present the main result of this paper, the Local Confluence Theorem for rules with ACs, based on our new critical pair notion. Roughly speaking, in order to show local confluence, we have to require that all critical pairs are confluent. However, as we have seen in Section 2, even in the case of plain graph transformation rules this is not sufficient to show local confluence [4,5]. Suppose that we have two one-step transformations $H_1 \Leftarrow G \Rightarrow H_2$. By completeness of critical pairs, there is a critical pair $P_1 \Leftarrow K \Rightarrow P_2$, with associated conditions ac_K and ac_K^*, which is embedded in $H_1 \Leftarrow G \Rightarrow H_2$ via an extension $m : K \rightarrow G$ satisfying ac_K and ac_K^*. If critical pairs are plain strictly confluent, we have transformations $P_1 \overset{*}{\Rightarrow} K' \overset{*}{\Leftarrow} P_2$, which can be embedded into transformations $H_1 \overset{*}{\Rightarrow} H \overset{*}{\Leftarrow} H_2$, if we disregard the ACs. However, if we consider the ACs, we may be unable to apply some rule in these derivations if the corresponding match fails to satisfy the ACs of that rule. Now, for every AC-disregarding transformation $\bar{t} : K \overset{*}{\Rightarrow} K'$ we may compute, as described below, a condition $\text{ac}(\bar{t})$ which is equivalent to all the AC's in the transformation, in the sense that for every extension $m : K \rightarrow G$ we have that m satisfies $\text{ac}(\bar{t})$ if all the match morphisms in the extended transformation $G \overset{*}{\Rightarrow} G'$ satisfy the ACs of the corresponding rule. Then, if we can prove *AC-compatibility*, i.e. that ac_K and ac_K^* imply $\text{ac}(\bar{t})$ and $\text{ac}(\bar{t'})$, where $\bar{t} : K \Rightarrow P_1 \overset{*}{\Rightarrow} K' \overset{*}{\Leftarrow} P_2 \Leftarrow K : \bar{t'}$, we would know that all the rule applications in the transformations $H_1 \overset{*}{\Rightarrow} H \overset{*}{\Leftarrow} H_2$ satisfy the corresponding ACs, since $m : K \rightarrow G$ satisfies ac_K and ac_K^*. AC-compatibility plus plain strict confluence is called *strict AC-confluence*.

Let us now describe how we can compute $\text{ac}(t)$ for a transformation $P_1 \Rightarrow_{\rho_1, m_1} P_2 \Rightarrow_{\rho_2, m_2} \cdots \Rightarrow_{\rho_n, m_n} P_n$. First, we know that using the shift construction in Lemma 1, we may translate the AC, ac_{L_i}, on the rule ρ_i into an equivalent condition $\text{Shift}(m_i, \text{ac}_{L_i})$ on P_i. On the other hand, we also know that, applying repeatedly the construction in Lemma 2, we can transform every condition $\text{Shift}(m_i, \text{ac}_{L_i})$ on P_i into an equivalent condition ac'_i on P_1. Then, $\text{ac}(t)$ is just the conjunction of all these conditions, i.e. $\text{ac}(t) = \bigwedge_i \text{ac}'_i$.

Theorem 2 (Local Confluence with ACs [10]). A transformation system with ACs is locally confluent, if all critical pairs are strictly AC-confluent, i.e. AC-compatible and plain strict confluent.

Remark 1. As proven in [8], in the case of graphs, nested application conditions are expressively equivalent to first order graph formulas. This means that the satisfiability problem for ACs is undecidable and, as a consequence, constructing the set of critical pairs for a transformation system with arbitrary ACs would be a non-computable problem. Similarly, showing logical consequence, and in particular AC-compatibility, and therefore showing strict confluence, is also undecidable. However in [16,17,18], techniques are presented to tackle the satisfiability and the deduction problems in pratice. Obviously, this kind of techniques would be important in this context. Nevertheless, it must be taken into account that, as shown in [5], checking local confluence for terminating graph transformation systems is undecidable, even in the case of rules without ACs.

Example 7 (Local Confluence with ACs). In order to apply Theorem 2 we show that the critical pair in Example 4 and 5 is strictly AC-confluent. First of all, rule process_stop_down and then rule move_down can be applied to P_2 leading to $K' = P_1$ (see Fig. 5, where $G' = H_1 \Leftarrow_{\rho_1,m_1} G \Rightarrow_{\rho_2,m_2} H_2$ is shown explicitly in Fig. 3). This is a strict solution, since it deletes no floor, nor the elevator, the only structures which are preserved by the critical pair. Moreover, we can see that this solution is AC-compatible. Let $\bar{t}_1 : K \Rightarrow P_1 = K'$ and $\bar{t}_2 : K \Rightarrow P_2 \Rightarrow P_3 \Rightarrow K'$. At first, we can conclude that $\mathrm{ac}_K \Rightarrow \mathrm{ac}(\bar{t}_2)$ (see ac_K in Fig. 4), because $\mathrm{ac}(\bar{t}_2)$ is, in particular, equivalent to ac_K. Therefore, also $\mathrm{ac}_K \wedge \mathrm{ac}_K^* \Rightarrow \mathrm{ac}(\bar{t}_1) \wedge \mathrm{ac}(\bar{t}_2)$, since $\mathrm{ac}(\bar{t}_1) = \mathrm{Shift}(o_1, \mathrm{ac}_{L_1})$, where $\mathrm{ac}_K = \mathrm{Shift}(o_1, \mathrm{ac}_{L_1}) \wedge \mathrm{Shift}(o_2, \mathrm{ac}_{L_2})$, and as mentioned in Example 4 ac_K^* is equivalent to true. By Theorem 2 we can conclude that the pair $G' = H_1 \Leftarrow_{\rho_1,m_1} G \Rightarrow_{\rho_2,m_2} H_2$ is locally confluent as shown in the outer diagram of Fig. 5. This means that the elevator in downward mode in the four-storey building with a request on the lowest floor can first process the generated stop request and then continue moving downward instead of moving downward immediately.

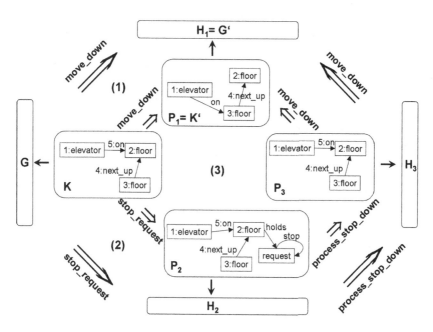

Fig. 5. strictly AC-confluent critical pair of *Elevator*

6   Conclusion, Related and Future Work

In this paper we have presented for the first time a new method for checking local confluence for graph transformation systems with ACs. This kind of ACs provide a considerable increase of expressive power with respect to the NACs that were used up to now. Moreover, as said in the introduction, all the results presented apply not only to the categories of graphs or typed graphs, but to arbitrary adhesive categories and, more generally, to (weak) adhesive high-level replacement

(HLR) categories with some additional properties. An example describing the specification of an elevator system shows that this increase of descriptive power is really necessary. The new method is not just a generalization of the critical pair method introduced in [19,7] to prove local confluence of GTS with NACs. In those papers the critical pairs are defined using graph transformations that satisfy the given NACs. This cannot be done when dealing with nested conditions, because they may include positive conditions which may not be satisfied by the critical pair but only by the embedding of these transformations into a larger context. As a consequence, a new kind of critical pairs had to be defined. As main results we have shown a Completeness and Local Confluence Theorem for the case with ACs.

Graph conditions with a similar expressive power as nested conditions were introduced in [20], while nested conditions were introduced in [8], where we can find a detailed account of many results on them, including their expressive power.

The use of critical pairs to check confluence in GTS was introduced in [4]. On the other hand, graph transformation with the important special kind of negative application conditions was introduced in [6], where it was shown how to transform right application conditions into left application conditions and graph constraints into application conditions. These results were generalized to arbitrary adhesive HLR categories [14] and a critical pair method to study the local confluence of a GTS, and in general of an adhesive rewriting system, with this kind of NACs was presented in [19,7].

Since the construction of critical pairs with NACs is implemented already in the AGG system, an extension to the case with nested application conditions is planned. However, before starting this kind of implementation, some aspects of our techniques need some additional refinement. In particular, additional work is needed on suitable implementations of satisfiability (resp. implication) solvers [16,17] for nested application conditions in order to prove AC-compatibility.

References

1. Baader, F., Nipkow, T.: Term Rewriting and All That. Cambridge University Press, Cambridge (1998)
2. Ehrig, H., Ehrig, K., Prange, U., Taentzer, G.: Fundamentals of Algebraic Graph Transformation. In: EATCS Monographs of Theoretical Computer Science. Springer, Heidelberg (2006)
3. Knuth, N.E., Bendix, P.B.: Simple word problems in universal algebra. In: Leech, J. (ed.) Computational Problems in Abstract Algebra, pp. 263–297 (1970)
4. Plump, D.: Hypergraph rewriting: Critical pairs and undecidability of confluence. In: Term Graph Rewriting: Theory and Practice, pp. 201–213. John Wiley, Chichester (1993)
5. Plump, D.: Confluence of graph transformation revisited. In: Middeldorp, A., van Oostrom, V., van Raamsdonk, F., de Vrijer, R. (eds.) Processes, Terms and Cycles: Steps on the Road to Infinity. LNCS, vol. 3838, pp. 280–308. Springer, Heidelberg (2005)
6. Habel, A., Heckel, R., Taentzer, G.: Graph grammars with negative application conditions. Fundamenta Informaticae 26, 287–313 (1996)

7. Lambers, L., Ehrig, H., Prange, U., Orejas, F.: Embedding and confluence of graph transformations with negative application conditions. In: Ehrig, H., Heckel, R., Rozenberg, G., Taentzer, G. (eds.) ICGT 2008. LNCS, vol. 5214, pp. 162–177. Springer, Heidelberg (2008)

8. Habel, A., Pennemann, K.H.: Correctness of high-level transformation systems relative to nested conditions. Mathematical Structures in Computer Science 19, 245–296 (2009)

9. Courcelle, B.: The expression of graph properties and graph transformations in monadic second-order logic. In: Handbook of Graph Grammars and Computing by Graph Transformation, vol. 1, pp. 313–400. World Scientific, Singapore (1997)

10. Lambers, L., Ehrig, H., Habel, A., Orejas, F., Golas, U.: Local Confluence for Rules with Nested Application Conditions based on a New Critical Pair Notion. Technical Report 2010-7, Technische Universität Berlin (2010)

11. Lack, S., Sobociński, P.: Adhesive categories. In: Walukiewicz, I. (ed.) FOSSACS 2004. LNCS, vol. 2987, pp. 273–288. Springer, Heidelberg (2004)

12. Newman, M.H.A.: On theories with a combinatorial definition of "equivalence". Annals of Mathematics (43,2), 223–243 (1942)

13. Koch, M., Mancini, L.V., Parisi-Presicce, F.: Graph-based specification of access control policies. Journal of Computer and System Sciences 71, 1–33 (2005)

14. Ehrig, H., Ehrig, K., Habel, A., Pennemann, K.H.: Theory of constraints and application conditions: From graphs to high-level structures. Fundamenta Informaticae 74(1), 135–166 (2006)

15. Ehrig, H., Habel, A., Lambers, L.: Parallelism and concurrency theorems for rules with nested application conditions. In: Essays Dedicated to Hans-Jörg Kreowski on the Occasion of His 60th Birthday. Electronic Communications of the EASST, vol. 26 (2010)

16. Orejas, F., Ehrig, H., Prange, U.: A logic of graph constraints. In: Fiadeiro, J.L., Inverardi, P. (eds.) FASE 2008. LNCS, vol. 4961, pp. 179–198. Springer, Heidelberg (2008)

17. Pennemann, K.H.: Resolution-like theorem proving for high-level conditions. In: Ehrig, H., Heckel, R., Rozenberg, G., Taentzer, G. (eds.) ICGT 2008. LNCS, vol. 5214, pp. 289–304. Springer, Heidelberg (2008)

18. Pennemann, K.H.: An algorithm for approximating the satisfiability problem of high-level conditions. In: Proc. Int. Workshop on Graph Transformation for Verification and Concurrency (GT-VC 2007). ENTCS, vol. 213, pp. 75–94 (2008), http://formale-sprachen.informatik.uni-oldenburg.de/pub/index.html

19. Lambers, L., Ehrig, H., Orejas, F.: Conflict detection for graph transformation with negative application conditions. In: Corradini, A., Ehrig, H., Montanari, U., Ribeiro, L., Rozenberg, G. (eds.) ICGT 2006. LNCS, vol. 4178, pp. 61–76. Springer, Heidelberg (2006)

20. Rensink, A.: Representing first-order logic by graphs. In: Ehrig, H., Engels, G., Parisi-Presicce, F., Rozenberg, G. (eds.) ICGT 2004. LNCS, vol. 3256, pp. 319–335. Springer, Heidelberg (2004)

Multi-Amalgamation
in Adhesive Categories

Ulrike Golas[1], Hartmut Ehrig[1], and Annegret Habel[2]

[1] Technische Universität Berlin, Germany
{ugolas,ehrig}@cs.tu-berlin.de
[2] Universität Oldenburg, Germany
annegret.habel@informatik.uni-oldenburg.de

Abstract. Amalgamation is a well-known concept for graph transformations in order to model synchronized parallelism of rules with shared subrules and corresponding transformations. This concept is especially important for an adequate formalization of the operational semantics of statecharts and other visual modeling languages, where typed attributed graphs are used for multiple rules with general application conditions. However, the theory of amalgamation for the double pushout approach has been developed up to now only on a set-theoretical basis for pairs of standard graph rules without any application conditions.

For this reason, we present the theory of amalgamation in this paper in the framework of adhesive categories for a bundle of rules with (nested) application conditions. In fact, it is also valid for weak adhesive HLR categories. The main result is the Multi-Amalgamation Theorem, which generalizes the well-known Parallelism and Amalgamation Theorems to the case of multiple synchronized parallelism.

The constructions are illustrated by a small running example. A more complex case study for the operational semantics of statecharts based on multi-amalgamation is presented in a separate paper.

1 Introduction

1.1 Historical Background of Amalgamation

The concepts of adhesive [1] and (weak) adhesive high-level replacement (HLR) [2] categories have been a break-through for the double pushout approach of algebraic graph transformations [3]. Almost all main results could be formulated and proven in these categorical frameworks and instantiated to a large variety of HLR systems, including different kinds of graph and Petri net transformation systems [2]. These main results include the Local Church–Rosser, Parallelism, and Concurrency Theorems, the Embedding and Extension Theorem, completeness of critical pairs, and the Local Confluence Theorem.

However, at least one main result is missing up to now. The Amalgamation Theorem in [4] has been developed only on a set-theoretical basis for a pair of standard graph rules without application conditions. In [4], the Parallelism Theorem of [5] is generalized to the Amalgamation Theorem, where the assumption

H. Ehrig et al. (Eds.): ICGT 2010, LNCS 6372, pp. 346–361, 2010.

of parallel independence is dropped and pure parallelism is generalized to synchronized parallelism. The synchronization of two rules p_1 and p_2 is expressed by a common subrule p_0, which we call *kernel rule* in this paper. The subrule concept is formalized by a rule morphism $s_i : p_0 \to p_i$, called *kernel morphism* in this paper, based on pullbacks and a pushout complement property. p_1 and p_2 can be glued along p_0 leading to an amalgamated rule $\tilde{p} = p_1 +_{p_0} p_2$. The Amalgamation Theorem states that each amalgamable pair of direct transformations $G \xrightarrow{p_i, m_i} G_i (i = 1, 2)$ via p_1 and p_2 leads to an amalgamated transformation $G \xrightarrow{\tilde{p}, \tilde{m}} H$ via $\tilde{p}$, and vice versa yielding a bijective correspondence.

Moreover, the Complement Rule Theorem in [4] allows to construct a complement rule $\overline{p}$ of a kernel morphism $s : p_0 \to p$ leading to a concurrent rule $p_0 *_E \overline{p}$ which is equal to p. Now the Concurrency Theorem allows to decompose each amalgamated transformation $G \xrightarrow{\tilde{p}, \tilde{m}} H$ into sequences $G \xrightarrow{p_i} G_i \xrightarrow{q_i} H$ for $i = 1, 2$ and vice versa, where q_i is the complement rule of $t_i : p_i \to \tilde{p}$.

The concepts of amalgamation are applied to communication based systems and visual languages in [4, 6, 7, 8] and transferred to the single-pushout approach of graph transformation in [9]. Other approaches dealing with the problem of similar parallel actions use a collection operator [10] or multi-objects for cloning the complete matches [11]. In [12], an approach based on nested graph predicates combined with amalgamation is introduced which define a relationship between rules and matches. While nesting extends the expressiveness of these transformations, it is quite complicated to write and understand these predicates and, as for the other approaches, it seems to be difficult to relate or integrate them to the theoretical results for graph transformation.

1.2   The Aim of This Paper

The concept of amalgamation plays a key role in the application of parallel graph transformation to communication-based systems [7] and in the modeling of the operational semantics for visual languages [8]. However, in most of these applications we need amalgamation for n rules, called multi-amalgamation, based not only on standard graph rules, but on different kinds of typed and attributed graph rules including (nested) application conditions.

The main idea of this paper is to fill this gap between theory and applications. For this purpose, we have developed the theory of multi-amalgamation for adhesive and adhesive HLR systems based on rules with application conditions. This allows to instantiate the theory to a large variety of graphs and corresponding graph transformation systems and, using weak adhesive HLR categories, also to typed attributed graph transformation systems [2]. A complex case study for the operational semantics of statecharts based on typed attributed graphs and multi-amalgamation is presented in [13]. For simplicity and due to space limitations, we present the theory in this paper for adhesive categories, while weak adhesive HLR categories are considered in [14].

1.3   Review of Basic Notions

The basic idea of adhesive categories [1] is to have a category with pushouts along monomorphisms and pullbacks satisfying the van Kampen property. Intuitively, this means that pushouts along monomorphisms and pullbacks are compatible with each other. This holds for sets and various kinds of graphs (see [1, 2]), including the standard category of graphs which is used as a running example in this paper.

In the double pushout approach to graph transformation, a rule is given by a span $p = (L \xleftarrow{l}$
$K \xrightarrow{r} R)$ with objects L, K, and R, called left-
hand side, interface, and right-hand side, respec-

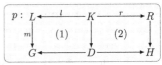

tively, and monomorphisms l and r. An application of such a rule to a graph G via a match $m : L \to G$ is constructed as two gluings (1) and (2), which are pushouts in the corresponding graph category, leading to a direct transformation $G \xRightarrow{p,m} H$.

An important extension is the use of rules with suitable application conditions. These include positive application conditions of the form $\exists a$ for a morphism $a : L \to C$, demanding a certain structure in addition to L, and also negative application conditions $\neg \exists a$, forbidding such a structure. A match $m : L \to G$ satisfies $\exists a$ ($\neg \exists a$) if there is a (no) monomorphism $q : C \to G$ satisfying $q \circ a = m$. In more detail, we use nested application conditions [15], short application conditions. In particular, true is an application condition which is always satisfied. For a basic application condition $\exists(a, \mathrm{ac}_C)$ on L with an application condition ac_C on C, in addition to the existence
of q it is required that q satisfies ac_C. In particular, we

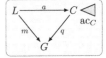

have $\exists a = \exists(a, \mathrm{true})$ for $\mathrm{ac}_C = \mathrm{true}$. In general, we write
$m \models \exists(a, \mathrm{ac}_C)$ if m satisfies $\exists(a, \mathrm{ac}_C)$, and application conditions are closed under boolean operations. Moreover, $\mathrm{ac}_C \cong \mathrm{ac}'_C$ denotes the semantical equivalence of ac_C and ac'_C on C.

In this paper we consider rules of the form $p = (L \xleftarrow{l} K \xrightarrow{r} R, \mathrm{ac})$, where $(L \xleftarrow{l} K \xrightarrow{r} R)$ is a (plain) rule and ac is an application condition on L. In order to handle rules with application conditions there are two important concepts, called the shift of application conditions over morphisms and rules ([15, 16]):

1. Given an application condition ac_L on L and a morphism $t : L \to L'$ then there is an application condition $\mathrm{Shift}(t, \mathrm{ac}_L)$ on L' such that for all $m' : L' \to G$ holds: $m' \models \mathrm{Shift}(t, \mathrm{ac}_L) \iff m = m' \circ t \models \mathrm{ac}_L$. For a basic application condition $\mathrm{ac} = \exists(a, \mathrm{ac}')$ we define $\mathrm{Shift}(t, \mathrm{ac}) = \bigvee_{(a',t') \in \mathcal{F}} \exists(a', \mathrm{Shift}(t', \mathrm{ac}'))$ with $\mathcal{F} = \{(a', t') \mid (a', t') \text{jointly epimorphic}, t' \in \mathcal{M}, t' \circ a = a' \circ t\}$,

2. Given a plain rule $p = (L \xleftarrow{l} K \xrightarrow{r} R)$ and an application condition ac_R on R then there is an application condition $\mathrm{L}(p, \mathrm{ac}_R)$ on L such that for all transformations $G \xRightarrow{p,m,n} H$ with match m and comatch n holds: $m \models \mathrm{L}(p, \mathrm{ac}_R) \iff n \models \mathrm{ac}_R$. For a basic application condition $\mathrm{ac}_R = \exists(a, \mathrm{ac}'_R)$ we define $\mathrm{L}(p, \mathrm{ac}_R) = \exists(b, \mathrm{L}(p^*, \mathrm{ac}'_R))$ if $a \circ r$ has a pushout complement (1)

and $p^* = (Y \xleftarrow{l^*} Z \xrightarrow{r^*} X)$ is the derived rule by constructing pushout (2), and $L(p, \exists(a, ac'_R)) = \text{false}$ otherwise. Vice versa, there is also a construction $R(p, ac_L)$ shifting an application condition ac_L on L over p to R.

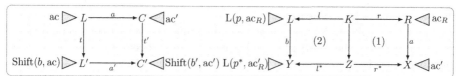

One of the main results for graph transformation is the Concurrency Theorem, which is concerned with the execution of transformations which may be sequentially dependent. Given an arbitrary sequence $G \xrightarrow{p_1, m_1} H \xrightarrow{p_2, m_2} G'$ of di-

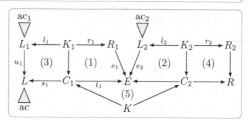

rect transformations it is possible to construct an E-concurrent rule $p_1 *_E p_2 = (L \leftarrow K \rightarrow R, ac)$ by pushouts (1), (2), (3), (4), and pullback (5). The object E is an overlap of R_1 and L_2, where the two overlapping morphisms have to be in a class $\mathcal{E}'$ of pairs of morphisms with the same codomain. The construction of the concurrent application condition $ac = \text{Shift}(u_1, ac_1) \wedge L(p^*, \text{Shift}(e_2, ac_2))$ and $p^* = (L \xleftarrow{s_1} C_1 \xrightarrow{t_1} E)$ is again based on the two shift constructions. The Concurrency Theorem states that for the transformation $G \xrightarrow{p_1, m_1} H \xrightarrow{p_2, m_2} G'$ the E-concurrent rule $p_1 *_E p_2$ allows us to construct a direct transformation $G \xrightarrow{p_1 *_E p_2} G'$ via $p_1 *_E p_2$, and vice versa, each direct transformation $p_1 *_E p_2$ can be sequentialized.

1.4 General Assumptions

In this paper we assume to have an adhesive category [1] with binary coproducts, epi-mono-factorization, and initial pushouts [2]. We consider rules with (nested) application conditions [15] as explained above. In the following, a *bundle* represents a family of morphisms or transformation steps with the same domain, which means that a bundle of things always starts at the same object. Note that the theory is also valid for weak adhesive HLR categories with a suitable class $\mathcal{M}$ of monomorphisms [2, 14].

1.5 Organization of This Paper

This paper is organized as follows: in Section 2, we introduce kernel rules, multi rules, and kernel morphisms leading to the Complement Rule Theorem as first main result. In Section 3, we construct multi-amalgamated rules and transformations and show as second main result the Multi-Amalgamation Theorem. In Section 4, we present a summary of our results and discuss future work. In the main part of the paper, we mainly give proof ideas, the full proofs can be found in [14].

2   Decomposition of Direct Transformations

In this section, we show how to decompose a direct transformation in adhesive categories into transformations via a kernel and a complement rule leading to the Complement Rule Theorem.

A kernel morphism describes how a smaller rule, the kernel rule, is embedded into a larger rule, the multi rule, which has its name because it can be applied multiple times for a given kernel rule match as described in Section 3. We need some more technical preconditions to make sure that the embeddings of the L-, K-, and R-components as well as the application conditions are consistent and allow to construct a complement rule.

Definition 1 (Kernel morphism). *Given rules* $p_0 = (L_0 \xleftarrow{l_0} K_0 \xrightarrow{r_0} R_0, \mathrm{ac}_0)$ *and* $p_1 = (L_1 \xleftarrow{l_1} K_1 \xrightarrow{r_1} R_1, \mathrm{ac}_1)$, *a kernel morphism* $s : p_0 \to p_1$, $s = (s_L, s_K, s_R)$ *consists of monomorphisms* $s_L : L_0 \to L_1$, $s_K : K_0 \to K_1$, *and* $s_R : R_0 \to R_1$ *such that in the following diagram* (1) *and* (2) *are pullbacks,* (1) *has a pushout complement* $(1')$ *for* $s_L \circ l_0$, *and* ac_0 *and* ac_1 *are* complement-compatible *w.r.t.* s, *i.e. given pushout* (3) *then* $\mathrm{ac}_1 \cong \mathrm{Shift}(s_L, \mathrm{ac}_0) \wedge \mathrm{L}(p_1^*, \mathrm{Shift}(v, \mathrm{ac}_1'))$ *for some* ac_1' *on* L_{10} *and* $p_1^* = (L_1 \xleftarrow{u} L_{10} \xrightarrow{v} E_1)$. *In this case,* p_0 *is called* kernel rule *and* p_1 multi rule.

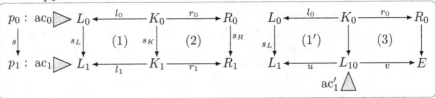

Remark 1. The complement-compatibility of the application conditions makes sure that there is a decomposition of ac_1 into parts on L_0 and L_{10}, where the latter ones are used later for the application conditions of the complement rule. L_{10} represents in addition to K_0 all these elements that are new in the multi rule and thus have to be present in the complement rule. This ensures that the complement rule is applicable after the kernel rule if and only if the multi rule can be applied. Otherwise, ac_1 uses combined elements from the kernel and complement rules such that the multi rule cannot be decomposed.

Example 1. To explain the concept of amalgamation, in our example we model a small transformation system for switching the direction of edges in labeled graphs, where we only have different labels for edges – black and dotted edges. The kernel rule p_0 is depicted in the top of Fig. 1. It selects a node with a black loop, deletes this loop, and adds a dotted loop, all of this if no dotted loop is already present. The matches are defined by the numbers at the nodes and can be induced for the edges by their position.

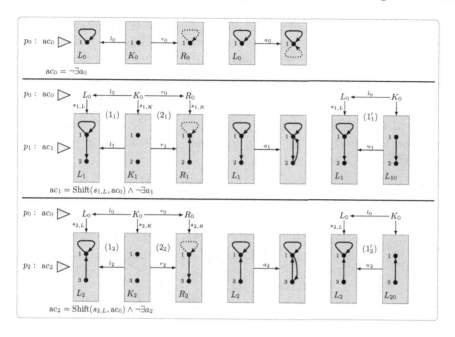

Fig. 1. The kernel rule p_0 and the multi rules p_1 and p_2

In the middle and bottom of Fig. 1, two multi rules p_1 and p_2 are shown, which extend the rule p_0 and in addition reverse an edge if no backward edge is present. They also inherit the application condition of p_0 forbidding a dotted loop at the selected node.

There is a kernel morphism $s_1 : p_0 \rightarrow p_1$ as shown in the top of Fig. 1 with pullbacks (1_1) and (2_1), and pushout complement $(1'_1)$. For the application conditions, $\text{ac}_1 = \text{Shift}(s_{1,L}, \text{ac}_0) \wedge \neg\exists a_1 \cong \text{Shift}(s_{1,L}, \text{ac}_0) \wedge \text{L}(p_1^*, \text{Shift}(v_1, \neg\exists a'_1))$ with a'_1 as shown in Fig. 2. We have that $\text{Shift}(v_1, \neg\exists a'_1) = \neg\exists a_{11}$, because square $(*)$ is the only possible commuting square leading to a_{11}, b_{11} jointly surjective and b_{11} injective. More-

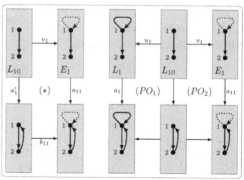

Fig. 2. Application condition construction

over, $\text{L}(p_1^*, \neg\exists a_{11}) = \neg\exists a_1$ as shown by the two pushout squares (PO_1) and (PO_2) in Fig. 2. Thus $\text{ac}'_1 = \neg\exists a'_1$, and ac_0 and ac_1 are complement compatible.

Similarly, there is a kernel morphism $s_2 : p_0 \rightarrow p_2$ as shown in the bottom of Fig. 1 with pullbacks (1_2) and (2_2), pushout complement $(1'_2)$, and ac_0 and ac_2 are complement compatible.

For a given kernel morphism, the complement rule is the remainder of the multi rule after the application of the kernel rule, i.e. it describes what the multi rule does in addition to the kernel rule.

Theorem 1 (Existence of complement rule). *Given rules $p_0 = (L_0 \xleftarrow{l_0} K_0 \xrightarrow{r_0} R_0, \mathrm{ac}_0)$ and $p_1 = (L_1 \xleftarrow{l_1} K_1 \xrightarrow{r_1} R_1, \mathrm{ac}_1)$, and a kernel morphism $s : p_0 \to p_1$ then there exists a rule $\overline{p_1} = (\overline{L_1} \xleftarrow{\overline{l_1}} \overline{K_1} \xrightarrow{\overline{r_1}} \overline{R_1}, \overline{\mathrm{ac}_1})$ and a jointly epimorphic cospan $R_0 \xrightarrow{e_1} E \xleftarrow{e_2} \overline{L_1}$ such that the E-concurrent rule $p_0 *_E \overline{p_1}$ exists and $p_1 = p_0 *_E \overline{p_1}$.*

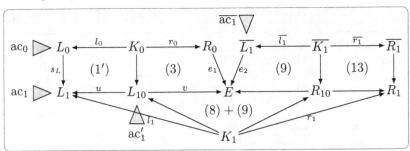

Proof idea. We consider the construction without application conditions. First, S is constructed as follows: we construct the initial pushout (4) over s_R, P as the pullback object of r_0 and b, and then the pushout (5). After the construction of different pushouts and pullbacks and using various properties of adhesive categories, we obtain

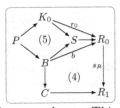

the diagram below where all squares (6) to (11) and (13) are pushouts. This leads to the required rule $\overline{p_1} = (\overline{L_1} \xleftarrow{\overline{l_1}} \overline{K_1} \xrightarrow{\overline{r_1}} \overline{R_1})$ and $p_1 = p_0 *_E \overline{p_1}$ for rules without application conditions.

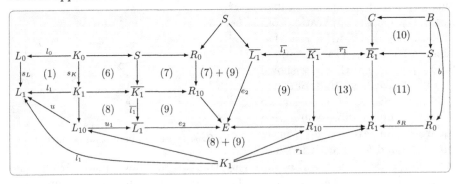

For the application conditions, suppose $\mathrm{ac}_1 \cong \mathrm{Shift}(s_L, \mathrm{ac}_0) \wedge L(p_1^*, \mathrm{Shift}(v, \mathrm{ac}_1'))$ for $p_1^* = (L_1 \xleftarrow{u} L_{10} \xrightarrow{v} E)$ with $v = e_2 \circ u_1$ and ac_1' on L_{10}. Now define $\overline{\mathrm{ac}_1} = \mathrm{Shift}(u_1, \mathrm{ac}_1')$, which is an application condition on $\overline{L_1}$.

We have to show that $(p_1, \mathrm{ac}_{p_0 * _E \overline{p_1}}) \cong (p_1, \mathrm{ac}_1)$. By construction of the E-concurrent rule we have that $\mathrm{ac}_{p_0 *_E \overline{p_1}} \cong \mathrm{Shift}(s_L, \mathrm{ac}_0) \wedge L(p_1^*, \mathrm{Shift}(e_2, \overline{\mathrm{ac}_1})) \cong$

$\text{Shift}(s_L, ac_0) \wedge L(p_1^*, \text{Shift}(e_2, \text{Shift}(u_1, ac_1'))) \cong \text{Shift}(s_L, ac_0) \wedge L(p_1^*, \text{Shift}(e_2 \circ u_1, ac_1')) \cong \text{Shift}(s_L, ac_0) \wedge L(p_1^*, \text{Shift}(v, ac_1')) \cong ac_1.$ □

Remark 2. Note, that by construction the interface K_0 of the kernel rule has to be preserved in the complement rule. The construction of $\overline{p_1}$ is not unique w.r.t. the property $p_1 = p_0 *_E \overline{p_1}$, since other choices for S with monomorphisms from K_0 and B also lead to a well-defined construction. In particular, one could choose $S = R_0$ leading to $\overline{p_1} = E \leftarrow R_{10} \rightarrow R_1$. Our choice represents a smallest possible complement, which should be preferred in most application areas.

Definition 2 (Complement rule). *Given rules* $p_0 = (L_0 \xleftarrow{l_0} K_0 \xrightarrow{r_0} R_0, ac_0)$ *and* $p_1 = (L_1 \xleftarrow{l_1} K_1 \xrightarrow{r_1} R_1, ac_1)$, *and a kernel morphism* $s : p_0 \rightarrow p_1$ *then the rule* $\overline{p_1} = (\overline{L_1} \xleftarrow{\overline{l_1}} \overline{K_1} \xrightarrow{\overline{r_1}} \overline{R_1}, \overline{ac_1})$ *constructed in Thm. 1 is called* complement rule *(of s).*

Example 2. Consider the kernel morphism s_1 depicted in Fig. 1. Using the construction in Thm. 1 we obtain the diagrams in Fig. 3 leading to the complement rule in the top row in Fig. 4 with the application condition $\overline{ac_1} = \neg \exists \overline{a_1}$. Similarly, we obtain a complement rule for the kernel morphism $s_2 : p_0 \rightarrow p_2$ in Fig. 1, which is depicted in the bottom row of Fig. 4.

Each direct transformation via a multi rule can be decomposed into a direct transformation via the kernel rule followed by a direct transformation via the complement rule.

Fact 1. *Given rules* $p_0 = (L_0 \xleftarrow{l_0} K_0 \xrightarrow{r_0} R_0, ac_0)$ *and* $p_1 = (L_1 \xleftarrow{l_1} K_1 \xrightarrow{r_1} R_1, ac_1)$, *a kernel morphism* $s : p_0 \rightarrow p_1$, *and a direct transformation* $t : G \xrightarrow{p_1, m} G_1$ *then* t *can be decomposed into the transformation* $G \xrightarrow{p_0, m_0} G_0 \xrightarrow{\overline{p_1}, \overline{m}} G_1$ *with* $m_0 = m \circ s_L$ *where* $\overline{p_1}$ *is the complement rule of s.*

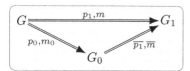

Proof. We have that $p_1 \cong p_0 *_E \overline{p_1}$. The analysis part of the Concurrency Theorem [16] now implies the decomposition into $G \xrightarrow{p_0, m_0} G_0 \xrightarrow{\overline{p_1}, \overline{m}} G_1$ with $m_0 = m \circ s_L$.   □

3 Multi-Amalgamation

In [4], an Amalgamation Theorem for a pair of graph rules without application conditions has been developed. It can be seen as a generalization of the Parallelism Theorem [5], where the assumption of parallel independence is dropped and pure parallelism is generalized to synchronized parallelism. In this section, we present an Amalgamation Theorem for a bundle of rules with application conditions, called Multi-Amalgamation Theorem, over objects in an adhesive category.

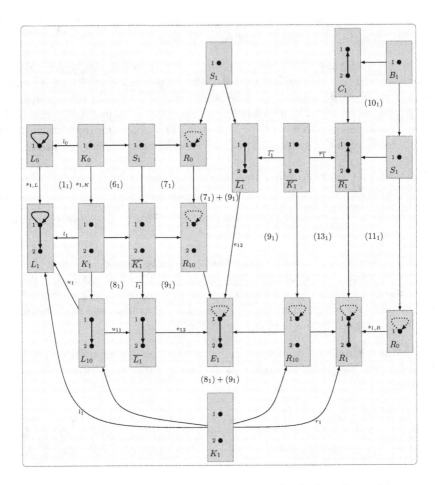

Fig. 3. The construction of the complement rule for the kernel morphism s_1

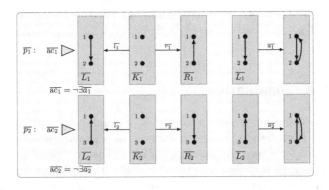

Fig. 4. The complement rules for the kernel morphisms

We consider not only single kernel morphisms, but bundles of them over a fixed kernel rule. Then we can combine the multi rules of such a bundle to an amalgamated rule by gluing them along the common kernel rule.

Definition 3 (Multi-amalgamated rule). *Given rules* $p_i = (L_i \xleftarrow{l_i} K_i \xrightarrow{r_i} R_i, ac_i)$ *for* $i = 0, \ldots, n$ *and a bundle of kernel morphisms* $s = (s_i : p_0 \to p_i)_{i=1,\ldots,n}$, *then the* (multi-)*amalgamated rule* $\tilde{p}_s = (\tilde{L}_s \xleftarrow{\tilde{l}_s} \tilde{K}_s \xrightarrow{\tilde{r}_s} \tilde{R}_s, \tilde{ac}_s)$ *is constructed as the componentwise colimit of the kernel morphisms:*

- $\tilde{L}_s = Col((s_{i,L})_{i=1,\ldots,n})$,
- $\tilde{K}_s = Col((s_{i,K})_{i=1,\ldots,n})$,
- $\tilde{R}_s = Col((s_{i,R})_{i=1,\ldots,n})$,
- $\tilde{l}_s$ *and* $\tilde{r}_s$ *are induced by* $(t_{i,L} \circ l_i)_{i=0,\ldots,n}$ *and* $(t_{i,R} \circ r_i)_{0=1,\ldots,n}$, *respectively,*
- $\tilde{ac}_s = \bigwedge_{i=1,\ldots,n} \text{Shift}(t_{i,L}, ac_i)$.

$$
\begin{array}{ccccccc}
p_0: & ac_0 \rhd & L_0 & \xleftarrow{l_0} & K_0 & \xrightarrow{r_0} & R_0 \\
& s_i \downarrow & s_{i,L} \downarrow & (1_i) & s_{i,K} \downarrow & (2_i) & s_{i,R} \downarrow \\
p_i: & ac_i \rhd & L_i & \xleftarrow{l_i} & K_i & \xrightarrow{r_i} & R_i \\
& t_i \downarrow & t_{i,L} \downarrow & (14_i) & t_{i,K} \downarrow & (15_i) & t_{i,R} \downarrow \\
\tilde{p}_s: & \tilde{ac}_s \rhd & \tilde{L}_s & \xleftarrow{\tilde{l}_s} & \tilde{K}_s & \xrightarrow{\tilde{r}_s} & \tilde{R}_s
\end{array}
$$

Fact 2. *The amalgamated rule is well-defined and we have kernel morphisms* $t_i = (t_{i,L}, t_{i,K}, t_{i,R}) : p_i \to \tilde{p}_s$ *for* $i = 0, \ldots, n$.

Proof idea. The colimit of a bundle of n morphisms can be constructed by iterated pushout constructions, which means that we only have to require pushouts over monomorphisms. Since pushouts are closed under monomorphisms, the iterated pushout construction leads to t be-

$$
\begin{array}{ccccccc}
p_0: & ac_0 \rhd & L_0 & \xleftarrow{l_0} & K_0 & \xrightarrow{r_0} & R_0 \\
& s_{i,L} \downarrow & (1'_i) & w_i \downarrow & (3_i) & e_{i1} \downarrow \\
p_i^*: & ac_i \rhd & L_i & \xleftarrow{u_i} & L_{i0} & \xrightarrow{v_i} & E_i \\
& t_{i,L} \downarrow & (17_i) & \tilde{l}_i \downarrow & (18_i) & \tilde{k}_i \downarrow \\
\tilde{p}_s^*: & \tilde{ac}_s \rhd & \tilde{L}_s & \xleftarrow{\tilde{u}} & \tilde{L}_0 & \xrightarrow{\tilde{v}} & \tilde{E}
\end{array}
$$

ing a monomorphism. In addition, it can be shown by induction that that (14_i) resp. $(14_i) + (1_i)$ and (15_i) resp. $(15_i) + (2_i)$ are pullbacks, and (14_i) resp. $(14_i) + (1_i)$ has a pushout complement (17_i) resp. $(17_i) + (1'_i)$ for $t_{i,L} \circ l_i$.

Since ac_0 and ac_i are complement-compatible for all i we have that $ac_i \cong \text{Shift}(s_{i,L}, ac_0) \wedge \text{L}(p_i^*, \text{Shift}(v_i, ac'_i))$. For any ac'_i, it holds that $\text{Shift}(t_{i,L}, \text{L}(p_i^*, \text{Shift}(v_i, ac'_i)))) \cong \text{L}(\tilde{p}_s^*, \text{Shift}(\tilde{k}_i \circ v_i, ac'_i)) \cong \text{L}(\tilde{p}_s^*, \text{Shift}(\tilde{v}, \text{Shift}(\tilde{l}_i, ac'_i)))$, since all squares are pushouts by pushout-pullback decomposition and the uniqueness of pushout complements. Define $ac_i^* := \text{Shift}(\tilde{l}_i, ac'_i)$ as an application condition on $\tilde{L}_0$. It follows that $\tilde{ac}_s = \bigwedge_{i=1,\ldots,n} \text{Shift}(t_{i,L}, ac_i) \cong \bigwedge_{i=1,\ldots,n}(\text{Shift}(t_{i,L} \circ s_{i,L}, ac_0) \wedge \text{Shift}(t_{i,L}, \text{L}(p_i^*, \text{Shift}(v_i, ac'_i)))) \cong \text{Shift}(t_{0,L}, ac_0) \wedge \bigwedge_{i=1,\ldots,n} \text{L}(\tilde{p}_s^*, \text{Shift}(\tilde{v}, ac_i^*))$.

For $i = 0$ define $ac'_{s0} = \bigwedge_{j=1,\ldots,n} ac_j^*$, and hence $\tilde{ac}_s = \text{Shift}(t_{0,L}, ac_0) \wedge \text{L}(\tilde{p}_s^*, \text{Shift}(\tilde{v}, ac'_{s0}))$ implies the complement-compatibility of ac_0 and $\tilde{ac}_s$. For $i > 0$, we have that $\text{Shift}(t_{0,L}, ac_0) \wedge \text{L}(\tilde{p}_s^*, \text{Shift}(\tilde{v}, ac_i^*)) \cong \text{Shift}(t_{i,L}, ac_i)$. Define $ac'_{si} = \bigwedge_{j=1,\ldots,n\setminus i} ac_j^*$, and hence $\tilde{ac}_s = \text{Shift}(t_{i,L}, ac_i) \wedge \text{L}(\tilde{p}_s^*, \text{Shift}(\tilde{v}, ac'_{si}))$ implies the complement-compatibility of ac_i and $\tilde{ac}_s$. □

The application of an amalgamated rule yields an amalgamated transformation.

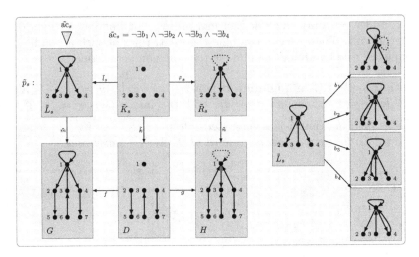

Fig. 5. An amalgamated transformation

Definition 4 (Amalgamated transformation). *The application of an amalgamated rule to a graph G is called an* amalgamated transformation.

Example 3. Consider the bundle $s = (s_1, s_2, s_3 = s_1)$ of the kernel morphisms depicted in Fig. 1. The corresponding amalgamated rule $\tilde{p}_s$ is shown in the top row of Fig. 5. This amalgamated rule can be applied to the graph G leading to the amalgamated transformation depicted in Fig. 5, where the application condition $\tilde{ac}_s$ is obviously fulfilled by the match $\tilde{m}$.

If we have a bundle of direct transformations of a graph G, where for each transformation one of the multi rules is applied, we want to analyze if the amalgamated rule is applicable to G combining all the single transformation steps. These transformations are compatible, i.e. multi-amalgamable, if the matches agree on the kernel rules, and are independent outside.

Definition 5 (Multi-amalgamable). *Given a bundle of kernel morphisms $s = (s_i : p_0 \rightarrow p_i)_{i=1,...,n}$, a bundle of direct transformations steps $(G \xrightarrow{p_i, m_i} G_i)_{i=1,...n}$ is s-multi-amalgamable, or short s-amalgamable, if*

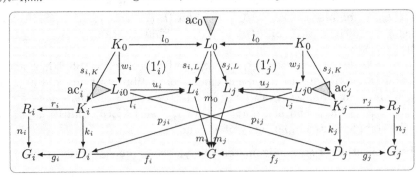

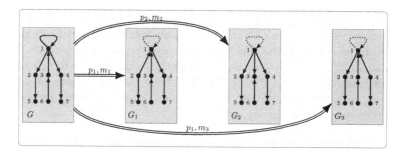

Fig. 6. An s-amalgamable bundle of direct transformations

– *it has* consistent matches, *i.e.* $m_i \circ s_{i,L} = m_j \circ s_{j,L} =: m_0$ *for all* $i, j = 1, \ldots, n$ *and*
– *it has* weakly independent matches, *i.e. for all* $i \neq j$ *consider the pushout complements* $(1'_i)$ *and* $(1'_j)$, *and then there exist morphisms* $p_{ij} : L_{i0} \to D_j$ *and* $p_{ji} : L_{j0} \to D_i$ *such that* $f_j \circ p_{ij} = m_i \circ u_i$, $f_i \circ p_{ji} = m_j \circ u_j$, $g_j \circ p_{ij} \models \mathrm{ac}'_i$, *and* $g_i \circ p_{ji} \models \mathrm{ac}'_j$.

Similar to the characterization of parallel independence in [2] we can give a set-theoretical characterization of weak independence.

Fact 3. *For graphs and other set-based structures, weakly independent matches means that* $m_i(L_i) \cap m_j(L_j) \subseteq m_0(L_0) \cup (m_i(l_i(K_i)) \cap m_j(l_j(K_j)))$ *for all* $i \neq j = 1, \ldots, n$, *i.e. the elements in the intersection of the matches* m_i *and* m_j *are either preserved by both transformations, or are also matched by* m_0.

Proof. We have to proof the equivalence of $m_i(L_i) \cap m_j(L_j) \subseteq m_0(L_0) \cup (m_i(l_i(K_i)) \cap m_j(l_j(K_j)))$ for all $i \neq j = 1, \ldots, n$

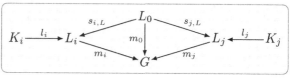

with the definition of weakly independent matches.

"$\Leftarrow$" Let $x = m_i(y_i) = m_j(y_j)$, and suppose $x \notin m_0(L_0)$. Since $(1'_i)$ is a pushout we have that $y_i = u_i(z_i) \in u_i(L_{i0} \backslash w_i(K_0))$, and $x = m_i(u_i(z_i)) = f_j(p_i(z_i)) = m_j(y_j)$, and by pushout properties $y_j \in l_j(K_j)$ and $x \in m_j(l_j(K_j))$. Similarly, $x \in m_i(l_i(K_i))$.

"$\Rightarrow$" For $x \in L_{i0}$, $x = w_i(k)$ define $p_{ij}(x) = k_j(s_{j,K}(k))$, then $f_j(p_{ij}(x)) = f_j(k_j(s_{j,K}(k))) = m_j(l_j(s_{j,K}(k))) = m_j(s_{j,L}(l_0(k))) = m_i(s_{i,L}(l_0(k))) = m_i(n_i(w_i(k))) = m_i(u_i(x))$. Otherwise, $x \notin w_i(K_0)$, i.e. $u_i(x) \notin s_{i,L}(L_0)$, and define $p_{ij}(x) = y$ with $f_j(y) = m_i(u_i(x))$. This y exists, because either $m_i(u_i(x)) \notin m_j(L_j)$ or $m_i(u_i(x)) \in m_j(L_j)$ and then $m_i(u_i(x)) \in m_j(l_j(K_j))$, and in both cases $m_i(u_i(x)) \in f_j(D_j)$. Similarly, we can define p_{ji} with the required property. □

Example 4. Consider the bundle $s = (s_1, s_2, s_3 = s_1)$ of kernel morphisms from Ex. 3. For the graph G given in Fig. 5 we find matches $m_0 : L_0 \to G$, $m_1 : L_1 \to$

G, $m_2 : L_2 \to G$, and $m_3 : L_1 \to G$ mapping all nodes from the left hand side to their corresponding nodes in G, except for m_3 mapping node 2 in L_1 to node 4 in G. For all these matches, the corresponding application conditions are fulfilled and we can apply the rules p_1, p_2, p_1, respectively, leading to the bundle of direct transformations depicted in Fig. 6. This bundle is s-amalgamable, because the matches m_1, m_2, and m_3 agree on the match m_0, and are weakly independent, because they only overlap in m_0.

For an s-amalgamable bundle of direct transformations, each single transformation step can be decomposed into an application of the kernel rule followed by an application of the complement rule. Moreover, all kernel rule applications lead to the same object, and the following applications of the complement rules are parallel independent.

Fact 4. *Given a bundle of kernel morphisms* $s = (s_i : p_0 \to p_i)_{i=1,\ldots,n}$ *and an s-amalgamable bundle of direct transformations* $(G \xrightarrow{p_i,m_i} G_i)_{i=1,\ldots,n}$ *then each direct transformation* $G \xrightarrow{p_i,m_i} G_i$ *can be decomposed into a transformation* $G \xrightarrow{p_0,m_0} G_0 \xrightarrow{\overline{p_i},\overline{m_i}} G_i$. *Moreover, the transformations* $G_0 \xrightarrow{\overline{p_i},\overline{m_i}} G_i$ *are pairwise parallel independent.*

Proof idea. From Fact 1 it follows that each single direct transformation $G \xrightarrow{p_i,m_i} G_i$ can be decomposed into a transformation $G \xrightarrow{p_0,m_0^i} G_0^i \xrightarrow{\overline{p_i},\overline{m_i}} G_i$ with $m_0^i = m_i \circ s_{i,L}$, and since the bundle is s-amalgamable, $m_0 = m_i \circ s_{i,L} = m_0^i$ and $G_0 := G_0^i$ for all

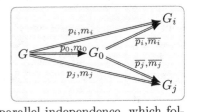

$i = 1, \ldots, n$. It remains to show the pairwise parallel independence, which follows from consistent and weakly independent matches. □

If a bundle of direct transformations of a graph G is s-amalgamable, then we can apply the amalgamated rule directly to G leading to a parallel execution of all the changes done by the single transformation steps.

Theorem 2 (Multi-Amalgamation). *Consider a bundle of kernel morphisms* $s = (s_i : p_0 \to p_i)_{i=1,\ldots,n}$.

1. *Synthesis. Given an s-amalgamable bundle of direct transformations* $(G \xrightarrow{p_i,m_i} G_i)_{i=1,\ldots,n}$ *then there is an amalgamated transformation* $G \xrightarrow{\tilde{p}_s,\tilde{m}} H$ *and transformations* $G_i \xRightarrow{q_i} H$ *over the complement rules q_i of the kernel morphisms* $t_i : p_i \to \tilde{p}_s$ *such that* $G \xrightarrow{p_i,m_i} G_i \xRightarrow{q_i} H$ *is a decomposition of* $G \xrightarrow{\tilde{p}_s,\tilde{m}} H$.

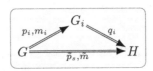

2. Analysis. *Given an amalgamated transformation* $G \xrightarrow{\tilde{p}_s, \tilde{m}} H$ *then there are* s_i*-related transformations* $G \xrightarrow{p_i, m_i} G_i \xRightarrow{q_i} H$ *for* $i = 1, \ldots, n$ *such that* $G \xrightarrow{p_i, m_i} G_i$ *is s-amalgamable.*

3. Bijective Correspondence. *The synthesis and analysis constructions are inverse to each other up to isomorphism.*

Proof idea.

1. We can show that $\tilde{p}_s$ is applicable to G leading to an amalgamated transformation $G \xrightarrow{\tilde{p}_s, \tilde{m}} H$ with $m_i = \tilde{m} \circ t_{i,L}$, where $t_i : p_i \to \tilde{p}_i$ is the kernel morphism constructed in Fact 2. Then we can apply Fact 1 which implies the decomposition of $G \xrightarrow{\tilde{p}_s, \tilde{m}} H$ into $G \xrightarrow{p_i, m_i} G_i \xRightarrow{q_i} H$, where q_i is the complement rule of the kernel morphism t_i.

2. Using the kernel morphisms t_i we obtain transformations $G \xrightarrow{p_i, m_i} G_i \xRightarrow{q_i} H$ from Fact 1 with $m_i = \tilde{m} \circ t_{i,L}$. We have to show that this bundle of transformation is s-amalgamable. Applying again Fact 1 we obtain transformations $G \xrightarrow{p_0, m_0^i} G_0^i \xrightarrow{\overline{p_i}} G_i$ with $m_0^i = m_i \circ s_{i,L}$. It follows that $m_0^i = m_i \circ s_{i,L} = \tilde{m} \circ t_{i,L} \circ s_{i,L} = \tilde{m} \circ t_{0,L} = \tilde{m} \circ t_{j,L} \circ s_{j,L} = m_j \circ s_{j,L}$ and thus we have consistent matches with $m_0 := m_0^i$ well-defined and $G_0 = G_0^i$. Moreover, the matches are weakly independent.

3. Because of the uniqueness of the used constructions, the above constructions are inverse to each other up to isomorphism. □

Remark 3. Note, that q_i can be constructed as the amalgamated rule of the kernel morphisms $(p_{K_0} \to \overline{p_j})_{j \neq i}$, where $p_{K_0} = (K_0 \xleftarrow{id_{K_0}} K_0 \xrightarrow{id_{K_0}} K_0, \text{true}))$ and $\overline{p_j}$ is the complement rule of p_j.

For $n = 2$ and rules without application conditions, the Multi-Amalgamation Theorem specializes to the Amalgamation Theorem in [4]. Moreover, if p_0 is the empty rule, this is the Parallelism Theorem in [16], since the transformations are parallel independent for an empty kernel match.

Example 5. As already stated in Example 4, the transformations $G \xrightarrow{p_1, m_1} G_1$, $G \xrightarrow{p_2, m_2} G_2$, and $G \xrightarrow{p_1, m_3} G_3$ shown in Fig. 6 are s-amalgamable for the bundle $s = (s_1, s_2, s_3 = s_1)$ of kernel morphisms. Applying Fact 4, we can decompose these transformations into a transformation $G \xrightarrow{p_0, m_0} G_0$ followed by transformations $G_0 \xrightarrow{\overline{p_1}, \overline{m_1}} G_1$, $G_0 \xrightarrow{\overline{p_2}, \overline{m_2}} G_2$, and $G_0 \xrightarrow{\overline{p_1}, \overline{m_3}} G_3$ via the complement rules, which are pairwise parallel independent. These transformations are depicted in Fig. 7. Moreover, Thm. 2 implies that we obtain for this bundle of direct transformations an amalgamated transformation $G \xrightarrow{\tilde{p}_s, \tilde{m}} H$, which is the transformation already shown in Fig. 5. Vice versa, the analysis of this amalgamated transformation leads to the s-amalgamable bundle of transformations $G \xrightarrow{p_1, m_1} G_1$, $G \xrightarrow{p_2, m_2} G_2$, and $G \xrightarrow{p_1, m_3} G_3$ from Fig. 6.

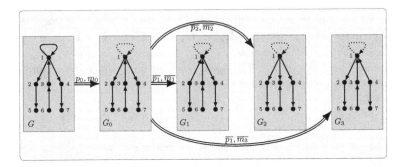

Fig. 7. The decomposition of the s-amalgamable bundle

Extension to Multi-Amalgamation with Maximal Matchings

An important extension of the presented theory is the introduction of interaction schemes and maximal matchings. An interaction scheme defines a bundle of kernel morphisms. In contrast to a concrete bundle, for the application of such an interaction scheme all possible matches for the multi rules are computed that agree on a given kernel match and lead to an amalgamable bundle of transformations. In our example, the interaction scheme $is = \{s_1, s_2\}$ contains the two kernel morphisms from Fig. 1. For the kernel match m_0, the matches m_1, m_2, m_3 are maximal: they are s-amalgamable, and any other match for p_1 or p_2 that agrees an m_0 would hold only already matched elements. This technique is very useful for the definition of the semantics of visual languages. For our example concerning statecharts [13], an unknown number of state transitions triggered by the same event, which is highly dependent on the actual system state, can be handled in parallel.

4   Conclusion

In this paper, we have generalized the theory of amalgamation in [4] to multi-amalgamation in adhesive categories. More precisely, the Complement Rule and Amalgamation Theorems in [4] are presented on a set-theoretical basis for pairs of plain graph rules without any application conditions. The Complement Rule and Multi-Amalgamation Theorems in this paper are valid in adhesive and weak adhesive HLR categories for n rules with application conditions [15]. These generalizations are non-trivial and important for applications of parallel graph transformations to communication-based systems [7], to model transformations from BPMN to BPEL [17], and for the modeling of the operational semantics of visual languages [8], where interaction schemes are used to generate multi-amalgamated rules and transformations based on suitable maximal matchings.

The theory of multi-amalgamation is a solid mathematical basis to analyze interesting properties of the operational semantics, like termination, local confluence, and functional behavior. However, it is left open for future work to generalize the corresponding results in [2], like the Local Church–Rosser, Parallelism, and Local Confluence Theorems, to the case of multi-amalgamated rules, especially to the operational semantics of statecharts based on amalgamated graph transformation with maximal matchings in [13].

References

[1] Lack, S., Sobociński, P.: Adhesive Categories. In: Walukiewicz, I. (ed.) FOSSACS 2004. LNCS, vol. 2987, pp. 273–288. Springer, Heidelberg (2004)

[2] Ehrig, H., Ehrig, K., Prange, U., Taentzer, G.: Fundamentals of Algebraic Graph Transformation. EATCS Monographs. Springer, Heidelberg (2006)

[3] Rozenberg, G. (ed.): Handbook of Graph Grammars and Computing by Graph Transformation. Foundations, vol. 1. World Scientific, Singapore (1997)

[4] Böhm, P., Fonio, H.R., Habel, A.: Amalgamation of Graph Transformations: A Synchronization Mechanism. JCSS 34(2-3), 377–408 (1987)

[5] Ehrig, H., Kreowski, H.J.: Parallelism of Manipulations in Multidimensional Information Structures. In: Mazurkiewicz, A. (ed.) MFCS 1976. LNCS, vol. 45, pp. 285–293. Springer, Heidelberg (1976)

[6] Taentzer, G., Beyer, M.: Amalgamated Graph Transformations and Their Use for Specifying AGG - an Algebraic Graph Grammar System. In: Ehrig, H., Schneider, H.-J. (eds.) Dagstuhl Seminar 1993. LNCS, vol. 776, pp. 380–394. Springer, Heidelberg (1994)

[7] Taentzer, G.: Parallel and Distributed Graph Transformation: Formal Description and Application to Communication Based Systems. PhD thesis, TU Berlin (1996)

[8] Ermel, C.: Simulation and Animation of Visual Languages based on Typed Algebraic Graph Transformation. PhD thesis, TU Berlin (2006)

[9] Löwe, M.: Algebraic Approach to Single-Pushout Graph Transformation. TCS 109, 181–224 (1993)

[10] Grønmo, R., Krogdahl, S., Møller-Pedersen, B.: A Collection Operator for Graph Transformation. In: Paige, R.F. (ed.) ICMT 2009. LNCS, vol. 5563, pp. 67–82. Springer, Heidelberg (2009)

[11] Hoffmann, B., Janssens, D., Eetvelde, N.: Cloning and Expanding Graph Transformation Rules for Refactoring. ENTCS 152, 53–67 (2006)

[12] Rensink, A., Kuperus, J.: Repotting the Geraniums: On Nested Graph Transformation Rules. ECEASST 18, 1–15 (2009)

[13] Golas, U., Biermann, E., Ehrig, H., Ermel, C.: A Visual Interpreter Semantics for Statecharts Based on Amalgamated Graph Transformation (Submitted 2010), http://tfs.cs.tu-berlin.de/publikationen/Papers10/GBEE10.pdf

[14] Golas, U.: Multi-Amalgamation in M-Adhesive Categories: Long Version. Technical Report 2010/05, Technische Universität Berlin (2010)

[15] Habel, A., Pennemann, K.H.: Correctness of High-Level Transformation Systems Relative to Nested Conditions. MSCS 19(2), 245–296 (2009)

[16] Ehrig, H., Habel, A., Lambers, L.: Parallelism and Concurrency Theorems for Rules with Nested Application Conditions. ECEASST 26, 1–23 (2010)

[17] Biermann, E., Ehrig, H., Ermel, C., Golas, U., Taentzer, G.: Parallel Independence of Amalgamated Graph Transformations Applied to Model Transformation. In: Engels, G., Lewerentz, C., Schäfer, W., Schürr, A., Westfechtel, B. (eds.) Essays Dedicated to M. Nagl. LNCS, vol. 5765. Springer, Heidelberg (2010)

Amalgamating Pushout and Pullback Graph Transformation in Collagories

Wolfram Kahl*

McMaster University, Hamilton, Ontario, Canada
`kahl@cas.mcmaster.ca`

Abstract. The relation-algebraic approach to graph transformation replaces the universal category-theoretic characterisations of pushout and pullbacks with the local characterisations of tabulations and co-tabulations. The theory of collagories is a weak axiomatisation of relation-algebraic operations that closely corresponds to adhesive categories.

We show how to amalgamate double-pushout and double-pullback rewriting steps into a fused rewriting concept where rules can contain subgraph variables in a natural and flexible way, and rewriting can delete or duplicate the matched instances of such variables.

1 Introduction

We set out to show how the relation-algebraic approach to graph transformation [Kaw90, Kah01, Kah04, Kah09] can not only produce high-level descriptions of crucial concepts in conventional categorical approaches to graph transformation (Sect. 5), but also enables more complex rewriting concepts, in particular an elegant and flexible way to allow proper subgraph variables in rules, defined as amalgamation of DPO and DPB rewriting (Sect. 6).

For this, we use the recently developed setting of collagories [Kah09, Kah10], which can be understood as "relation algebras without complement, and without zero laws for empty relations". We give a brief overview over the necessary foundations in Sections 2–4. Collagories admit as models not only relations in any topos, in particular relational homomorphisms (bisimulations) between graph structures, but also relations in comma categories (like that of pointed sets) that only form adhesive categories (introduced by [LS04]).

The development of Sections 5 and 6 was previously performed in the stronger setting of Dedekind categories in [Kah01], but only published as a short informal overview [Kah02], without the details given in the current paper, which also adds the adaptation to the weaker collagory setting, and therewith widens the scope of possible applications.

2 Categories, Collagories: Definitions and Notation

This section only serves to fix notation and terminology for standard concepts, see [FS90, SS93, Kah04], and for collagories [Kah09]. Like Freyd and Scedrov

---

* This research is supported by the National Science and Engineering Research Council of Canada (NSERC).

H. Ehrig et al. (Eds.): ICGT 2010, LNCS 6372, pp. 362–378, 2010.

and a slowly increasing number of categorists, we denote composition in "diagram order" not only in relation-algebraic contexts, where this is customary, but also in the context of categories. We will always use the infix operator ";" to make composition explicit, $R \mathbin{;} S = \mathcal{A} \xrightarrow{R} \mathcal{B} \xrightarrow{S} \mathcal{C}$, assign ";" higher priority than other binary operators, and assign unary operators higher priority than all binary operators.

Definition 2.1. A *category* $\mathbf{C}$ is a tuple $(\mathsf{Obj_C}, \mathsf{Mor_C}, \mathsf{src}, \mathsf{trg}, \mathbb{I}, \mathbin{;})$ where
- $\mathsf{Obj_C}$ is a collection of *objects*.
- $\mathsf{Mor_C}$ is a collection of *arrows* or *morphisms*.
- src (resp. trg) maps each morphism to its source (resp. target) object. Instead of $\mathsf{src}(f) = \mathcal{A} \wedge \mathsf{trg}(f) = \mathcal{B}$ we write $f : \mathcal{A} \to \mathcal{B}$. The collection of all morphisms f with $f : \mathcal{A} \to \mathcal{B}$ is denoted as $\mathsf{Mor_C}[\mathcal{A}, \mathcal{B}]$ and also called a *homset*.
- ";" is the binary *composition* operator, and composition of two morphisms $f : \mathcal{A} \to \mathcal{B}$ and $g : \mathcal{B}' \to \mathcal{C}$ is defined iff $\mathcal{B} = \mathcal{B}'$, and then $(f \mathbin{;} g) : \mathcal{A} \to \mathcal{C}$; composition is associative.
- $\mathbb{I}$ associates with every object $\mathcal{A}$ a morphism $\mathbb{I}_\mathcal{A}$ which is both a right and left unit for composition. $\qquad\Box$

Collagories [Kah09] extend the allegories of [FS90] with binary joins, but don't postulate the least morphisms and zero laws required for distributive allegories.

Definition 2.2. A *collagory* is a category $\mathbf{C}$ such that
- for each two objects $\mathcal{A}$ and $\mathcal{B}$, the homset $\mathsf{Mor_C}[\mathcal{A}, \mathcal{B}]$ is a distributive lattice with ordering $\sqsubseteq_{\mathcal{A},\mathcal{B}}$, binary meet $\sqcap_{\mathcal{A},\mathcal{B}}$, and binary join $\sqcup_{\mathcal{A},\mathcal{B}}$ (the indices will usually be omitted),
- composition is monotonic with respect to $\sqsubseteq$ in both arguments, and distributes over $\sqcup$ from both sides,
- each morphism $R : \mathcal{A} \to \mathcal{B}$ has a *converse* $R^\smile : \mathcal{B} \to \mathcal{A}$,
- the *involution equations* hold for all $R : \mathcal{A} \to \mathcal{B}$ and $S : \mathcal{B} \to \mathcal{C}$:
$$(R^\smile)^\smile = R \qquad\qquad \mathbb{I}_\mathcal{A}^\smile = \mathbb{I}_\mathcal{A} \qquad\qquad (R \mathbin{;} S)^\smile = S^\smile \mathbin{;} R^\smile$$
- conversion is monotonic with respect to $\sqsubseteq$,
- for all $Q : \mathcal{A} \to \mathcal{B}$, $R : \mathcal{B} \to \mathcal{C}$, and $S : \mathcal{A} \to \mathcal{C}$, the *modal rule* holds:
$$Q \mathbin{;} R \sqcap S \sqsubseteq (Q \sqcap S \mathbin{;} R^\smile) \mathbin{;} R \ . \qquad\qquad\Box$$

Many relation properties can be characterised in the context of collagories:

Definition 2.3. A morphism $R : \mathcal{A} \to \mathcal{B}$ in a collagory is called:
- *univalent* iff $R^\smile \mathbin{;} R \sqsubseteq \mathbb{I}_\mathcal{B}$,   - *surjective* iff $\mathbb{I}_\mathcal{B} \sqsubseteq R^\smile \mathbin{;} R$,
- *total* iff $\mathbb{I}_\mathcal{A} \sqsubseteq R \mathbin{;} R^\smile$,
- *injective* iff $R \mathbin{;} R^\smile \sqsubseteq \mathbb{I}_\mathcal{A}$,   - *a mapping* iff it is univalent and total,
     - *bijective* iff it is injective and surjective. $\quad\Box$

For a collagory $\mathbf{C}$, we write $\mathsf{Map}\,\mathbf{C}$ for the sub-category of $\mathbf{C}$ that contains only the mappings as arrows.

Another important property is difunctionality (see [SS93, 4.4]):

Definition 2.4. A morphism R in a collagory is *difunctional* iff $R \,\hat{;}\, R^{\smile} \,\hat{;}\, R \sqsubseteq R$. $\Box$

A concrete relation, understood as a Boolean matrix, is difunctional iff it can be rearranged into "loose block-diagonal form", with full rectangular blocks such that there is no overlap between different blocks in either direction. If R is univalent or injective, it is a *fortiori* difunctional.

We shall need difunctional closures, which could be defined using Kleene star, but can also be characterised directly using axioms similar to Kozen's Kleene star axioms [Koz94]:

Definition 2.5. For a morphism $R : \mathcal{A} \to \mathcal{B}$ in a collagory, we define $R^{\boxplus} : \mathcal{A} \to \mathcal{B}$ as satisfying, for all $Q : \mathcal{C} \to \mathcal{A}$, and $Q' : \mathcal{C} \to \mathcal{B}$, and $S : \mathcal{B} \to \mathcal{C}$, and $S' : \mathcal{A} \to \mathcal{C}$:

$$R^{\boxplus} = R \sqcup R^{\boxplus} \,\hat{;}\, (R^{\boxplus})^{\smile} \,\hat{;}\, R^{\boxplus} \qquad \text{(recursive definition)},$$
$$Q \,\hat{;}\, R \sqsubseteq Q' \wedge Q' \,\hat{;}\, R^{\smile} \,\hat{;}\, R \sqsubseteq Q' \;\Rightarrow\; Q \,\hat{;}\, R^{\boxplus} \sqsubseteq Q' \qquad \text{(right induction)},$$
$$R \,\hat{;}\, S \sqsubseteq S' \wedge R \,\hat{;}\, R^{\smile} \,\hat{;}\, S' \sqsubseteq S' \;\Rightarrow\; R^{\boxplus} \,\hat{;}\, S \sqsubseteq S' \qquad \text{(left induction)}.$$

We further define $R^{\triangleright} : \mathcal{A} \to \mathcal{A}$ and $R^{\triangleleft} : \mathcal{B} \to \mathcal{B}$ as:

$$R^{\triangleright} := \mathbb{I}_{\mathcal{A}} \sqcup R^{\boxplus} \,\hat{;}\, (R^{\boxplus})^{\smile} \qquad \text{and} \qquad R^{\triangleleft} := \mathbb{I}_{\mathcal{B}} \sqcup (R^{\boxplus})^{\smile} \,\hat{;}\, R^{\boxplus} . \qquad \Box$$

If $R^{\boxplus}$ exists, it is uniquely determined and is the *difunctional closure* of R, that is, the least difunctional morphism containing R.

For endomorphisms, there are a few additional properties of interest:

Definition 2.6. A morphism $R : \mathcal{A} \to \mathcal{A}$ in a collagory is called:

- *reflexive* iff $\mathbb{I}_{\mathcal{A}} \sqsubseteq R$,
- a *subidentity* iff $R \sqsubseteq \mathbb{I}_{\mathcal{A}}$,
- *symmetric* iff $R^{\smile} \sqsubseteq R$,
- *transitive* iff $R \,\hat{;}\, R \sqsubseteq R$,
- *idempotent* iff $R \,\hat{;}\, R = R$,
- an *equivalence* iff it is symmetric, reflexive and transitive. $\Box$

We write $\mathsf{SubId}\,\mathcal{A}$ for the type containing all subidentities on the object $\mathcal{A}$.

In a collagory, subidentities are symmetric and idempotent [FS90, 2.12], and are the preferred means to identify "parts" of an object, as for example domain and range of a morphism:

Definition 2.7. For a morphism $R : \mathcal{A} \to \mathcal{B}$ in a collagory, we define $\mathsf{dom}\,R : \mathsf{SubId}\,\mathcal{A}$ and $\mathsf{ran}\,R : \mathsf{SubId}\,\mathcal{B}$:

$$\mathsf{dom}\,R = \mathbb{I}_{\mathcal{A}} \sqcap R \,\hat{;}\, R^{\smile} \qquad\qquad \mathsf{ran}\,R = \mathbb{I}_{\mathcal{B}} \sqcap R^{\smile} \,\hat{;}\, R \qquad\qquad \Box$$

3   The Collagory of Graphs

The category *Rel* with sets as objects and relations between sets as morphisms extends to a collagory by using standard relation-algebraic operations.

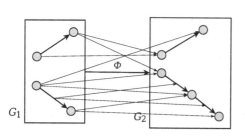

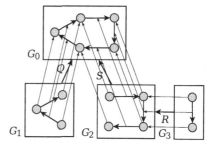

Fig. 1. A graph bisimulation   **Fig. 2.** Subgraph semi-/pseudo-compl

In [Kah09] it was shown that for many-sorted signatures with only zero-ary and unary function symbols, we obtain, from a base collagory $\mathbf{C}$, a new collagory with objects being algebras where the function symbols are interpreted as mappings in $\mathbf{C}$, and with bisimulations between these algebras as morphisms.

The signature for graphs has only two unary function symbols, assigning source respectively target nodes to edges:

$$\mathsf{sigGraph} := \langle \mathbf{sorts:}\ \mathsf{V}, \mathsf{E};\ \mathbf{ops:}\ \mathsf{s}, \mathsf{t} : \mathsf{E} \to \mathsf{V} \rangle$$

A graph G_i is a $\mathsf{sigGraph}$-algebra over Rel, that is, it consists of two sets V_i and E_i, and two mappings (total functions) $\mathsf{s}_i, \mathsf{t}_i : \mathsf{E}_i \to \mathsf{V}_i$.

Given two graphs G_1 and G_2, a bisimulation $\varPhi : G_1 \to G_2$ is a pair of relations $\varPhi = (\varPhi_\mathsf{V}, \varPhi_\mathsf{E})$ with $\varPhi_\mathsf{V} : \mathsf{V}_1 \to \mathsf{V}_2$ and $\varPhi_\mathsf{E} : \mathsf{E}_1 \to \mathsf{E}_2$ in Rel, satisfying

$$\varPhi_\mathsf{E} \,\mathbin{;}\, \mathsf{s}_2 \sqsubseteq \mathsf{s}_1 \,\mathbin{;}\, \varPhi_\mathsf{V} \qquad \text{and} \qquad \varPhi_\mathsf{E} \,\mathbin{;}\, \mathsf{t}_2 \sqsubseteq \mathsf{t}_1 \,\mathbin{;}\, \varPhi_\mathsf{V} \;.$$

In predicate logic, the first of these conditions can be expressed as follows:

$$\forall\, e_1 : \mathsf{E}_1;\ e_2 : \mathsf{E}_2 \ \bullet\ (e_1, e_2) \in \varPhi_\mathsf{E} \Rightarrow (\mathsf{s}_1(e_1), \mathsf{s}_2(e_2)) \in \varPhi_\mathsf{V}$$

Fig. 1 shows a graph bisimulation $\varPhi : G_1 \to G_2$, visualising $\varPhi_\mathsf{V}$ as thin arrows between nodes, and $\varPhi_\mathsf{E}$ as thin arrows between the thicker edge arrows.

Composition, identities, converse, meet, and join are defined component-wise on the two constituent relations of a bisimulation, and these operations satisfy the collagory laws due to the general algebra collagory construction from [Kah09].

In the resulting collagory of graphs, there is a one-to-one correspondence between subidentities and subgraphs. The subgraph lattice is not Boolean, and we will need both pseudo-complements and their dual, which we can define for general collagories (where both don't necessarily exist).

Definition 3.1. If $p : \mathsf{SubId}\,\mathcal{A}$ is a subidentity in a collagory, the *semi-complement* $p^\sim : \mathsf{SubId}\,\mathcal{A}$ of p is defined as the least subidentity such that $p \sqcup p^\sim = \mathbb{I}_\mathcal{A}$, that is, by $p^\sim \sqsubseteq q \Leftrightarrow p \sqcup q = \mathbb{I}_\mathcal{A}$ for all subidentities $q : \mathsf{SubId}\,\mathcal{A}$. □

The intersection $p^\sim \sqcap p$ is not necessarily empty. For a subgraph p of G, the semi-complement contains all nodes and edges of G that are not in p, and all nodes incident with those edges (some of which will be border nodes of p). For example, in the subgraph injection diagram in Fig. 2 we have $(\operatorname{ran} Q)^\sim = \operatorname{ran} S$, or, informally, $G_1^\sim = G_2$.

Definition 3.2. For subidentities $p, q : \operatorname{SubId} \mathcal{A}$, the *relative pseudo-complement* $p \to q$ is defined by $x \sqcap p \sqsubseteq q \Leftrightarrow x \sqsubseteq (p \to q)$ for all $p, q : \operatorname{SubId} \mathcal{A}$.   □

For subgraphs of G, the relative pseudo-complement $p \to q$ consists of all nodes of G that are in q or not in p, and all edges in q or not in p that are also only incident with nodes in q or not in p. Intuitively, $p \to q$ therefore is G with the parts of p outside q removed, and then also all dangling edges removed. A *pseudo-complement* is a relative pseudo-complement with respect to the least morphism $\bot\!\!\!\bot$; for example, in Fig. 2 we have $(\operatorname{ran} Q) \to \bot\!\!\!\bot = \operatorname{ran}(R \,\S\, S)$, or, informally. $G_1 \to \varnothing = G_3$.

4   Tabulations and Co-tabulations

Central to the connection between pullbacks and pushouts in categories of mappings on the one hand and constructions in relational theories on the other hand is the fact that a square of mappings commutes iff the "relation" induced by the source span is contained in that induced by the target co-span [FS90, 2.146]:

Lemma 4.1. Given a square of mappings in a collagory as drawn to the right, we have $P \,\S\, R = Q \,\S\, S$ iff $P^\sim \,\S\, Q \sqsubseteq R \,\S\, S^\sim$. □

This provides a first hint that in the relational setting, the precise choice of the two mappings P and Q does not matter when looking for a pushout of the span $\mathcal{B} \xleftarrow{\ P\ } \mathcal{A} \xrightarrow{\ Q\ } \mathcal{C}$ — we only need to consider the diagonal $P^\sim \,\S\, Q$. Dually, when looking for a pullback of the co-span $\mathcal{B} \xrightarrow{\ R\ } \mathcal{D} \xleftarrow{\ S\ } \mathcal{C}$, only $R \,\S\, S^\sim$ needs to be considered. The gap between the two ways of calculating the horizontal diagonal can be significant since $R \,\S\, S^\sim$ is always difunctional (because $R \,\S\, S^\sim \,\S\, S \,\S\, R^\sim \,\S\, R \,\S\, S^\sim \sqsubseteq R \,\S\, S^\sim$ from univalence of R and S.)

Producing the result span of a pullback (respectively the result co-span of a pushout) from the horizontal diagonal alone is, in some sense, a generalisation of Freyd and Scedrov's splitting of idempotents; [Kah04] contains more discussion of this aspect.

While Freyd and Scedrov introduced tabulations as spans of mappings, the following equivalent characterisation provided by [Kah04] has the advantage that it is fully equational, without the implicit inclusion conditions in the requirement that P and Q are mappings. This frequently facilitates calculations, and also eases amalgamation of tabulations and co-tabulations in Sect. 6.

Definition 4.2. In a collagory, the span $\mathcal{B} \xleftarrow{\ P\ } \mathcal{A} \xrightarrow{\ Q\ } \mathcal{C}$ is a tabulation of $V : \mathcal{B} \to \mathcal{C}$ if and only if the following equations hold:

$$P^\sim \,\S\, Q = V \qquad \begin{aligned} P^\sim \,\S\, P &= \mathbb{I}_\mathcal{B} \sqcap V \,\S\, V^\sim \\ Q^\sim \,\S\, Q &= \mathbb{I}_\mathcal{C} \sqcap V^\sim \,\S\, V \end{aligned} \qquad P \,\S\, P^\sim \sqcap Q \,\S\, Q^\sim = \mathbb{I}_\mathcal{A} \ .$$

The co-span $\mathcal{B} \xrightarrow{R} \mathcal{D} \xleftarrow{S} \mathcal{C}$ is a co-tabulation of $W : \mathcal{B} \to \mathcal{C}$ iff the following equations hold:

$$R \,;\, S^{\smile} = W \qquad \begin{aligned} R \,;\, R^{\smile} &= \mathbb{I}_{\mathcal{B}} \sqcup W \,;\, W^{\smile} \\ S \,;\, S^{\smile} &= \mathbb{I}_{\mathcal{C}} \sqcup W^{\smile} \,;\, W \end{aligned} \qquad R^{\smile} \,;\, R \sqcup S^{\smile} \,;\, S = \mathbb{I}_{\mathcal{D}} \,. \qquad \square$$

Together with the univalence of R and S following from the last equation, $R \,;\, S^{\smile} = W$ implies that if W has a co-tabulation, it must be difunctional.

Tabulations and co-tabulations are unique up to isomorphism. If a co-span $\mathcal{B} \xrightarrow{R} \mathcal{D} \xleftarrow{S} \mathcal{C}$ of mappings is given, then a tabulation of $R \,;\, S^{\smile}$ in the collagory $\mathbf{C}$ is a pullback for that co-span in $\mathsf{Map}\,\mathbf{C}$ [FS90, 2.147].

Dually, if a span $\mathcal{B} \xleftarrow{P} \mathcal{A} \xrightarrow{Q} \mathcal{C}$ of mappings is given, then a co-tabulation for $(P^{\smile} \,;\, Q)^{\boxplus}$ (if that exists) in the collagory $\mathbf{C}$ is a pushout for that span in $\mathsf{Map}\,\mathbf{C}$. More precisely, given a span $\mathcal{B} \xleftarrow{P} \mathcal{A} \xrightarrow{Q} \mathcal{C}$ of mappings, the bi-pushouts for this in the bicategory with cells defined by the local ordering $\sqsubseteq$ are exactly the co-tabulations for $(P^{\smile} \,;\, Q)^{\boxplus}$ (the existence of which follows from the existence of a bipushout), as shown in [Kah10].

A co-tabulation for $U^{\boxplus}$ satisfies the following equations:

$$R \,;\, S^{\smile} = U^{\boxplus} \qquad R \,;\, R^{\smile} = U^{\triangleright} \qquad S \,;\, S^{\smile} = U^{\triangleleft} \qquad R^{\smile} \,;\, R \sqcup S^{\smile} \,;\, S = \mathbb{I}_{\mathcal{D}} \,.$$

This was introduced as a *gluing for* U in [Kah01]. Kawahara is the first to have characterised pushouts relation-algebraically in essentially this way [Kaw90].

Definition 4.3. If a collagory has a tabulation for each morphism and a co-tabulation for each difunctional morphism, then we call it *bi-tabular*. $\square$

5   Categoric Graph Transformation in Collagories

We now summarise key aspects of conventional categorical approaches to graph transformation, mainly in order to be able to provide some base material for the amalgamated approach.

5.1   Gluing Condition and Pushout Complements

Since the identification condition is of a shape that will also be useful in other contexts, we first define:

Definition 5.1. Given a subidentity $u : \mathsf{SubId}\,\mathcal{A}$, a relation $R : \mathcal{A} \leftrightarrow \mathcal{B}$ is called *almost-injective besides* u iff $R \,;\, R^{\smile} \sqsubseteq \mathbb{I}_{\mathcal{A}} \sqcup u \,;\, R \,;\, R^{\smile} \,;\, u$. $\square$

Definition 5.2. In a collagory with semi-complement $\_^{\smile}$ on subidentities, Let $\varPhi : \mathcal{G} \to \mathcal{L}$ and $X : \mathcal{L} \to \mathcal{A}$ be given.

– The *identification condition* holds iff X is almost-injective besides $\mathsf{ran}\,\varPhi$.
– The *dangling condition* holds iff $X \,;\, (\mathsf{ran}\,X)^{\smile} \sqsubseteq (\mathsf{ran}\,\varPhi) \,;\, X$. $\square$

Kawahara [Kaw90] used a different dangling condition based on relative pseudo-complements; under the identification condition, this is equivalent to our semi-complement-based condition, which may be easier to understand: It specifies that the part of X that maps to the border of the image of X should start only from $\operatorname{ran} \Phi$, which corresponds closely to the usual item-based definitions of the dangling condition, e.g. in [CMR^{+}97].

Definition 5.3. Let two morphisms $\Phi : \mathcal{G} \to \mathcal{L}$ and $X : \mathcal{L} \to \mathcal{A}$ in a collagory with semi-complements and relative pseudo-complements on subidentities be given.

If $q : \operatorname{SubId} \mathcal{A}$ is a subidentity on $\mathcal{A}$, then a *subobject host construction for* $\mathcal{G} \xrightarrow{\Phi} \mathcal{L} \xrightarrow{X} \mathcal{A}$ *by* q is a diagram $\mathcal{G} \xrightarrow{\Xi} \mathcal{H} \xrightarrow{\Psi} \mathcal{A}$ where Ψ is a subobject injection for q and $\Xi := \Phi \,\mathbin{;}\, X \,\mathbin{;}\, \Psi^{\smile}$.

A *straight host construction* uses $q := \operatorname{ran} X \to \operatorname{ran}(\Phi \,\mathbin{;}\, X)$.

A *sloppy host construction* uses $q := (\operatorname{ran} X)^{\smile} \sqcup \operatorname{ran}(\Phi \,\mathbin{;}\, X)$. □

Theorem 5.4 (Pushout Complement). In a collagory with semi-complements and relative pseudo-complements on subidentities, if $\Phi : \mathcal{G} \to \mathcal{L}$ and $X : \mathcal{L} \to \mathcal{A}$ are two mappings such that the gluing condition holds, then straight and sloppy host constructions coincide, and a straight host construction $\mathcal{G} \xrightarrow{\Xi} \mathcal{H} \xrightarrow{\Psi} \mathcal{A}$ for $\mathcal{G} \xrightarrow{\Phi} \mathcal{L} \xrightarrow{X} \mathcal{A}$ in a collagory **C** is a pushout complement for $\mathcal{G} \xrightarrow{\Phi} \mathcal{L} \xrightarrow{X} \mathcal{A}$ in $\operatorname{\mathsf{Map}} \mathbf{C}$.

If Φ is injective, then the pushout complement is unique up to isomorphism. □

5.2 Pullback Complements

The double-pullback approach is, categorically, the strict dual to the double-pushout approach. Given two mappings $\mathcal{D} \xrightarrow{P} \mathcal{B} \xrightarrow{R} \mathcal{A}$, where R would be the rule's left-hand side and P the redex matching, the double-pullback approach therefore requires construction of a *pullback complement* $\mathcal{D} \xrightarrow{Q} \mathcal{C} \xrightarrow{S} \mathcal{A}$, that is, an object $\mathcal{C}$ and two mappings $Q : \mathcal{D} \to \mathcal{C}$ and $S : \mathcal{C} \to \mathcal{A}$ such that the resulting square is a pullback for R and S.

This has been studied by Bauderon and Jacquet [Jac99, BJ01] at the concrete graph level (reasoning about individual nodes and edges), and we follow some of their terminology here while summarising the results first obtained in [Kah01], which are formalised completely at the component-free level.

The crucial condition here is *coherence*:

Definition 5.5. Two mappings $\mathcal{D} \xrightarrow{P} \mathcal{B} \xrightarrow{R} \mathcal{A}$ are called *coherent* iff there exists an equivalence relation $\Theta : \mathcal{D} \to \mathcal{D}$ such that the following conditions hold:

1. $P \,\mathbin{;}\, P^{\smile} \sqcap \Theta \sqsubseteq \mathbb{I}_{\mathcal{D}}$,
2. $\Theta \,\mathbin{;}\, P = P \,\mathbin{;}\, R \,\mathbin{;}\, R^{\smile}$.

We also say then P and R are *coherent via* Θ. □

These coherence conditions imply that Θ is a complement of $P \, ; P^\smile$ in the lattice of all equivalence relations contained in $P \, ; R \, ; R^\smile \, ; P^\smile$. Since lattices of equivalence relations are not necessarily modular, such a complement is not uniquely determined, and not all complements satisfy (2). Now we have:

Proposition 5.6. If $\mathcal{B} \xleftarrow{\ P\ } \mathcal{D} \xrightarrow{\ Q\ } \mathcal{C}$ is a tabulation for $\mathcal{B} \xrightarrow{\ R\ } \mathcal{A} \xleftarrow{\ S\ } \mathcal{C}$, then:

1. P and R are coherent via $\Theta := Q \, ; Q^\smile$.
2. If R is injective, then $\Theta = \mathbb{I}_{\mathcal{D}}$ (and Q is injective, too).
3. If R is surjective, then Q is surjective, too.   □

If we know that Q is surjective, then such an equivalence relation Θ already uniquely determines Q (up to isomorphism) as a quotient projection for Θ.

If R is not surjective, then Q need not be surjective, either, and the pullback complement object is not uniquely determined. However, if Map $\mathcal{C}$ allows epi-mono-decompositions, then every pullback complement factors over a pullback complement with surjective Q. When looking for pullback complements, we shall therefore restrict our search to candidates with surjective Q.

With this restriction, surjectivity and univalence of Q together with commutativity then also determine S, as: $S = Q^\smile \, ; Q \, ; S = Q^\smile \, ; P \, ; R$. Showing that all this then gives rise to a pullback complement is routine:

Theorem 5.7 (Pullback Complement). If there exists an equivalence relation $\Theta : \mathcal{D} \to \mathcal{D}$ such that P and R are coherent via Θ, then a pullback complement $\mathcal{D} \xrightarrow{\ Q\ } \mathcal{C} \xrightarrow{\ S\ } \mathcal{A}$ for $\mathcal{D} \xrightarrow{\ P\ } \mathcal{B} \xrightarrow{\ R\ } \mathcal{A}$ is obtained as follows:

- Let $\mathcal{C}$ be a quotient of $\mathcal{D}$ for Θ, with projection $Q : \mathcal{D} \to \mathcal{C}$, that is, Q satisfies $Q \, ; Q^\smile = \Theta$ and $Q^\smile \, ; Q = \mathbb{I}_{\mathcal{C}}$.
- Define $S : \mathcal{C} \to \mathcal{A}$ as $S := Q^\smile \, ; P \, ; R$.   □

Löwe *et al.* gave an example that even for surjective R, different pullback complements do not need to be isomorphic [LKPS06, Fig. 3], and even where different pullback complement objects are isomorphic, these isomorphisms do not necessarily factor the two pullback complements [Kah01, p. 119].

So there are two reasons that make working with general pullback complements unsatisfactory: finding an equivalence relation determining the second projection Q is, in general, computationally inefficient, and the resulting diagram is not even unique up to isomorphism.

However, in Sect. 6 we will consider R as mapping variable occurrences in a rule's left-hand side L to variables in the gluing object G, and in that context, injectivity of R corresponds to the restriction to left-linear rules, and surjectivity of R corresponds to (part of) the restriction to rules with no new variables on the right-hand side. If R is injective, then Q is injective, too, so with the assumption that Q is also surjective, we may set $Q = \mathbb{I}$ when constructing a pullback complement.

In contrast to the general case, this trivial pullback complement is also more "well-behaved" in that it factors other candidates [Kah01, Prop. 5.5.6]:

Proposition 5.8. If $\mathcal{D} \xrightarrow{Q} C \xrightarrow{S} \mathcal{A}$ and $\mathcal{D} \xrightarrow{Q'} C' \xrightarrow{S'} \mathcal{A}$ are two pullback complements for $\mathcal{D} \xrightarrow{P} \mathcal{B} \xrightarrow{R} \mathcal{A}$, and if Q is injective and surjective, then $Y := Q^{\smile} \,\mathring{;}\, Q'$ is a mapping and $Q' = Q \,\mathring{;}\, Y$ and $S = Y \,\mathring{;}\, S'$.   □

6   Amalgamating Pushout and Pullback Graph Transformation

While the DPO approach is recognised to produce much more intuitive rule formulations than the DPB approach, one of the limitations of the expressiveness of DPO graph transformation is that it can delete only items explicitly mentioned in the rule, but can neither delete nor duplicate more loosely described subgraphs. One way to formulate this additional expressiveness is to include some concept of variables in the rule constituents, and allow such variables to match subgraphs. Such duplication and deletion can, however, be very easily expressed in DPB rules. To combine the best of both worlds, we now amalgamate the two approaches, using DPB rewriting on "subgraph variables", and DPO rewriting on the "constant context parts".

An example rule $L \xleftarrow{\varPhi_{\mathrm{L}}} (G, u_0) \xrightarrow{\varPhi_{\mathrm{R}}} R$ with redex $X_{\mathrm{L}} : L \to A$ is shown in Fig. 3, and, for the sake of readability, the remainder of the induced rewriting step is shown in the separate Figures 4 and 5 with rearranged rule graphs.

The subidentity u_0 on the gluing graph G marks its constant part; the semi-complement of that is then the *parameter* part — in the example in Fig. 3, the parameter part consists of a unit graph (a node with a loop edge) at the bottom, connected by an incoming edge with the constant part.

The rule's LHS $\varPhi_{\mathrm{L}}$ embeds the gluing graph into additional context (top left) which is also constant, and will be deleted as in DPO rewriting; the rule's RHS duplicates the parameter part.

Rewriting now essentially applies a DPO step to the constant part, and a DPB step to the parameter part. The unit graph of the parameter part can be

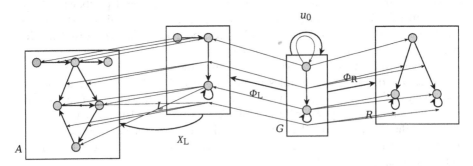

Fig. 3. Double-pullout rule with matching $X_{\mathrm{L}} : L \to A$

matched to an arbitrary subgraph; here it is matched to a triangular cycle. The connecting edge of the parameter part specifies that the image of the unit graph can only be connected with the context by edges starting from the context; here, there are two such edges. Since the rule here is left-linear, constructing the host graph is straight-forward, and for calculating the result, duplication of the image of the parameter part has to be performed as specified by the RHS.

We now first introduce the definitions necessary for working with an appropriate separation into parameter and constant parts. Then we amalgamate the tabulation and co-tabulation concepts into bi-tabulations. The corresponding square diagrams will be called "pullouts", and we also characterise and construct pullout complements.

6.1 Gluing Setups and Pullouts

As we have seen above, the gluing object in our approach is accompanied by a subidentity specifying its interface part:

Definition 6.1. Let an object G_0 and a partial identity $u_0 : \mathsf{SubId}\, G_0$ be given. The pair (G_0, u_0) is then called a *gluing object G_0 along u_0*, and u_0 is called the *interface component* of G_0.

Furthermore, we define $v_0 := \widetilde{u_0}$ and consider it as part of the gluing setup; we may now talk about a *gluing object G_0 along u_0 over v_0*, and we call v_0 the *parameter component* of G_0.   □

In the rule above, the interface component (u_0) consists only of the top node, while the parameter component v_0 is the whole graph. This shows an asymmetry between the two components: where u_0 is the whole graph, v_0 is empty.

We will restrict the interface between the rule and its application context to (the image of) the non-parameter part u_0. Furthermore, the border between u_0 and the parameter part v_0 will be the only interface allowed between the parameter part and its context.

All the morphisms starting at the gluing object will need to satisfy the sanity conditions of *interface preservation*:

Definition 6.2. Let a gluing object G_0 along u_0 over v_0 and a morphism Φ from G_0 to another object G_1 be given.

In such a context, we shall use the following abbreviations: Let $u_1 := \mathsf{ran}\,(u_0 \,\text{;}\, \Phi)$ and $v_1 := \mathsf{ran}\,(v_0 \,\text{;}\, \Phi)$ be the images of the gluing components; let $b_1 := v_1 \sqcap u_1$ be the border of the parameter image with the image of the constant part.

Φ is called *interface preserving* if the following conditions are satisfied:

– The border of the image of the parameter part is contained in the image of the interface component: $v_1 \sqcap \widetilde{v_1} \sqsubseteq u_1$

− Φ is total on the interface component, and relates only constant items of G_0 with non-parameter items of G_1: $\mathsf{dom}\,(\Phi \mathbin{;} (v_1^{\smile} \sqcup u_1)) = u_0$
− Φ is univalent on the border of the parameter image: $b_1 \mathbin{;} \Phi^{\smile} \mathbin{;} \Phi \sqsubseteq b_1$.   □

Both rule sides in the example rule in Fig. 3, and also the resulting host morphism in Fig. 4 are interface preserving.

Definition 6.3. A *gluing setup* $(G_0, u_0, v_0, \Xi, \Phi)$ consists of a gluing object G_0 along u_0 over v_0, and a span $G_2 \xleftarrow{\ \Xi\ } G_0 \xrightarrow{\ \Phi\ } G_1$ of interface-preserving morphisms, where G_2 is called the *host object* and G_1 is called the *rule side*.

In a gluing setup with names as above, we additionally define the following names for the interface, parameter, and target specific components in the target objects:

$$v_1 := \mathsf{ran}\,(v_0 \mathbin{;} \Phi) \qquad u_1 := \mathsf{ran}\,(u_0 \mathbin{;} \Phi) \qquad r_1 := (\mathsf{ran}\,\Phi)^{\sim}$$
$$v_2 := \mathsf{ran}\,(v_0 \mathbin{;} \Xi) \qquad u_2 := \mathsf{ran}\,(u_0 \mathbin{;} \Xi) \qquad h_2 := (\mathsf{ran}\,\Xi)^{\sim}$$

In addition, we define names for the borders between the interface and parameter parts, and for the whole non-parameter parts:

$$b_1 := u_1 \sqcap v_1 \qquad c_1 := u_1 \sqcup r_1$$
$$b_2 := u_2 \sqcap v_2 \qquad c_2 := u_2 \sqcup h_2 \qquad q_2 := b_2 \sqcup h_2 \qquad \square$$

A simple first approach to completing a gluing setup to a full square starts with producing a pushout for the interface parts $u_0 \mathbin{;} \Phi$ and $u_0 \mathbin{;} \Xi$, and a pullback for the converses of the parameter parts, i.e., for $\Phi^{\smile} \mathbin{;} v_0$ and $\Xi^{\smile} \mathbin{;} v_0$. To make these pushouts and pullbacks possible, Φ and Ξ have to behave "essentially like mappings" in the forward direction on the interface part, and in the backward direction on the parameter part:

Definition 6.4. Given a gluing object G_0 along u_0, a morphism $\Phi : G_0 \leftrightarrow G_1$ is called a *standard gluing morphism* iff the following conditions are satisfied:
− Φ is interface preserving,
− Φ is total on u_0, that is, $u_0 \sqsubseteq \mathsf{dom}\,\Phi$,
− Φ is univalent on u_0, that is, $u_0 \mathbin{;} \Phi$ is univalent, and
− Φ is almost-injective besides u_0.   □

The next step will now be to complete such a gluing setup to a commuting square as drawn to the right. Once we have the pushout of the interface part and the pullback of the parameter part, the two result objects then have to be glued together along a gluing relation induced by the restrictions of the pushout and pullback morphisms to the borders b_1 and b_2.

$$\begin{array}{ccc} G_0 & \xrightarrow{\ \Phi\ } & G_1 \\ {\scriptstyle\Xi}\downarrow & & \downarrow{\scriptstyle X} \\ G_2 & \xrightarrow{\ \Psi\ } & G_3 \end{array}$$

The following diagram should help with orientation:

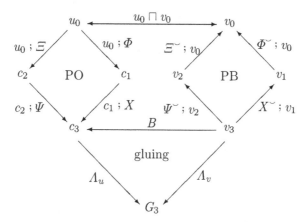

According to this intuition, the specification of this construction also has to join the specifications of pushout and pullback into a single specification. We have to be careful to restrict the occurrences of $\blacktriangleright$ and $\blacktriangleleft$ to those parts of G_1 and G_2 that are governed by the pushout construction, since otherwise the reflexive part would override the pullback domains. The final gluing is not directly reflected in these conditions since interface preservation keeps the border part inside the interface part, so the gluing components for the interface part already cover the final gluing.

Definition 6.5. Given a gluing setup $G_1 \xleftarrow{\Phi} G_0 \xrightarrow{\Xi} G_2$ along u_0 over v_0 in a Dedekind category $\mathbf{D}$, a cospan $G_1 \xrightarrow{X} G_3 \xleftarrow{\Psi}$ of relations in $\mathbf{D}$ is a *pullout* for $G_1 \xleftarrow{\Phi} G_0 \xrightarrow{\Xi} G_2$ along u_0 iff the following conditions are satisfied:

$$
\begin{aligned}
\Phi \,;\, X &= \Xi \,;\, \Psi \\
X \,;\, \Psi^\smile &= (\Phi^\smile \,;\, u_0 \,;\, \Xi)^{\circledast} \quad \sqcup \quad \Phi^\smile \,;\, v_0 \,;\, \Xi \\
X \,;\, X^\smile &= c_1 \,;\, (\Phi^\smile \,;\, u_0 \,;\, \Xi)^{\blacktriangleright} \,;\, c_1 \;\sqcup\; \mathrm{dom}\,(\Phi^\smile \,;\, v_0 \,;\, \Xi) \\
\Psi \,;\, \Psi^\smile &= c_2 \,;\, (\Phi^\smile \,;\, u_0 \,;\, \Xi)^{\blacktriangleleft} \,;\, c_2 \;\sqcup\; \mathrm{ran}\,(\Phi^\smile \,;\, v_0 \,;\, \Xi) \\
\mathbb{I} &= X^\smile \,;\, c_1 \,;\, X \sqcup \Psi^\smile \,;\, c_2 \,;\, \Psi \sqcup (X^\smile \,;\, v_1 \,;\, X \sqcap \Psi^\smile \,;\, v_2 \,;\, \Psi)
\end{aligned}
$$ □

We now have all the definitions in place to set up a *double-pullout rewriting approach* in analogy to the double-pushout approach.

Definition 6.6. In a collagory, a *double-pullout rule* $\mathcal{L} \xleftarrow{\Phi_L} (\mathcal{G}, u_0) \xrightarrow{\Phi_R} \mathcal{R}$ consists of a gluing object $\mathcal{G}$ along u_0, the left-hand side and right-hand side objects $\mathcal{L}$ and $\mathcal{R}$, and two standard gluing morphisms $\Phi_L : (\mathcal{G}, u_0) \to \mathcal{L}$ and $\Phi_R : (\mathcal{G}, u_0) \to \mathcal{R}$.

Such a double-pullout rule is called *left-linear* iff Φ_L is univalent.

Given such a rule, if for an application object $\mathcal{A}$ and a morphism $X_L : \mathcal{L} \leftrightarrow \mathcal{A}$ all preconditions of Theorem 6.11 hold, then the rule is *applicable to $\mathcal{A}$ via*

X_L, and *application* first calculates a pullout complement $\mathcal{G} \xrightarrow{\Xi} \mathcal{H} \xrightarrow{\Psi_L} \mathcal{A}$ for $(\mathcal{G}, u_0) \xrightarrow{\Phi_L} \mathcal{L} \xrightarrow{X_L} \mathcal{A}$, and then constructs a pullout $\mathcal{R} \xrightarrow{X_R} \mathcal{B} \xleftarrow{\Psi_R} \mathcal{H}$ for $\mathcal{R} \xleftarrow{\Phi_R} (\mathcal{G}, u_0) \xrightarrow{\Xi} \mathcal{H}$, where $\mathcal{B}$ is the *result* of the application. $\square$

$$\begin{array}{ccccc}
\mathcal{L} & \xleftarrow{\Phi_L} & (\mathcal{G}, u_0) & \xrightarrow{\Phi_R} & \mathcal{R} \\
{\scriptstyle X_L}\downarrow & & {\scriptstyle \Xi}\downarrow & & \downarrow{\scriptstyle X_R} \\
\mathcal{A} & \xleftarrow{\Psi_L} & \mathcal{H} & \xrightarrow{\Psi_R} & \mathcal{B}
\end{array}$$

For the completion of the redex of Fig. 3 to a double-pullout rewriting step, we show the host and result morphisms in Figures 4 and 5.

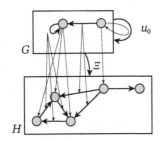

Fig. 4. Host morphism for Fig. 3

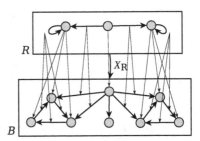

Fig. 5. Result morphism for Fig. 3

6.2   Bi-tabulations

Since the definition of pullouts obviously does not need all the details of Φ and Ξ, we now perform the step from square diagram to triangle, as previously in the move from pullback to tabulation, or from pushout to co-tabulation. While there, a single (relational) morphism was sufficient as a starting point equivalent to the original (co-)span of mappings, replacing the gluing setup requires treating parameter and constant parts separately. We therefore start from the following two relations:

$$U := (\Phi^{\smile} \,;\, u_0 \,;\, \Xi)^{\boxplus} \,, \qquad\qquad V := \Phi^{\smile} \,;\, v_0 \,;\, \Xi \,.$$

Definition 6.7. Let $U, V : G_1 \to G_2$ and $c_1 \sqsubseteq \mathbb{I}_{G_1}$ and $c_2 \sqsubseteq \mathbb{I}_{G_2}$ be given. Define:

$$v_1 := c_1^{\smallsmile} \,, \qquad\qquad v_2 := c_2^{\smallsmile} \,.$$

If U is difunctional and if we have *direct interface preservation*:

$$\begin{array}{lll}
\operatorname{dom} U \sqsubseteq c_1 & \operatorname{dom} V \sqsubseteq v_1 & c_1 \,;\, V \sqsubseteq U \\
\operatorname{ran} U \sqsubseteq c_2 & \operatorname{ran} V \sqsubseteq v_2 & V \,;\, c_2 \sqsubseteq U \,,
\end{array}$$

then a cospan $G_1 \xrightarrow{X} G_3 \xleftarrow{\Psi} G_2$ is called a *bi-tabulation for V along U on c_1 and c_2* iff the following conditions hold:

$$
\begin{aligned}
X \,;\, \Psi^\smile = &\quad U &\sqcup&\quad V \\
X \,;\, X^\smile = &\quad c_1 \,;\, (\mathbb{I} \sqcup U \,;\, U^\smile) \,;\, c_1 &\sqcup&\quad (\mathbb{I} \sqcap V \,;\, V^\smile) \\
\Psi \,;\, \Psi^\smile = &\quad c_2 \,;\, (\mathbb{I} \sqcup U^\smile \,;\, U) \,;\, c_2 &\sqcup&\quad (\mathbb{I} \sqcap V^\smile \,;\, V) \\
\mathbb{I} = &\, X^\smile \,;\, c_1 \,;\, X \sqcup \Psi^\smile \,;\, c_2 \,;\, \Psi \sqcup (X^\smile \,;\, v_1 \,;\, X \sqcap \Psi^\smile \,;\, v_2 \,;\, \Psi) &&\qquad\qquad \square
\end{aligned}
$$

Theorem 6.8. Bi-tabulations for V along difunctional U on c_1 and c_2 with direct interface preservation are unique up to isomorphism. $\qquad\square$

When actually constructing a bi-tabulation in a bi-tabular collagory, the only minor technical complication arises from the restriction of the non-parameter gluing to c_1 and c_2 which is necessary for preserving the possibility that the parameter parts of X and Ψ may be partial. For achieving this cleanly, we have to construct this first gluing from appropriate subobjects. For the parameter part, however, no such subobjects are necessary since the tabulation does not reach beyond the domain and range of V.

Theorem 6.9. Under the preconditions of Def. 6.7, we obtain a bi-tabulation for V along U on c_1 by defining the co-span $G_1 \xrightarrow{X} G_3 \xleftarrow{\Psi} G_2$ as follows:

- Let $\lambda_1 : C_1 \rightarrowtail G_1$ be a subobject for c_1, and let $\lambda_2 : C_2 \rightarrowtail G_2$ be a subobject for c_2.
- Let $C_1 \xrightarrow{X_u} G_u \xleftarrow{\Psi_u} C_2$ be a co-tabulation for $\lambda_1 \,;\, U \,;\, \lambda_2^\smile$ (which is difunctional).
- Let $G_1 \xleftarrow{X_v} G_v \xrightarrow{\Psi_v} G_2$ be a tabulation for V.
- Define $B := X_v \,;\, \lambda_1^\smile \,;\, X_u \sqcup \Psi_v \,;\, \lambda_2^\smile \,;\, \Psi_u$.
- Let $G_v \xrightarrow{\Lambda_v} G_3 \xleftarrow{\Lambda_u} G_u$ be a gluing for B, that is, a co-tabulation for $B^\boxdot$.
- Define: $X := \lambda_1^\smile \,;\, X_u \,;\, \Lambda_u \sqcup X_v^\smile \,;\, \Lambda_v$ and $\Psi := \lambda_2^\smile \,;\, \Psi_u \,;\, \Lambda_u \sqcup \Psi_v^\smile \,;\, \Lambda_v$ $\qquad\square$

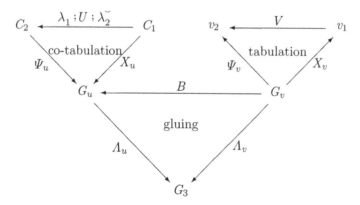

A bi-tabulation induced by a gluing setup with standard gluing morphisms is well-defined, and produces a pullout:

Theorem 6.10. Let a gluing setup $(G_0, u_0, v_0, \Xi, \Phi)$ be given where Φ and Ξ both are standard gluing morphisms. Defining $U := \Phi^\smile \,;\, u_0 \,;\, \Xi$ and $V := \Phi^\smile \,;\, v_0 \,;\, \Xi$ ensures direct interface preservation, and a bi-tabulation $G_1 \xrightarrow{X} G_3 \xleftarrow{\Psi} G_2$ for V along U on c_1 is also a pullout for the gluing setup $(G_0, u_0, v_0, \Xi, \Phi)$. $\qquad\square$

6.3   Pullout Complements

Theorem 6.11. Let a gluing object $\mathcal{G}$ along u_0 over v_0 and a standard gluing morphism $\Phi : \mathcal{G} \to \mathcal{L}$ be given, and a morphism $X : \mathcal{L} \to \mathcal{A}$. Define:

$$v_1 := \mathrm{ran}\,(v_0 \,\mathring{,}\, \Phi) \qquad\qquad u_1 := \mathrm{ran}\,(u_0 \,\mathring{,}\, \Phi) \qquad\qquad c_1 := u_1 \sqcup (\mathrm{ran}\,\Phi)^{\sim}$$
$$v_3 := \mathrm{ran}\,(v_1 \,\mathring{,}\, X) \qquad\qquad u_3 := \mathrm{ran}\,(u_1 \,\mathring{,}\, X) \qquad\qquad c_3 := u_3 \sqcup (\mathrm{ran}\,X)^{\sim}$$

If the following conditions are satisfied:
1. $c_1 \,\mathring{,}\, X$ is univalent, and $c_1 \sqsubseteq \mathrm{dom}\,X$,
2. X is almost-injective besides u_1,
3. $X \,\mathring{,}\, (\mathrm{ran}\,X)^{\sim} \sqsubseteq u_1 \,\mathring{,}\, X$,
4. there is a partial equivalence relation $\Theta : \mathcal{A} \leftrightarrow \mathcal{A}$ such that

$$\mathrm{ran}\,\Theta = v_3 \qquad\qquad X^{\smile} \,\mathring{,}\, v_1 \,\mathring{,}\, X \sqcap \Theta \sqsubseteq \mathbb{I}$$
$$u_3 \,\mathring{,}\, \Theta \sqsubseteq u_3 \qquad\qquad v_1 \,\mathring{,}\, X \,\mathring{,}\, \Theta = \Phi^{\smile} \,\mathring{,}\, v_0 \,\mathring{,}\, \Phi \,\mathring{,}\, X \ ,$$

then there is a pullout complement $\mathcal{A} \xrightarrow{\;\Xi\;} \mathcal{H} \xrightarrow{\;\Psi\;} \mathcal{L}$ constructed as follows:
– Let $\mathcal{V}$ contain the parameter instantiation, characterised as a combined quotient and subobject by $\nu : \mathcal{A} \leftrightarrow \mathcal{V}$ with:

$$\nu^{\smile} \,\mathring{,}\, \nu = \mathbb{I} \ , \qquad\qquad \nu \,\mathring{,}\, \nu^{\smile} = \Theta \ .$$

– Let $\mathcal{C}$ be the subobject of $\mathcal{A}$ containing the context:

$$\lambda : \mathcal{C} \leftrightarrow \mathcal{A} \qquad\qquad \lambda \,\mathring{,}\, \lambda^{\smile} = \mathbb{I} \qquad\qquad \lambda^{\smile} \,\mathring{,}\, \lambda = c_3$$

– Let $\mathcal{C} \xrightarrow{\;\iota\;} \mathcal{H} \xleftarrow{\;\kappa\;} \mathcal{V}$ be the gluing of $Z : \mathcal{C} \leftrightarrow \mathcal{V}$ with $Z := \lambda \,\mathring{,}\, \nu$.
– Define: $\Psi := \iota^{\smile} \,\mathring{,}\, \lambda \sqcup \kappa^{\smile} \,\mathring{,}\, \nu^{\smile}$ and $\Xi := \Phi \,\mathring{,}\, X \,\mathring{,}\, \Psi^{\smile}$.

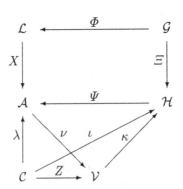

Ψ is injective by construction. Furthermore, Ξ is a standard gluing morphism if the following additional conditions are satisfied:
5. $v_3 \sqcap v_3^{\smile} \sqsubseteq u_3$, and
6. $X \,\mathring{,}\, u_3 \sqsubseteq u_1 \,\mathring{,}\, X$. $\qquad\qquad\qquad\qquad\qquad\qquad\qquad\qquad\qquad\qquad\square$

For left-linear rules, no search for a partial equivalence is necessary:

Theorem 6.12. If, in the setting of Theorem 6.11, Φ is univalent, then condition 6.11.(4) is satisfied by setting $\Theta := v_3$, and if conditions 6.11.(1)–6.11.(3) also hold, the straight host construction of Def. 5.3 produces a pullout complement that is isomorphic to that produced by the construction of Theorem 6.11. $\square$

7   Conclusion

Double-pullout graph rewriting allows to formulate rules that contain subgraph variables and may use these to specify duplication or deletion of the matched subgraphs. This is an expressive power not available to the double-pushout approach. Although this expressive power is present in the double-pullback approach, the equivalent rules there would be much less readable. Therefore, double-pullout graph rewriting combines the intuitive accessibility of the double-pushout approach with the expressive power of the double-pullback approach. It achieves this using an elegant amalgamation of the two approaches at the level of their abstract relation-algebraic characterisations, and we identified bi-tabular collagories with difunctional closure and pseudo- and semi-complements of subidentities as sufficient relation-categoric foundation.

References

[BJ01]   Bauderon, M., Jacquet, H.: Pullback as a Generic Graph Rewriting Mechanism. Applied Categorical Structures 9(1), 65–82 (2001)

[CMR+97] Corradini, A., Montanari, U., Rossi, F., Ehrig, H., Heckel, R., Löwe, M.: Algebraic Approaches to Graph Transformation, Part I: Basic Concepts and Double Pushout Approach. In: Rozenberg, G. (ed.) Handbook of Graph Grammars and Computing by Graph Transformation, ch. 3. Foundations, vol. 1, pp. 163–245. World Scientific, Singapore (1997)

[FS90]   Freyd, P.J., Scedrov, A.: Categories, Allegories, North-Holland Mathematical Library, vol. 39. North-Holland, Amsterdam (1990)

[Jac99]   Jacquet, H.: Une approche catégorique de la réécriture de sommets dans les graphes. PhD thesis, Université Bordeaux 1 (1999)

[Kah01]   Kahl, W.: A Relation-Algebraic Approach to Graph Structure Transformation, Habil. Thesis, Fakultät für Informatik, Univ. der Bundeswehr München, Techn. Report 2002-03 (2001), http://sqrl.mcmaster.ca/~kahl/Publications/RelRew/

[Kah02]   Kahl, W.: A Relation-Algebraic Approach to Graph Structure Transformation. In: de Swart, H. (ed.) RelMiCS 2001. LNCS, vol. 2561, pp. 1–14. Springer, Heidelberg (2002)

[Kah04]   Kahl, W.: Refactoring Heterogeneous Relation Algebras around Ordered Categories and Converse. J. Relational Methods in Comp. Sci. 1, 277–313 (2004)

[Kah09]   Kahl, W.: Collagories for Relational Adhesive Rewriting. In: Berghammer, R., Jaoua, A.M., Möller, B. (eds.) RelMiCS 2009. LNCS, vol. 5827, pp. 211–226. Springer, Heidelberg (2009)

[Kah10]   Kahl, W.: Collagory Notes, Version 1. SQRL Report 57, Software Quality
           Research Laboratory, Department of Computing and Software, McMaster
           University, 53 pages (2010),
           http://sqrl.mcmaster.ca/sqrl_reports.html

[Kaw90]   Kawahara, Y.: Pushout-Complements and Basic Concepts of Grammars in
           Toposes. Theoretical Computer Science 77, 267–289 (1990)

[Koz94]   Kozen, D.: A Completeness Theorem for Kleene Algebras and the Algebra
           of Regular Events. Inform. and Comput. 110(2), 366–390 (1994)

[LS04]   Lack, S., Sobociński, P.: Adhesive Categories. In: Walukiewicz, I. (ed.) FOS-
           SACS 2004. LNCS, vol. 2987, pp. 273–288. Springer, Heidelberg (2004)

[LKPS06]   Löwe, M., König, H., Peters, M., Schulz, C.: Refactoring Informations Sys-
           tems. In: Software Evolution through Transformations. Electronic Commu-
           nications of the EASST, vol. 3 (2006)

[SS93]   Schmidt, G., Ströhlein, T.: Relations and Graphs, Discrete Mathematics
           for Computer Scientists. EATCS-Monographs on Theoret. Comput. Sci.
           Springer, Heidelberg (1993)

ICGT 2010 Doctoral Symposium

Andrea Corradini[1] and Maarten de Mol[2]

[1] Dipartimento di Informatica, Università di Pisa, Italy
andrea@di.unipi.it
[2] University of Twente, The Netherlands
M.J.deMol@utwente.nl

Given the success of the Doctoral Symposium held for the first time during ICGT in 2008 in Leicester, also this year a specific event of the International Conference on Graph Transformation was explicitly dedicated to Ph.D. students. The Doctoral Symposium consisted of technical sessions dedicated to presentations by doctoral students, held during the main conference, giving them a unique opportunity to present their research project and to interact with established researchers of the graph transformation community and with other students.

Among the several submissions of high quality that we received, the following twelve three-page abstracts were accepted for presentation at the conference and are included in the ICGT 2010 proceedings:

1. Enrico Biermann, *EMF Model Transformation based on Graph Transformation: Formal Foundation and Tool Environment*
2. Christoph Blume, *Recognizable graph languages for the verification of dynamic systems*
3. Adwoa Donyina, *Stochastic Modelling and Simulation of Dynamic Resource Allocation*
4. Mathias Hülsbusch, *Bisimulation Theory for Graph Transformation Systems*
5. Eugen Jiresch, *Realizing Impure Functions in Interaction Nets*
6. Stefan Jurack, *Composite EMF Modeling based on Typed Graphs with Inheritance and Containment Structures*
7. Tony Modica, *Formal Modeling and Analysis of Communication Platforms like Skype based on Petri Net Transformation Systems*
8. Giacoma Monreale *LTS semantics for process calculi from their graphical encodings*
9. Fawad Qayum, *Automated Assistance for Search-Based Refactoring using Unfolding of Graph Transformation Systems*
10. Hendrik Radke, *Correctness of Graph Programs relative to HR+ Conditions*
11. Zoltán Ujhelyi, *Static Type Checking of Model Transformation Programs*
12. Eduardo Zambon, *Using Graph Transformations and Graph Abstractions for Software Verification*

H. Ehrig et al. (Eds.): ICGT 2010, LNCS 6372, pp. 379–380, 2010.

These contributions were selected on the basis of the submitted abstracts according to their originality, significance and general interest, by the members of the following Program Committee:

- Paolo Baldan
- Luciano Baresi
- Michel Bauderon
- Andrea Corradini (chair)
- Juan de Lara
- Maarten de Mol
- Hartmut Ehrig
- Gregor Engels
- Annegret Habel
- Reiko Heckel
- Dirk Janssens
- Barbara König
- Hans-Joerg Kreowski
- Mark Minas
- Ugo Montanari
- Mohamed Mosbah
- Fernando Orejas
- Francesco Parisi-Presicce
- Mauro Pezzè
- Detlef Plump
- Gabriele Taentzer
- Emilio Tuosto

After the conference, authors of selected contributions will be invited to submit a full paper for the refereed post-proceedings of the Doctoral Symposium, which will be published as a volume of the Electronic Communications of the European Association of Software Science and Technology (ECEASST) series.

EMF Model Transformation Based on Graph Transformation: Formal Foundation and Tool Environment

Enrico Biermann

Institut für Softwaretechnik und Theoretische Informatik
Technische Universität Berlin, Germany
`enrico@cs.tu-berlin.de`

Introduction. Model-driven software development is considered as a promising paradigm in software engineering [10].

Models are the central artifacts in model-driven development. Hence, inspecting and modifying models to reduce their complexity and improve their readability, maintainability and extensibility (i.e. by performing *model refactoring* [8]) are important issues of model development. Thus model transformation can be considered as one of the key activities in model-driven software development.

The Eclipse Modeling Framework (EMF) [5] has evolved to a de facto standard technology to define models and modeling languages. EMF provides a modeling and code generation framework for Eclipse applications based on structured data models. The modeling approach is similar to that of MOF, actually EMF supports Essential MOF (EMOF) as part of the OMG MOF 2.0 specification [9]. Several EMF model transformation approaches have been developed, focusing on different transformation aspects.

In this paper, an overview of the state of my doctoral thesis is presented. In the thesis, graph transformation will be used to transform EMF models. Therefore, several graph transformation concepts like amalgamated graph transformation and application conditions will be applied to EMF model transformations. Furthermore some results of graph transformation theory will be adapted to analyze EMF model transformations. In particular, the following topics are the main subjects of my thesis:

Consistent EMF Transformations. EMF model transformations are defined as a special kind of typed attributed graph transformations using node type inheritance. However, EMF models have several structural properties that do not have a direct correspondence in the domain of graphs. The most prominent one being containment relations (i.e. compositions in UML) define an ownership relation between objects. Thereby, they induce a hierarchical structure in model instantiations. In MOF and EMF, this structure is further used to implement a mapping to XML, known as XMI (XML Meta data Interchange) [9]. Containment always implies a number of constraints for model instantiations that must be ensured at run-time, e.g. no containment cycles and only a maximum of one container for each object. Furthermore, other EMF model properties like

H. Ehrig et al. (Eds.): ICGT 2010, LNCS 6372, pp. 381–383, 2010.

bidirectional edges or the non-existence of parallel edges are ensured by graph constraints. First results on consistent EMF transformations have been presented in [1].

Parallel and Amalgamated Graph Transformation. In order to increase the expressivness of graph transformations *parallel graph transformation* concepts were introduced in [6] and extended to synchronized, overlapping rules in [11]. The essence of parallel graph transformation is that a (possibly infinite) set of rules with some regularity, a so-called *rule scheme*, can be described by a finite set of *multi-rules* modeling elementary actions only. For the description of rule schemes, the concept of amalgamating rules at *kernel rules* [4] is used to describe the application of multi-rules in variable contexts. Thus, the transformation of multi-object structures can be described in a general way. To use this enhanced expressiveness for EMF model transformations, these concepts were lifted to amalgamted EMF transformations in [3] and [2].

Application Conditions. Apart from transformation units, application conditions that can be hierarchically nested, have recently been shown to raise the expressiveness of visual behavior modelling to the level of first-order logic. Hence, graph transformation rules can be enhanced by nested application conditions. There also exist theoretical results regarding local confluence of rules with nested application conditions [7]. In my thesis, the application conditions are used also for EMF transformations.

Tool Environment. The proposed approach to formal EMF model transformation has been implemented in the tool EMF Henshin[1], a subproject of the Eclipse Modeling Framework Technology Project. EMF Henshin contains visual modeling capabilities and a transformation framework supporting the definition and in-place execution of consistent EMF model transformations. The engine which is used to transform EMF models is implemented in Java. However consistent EMF model transformations modeled with EMF Henshin can be converted to AGG graph rules to analyze their behaviour wrt. termination and functional behavior.

Independence Analysis for EMF Model Transformations. In my thesis, the definition of consistent EMF transformations will be extended by the notion of independency between different rules. Even though individual rules might lead to corrrect transformation steps, one such step might lead to an EMF model on which a second rule is no longer applicable resulting in a conflict. In graph transformation theory such conflicts are detectable by critical pair analysis. However, the additional structural properties of EMF models are not taken into account during that analysis. One of the goals of my thesis is to include those properties towards a notion of independence for EMF model transformations. Thus, consistent EMF model transformations behave like algebraic graph transformations

---

[1] http://www.eclipse.org/modeling/emft/henshin/

and the rich theory of algebraic graph transformation can be applied to these EMF model transformations to show functional behavior and correctness. This part of my thesis is ongoing research.

Summary. The paper gives an overview about the use of graph transformation for the definition of EMF model transformations. In addition, ongoing research is presented concerning independence analysis for EMF rules based on local confluence.

References

1. Biermann, E., Ermel, C., Taentzer, G.: Precise semantics of EMF model transformations by graph transformation. In: Czarnecki, K., Ober, I., Bruel, J.-M., Uhl, A., Völter, M. (eds.) MODELS 2008. LNCS, vol. 5301, pp. 53–67. Springer, Heidelberg (2008), http://tfs.cs.tu-berlin.de/publikationen/Papers08/BET08.pdf
2. Biermann, E., Ermel, C., Taentzer, G.: Formal foundation of consistent EMF model transformations by algebraic graph transformation. In: Software and Systems Modeling, SoSyM (to appear 2010)
3. Biermann, E., Ermel, C., Taentzer, G.: Lifting parallel graph transformation concepts to model transformation based on the Eclipse modeling framework. ECE-ASST 26 (2010),
   http://journal.ub.tu-berlin.de/index.php/eceasst/issue/view/36
4. Böhm, P., Fonio, H.R., Habel, A.: Amalgamation of graph transformations: a synchronization mechanism. Computer and System Sciences (JCSS) 34, 377–408 (1987)
5. Eclipse Consortium: Eclipse Modeling Framework (EMF) – Version 2.4 (2008), http://www.eclipse.org/emf
6. Ehrig, H., Kreowski, H.J.: Parallel graph grammars. In: Lindenmayer, A., Rozenberg, G. (eds.) Automata, Languages, Development, pp. 425–447. North Holland, Amsterdam (1976)
7. Lambers, L., Ehrig, H., Habel, A., Orejas, F., Golas, U.: Local confluence for rules with nested application conditions based on a new critical pair notion. Tech. rep., Technische Universität Berlin (2010),
   http://www.eecs.tu-berlin.de/menue/forschung/forschungsberichte/2010
8. Mens, T., Tourwé, T.: A survey of software refactoring. Transactions on Software Engineering 30(2), 126–139 (2004)
9. Object Management Group: Meta Object Facility (MOF) Core Specification Version 2.0 (2008),
   http://www.omg.org/technology/documents/modeling_spec_catalog.htm#MOF
10. Schmidt, D.C.: Model-driven engineering. IEEE Computer 39(2), 25–31 (2006)
11. Taentzer, G.: Parallel and Distributed Graph Transformation: Formal Description and Application to Communication-Based Systems. Ph.D. thesis, TU Berlin, Shaker Verlag (1996)

Recognizable Graph Languages for the Verification of Dynamic Systems

Christoph Blume

Universität Duisburg-Essen, Germany
christoph.blume@uni-due.de

The theory of regular word languages has a large number of applications in computer science, especially in verification. The notion of regularity can be straightforwardly generalized to trees and tree automata. Therefore it is natural to ask for a theory of regular graph languages.

There exist several notions of regular graph languages [10,7,8] – in this context called *recognizable* graph languages – which all turned out to be equivalent. Especially the notion of Courcelle is widely accepted. Very roughly, one can say that a property (or language) of graphs is recognizable whenever it can be derived inductively via an (arbitrary) decomposition of the graph. Alternatively recognizability can be defined via a family of Myhill-Nerode style congruences of finite index, i.e., congruences with finitely many equivalence classes.

The notion of recognizability by Bruggink and König [8] is based on a categorical definition of recognizability in terms of so-called *automaton functors (AF)*, which are a generalization of non-deterministic finite automata. An advantage of this notion of recognizability is that many familiar constructions on finite automata can be straightforwardly generalized to automaton functors.

Let $Cospan(Graph)$ be the category which has arbitrary graphs as objects and cospans of graphs which can be seen as a graph with a left and a right interface as arrows. If automaton functors from the category $Cospan(Graph)$ are considered this yields exactly the notion of recognizable graph languages mentioned above. Cospans of graphs are closely related to GTSs, in particular to the double-pushout (DPO) approach to graph transformation [13]. A DPO rule $\rho\colon L \xrightarrow{\rho_L} I \xleftarrow{\rho_R} R$ can be seen as a pair of cospans $\ell\colon \emptyset \to L \xrightarrow{\rho_L} I$, $r\colon \emptyset \to R \xrightarrow{\rho_R} I$. Then it holds that $G \Rightarrow_\rho H$ iff $\emptyset \to G \leftarrow \emptyset = c \circ \ell$ and $\emptyset \to H \leftarrow \emptyset = c \circ r$, for some cospan c.

In the following I will give a short overview of my research topics. The focus is on verification techniques based on recognizable graph languages for dynamic systems. Below I will briefly present some projects on which I am working:

Recognizability and Invariant Checking: One of the most straightforward approaches to verification is to provide an invariant and to show that it is preserved by all transformation rules. In the case of words, a language is an invariant for a rule $\ell \to r$ if it holds for all words u and v that $u\ell v \in L$ implies $urv \in L$. In the case of regular (word) languages the rule $\ell \to r$ preserves the language L iff ℓ, r are ordered w.r.t. a monotone well-quasi order such that L is upward-closed w.r.t this well-quasi order [12]. The coarsest such order is the Myhill-Nerode quasi order of a language L which relates arbitrary words v and w iff it holds for

H. Ehrig et al. (Eds.): ICGT 2010, LNCS 6372, pp. 384–387, 2010.

all words u and x that $uvx \in L$ implies $uwx \in L$. This is the coarsest monotone quasi order such that L is upward-closed w.r.t. this quasi order and it can be computed by a fix-point iteration similar to the computation of the minimal finite automaton.

The notion of the Myhill-Nerode quasi order and the result that a rule $\ell \to r$ preserves a languages iff ℓ, r are ordered w.r.t. the Myhill-Nerode quasi order can be lifted to recognizable graph languages (based on *Cospan(Graph)*) [4]. The algorithm for computing the Myhill-Nerode quasi order can also be adapted to the more general setting and there exists a prototype implementation to check whether the language of all graphs containing a given subgraph is an invariant according to a given graph transformation rule [3].

Regular Model Checking: Another approach for the verification of distributed and infinite-state systems is regular model checking [6]. The main idea is to describe (infinite) sets of states as regular languages, i.e. every state is represented by a word, and transitions as regular relations which are represented by finite-state transducers. Verification can then be done by performing a forward or backward analysis. Note that in general the transitive closure of the application of transitions is not guaranteed to be a regular language, therefore it can be necessary to overapproximate the transitive closure in order to use this technique.

Since this approach has been extended to the setting of (regular) tree languages [5] it is a logical step to generalize regular model checking to (recognizable) graph languages. There already exists the notion of MSO-definable transductions invented by Courcelle [11], but this notion does not seem to be that useful, since these transductions are very complex and do not guarantee to preserve the recognizability in general which is required for a forward analysis. It has to be investigated how the notion of finite-state transducer can be generalized to (some kind of) "transduction functors" similar to the generalization of finite-state automata to automaton functors. The goal is to have a notion which is equivalent to finite-state transducers when restricted to word languages. In the case of words, there exists a characterization of transductions by Nivat [2, Chap. III, Thm. 3.2] in terms of regular languages and morphisms of free monoids. This is a possible starting point for the generalization, but it is not that obvious since a graph – unlike a word – can be decomposed in several ways. I have already invented a notion of transduction functor which is based on a category of sets and labeled relations. In this category every tuple is labeled by an arrow which indicates the output of the transduction functor according to the input. However, in order to get a better understanding of these transductions functors, the study of transductions between monoids which are not free is interesting.

Efficient Implementation of Automaton Functors: In general an automaton functor consists of infinitely many finite state sets. But if only recognizable graph languages of bounded path-width are allowed, it is possible to use automaton functors of bounded size. However, the automaton functors might still be very large. This is an important problem that has to be attacked in order to provide tools based on recognizability.

There already exists a prototype implementation of an automaton functor [3] (which is used for invariant checking) that uses an explicit representation of the automaton functors leading to a high memory consumption. A possible solution to this problem is to find a good representation of the transition relations of the automaton functors. One kind of data structure which is very suitable for the compact representation of large relations are Binary Decision Diagrams (BDDs) [1]. An implementation using BDDs is currently under development and the first experiments have been very promising. One example which has been tested, is the AF accepting all graphs containing a specific subgraph. Using the explicit-state implementation it is only possible to compute this AF for interfaces of size up to 8. If one uses the BDD-based implementation it is possible to compute this automaton functor for interfaces of size up to 100.

Another problem is the determinization of automaton functors, which is required for many constructions, since the direct computation of deterministic automaton functors is not possible in practice due to the state explosion problem. A possible solution is a technique which uses BDDs for the search in powerset automata [9]. How this technique can be adapted to AFs has to be investigated.

The long-term goal is to provide a tool suite for the representation and manipulation of (bounded) automaton functors. Moreover, this tool suite should be usable to verify dynamic systems represented as graphs and GTSs.

The goal of my research is to suggest new directions in the verification of dynamic systems based on recognizable graph languages as well as to investigate how established analysis techniques for regular languages can be adapted to recognizable graph languages. Furthermore, the results will be used for an implementation to provide tools for verification based on automaton functors.

References

1. Andersen, H.R.: An introduction to binary decision diagrams. Course Notes (1997)
2. Berstel, J.: Transductions and Context-Free Languages. Teubner Verlag (1979)
3. Blume, C.: Graphsprachen für die Spezifikation von Invarianten bei verteilten und dynamischen Systemen. Master's thesis, Universität Duisburg-Essen (2008)
4. Blume, C., Bruggink, S., König, B.: Recognizable graph languages for checking invariants. In: Proc. of GT-VMT 2010. Elec. Communications of the EASST (2010)
5. Bouajjani, A., Habermehl, P., Rogalewicz, A., Vojnar, T.: Abstract regular tree model checking of complex dynamic data structures. In: Yi, K. (ed.) SAS 2006. LNCS, vol. 4134, pp. 52–70. Springer, Heidelberg (2006)
6. Bouajjani, A., Jonsson, B., Nilsson, M., Touili, T.: Regular model checking. In: Emerson, E.A., Sistla, A.P. (eds.) CAV 2000. LNCS, vol. 1855. Springer, Heidelberg (2000)
7. Bozapalidis, S., Kalampakas, A.: Graph automata. Theor. Comp. Sci. 393 (2008)
8. Bruggink, S., König, B.: On the recognizability of arrow and graph languages. In: Ehrig, H., Heckel, R., Rozenberg, G., Taentzer, G. (eds.) ICGT 2008. LNCS, vol. 5214, pp. 336–350. Springer, Heidelberg (2008)
9. Cimatti, A., Roveri, M., Bertoli, P.: Searching powerset automata by combining explicit-state and symbolic model checking. In: Margaria, T., Yi, W. (eds.) TACAS 2001. LNCS, vol. 2031, p. 313. Springer, Heidelberg (2001)

10. Courcelle, B.: The monadic second-order logic of graphs. I. recognizable sets of finite graphs. Inf. Comput. 85(1) (1990)
11. Courcelle, B.: The expression of graph properties and graph transformations in monadic second-order logic. In: Handbook of Graph Grammars and Computing by Graph Transformation, ch. 5. Foundations, vol. 1. World Scientific, Singapore (1997)
12. de Luca, A., Varricchio, S.: Well quasi-orders and regular languages. Acta Inf. 31(6) (1994)
13. Sassone, V., Sobociński, P.: Reactive systems over cospans. In: LICS (2005)

Stochastic Modelling and Simulation of Dynamic Resource Allocation

Adwoa Donyina

Department of Computer Science
University of Leicester
Leicester, United Kingdom
add7@le.ac.uk

1 Introduction

In contrast to computer systems, human behaviour is only predictable to a certain degree of probability. In semi-automated business processes human actors are guided by predetermined policies and regulations but retain the freedom to react to unforeseen events. For example, if an urgent prescription has to be dispensed by a pharmacist and the current pharmacist on duty is busy, it is likely that they would interrupt their current activity. The assignment policies in a model of this process should accurately define exception handling procedures in order to realistically simulate business processes. Other policies may require task assignment to the least-qualified person available to do the job if the people involved have different levels of access rights and qualifications.

Problem statement. It is difficult to accurately model and simulate the dynamic behaviour of humans in business processes without taking into account the following requirements: 1) access control; 2) dynamic (re)-assignment; 3) role promotion ; 4) resource assignment; 5) assignment policies; 6) process scheduling influenced by deadlines and priorities; 7) escalation handling; 8) probability of the resource performing/selecting tasks; 9) non-deterministic duration of tasks.

2 Proposed Solutions

A new modelling approach for human actors in business processes that replaces the rigid control flow structure with a more flexible rule-based approach is being proposed. A metamodel is used to define the abstract syntax influenced by Role-Based Access Control (RBAC) [9], whereas concrete syntax extends the UML notation of class and use case diagrams (R1). A Domain Specific Language (DSL) will be used to define the orchestration between participants and their corresponding roles and responsibilities in terms of standard organisational modeling [7] techniques for assignment of activities to resources in organisational structures (R3-5). Graphs will be used to represent system states and stochastic graph transformation will be used to model state changes with non-deterministic timing, such as the execution of a business action with a known

H. Ehrig et al. (Eds.): ICGT 2010, LNCS 6372, pp. 388–390, 2010.

average delay or the assignment of an actor to a role (R2,9). This allows us to model semi-structured processes, where actions are not chosen based on a fixed control flow but non-deterministically, influenced by deadlines, priorities and escalation events (R7-8). The operational semantics of stochastic graph trans-formation (GT) allows for simulation and analysis of dynamic reconfiguration (R2). The stochastic simulation will use probability distribution to predict the timing of operations, which will provide analysis facilities for gathering service level guarantee metrics and policy comparisons (R9).

At the same time the visual, rule-based approach provides an intuitive no-tation for structural changes. The approach also distinguishes between domain-specific and domain-independent GT rules. This enables specification of generic human resource allocation policies, which are defined using application condi-tions and constraints [6] (R5).

Current status of the PhD project. The syntax and semantics for the visual mod-elling language has been developed, which allows us to model human resources as part of business processes using a rule-based approach. The model will be implemented in an existing stochastic simulation GT tool [10] to validate per-formance aspects of the model, such as the probability for cases to finish within their deadlines, comparing different scheduling policies.

3   State of the Art

Standard business process modelling notations, such as UML or BPMN [8] do not address the dynamic assignment of roles to individual actors. BPMN represents business roles as participants, but is constrained by high-level concepts such as swimlanes to a static partitioning of activities, whereas WS-HumanTask and BPEL4People specify humans as part of service-oriented systems or processes by capturing task priority, assignment, and timeouts, and triggers appropriate escalation actions [1]. However, these xml-based languages lack visual represen-tations suitable for domain and business experts and were not built with the intention of use in simulation engines.

There are various simulation approaches that are currently used in industry and research environments, which are based on the flow oriented style of mod-elling and provide functionality for modelling, analysis and design for business process management. Little-Jil [5] is a domain independent agent coordination visual language used for modelling and simulating process steps in a control flow manner. It captures key aspects of exception handling and deadlines; however its focus is primarily on the process step with very little on human resource allocation, whereas ADONIS [2] is a simulation tool that captures general as-pects of resource allocation; however it is missing the requirements of escalation handling, deadlines, priorities, and assignment policies.

On the other hand, the rule-based modelling style has been used in agent-based systems to specify agent operations in a way comparable to the specifica-tion of our business activities [3]. While humans can be regarded as autonomous agents, our approach adds the element of dynamic reassignment of human agents to roles as well as the modelling of non-deterministic timing of actions. This relation is explored in more detail in a previous paper [4]. In summary, none of the approaches completely satisfies the requirements laid out in Section 1.

4   Evaluation

The model will be evaluated in terms of usability, flexibility and scalability. Usability testing will be accomplished by experiments with computer science students developing models in the new language in comparison to existing languages using standard business process model examples. Scalability will be verified through simulation experiments on larger models, whereas flexibility will be verified by modelling experiments and analyzing their resulting models. The evaluation's test results would provide data on the models: ease-of-use, response to uncertainty and load impact. These results will be used to verify the correctness of the modelling approach and for future language improvements.

References

1. Agrawal, A., et al.: Web Service Human Task (WS-HumanTask), version 1.0. Tech. rep., Adobe and BEA and Oracle and Active Endpoints and IBM and SAP (June 2007)
2. BOC-Group: ADONIS: Community Edition (2010),
   http://www.adonis-community.com/
3. Depke, R., Heckel, R., Küster, J.M.: Formal agent-oriented modeling with UML and graph transformation. In: Sci. Comput. Program, vol. 44, pp. 229–252. Elsevier North-Holland, Inc., Amsterdam (2002)
4. Donyina, A., Heckel, R.: Formal visual modeling of human agents in service oriented systems. In: Fourth South-East European Workshop on Formal Methods (SEEFM 2009), pp. 25–32. IEEE Computer Society, Los Alamitos (2009)
5. Group, L.P.W.: Little-JIL 1.5 Language Report. Tech. rep., Laboratory for Advanced Software Engineering Research, University of Massachusetts, Amherst (1997-2006)
6. Heckel, R., Wagner, A.: Ensuring consistency of conditional graph grammars - a constructive approach. In: Proc. of SEGRAGRA 1995 "Graph Rewriting and Computation". ENTCS, p. 2 (1995)
7. Muehlen, M.z.: Workflow-based Process Controlling: Foundation, Design and Application of Workflow-driven Process Information Systems. Logos Verlag, Berlin (2002)
8. omg.org: Business process modeling notation (BPMN), version 1.2. Tech. rep., Object Managment Group(OMG) (January 2009)
9. Sandhu, R.S.: Role-based access control. In: Adv. in Computers, vol. 46, pp. 237–286. Academic Press, London (1998)
10. Torrini, P., Heckel, R., Ráth, I.: Stochastic simulation of graph transformation systems. In: Rosenblum, D.S., Taentzer, G. (eds.) Fundamental Approaches to Software Engineering. LNCS, vol. 6013, pp. 154–157. Springer, Heidelberg (2010)

Bisimulation Theory for Graph Transformation Systems

Mathias Hülsbusch

Universität Duisburg-Essen, Germany
mathias.huelsbusch@uni-due.de

1 DPO, Reactive Systems and Borrowed Context

The main focus of our work is to analyze and compare the behavior of graph transformation systems (GTSs). Therefore, we use the borrowed context technique to specify the behavior of a single graph with interface in different contexts. This mechanism is used to construct an LTS with graphs as states and rule applications as transitions. Each transition is labeled with the rule name and the borrowed context that is needed to perform the step. This makes the ideas of behavior comparison from LTS theory applicable to GTSs [1,9,6].

It is a well known fact that *DPO* rewriting in an adhesive category $\mathbb{C}$ coincides with a reactive system over the category cospan($\mathbb{C}$), since cospan composition is done via pushout [11]. This idea is depicted in the following image (on the left). The thick black arrows denote the reactive system diagram. The thin gray arrows show the *DPO* diagram and the dotted arrows are additional morphisms in $\mathbb{C}$. The objects that are rewritten in $\mathbb{C}$ are the inner objects of the cospans a and b (G is rewritten to H via the rule (l', r'). 0 is the initial object in $\mathbb{C}$). In the reactive system, a is rewritten to b via the rule (l, r) in the context c.

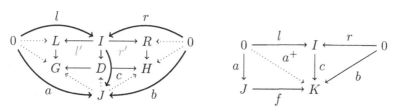

For reactive systems, the concept of borrowed contexts [1,9] is shown in the right diagram above, where the square, spanned by l and a, spelled out explicitly in $\mathbb{C}$, is the original borrowed context diagram. The idea is that a can be rewritten if one adds f to a [6]. If the square behaves like an IPO, f is minimal.

Application Conditions: When modeling real-world processes, one wants to be able to limit the applicability of rules by application conditions. The literature [8] so far has mainly focused on negative application conditions (NAC). Nevertheless, nested application conditions (or **G**eneralized **A**pplication **C**onditions) have been studied for *DPO* rewriting [7], but not for reactive systems.

Here is a short example how useful application conditions are to model real world processes: In a filesystem, a user U deletes a file F. This is forbidden if

H. Ehrig et al. (Eds.): ICGT 2010, LNCS 6372, pp. 391–393, 2010.

F is marked as protected (classical NAC). But even if F is protected, U may delete it, if U has root permissions (this is a GAC). In our approach, the GACs describe those conditions which the context c has to fulfill in order to make the rule applicable. Since l (the left-hand-side of the rule) is completely known, we only need rules for the context c. A NAC is modeled by an arrow $I \overset{nac}{\rightarrow} T$. A context $I \overset{c}{\rightarrow} J$ fulfills this NAC iff there is *no* morphism $T \overset{\varphi}{\rightarrow} J$ such that nac, φ and c form a commuting triangle in cospan($\mathbb{C}$). GACs are build as tree-like structures with root node I [10]. Application conditions for *DPO* rewriting can be translated into GACs for reactive systems under very mild conditions (e.g. if $L \leftarrow I$ is a mono). We worked on the label generation via borrowed context for rewriting rules with this kind of GACs, providing the conditions on the minimal context for each step as a tree of morphisms. The aim is to get a congruence result, in order to give a finite representation of the bisimulation-relation.

2   Context-Aware Bisimulations

The behavior of two systems might be different if both systems are in a certain context, but may be bisimilar in other contexts (due to application conditions on the rules). To handle this situation, there are several problems to solve:

1. Is the context of a system fixed for all times or may the context change over time? Could it be that the rule is sometimes applicable, sometimes not?
2. Is the whole context visible to the system, or does the system only see the context that actually matters for the next step (saturated bisimulation)?
3. How can the difference between external and internal non-determinism be modeled? External non-determinism is caused by the actual context, the internal non-determinism is caused by decisions of the system.

Our aim is to develop a theory for bisimulation of graph transformation systems, that allows to handle influences of the actual context on the behavior of a state. We are currently working on this setting: Hide the context-specific informations in a monad and apply techniques from the field of coalgebra theory to the resulting Kleisli category. As a first result we have sketched an algorithm that does not compute the complete final coalgebra, but only the part that is needed to compare the behavior of the given states. Two states are bisimilar in a special context, iff they are mapped to the same object in the final coalgebra if the mapping is restricted to the actual context (the mapping is an arrow in the Kleisli-category). The next step is to generalize these concepts to graph transformation systems, where the restrictions on the contexts may be modeled via GACs (see Sct. 1)

3   Verification of Model Transformations

The verification of model transformations (see [9,2]) deals with the question whether a model transformation preserves the behavior. This is an interesting application for the techniques presented here. Our approach is to decide whether the source model is (weakly) bisimilar to the target model. So far, we concentrated on *in-situ* transformations, where a part of the source model is replaced by a part of the target model. Other approaches, such as TGGs, construct the target model while parsing the source model, but do not destroy the source model (See [5,4]). For the verification of a model transformation, the following sets of rules are needed: The semantic rules of the source model, of the target model and the set of transformation rules. Our ideas can be used to handle: NACs on the semantic rules [3], GACs on the semantic rules (both using ideas from Sct. 1) and GACs on the transformation rules (see Sct. 2). This approach does not only cover classical model transformation, but also system migration, where subsystems of a model are replaced step by step. The in-situ transformation fits perfectly into this setting, since it directly supports step by step transformations. Several actual IT problems occur during a system migration (switch from IPv4 to IPv6, from DNS to DNSSEC, etc.). Thus the analysis of in situ transformations is not only interesting for bisimulation theory, but could have an influence on practical applications as well.

We are working on a Mathematica^{TM}-based implementation of these concepts to provide tools that make our results applicable to larger problems.

References

1. Ehrig, H., König, B.: Deriving bisimulation congruences in the DPO approach to graph rewriting with borrowed contexts. MSCS 16(6) (2006)
2. Engels, G., Kleppe, A., Rensink, A., Semenyak, M., Soltenborn, C., Wehrheim, H.: From UML Activities to TAAL - Towards Behaviour-Preserving Model Transformations. In: Schieferdecker, I., Hartman, A. (eds.) ECMDA-FA 2008. LNCS, vol. 5095, pp. 94–109. Springer, Heidelberg (2008)
3. Hermann, F., Hülsbusch, M., König, B.: Specification and verification of model transformations. In: Proc. of GraMoT 2010. Electronic Communications of the EASST, vol. 30 (to appear 2010)
4. Hülsbusch, M., König, B., Rensink, A., Semenyak, M., Soltenborn, C., Wehrheim, H.: Verifying full semantic preservation of model transformation is hard (to appear)
5. Königs, A.: Model transformation with triple graph grammars. In: Workshop on Model Transformations in Practice (2005)
6. Leifer, J., Milner, R.: Deriving bisimulation congruences for reactive systems. In: Palamidessi, C. (ed.) CONCUR 2000. LNCS, vol. 1877, p. 243. Springer, Heidelberg (2000)
7. Pennemann, K.H.: Resolution-like theorem proving for high-level conditions. In: Ehrig, H., Heckel, R., Rozenberg, G., Taentzer, G. (eds.) ICGT 2008. LNCS, vol. 5214. Springer, Heidelberg (2008)
8. Rangel, G., König, B., Ehrig, H.: Deriving bisimulation congruences in the presence of negative application conditions. In: Amadio, R.M. (ed.) FOSSACS 2008. LNCS, vol. 4962, pp. 413–427. Springer, Heidelberg (2008)
9. Rangel, G., Lambers, L., König, B., Ehrig, H., Baldan, P.: Behavior preservation in model refactoring using DPO transformations with borrowed contexts. In: Ehrig, H., Heckel, R., Rozenberg, G., Taentzer, G. (eds.) ICGT 2008. LNCS, vol. 5214. Springer, Heidelberg (2008)
10. Rensink, A.: Representing first-order logic using graphs. In: Ehrig, H., Engels, G., Parisi-Presicce, F., Rozenberg, G. (eds.) ICGT 2004. LNCS, vol. 3256, pp. 319–335. Springer, Heidelberg (2004)
11. Sassone, V., Sobociński, P.: Reactive systems over cospans. In: LICS 2005. IEEE, Los Alamitos (2005)

Realizing Impure Functions in Interaction Nets

Eugen Jiresch*

Institute of Computer Languages, Vienna University of Technology

Abstract. We propose first steps towards an extension of interaction nets for handling functions with side effects via monads.

1 Introduction and Overview

Models of computation are the basis for many programming languages, e.g., for reasoning on formal properties of programs such as correctness and termination. *Interaction nets* are a model of computation based on graph rewriting. They enjoy several useful properties which makes them a promising candidate for a future programming language. *Interaction nets* were first introduced in [4]. A *net* is a graph consisting of *agents* (nodes) and
ports (edges). Agent labels denote data or function
symbols. Computation is modeled by rewriting
the graph, which is based on *interaction rules*.

These rules apply to two nodes which are connected by their *principal ports* (indicated by the arrows). For example, the following rules model the addition of natural numbers (encoded by 0 and a successor function S):

$$(1) \; \boxed{0} \bowtie \boxed{+} \; y \;\; \Rightarrow \;\; \diagup y \quad\quad (2) \; x \; \boxed{S} \bowtie \boxed{+} \; y \;\; \Rightarrow x \vartriangleleft \boxed{+} \; \boxed{S} \; y$$

This simple system allows for parallel evaluation of programs: If several rules are applicable at the same time, they can be applied in parallel without interfering with each other. In addition, reducible expressions in a program cannot be duplicated: They are evaluated only once, which allows for sharing of computation.

Context, Overall Goal and Current Contribution. Functions that incorporate side effects such as I/O or exception handling generally destroy the above properties: Due to parallelism, side effects can occur in any order, causing nondeterministic results. Yet, these impure functions are a crucial part of any programming language. Our goal is to extend interaction nets in order to support generic computations with side effects. To our knowledge, very little work has been done on this topic. In this abstract, we make first steps towards a solution of this problem based on monads.

---

* The author was supported by the Austrian Academy of Sciences (ÖAW) under grant no. 22932 and the Vienna PhD School of Informatics.

H. Ehrig et al. (Eds.): ICGT 2010, LNCS 6372, pp. 394–396, 2010.

2 Towards a Monad Approach to Side Effects

We base our solution to impure functions in interaction nets on *monads*, a model to structure computation. Monads have been used with great success in functional programming: Monadic functions can encode side effects in a pure language and determine the order in which they occur [5]. A monad consists of an abstract datatype M a and two (higher-order) functions operating on this type:

```
data M a
return        :: a -> M a
>>= (bind) :: M a -> (a -> M b) -> M b
```

Monads are used to augment values of some type with computations that contain potential side effects. Intuitively, M adds a sort of wrapper to some value x of type a, potentially containing additional data or functions. *return* x wraps a value x, yielding a monadic object $M\ x$. *bind* handles sequentialization or ordering of function applications and their potential side effects.

A monad needs to satisfy the following laws:

```
(1)    return a >>= f    = f a
(2)    M a >>= return    = M a
(3)    (M a >>= f) >>= g = M a >>= (\x -> (f x >>= g))
```

Intuitively, *return* performs no side effects, and *bind* is associative (\x -> denotes lambda abstraction, i.e., function application).

Interaction nets as such do not support abstract datatypes or higher-order functions, which are essential ingredients of monads. This, together with the restricted shape of rules, makes an adaptation non-trivial. However, we can define a monad just by the basic features of interaction nets, agents and rules. We illustrate the basic idea and approach by a typical example, namely the monad *Maybe* which is used for exception handling:

```
         data Maybe a    = Just a | Nothing
(1)      return x        = Just x
(2)      (Just x) >>= f = f x
(3)      Nothing  >>= f = Nothing
```

The rules below model this monad for functions f on natural numbers:

The correspondence of interaction rules with the original definition of the *Maybe* monad is indicated by the rule labels. To qualify as a monad, these rules need to satisfy the monad laws. This can be shown by reduction of nets, such that both sides of the respective equation have a common reduct. The following reduction sequence proves law (1) (return a >>= f) = f a:

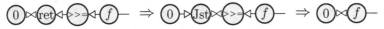

We only show the case a = 0. For a > 0 (i.e., substituting 0 with S), law (1) can be shown analogously. The sequence to show law (2) is very similar.

To prove law (3), we can prove that the sequentialization of two functions via *bind* agents always yields the same result, independently of the order of evaluation. This corresponds to the associativity property expressed in law (3), and can again be verified by giving reduction sequences of interaction nets.

Discussion. We have defined agents and rules that have the same functionality as the *Maybe* monad: These rules can effectively be used to model exception handling in interaction nets. Furthermore, the monad laws hold. We conclude that despite their simplicity, interaction nets can be used to define monads. This is an important step towards realizing impure functions: The *Maybe* monad example acts as a proof of concept.

Yet, this example constitutes an ad-hoc solution and is not general: First, only one type of computational side effect is covered by this monad. For other features (I/O, state,...), different monads need to be defined. Second, the *Maybe* monad in the example is only defined on natural numbers. This is mainly due to a current restriction of interaction nets: Existing type systems do not support abstract datatypes with type variables. Tackling these two issues will be subject to further work. Besides *Maybe*, we have already successfully defined another monad in interaction nets, namely a *Writer monad* (for logging / profiling), and plan to extend these results to other impure computations. Our overall goal is to define a generic, abstract and parametrized framework of side effect handling. To achieve this, we plan to extend the existing type systems for interaction nets, e.g., the intersection types approach of [1].

Related Work. Extensions to interaction nets have been proposed in several ways. For example, a system for more complex rule patterns is developed in [2].

Monads have been applied in various settings. In a recent application, monads are used to express the semantics of procedural proof languages [3].

References

1. Fernández, M.: Type assignment and termination of interaction nets. Mathematical Structures in Computer Science 8(6), 593–636 (1998)
2. Hassan, A., Jiresch, E., Sato, S.: An implementation of nested pattern matching in interaction nets. EPTCS 21, 13–25 (2010)
3. Kircher, F., Munoz, C.: The proof monad. Journal of Logic and Algebraic Programming 79, 264–277 (2010)
4. Lafont, Y.: Interaction nets. In: Proceedings, 17th ACM Symposium on Principles of Programming Languages, pp. 95–108 (1990)
5. Wadler, P.: How to declare an imperative. ACM Comp. Surv. 29(3), 240–263 (1997)

Composite EMF Modeling Based on Typed Graphs with Inheritance and Containment Structures

Stefan Jurack

Philipps-Universität Marburg, Germany
sjurack@mathematik.uni-marburg.de

1 Introduction

The rising paradigm of model-driven development (MDD) promises to be a new technique to control the complexity of today's software systems. However, increasing complexity may lead to large models which may justify a logical and/or physical distribution into several component models for better manageability. The concurrent development by distributed teams is desirable and already common practice in conventional software development. Naive solutions as storing all models in a central repository may not be adequate for distributed software development. E.g. Open Source Development is performed by distributed developer teams working in independent projects.

This PhD project investigates in an approach for composite modeling, and can be divided into two essential parts: Since graphs can be understood as abstract representation of visual models, the first part aims at providing a formal foundation of a composition concept for models based on graphs and graph morphisms. Moreover, composite model transformations shall specify distributed development steps. The second part of this thesis shall realize a framework for composite modeling. In particular, this thesis focuses on the Eclipse Modeling Framework (EMF) [3] as well-known and broadly used technology for modeling language definitions, providing modeling and code generation facilities. Composite EMF model transformation shall be implemented utilizing Henshin [4,1], an EMF model transformation tool basing on graph transformation concepts. It comes with an own transformation language including enhanced concepts as control structures. The author is involved in the Henshin project.

In the following, the approach, existing and future work are briefly outlined.

2 Composite Graph Concepts and Their Transformation

Alternatively to a central model repository as discussed above, this thesis proposes a (distributed) set of interconnected *component models* constituting a *composite model* (cf. [6]), based on the category of graphs and graph morphisms. The incorporation of distributed graphs (cf. [7]) introduces a twofold abstraction: A *network* graph specifies the distribution structure while the *object layer* is represented by local graph structures.

Each component has a body graph and an arbitrary number of explicit export and import interfaces being graphs as well. Export and imports graphs specify what is provided to and required from the environment, respectively. They refer to their component's body by total morphisms, identifying selected body elements. Connections between

H. Ehrig et al. (Eds.): ICGT 2010, LNCS 6372, pp. 397–399, 2010.

components are specified by partial morphisms between export and import interfaces only. Real partiality can be interpreted as inconsistency with imports not being served. Besides encapsulation and information hiding, explicit interfaces allow to define and develop component models independently such that they can be connected later.

Building up on the work by Biermann et al. (cf. [2]), we formalize EMF models by means of typed graphs with *in*heritance and *c*ontainment structures, short ICgraphs, and model relations by IC-morphisms (cf. [5]), leading to the category ICGRAPH. Moreover, this category has been combined with our composition concept resulting in the category of composite IC-graphs, called COMPICGRAPH. Figure 1

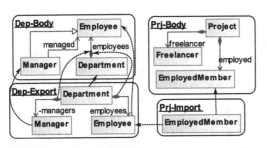

Fig. 1. Composite model with two component models

shows a simple composite model of two component models, with related IC-graphs and mappings in between. IC-morphisms support a flexible mapping along inheritance structures such that interface graphs may appear simpler than the structure in its mapped body graph, e.g. edge managers in interface Dep-Export is mapped to a different edge in the body graph. Typed composite graphs are considered in addition by utilizing composite IC-morphisms, i.e. interfaces and bodies on the instance level have to comply to interfaces and bodies on the type level. This allows to define exportable and importable elements on the type level already.

While the formal foundation is pretty far developed already wrt. to the static structures, dynamic aspects still require research. As model-driven development relies on model transformation, the use of graph transformation is an obvious choice in our case. In [6] we give a general definition of composite graph transformation on the one hand, and identify different kinds of such transformations on the other hand, having a certain amount of impact on network and object layer. All in all, this is pretty vague and has to be worked out in the near future.

3   Towards Tool Support and Case Studies

Our approach to composite graphs shall be implemented based on EMF, while the composite graph transformation concepts shall be realized with Henshin.

The Eclipse Modeling Framework provides means to define modeling languages by meta models and to generate model and editor code out of them by predefined templates. EMF already provides a distribution concept for models in separate resources, e.g. files, where remote resources can be addressed in a unique and uniform manner using so-called uniform resource identifier (URI). The author believes, that this concept of physical structuring is not relevant for the approach in this thesis which deals with a logical distribution of models.

A first approach towards an implementation of composite EMF models may be the use of annotations in EMF models to introduce an additional layer for import and export specifications. In this case, each annotation has to include a variety of information e.g. the related interface and mapping information. Moreover, it has to be investigated how elements of such component model parts can be referred to. One idea is to utilize URIs, already used by EMF. Furthermore it has to be investigated how far model code templates have to be adapted in order to generate composite model code.

To support (distributed) model-driven development in particular, an appropriate visual editor for composite EMF models has to be developed as well. It may read annotations and provide appropriate views, e.g. on interfaces, mappings and the overall modularization structure. Among standard editing capabilities on component models such as create, read, update and delete, the editor may also provide additional features for checking the compatibility between export and import interfaces, connecting them and revealing non-fully mapped imports resulting in inconsistencies.

Evolution steps of composite models such as synchronizations and network reconfigurations shall be based on composite EMF model transformation in order to ensure validity of composite EMF models. However, this deserves a dedicated transformation language. Being based on Henshin, we believe that there are mainly two conceivable approaches: The first one is the extension of Henshin's transformation language, while the alternative is to define an independent composite EMF model transformation language and specify its translation into Henshin's transformation language.

To complete the thesis, an elaboration of meaningful development scenarios is conceivable. On the one hand, the interconnected models for graphical editor development with the Eclipse Graphical Modeling Framework (GMF) seem to be a good choice for composite modeling. On the other hand, interacting model-driven development of open source software seems to be an instructive application scenario for composite modeling.

References

1. Arendt, T., Biermann, E., Jurack, S., Krause, C., Taentzer, G.: Henshin: Advanced Concepts and Tools for In-Place EMF Model Transformation. In: Proc. of 13th Int. Conference on Model Driven Engineering Languages and Systems (MoDELS 2010). LNCS, vol. 6394. Springer, Heidelberg (2010)
2. Biermann, E., Ermel, C., Taentzer, G.: Precise Semantics of EMF Model Transformations by Graph Transformation. In: Czarnecki, K., Ober, I., Bruel, J.-M., Uhl, A., Völter, M. (eds.) MODELS 2008. LNCS, vol. 5301, pp. 53–67. Springer, Heidelberg (2008)
3. EMF: Eclipse Modeling Framework (2010), http://www.eclipse.com/emf
4. Henshin (2010), http://www.eclipse.org/modeling/emft/henshin
5. Jurack, S., Taentzer, G.: A Component Concept for Typed Graphs with Inheritance and Containment Structures. In: Ehrig, H., et al. (eds.) ICGT 2010. LNCS, vol. 6372, pp. 186–200. Springer, Heidelberg (2010)
6. Jurack, S., Taentzer, G.: Towards Composite Model Transformations Using Distributed Graph Transformation Concepts. In: Schürr, A., Selic, B. (eds.) MODELS 2009. LNCS, vol. 5795, pp. 226–240. Springer, Heidelberg (2009)
7. Taentzer, G.: Distributed Graphs and Graph Transformation. Applied Categorical Structures 7(4) (1999)

Formal Modeling and Analysis of Communication Platforms Like Skype Based on Petri Net Transformation Systems

Tony Modica

Integrated Graduate Program Human-Centric Communication
Technische Universität Berlin
modica@cs.tu-berlin.de

The aim of this PhD thesis is to use an extension of Petri nets and Petri net transformation systems called AHOI nets in order to allow formal modeling and analysis of communication platforms (CP) like Skype. In the following, we explain the main ideas of AHOI nets and discuss how they can be used to model features of Skype. We give an overview of results achieved so far and remaining open problems to be solved in this PhD thesis.

1 AHOI Net Transformation Based on Algebraic Graph Transformation

An adequate formal modeling approach for CP would have to cover: 1. data, content, and the knowledge of actors, 2. the communication structure, which actors are connected to each other (in short, the topology), and 3. interactions within and between different applications, transmission of data.

For covering these main aspects, I propose an integrated approach that integrates Petri nets (topology), abstract data types (content spaces), and transformation of Petri nets based on graph transformation (interaction) into one formal technique. In detail, we extend classical Petri nets by the following features and call the integrated approach *reconfigurable algebraic higher-order Petri nets with individual tokens*, short (AHOI nets).:

Algebraic high-level (AHL) Petri nets integrate the classical (low-level) Petri nets with algebras for using algebraic data values as tokens and algebraic expressions as firing conditions [9].

Higher-order (HO) nets are AHL nets with a suitable algebra allowing Petri nets and transformation rules as tokens together with operations for firing net tokens and applying rule tokens on them [5].

Reconfiguration allows us to modify AHOI nets' structure or their tokens at runtime. For this, we use weak adhesive high-level replacement (waHLR) systems with double pushout transformation [4].

Individual Tokens have been proposed in [3] as a semantics for firing steps, i.e. marking tokens are considered distinguishable, in contrast to the classical "collective" paradigm. But for a graph-based transformation approach we need individual tokens already in the syntax instead of only interpreting

H. Ehrig et al. (Eds.): ICGT 2010, LNCS 6372, pp. 400–402, 2010.

marked Petri nets semantically under the individual viewpoint. For this, we have developed a formal theory for firing and transformation of algebraic high-level nets with markings of individual tokens and compared it with the classical collective approach [8]. A considerable advantage to the Petri net transformation in [9] is that markings can be changed without manipulating the net's structure and we can simulate firing steps with equivalent firing transformation rules.

2   Modeling Skype with AHOI Nets

As a representative modeling case study, I chose Skype as it has many aspects and features that are prominent in many CP as contact lists, privacy management, conferences, and group chats. In contrast to other work as [6], which aims at the technical design of network protocols, I focus on modeling user behavior and observable system reactions. We successfully used reconfigurable AHOI nets to model many features of Skype relying on the following important formal concepts:

High-level Petri nets. We represent each Skype client as a component of a single system net. The possible actions of a user are given by the activated transitions of its client component. We need algebraic values a.o. for user identities in the contact lists and for the data to be transmitted.

Net transformation. We follow strictly the principle of separation of user and system behavior. In its initial configuration, a client component can only change its state, e.g. to offline, online, or not available, and announce requests for contact exchange, establishing communication channels, or transmission of data. To extend a user's possible behavior, e.g. in conferences to invite more people or while a contact exchange to accept or to refuse, reconfiguring rule applications are triggered by requests (produced by transition firings) that add the structure corresponding to the new possible actions to the clients (or remove it, respectively). For ensuring that rules are only applied at reasonable matches, we employ rules with application conditions [4]. The individual tokens are vital for the transformations because we need rules that can change a net's marking without necessarily manipulating its structure.

Amalgamated Rules. A particular problem in Petri nets is multicasting of data, i.e. selective distribution of data tokens to participants of a conference or a chat. The challenge of multicasting is the a priori not known number of recipients in a dynamic system or conference at a certain time. For handling multicasting we have proposed to use amalgamated rules [2] and individual tokens.

Higher-order Markings. To control firing and reconfiguration of a Skype AHOI net, we use the AHOI formalism itself by considering the Skype net and the reconfiguring rules to be tokens in a control AHOI net. The control net binds transition firings (requests) with appropriate rule applications (request handling).

3 Results and Future Work

We have introduced algebraic high-level nets with individual tokens and developed a transformation theory based on waHLR systems. AHOI nets have been defined as a special instance of this framework. We examined requirements of CP and elaborated a case study on modeling Skype with AHOI nets [2,8,7]. It remains to examine how the structure of the higher-order net can further be used to improve modeling of CP.

Currently, we are working on the analysis of security-related properties of AHOI net models for CP. Besides the results for waHLR systems from [4], e.g. Local-Church-Rosser, Concurrency, and Parallelism Theorem, we try to employ the technique of invariants for high-level Petri nets [10] and combine it with net transformation. With these, we aim to prove properties like that a conference consists of invited clients only, that only specific actions can delete contacts, and that during conferences the participants' histories do not lose information.

Till the end of the year, we plan to extend an existing Eclipse-based tool [1] for supporting modeling, simulation, and analysis of reconfigurable AHOI nets.

References

1. Biermann, E., Ermel, C., Hermann, F., Modica, T.: A Visual Editor for Reconfigurable Object Nets based on the ECLIPSE Graphical Editor Framework. In: Proc. 14th Workshop on Algorithms and Tools for Petri Nets, AWPN (2007)
2. Biermann, E., Ehrig, H., Ermel, C., Hoffmann, K., Modica, T.: Modeling Multicasting in Dynamic Communication-based Systems by Reconfigurable High-level Petri Nets. In: IEEE Symp. on Visual Languages and Human-Centric Computing, VL/HCC (2009)
3. Bruni, R., Meseguer, J., Montanari, U., Sassone, V.: Functorial semantics for petri nets under the individual token philosophy. In: Category Theory and Computer Science, CTCS 1999 (1999)
4. Ehrig, H., Ehrig, K., Prange, U., Taentzer, G.: Fundamentals of Algebraic Graph Transformation. In: EATCS Monographs. Springer, Heidelberg (2006)
5. Hoffmann, K., Mossakowski, T., Ehrig, H.: High-Level Nets with Nets and Rules as Tokens. In: Ciardo, G., Darondeau, P. (eds.) ICATPN 2005. LNCS, vol. 3536, pp. 268–288. Springer, Heidelberg (2005)
6. Khan, A., Heckel, R., Torrini, P., Ráth, I.: Model-based stochastic simulation of p2p voip using graph transformation system. In: Proc. 17th Int. Conf. Analytical and Stochastic Modeling Techniques and Applications, ASMTA (2010)
7. Modica, T., Ermel, C., Ehrig, H., Hoffmann, K., Biermann, E.: Modeling communication spaces with higher-order petri nets. In: Advances in Multiagent Systems, Robotics and Cybernetics: Theory and Practice, vol. III (to appear 2010)
8. Modica, T., Gabriel, K., Ehrig, H., Hoffmann, K., Shareef, S., Ermel, C.: Low and High-Level Petri Nets with Individual Tokens. Tech. Rep. 2009/13, Technische Universität Berlin (2010)
9. Padberg, J., Ehrig, H., Ribeiro, L.: Algebraic high-level net transformation systems. Mathematical Structures in Computer Science 5, 217–256 (1995)
10. Schmidt, K.: Symbolische Analysemethoden für algebraische Petri-Netze. Dissertation, Humboldt-Universität zu Berlin, Berlin, Germany (1996)

LTS Semantics for Process Calculi from Their Graphical Encodings

Giacoma Valentina Monreale

Dipartimento di Informatica, Università di Pisa

The behaviour of a computational device is often naturally defined by means of reduction semantics: a set representing the possible states of the device, and an unlabelled relation among them, usually inductively defined, representing the possible evolutions of the device. Despite the advantage of conveying the semantics with few compact rewriting rules, freely instantiated and contextualized, the main drawback of reduction-based solutions is that the dynamic behaviour of a system is described in a monolithic way, and so it can be interpreted only by inserting the system in appropriate contexts, where a reduction may take place.

To make simpler the analysis of systems, it is often necessary to consider descriptions allowing the analysis of the behaviour of each single subcomponent, thus increasing modularity and enhancing the capabilities of verification. In such a context, labelled transition systems (LTSs) represent the mostly used tool. It usually leads to the definition of suitable observational equivalences, abstractly characterising when two systems have the same behaviour, thus allowing the possibility of verifying properties of system composition. However, the identification of the "right" labels is a difficult task and it is usually left to the ingenuity of the researcher. A case at hand is the calculus of *mobile ambients* (MAs) [5], for which only recently suitable labelled semantics were proposed [12,13].

A series of papers recently addressed the need to derive LTSs starting from a reduction semantics, mostly focusing on process calculi equipped with a structural congruence (equating processes intuitively denoting the same system). The most successful technique is represented by the theory of reactive systems (RSs) [11], based on the notion of relative pushout (RPO): it captures in an abstract setting the intuitive notion of "minimal" environment into which a process has to be inserted to allow a reduction to occur, which is then used as the label for a transition outgoing from that process. However, proving that a calculus satisfies the requirements needed for applying the RPOs technique is often quite a daunting task, due to the intricacies of the structural congruence. A way to overcome this problem lies with graphical encodings: processes are mapped into graphs with interfaces (roughly, with a distinguished set of nodes) such that process congruence is turned into graph isomorphism [1,7,8,9]. The reduction relation over processes is modeled via a graph transformation system (GTS), and the LTS on (processes encoded as) graphs with interfaces is distilled by applying the *borrowed context* (BC) mechanism [6], an instance of the RPO technique.

In order to obtain encodings preserving the process congruence, the graphical implementation can not simply reflect the syntactical structure of processes. For example, the implementations of calculi into (hyper-)graphs presented in the

H. Ehrig et al. (Eds.): ICGT 2010, LNCS 6372, pp. 403–406, 2010.

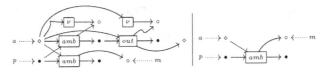

Fig. 1. Encoding and barb for the process $(\nu n)n[(\nu k)out\ k]|m[0]$ (left to right)

above papers do not provide a graphical operator for simulating parallel composition, which is instead modeled by the disjoint union of the graphs representing two processes, and by gluing the roots and the leaves of the two graphs. The left graph of Figure 1 shows the encoding for a MAs process having the parallel operator at top level. Also the restriction operator can not be modeled straightforwardly, and the search of a proper encoding for it becomes more challenging for topologically richer calculi such as MAs, as shown in [8,9]. The decentralized implementation proposed in [8] distinguishes the syntactic structure of a process from the activation order of the process components. Therefore, each graphical operator has an activation root node ◇, which represents the activating point for the reductions of the process that has this operator at top level. The proposed solution, as in [9], just drops the explicit graphical operator for simulating the restriction, and it deals with the lack of this operator simply by manipulating the "name nodes" ○ in the graph interfaces. The introduction of a decentralized implementation based on activation nodes suggests however an alternative solution: the addition of an explicit graphical operator simulating the restriction that is linked to the activation node ◇ of the graph representing the process where it occurs. This solution allows us both to capture the commutativity of sequential restricted names and to extend the scope of a restriction operator. Indeed, the encoding on the left of Figure 1 represents all the processes $(\nu n)n[(\nu k)out\ k]|m[0]$, $(\nu n)(\nu k)n[out\ k]|m[0]$, $(\nu n)(\nu k)(n[out\ k]|m[0])$ and $(\nu k)(\nu n)(n[out\ k]|m[0])$. This encoding may appear counter-intuitive, since, unlike the other operators, the restriction has as root only an activation node ◇ and has no process node •. Nevertheless, this solution, unlike the one proposed in [8], allows to precisely capture the standard structural congruence of MAs.

As said before, graphical encodings as the one above can be used to distill a LTS on (processes as) graphs. Graphs with interfaces are indeed amenable to the BC technique, which offers a constructive approach to calculate the minimal contexts enabling a graph transformation rule, allowing the construction of a LTS such that the associated bisimilarity is a congruence. The earliest proposal we are aware of for the application of the BC technique to establish formalisms is for CCS in [1]. In [3], the approach is instead used to synthesise a LTS for MAs. Here the encoding introduced in [8] is exploited to distill a LTS on graphs, that is used to infer a LTS directly defined on processes. Then, in [2], the adequacy of the derived LTS is proved by showing that it is the same as the one in [13].

Unfortunately, for most calculi, the semantics via minimal contexts is not the preferred equivalence, since too discriminating. Therefore, as proved e.g. for MAs in [2], barbed saturated semantics (considering all contexts as labels, and adding the check of state predicates) has to be taken into account. However, in

[2] the BC mechanism is just used to derive the labels, then an RS is reverse-engineered, and the strong and weak barbed saturated semantics for RSs are applied to MAs, proving that they can respectively capture the strong [13] and the weak [12] reduction barbed congruence for MAs.

In order to constructively reason directly on GTSs, [4] introduces the notion of saturated LTS for adhesive rewriting systems (ARSs), a generalization of classical GTSs. It also introduces barbs and barbed semantics for ARSs. In general, a barb is a predicate over the states of a system, with $P \downarrow o$ denoting that P satisfies o. However, in defining barbs for ARSs, in [4] authors were guided by the graphical encodings of calculi, and by the nature of barbs in that setting, where basically barbs check the presence of a suitable subsystem, such that it is needed to perform an interaction with the environment. So, a barb for a system G is defined as a subsystem occurring in it, also occurring in the left-hand-side of some production. Moreover, the interface of the subsystem must be contained in the interface of G. This assumption derives once more from an observation on the encodings for process calculi: only the relevant names for a process must occur in the interface of its encoding, so only free names can be observed.

In MAs, $P \downarrow n$ denotes the presence at top-level of a unrestricted ambient n. So, the barb m for the encoding on the left of Figure 1 is the graph on the right of the same figure. We do believe that this general mechanism to define barbs for ARSs may help us to solve the problem of automatically derive suitable barbs for RSs, along the line of the solution proposed in [10] for bigraphical RSs.

References

1. Bonchi, F., Gadducci, F., König, B.: Process bisimulation via a graphical encoding. In: Corradini, A., Ehrig, H., Montanari, U., Ribeiro, L., Rozenberg, G. (eds.) ICGT 2006. LNCS, vol. 4178, pp. 168–183. Springer, Heidelberg (2006)
2. Bonchi, F., Gadducci, F., Monreale, G.: Reactive systems, barbed semantics, and the mobile ambients. In: de Alfaro, L. (ed.) FOSSACS 2009. LNCS, vol. 5504, pp. 272–287. Springer, Heidelberg (2009)
3. Bonchi, F., Gadducci, F., Monreale, G.V.: Labelled transitions for mobile ambients. In: EXPRESS 2008. ENTCS, vol. 242(1), pp. 73–98 (2009)
4. Bonchi, F., Gadducci, F., Monreale, G.V., Montanari, U.: Saturated LTSs for Adhesive Rewriting Systems. In: Ehrig, H., et al. (eds.) ICGT 2010. LNCS, vol. 6372. Springer, Heidelberg (2010)
5. Cardelli, L., Gordon, A.: Mobile ambients. TCS 240(1), 177–213 (2000)
6. Ehrig, H., König, B.: Deriving bisimulation congruences in the DPO approach to graph rewriting with borrowed contexts. MSCS 16(6), 1133–1163 (2006)
7. Gadducci, F.: Graph rewriting for the π-calculus. MSCS 17(3), 407–437 (2007)
8. Gadducci, F., Monreale, G.V.: A decentralized implementation of mobile ambients. In: Ehrig, H., Heckel, R., Rozenberg, G., Taentzer, G. (eds.) ICGT 2008. LNCS, vol. 5214, pp. 115–130. Springer, Heidelberg (2008)
9. Gadducci, F., Montanari, U.: A concurrent graph semantics for mobile ambients. In: MFPS 2001. ENTCS, vol. 45. Elsevier, Amsterdam (2001)
10. Grohmann, D., Miculan, M.: Deriving barbed bisimulations for bigraphical reactive systems. In: ICGT 2008 - Doctoral Symposium. ECEASST, EASST, vol. 16 (2008)

11. Leifer, J., Milner, R.: Deriving bisimulation congruences for reactive systems. In: Palamidessi, C. (ed.) CONCUR 2000. LNCS, vol. 1877, pp. 243–258. Springer, Heidelberg (2000)
12. Merro, M., Zappa Nardelli, F.: Behavioral theory for mobile ambients. JACM 52(6), 961–1023 (2005)
13. Sobociński, P., Rathke, J.: Deriving structural labelled transitions for mobile ambients. In: van Breugel, F., Chechik, M. (eds.) CONCUR 2008. LNCS, vol. 5201, pp. 462–476. Springer, Heidelberg (2008)

Automated Assistance for Search-Based Refactoring Using Unfolding of Graph Transformation Systems

Fawad Qayum

Department of Computer Science, University of Leicester, Leicester, UK
fq7@mcs.le.ac.uk

1 Motivation

Refactoring has emerged as a successful technique to enhance the internal structure of software by a series of small, behaviour-preserving transformations [4]. However, due to complex dependencies and conflicts between the individual refactorings, it is difficult to choose the best sequence of refactoring steps in order to effect a specific improvement. In the case of large systems the situation becomes acute because existing tools offer only limited support for their automated application [8]. Therefore, search-based approaches have been suggested in order to provide automation in discovering appropriate refactoring sequences [6,11]. The idea is to see the design process as a combinatorial optimization problem, attempting to derive the best solution (with respect to a quality measure called objective function) from a given initial design [9].

Two obvious problems with search-based approaches are scalability, i.e., the ability to apply to large models [10], and traceability, i.e., ability on behalf of the developer to understand the changes suggested by the optimisation [6]. Heavy modifications make it difficult to relate the improvement to the original design, so that developers will struggle to understand the new structure. Therefore, we would prefer changes to be focussed, addressing specific problems through local changes only.

2 Approach

In this paper, we use a representation of object-oriented designs as graphs and refactoring operations as graph transformation rules [8]. Such rules provide a local description, identifying and changing a specific part of the design graph only. This enables us to analyse refactoring processes based on graph transformation [3] techniques and tools, in particular the approximated unfolding of a graph transformation system and its implementation in Augur2 [7] to identify conflicts and dependencies. Unfolding analyses a hypergraph grammar, starting with the initial hypergraph and producing a branching structure by applying all possible rules on the system. The result is a structure called Petri graph, presenting the behaviour in terms of an over-approximation of its transformations

H. Ehrig et al. (Eds.): ICGT 2010, LNCS 6372, pp. 407–409, 2010.

and dependencies [2]. The Petri graph is used as input to a search problem. The desired result is a sequence of transformations leading from the given design to a design of high(er) quality, using only transformation steps that are necessary to achieve that quality improvement.

We employ Ant Colony Optimisation (ACO) [1] meta-heuristic search to find such a solution. ACO is inspired by the behaviour of foraging ants, which search for food individually and concurrently, but share information about food sources and paths leading to them by leaving pheromone trails. Rather than representing the search space of designs and refactorings explicitly, we use the unfolding as a more scalable representation, where designs (states) are given by sets of transformations downward closed under causal dependencies. We can thus reconstruct states when needed, for example in order to evaluate the objective function, but will deal with the more compact representation when navigating the search space. As a further tribute to scalability, we are using the approximated unfolding. The resulting over-approximation of transformations and dependencies leads to spurious solutions, which have to be filtered out later. In addition, we are following a hybrid approach [5] where the ACO metaheuristic is augmented with local search to improve its performance.

In order to formalise a notion of quality of designs we use probe rules representing (anti) patterns to recognise situations that are desirable or to be avoided. Then, we will seek to maximise the number of desirable and minimise the number of undesirable situations. Using the unfolding as the underlying data structure, such information about patterns occurring is available at little extra cost. In particular, the implicit representation of states (by a sets of transformations closed under causality and without conflicts) will allow us to scale the search to larger problems, avoiding state-space explosion.

3 Discussion and Evaluation

Our approach involves using a combination of graph transformation theory and ACO meta heuristic, aiming to improve performance/scalability and traceability/understandability of search-based refactoring. We believe that scalability is achieved through the use of approximated unfolding, combining its implicit representation of graphs as states with the use of approximation, while hybrid ACO has been shown to be effective in situations of large and rugged search spaces with complex constraints on solutions. An experimental evaluation will have to confirm this belief.

To improve traceability, we will use the causal relation to explain refactorings in terms of their interdependency. For example, if the programmer accepts that the last step performed represents an improvement, they will implicitly accept the relevance of all changes up to that point that the final step depends on. We can also specify constraints on the sets of transformations to ensure that every step contributes directly or indirectly to the last step in the sequence.

Traceability will be evaluated through experiments with smaller models, assessing the effort it takes a human developer to understand the changes proposed by the search-based approach. The use of dependency information between trans-formations allows us to remove steps that ate unrelated to the intended change, making each change relevant and therefore easier to interpret.

References

1. Dorigo, M., Stützle, T.: Ant Colony Optimization. MIT Press, Cambridge (2004)
2. Baldan, P., Corradini, A., König, B.: A static analysis technique for graph trans-formation systems. In: Larsen, K.G., Nielsen, M. (eds.) CONCUR 2001. LNCS, vol. 2154, pp. 381–395. Springer, Heidelberg (2001)
3. Baldan, P., Corradini, A., Montanari, U.: Unfolding and event structure semantics for graph grammars. In: Thomas, W. (ed.) FOSSACS 1999. LNCS, vol. 1578, pp. 73–89. Springer, Heidelberg (1999)
4. Fowler, M.: Refactoring: Improving the Design of Existing Code. Addison-Wesley, Boston (1999)
5. Gambardella, L.M., Dorigo, M.: Has-sop: Hybrid ant system for the sequential ordering problem (1997)
6. Harman, M., Ph, U., Jones, B.F.: Search-based software engineering. Information and Software Technology 43, 833–839 (2001)
7. König, B., Kozioura, V.: Augur 2—a new version of a tool for the analysis of graph transformation systems. In: Proc. of GT-VMT 2006 (Workshop on Graph Transformation and Visual Modeling Techniques). ENTCS, vol. 211, pp. 201–210. Elsevier, Amsterdam (2006)
8. Mens, T., Taentzer, G., Runge, O.: Analysing refactoring dependencies using graph transformation. Software and System Modeling 6(3), 269–285 (2007)
9. O'Keeffe, M., Cinnéide, M.O.: A stochastic approach to automated design im-provement. In: PPPJ 2003: Proceedings of the 2nd International Conference on Principles and Practice of Programming in Java, pp. 59–62. Computer Science Press, Inc., New York (2003)
10. O'Keeffe, M., Cinnéide, M.O.: Search-based refactoring: an empirical study. J. Softw. Maint. E 20(5), 345–364 (2008)
11. Seng, O., Stammel, J., Burkhart, D.: Search-based determination of refactorings for improving the class structure of object-oriented systems. In: GECCO 2006: Pro-ceedings of the 8th Annual Conference on Genetic and Evolutionary Computation, pp. 1909–1916. ACM, New York (2006)

Correctness of Graph Programs
Relative to HR$^+$ Conditions*

Hendrik Radke

Carl v. Ossietzky Universität Oldenburg, Germany
hendrik.radke@informatik.uni-oldenburg.de

Abstract. In (Pennemann 2009), the correctness of graph programs relative to nested graph conditions is considered. Since these conditions are expressively equivalent to first-order graph formulas, non-local graph properties in the sense of Gaifman are not expressible by nested graph conditions. We generalize the concept of nested graph conditions to so-called HR$^+$ conditions and investigate the correctness for graph programs relative to these generalized conditions.

Modeling of system states. As software systems increase in complexity, there is a growing need for design concepts that allow an intuitive overview of a system, usually by visual modeling techniques. Graph transformation systems are a visual modeling approach that emphasizes the interconnections of data structures. The states of a regarded real-world system are modeled by graphs, and changes to the system state are described by graph programs. Structural properties of the system are described by graph conditions.

In [6], nested graph conditions are introduced. These conditions enhance first-order logic on graphs with a graphical representation of the nodes and edges involved. Nested conditions are expressively equivalent to first-order graph properties [10]. As such, they can express only local properties in the sense of Gaifman [4]. However, many real-world properties are non-local, i.e. they cannot be expressed by nested graph conditions. For instance, it is not possible to express the property "there is a path from node 1 to node 2", the connectedness or circle-freeness of a graph with these conditions, as these properties go beyond the k-neighbourhood for any node and any fixed k. Therefore, an extension is desired that can capture such properties.

HR$^+$ *conditions.* We propose to integrate hyperedge replacement systems with the nested graph conditions to form HR conditions [8]. The graphs in HR conditions are enriched with hyperedge variables, which are then replaced by graphs according to a hyperedge replacement system. Further enhancement to deal with subgraphs leads to HR$^+$ conditions. This way, non-local properties can be expressed by hyperedge replacement. In fact, HR$^+$ conditions are more expressive than monadic second-order formulas over graphs. The example HR$^+$ condition

---

* This work is supported by the German Research Foundation (DFG), grants GRK 1076/1 (Graduate School on Trustworthy Software Systems).

H. Ehrig et al. (Eds.): ICGT 2010, LNCS 6372, pp. 410–412, 2010.

$\exists(\overset{1}{\bullet}\overset{+}{\longrightarrow}\overset{2}{\bullet})$ with the hyperedge replacement system $+ ::= \overset{1}{\bullet}\longrightarrow\overset{2}{\bullet}\,|\,\overset{1}{\bullet}\longrightarrow\bullet\overset{+}{\longrightarrow}\overset{2}{\bullet}$ is satisfied for all graphs with a path between two nodes 1 and 2.

The following "car platooning" example after [9] may further illustrate the need for increased expressiveness. In order to save space and gas on highway lanes, cars travelling in the same direction drive in platoons, i.e. a tight row of cars with little distance in between. To ensure safety, the cars are partially controlled by a system adhering to a car platooning protocol. Each platoon has one and only one leader (designated by a small α) and an arbitrary number of followers. This property is represented by the following condition:

$$\forall\left(\boxed{\text{car}}\,,\exists\left(\boxed{\text{car}}\xrightarrow{+}\boxed{\text{car}_\alpha}\right)\right)\land\nexists\left(\boxed{\text{car}_\alpha}\xrightarrow{+}\boxed{\text{car}_\alpha}\right)$$

Model checking for HR$^+$ *conditions.* For finite graphs, the problem whether a graph G satisfies a HR$^+$ condition c is still decidable. In a naïve approach, one could, for every hyperedge in c, derive every graph permitted by the corresponding replacement system. The resulting conditions would be nested graph conditions, for which the satisfaction problem is decidable [10]. Since every hyperedge replacement system can be transformed into a montonous one [5] and no condition with a generated graph larger than G may be satisfied, the number of nested graph conditions to check is finite.

Correctness of graph programs relative to HR$^+$ *conditions.* HR$^+$ conditions can be used together with graph programs to build graph specifications in form of a Hoare triple {pre, Prog, post}. Our goal is to check the correctness of this triple, i.e. whether for all graphs G satisfying pre and all graphs H resulting from application of Prog on G, H satisfies post. Following an approach of Dijkstra [2], we construct a weakest precondition out of the program and the postcondition. The weakest precondition is constructed by first transforming the postcondition into a right HR$^+$ application condition for the program, then a transformation from the right to a left application condition and finally, from the left application condition to the weakest precondition. This is a generalization of the transformations from [7], where the hyperedge variables and the corresponding replacement systems have to be regarded. The correctness problem can thus be reduced to the problem whether the precondition implies the weakest precondition. It is planned to extend the ENFORCE framework [1] for nested conditions and its theorem prover component to support HR$^+$ conditions.

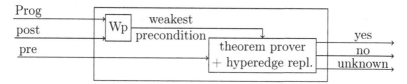

Case studies. In order to show the practical merit of the results, several case studies will be performed.

- HR^+ conditions and graph programs shall be used to verify the car platooning protocol mentioned earlier [9].
- HR^+ conditions shall be applied to the problem of C++ template instantiation. Type checking of the templates is done on graphs, and type errors are output as graphs, including suggestions for remedies. This may help developers dealing with templates by giving clearer error messages for type errors in templates.
- HR^+ conditions shall be used to express and check OCL constraints for UML diagrams. Together with a transformation of UML metamodels into graph grammars [3], this allows the generation of instances of a metamodel with constraints.

References

1. Azab, K., Habel, A., Pennemann, K.H., Zuckschwerdt, C.: ENFORCe: A system for ensuring formal correctness of high-level programs. In: Proc. of the Third Int. Workshop on Graph Based Tools (GraBaTs 2006). Electronic Communications of the EASST, vol. 1, pp. 82–93 (2007)
2. Dijkstra, E.W.: A Discipline of Programming. Prentice-Hall, Englewood Cliffs (1976)
3. Ehrig, K., Küster, J.M., Taentzer, G.: Generating instance models from meta models. Software and System Modeling 8(4), 479–500 (2009)
4. Gaifman, H.: On local and non-local properties. In: Stern, J. (ed.) Proceedings of the Herbrand Symposium: Logic Colloquium 1981, pp. 105–135. North Holland Pub. Co., Amsterdam (1982)
5. Habel, A.: Hyperedge replacement: grammars and languages. Springer, Heidelberg (1992)
6. Habel, A., Pennemann, K.H.: Correctness of high-level transformation systems relative to nested conditions. In: Mathematical Structures in Computer Science, pp. 1–52 (2009)
7. Habel, A., Pennemann, K.H., Rensink, A.: Weakest preconditions for high-level programs. In: Corradini, A., Ehrig, H., Montanari, U., Ribeiro, L., Rozenberg, G. (eds.) ICGT 2006. LNCS, vol. 4178, pp. 445–460. Springer, Heidelberg (2006)
8. Habel, A., Radke, H.: Expressiveness of graph conditions with variables. In: Int. Colloquium on Graph and Model Transformation on the occasion of the 65th birthday of Hartmut Ehrig., vol. 30 (to appear 2010)
9. Hsu, A., Eskafi, F., Sachs, S., Varaiya, P.: The design of platoon maneuver protocols for IVHS. Tech. rep., Institute of Transportation Studies, University of California at Berkeley (1991)
10. Pennemann, K.H.: Development of Correct Graph Transformation Systems. Ph.D. thesis, Universität Oldenburg (2009)

Static Type Checking
of Model Transformation Programs

Zoltán Ujhelyi

Budapest University of Technology and Economics,
Department of Measurement and Information Systems,
1117 Budapest, Magyar tudósok krt. 2
`ujhelyiz@mit.bme.hu`

1 Introduction

Model transformations, utilized for various tasks, such as formal model analysis or code generation are key elements of model-driven development processes. As the complexity of developed model transformations grows, ensuring the correctness of transformation programs becomes increasingly difficult. Nonetheless, error detection is critical as errors can propagate into the target application.

Various analysis methods are being researched for the validation of model transformations. Theorem proving based approaches, such as [5] show the possibility to prove statement validity over graph-based models. For the verification of dynamic properties model checking seems promising, but abstractions are needed to overcome the challenge of infinite state spaces [6].

In addition to model checking, static analysis techniques have been used in the verification of static properties. They provide efficiently calculable approximations of error-free behaviour, such as unfolding graph transformation systems into Petri nets [1], or using a two-layered abstract interpretation introduced in [2].

The current paper presents a static analysis approach for early detection of typing errors in partially typed model transformation programs. Transformation languages, such as the one of VIATRA2 [9], are often partially typed, e.g. it is common to use statically typed (checked at compile time) graph transformation rules with a dynamically typed (checked during execution) control structure.

The lack of static type enforcements in dynamically typed parts makes typing errors common and hard to trace, on the other hand the static parts allow efficient type inference. Our type checker approach uses constraint satisfaction problems (CSP) to propagate information between the different parts of the transformation program. Error messages are generated by a dedicated back-annotation method from the constraint domain.

2 Overview of the Approach

2.1 Type Safety as Constraint Satisfaction Problems

To evaluate type safety of transformation programs as CSPs, we create a CSP variable for each potential use of every transformation program variable with

H. Ehrig et al. (Eds.): ICGT 2010, LNCS 6372, pp. 413–415, 2010.

the domain of the elements of the type system. The constraints are created from the statements of the transformation program, expressing the type information inferrable from the language specification (e.g. conditions have a boolean type). For details about the constraint generation process see [8].

After the CSPs are evaluated, we check for two kinds of errors: (1) typing errors appear as *CSP violations*, (2) while variable type changes can be identified by comparing the different CSP variable representations of transformation program variables, looking for *inconsistencies* (although this may be valid, it is often erroneous, thus a warning is issued).

2.2   The Analysis Process

Our constraint-based type checking process is depicted in Fig. 1. The input of the static analysis is the transformation program and the metamodel(s) used by the program, while its output is a list of found errors. It is important to note that the instance models (that form the input of the transformation program) are not used at all in the static analysis process.

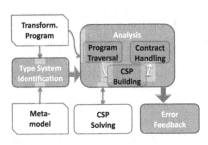

Fig. 1. Overview of the Approach

At first, we collect every possible type used in the transformation program (*Type System Identification*). The type system consists of the metamodel elements and built-in types (e.g. string, integer). To reduce the size of the type system, we prune the metamodel to provide a superset of the types used in the transformation program as described in [7]. The type system contains all elements referenced directly from the transformation program, all their parents, and in case of relations their endpoints and inverses.

To prepare the type system for the CSP based analysis, a unique integer set is assigned to each element of the type system as proposed in the algorithm in [3] to allow efficient calculation of subtype relationships.

Then in the *Analysis* step we traverse the transformation program, building and evaluating constraint satisfaction problems. Multiple traversal iterations are used to cover the different execution paths of the transformation problem.

For performance considerations a modular traversal is used: graph transformation rules and graph patterns are traversed separately. The partial analysis results of the rules (or patterns) are described and stored as pre- and postconditions based on the "design by contract" [4] methodology. After a contract is created for every reference to its corresponding rule (or pattern) the contract is used to generate the relevant constraints instead of re-traversing the referenced rule.

This modular approach is also used for the traversal of the control structure: it is similarly divided into smaller parts that are traversed and contracted separately.

The created CSPs can be solved using arbitrary finite domain CSP solver tools. The results are both used to build type contracts and to provide error feedback. The type contract of a rule (or pattern) holds the calculated types of the parameters at the beginning and the end of the method (the difference in the pre- and postcondition implies a type change in the parameter variable).

Finally, we look for typing errors and type changes to back-annotate them to the transformation program (*Error Feedback*). CSPs are evaluated together with the traversal to make context information also available to associate related code segment(s) to the found errors.

3   Implementation and Future Work

We have presented a static type checker approach for model transformation programs. It was implemented for the VIATRA2 transformation framework, and evaluated using transformation programs of various size.

In our inital evaluation the type checker seems useful for early error identification as it identified errors related to swapped variables or pattern calls.

As for the future, we plan to evaluate the possible usage of static program slicing methods for model transformation programs. This would allow to generate meaningful traces for reaching possibly erroneous parts of the transformation programs, thus helping more precise error identification. The generated slices are also usable to extend the system with additional validation options such as *dead code analysis* to detect unreachable code segments or *use-definition analysis* to detect the use of uninitialized or deleted variables.

References

1. Baldan, P., Corradini, A., Heindel, T., Knig, B., Sobociski, P.: Unfolding Grammars in Adhesive Categories. In: Kurz, A., Lenisa, M., Tarlecki, A. (eds.) CALCO 2009. LNCS, vol. 5728, pp. 350–366. Springer, Heidelberg (2009)
2. Bauer, J., Wilhelm, R.: Static Analysis of Dynamic Communication Systems by Partner Abstraction. In: Static Analysis, pp. 249–264. Springer, Heidelberg (2007)
3. Caseau, Y.: Efficient handling of multiple inheritance hierarchies. In: OOPSLA 1993: Proceedings of the Eighth Annual Conference on Object-Oriented Programming Systems, Languages, and Applications, pp. 271–287. ACM, New York (1993)
4. Meyer, B.: Applying 'design by contract'. Computer 25(10), 40–51 (1992)
5. Pennemann, K.: Resolution-Like theorem proving for High-Level conditions. In: Graph Transformations, pp. 289–304. Springer, Heidelberg (2008)
6. Rensink, A., Distefano, D.: Abstract Graph Transformation. Electronic Notes in Theoretical Computer Science 157(1), 39–59 (2006)
7. Sen, S., Moha, N., Baudry, B., Jézéquel, J.: Meta-model Pruning. In: Model Driven Engineering Languages and Systems, pp. 32–46. Springer, Heidelberg (2009)
8. Ujhelyi, Z., Horváth, A., Varró, D.: Static Type Checking of Model Transformations by Constraint Satisfaction Programming. Technical Report TUB-TR-09-EE20, Budapest University of Technology and Economics (June 2009)
9. Varró, D., Balogh, A.: The model transformation language of the VIATRA2 framework. Sci. Comput. Program. 68(3), 214–234 (2007)

Using Graph Transformations and Graph Abstractions for Software Verification[*]

Eduardo Zambon

Formal Methods and Tools Group
University of Twente, The Netherlands
zambon@cs.utwente.nl

Introduction. In this abstract we present an overview of our intended approach for the verification of software written in imperative programming languages. This approach is based on model checking of graph transition systems (GTS), where each program state is modeled as a graph and the exploration engine is specified by graph transformation rules. We believe that graph transformation [13] is a very suitable technique to model the execution semantics of languages with dynamic memory allocation. Furthermore, such representation provides a clean setting to investigate the use of graph abstractions, which can mitigate the space state explosion problem that is inherent to model checking techniques.

Overview. Figure 1 provides a picture of the whole verification cycle. The input is a program source code. The code is analysed by a compiler that produces as output an abstract syntax graph (ASG). This ASG is essentially the usual abstract syntax tree produced by a language parser enriched with type and variable bindings. The ASG, together with definitions of the language control flow semantics, is the input of a flow construction mechanism, that builds a flow graph for the given ASG. This flow graph represents how the execution point of the program should flow through the ASG, according to the rules of the programming language in use. Together, an ASG and a flow graph form a program graph, an executable representation of the program code as a graph. A program graph serves as input to an exploratory graph transformation system, composed by graph transformation rules that capture the execution semantics of elements of the programming language. This exploration mechanism generates the state space of the given program graph as a GTS that (eventually) captures all possible paths of execution of the program. Usually, these generated GTS are prohibitively large, or even infinite. At this point abstraction techniques come into play, in order to produce a finite over-approximation of the original GTS. After producing a GTS, we can perform model checking against a given set of correctness properties that the program is expected to have. This check produces either a verdict that the program is indeed correct, or a counter-example of an execution path that produces an error. This counter-example can then be traced back to the ASG, or, better yet, the input code, so that the user can inspect the error. As an exploration/model checking engine we use GROOVE [10], a tool specifically developed to perform model checking of graph production systems.

---

[*] The work reported herein is being carried out as part of the GRAIL project, funded by NWO (Grant 612.000.632).

H. Ehrig et al. (Eds.): ICGT 2010, LNCS 6372, pp. 416–418, 2010.

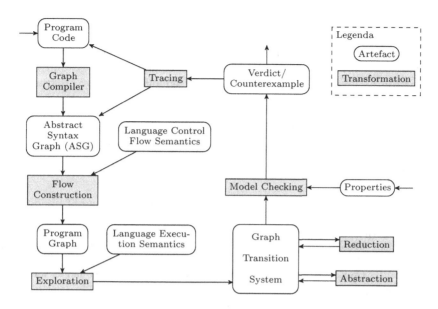

Fig. 1. Overview of the verification cycle proposed

Graph Abstractions. The key point of our verification method is the graph abstraction mechanism. In the context of graph transformation we have seen theoretical studies on suitable abstractions [12,1,6]. However, to the best of our knowledge only the last of these is backed up by an implementation, namely AUGUR [7]. Our graph abstractions are based on the concepts of shape analysis, proposed by Sagiv et al. [14], and of abstract interpretation developed by Cousot and Cousot [3]. A *graph shape* is an abstraction that captures the underlying structure of a set of concrete graphs, acting as their representative in the abstracted domain. Each node (resp. edge) of a graph shape is marked with a *multiplicity*, indicating how many nodes (resp. edges) must (or may) be present in a concrete graph. Previous work on graph abstractions were proposed by our group in [8,11,2]. One main issue with these abstractions is that they are unable to preserve important structural properties such as connectivity between nodes. This is a necessary information when analysing programs with heap-based data structures. Our current research is focused on elaborating good graph abstractions that keep the state space explosion under control while still allowing the verification of interesting properties on realistic programs.

Development. We chose Java as an initial programming language to handle, due to its wide-spread use. At the time of this writing we have a graph compiler that produces an abstract syntax graph from any legal Java program[1]. The details of the construction of this compiler are presented in [9]. The flow construction transformation is performed in GROOVE using a graph grammar that

---

[1] Available at http://groove.cs.utwente.nl/downloads/java2groove/

defines the Java control flow semantics. The next step is the elaboration of the execution semantics of Java also in terms of graph transformation rules. This is future work. It should be noted that all the ingredients of our proposed approach were previously investigated and their feasibility analysed. How graph transformations can be used to capture the execution semantics of a toy programming language was shown in [5]. The construction of a control flow semantics specification for a part of Java was given in [15]. Nevertheless, whether the combination of these techniques will indeed provide good practical results when applied to reasonable sized programs is still to be seen.

References

1. Baldan, P., König, B., König, B.: A logic for analyzing abstractions of graph transformation systems. In: Cousot, R. (ed.) SAS 2003. LNCS, vol. 2694, pp. 255–272. Springer, Heidelberg (2003)
2. Bauer, J., Boneva, I.B., Kurban, M.E., Rensink, A.: A modal-logic based graph abstraction. In: Ehrig, et al. (eds.) [4], pp. 321–335
3. Cousot, P., Cousot, R.: Abstract interpretation: A unified lattice model for static analysis of programs by construction or approximation of fixpoints. In: Principles of Programming Languages (POPL), pp. 238–252 (1977)
4. Ehrig, H., Heckel, R., Rozenberg, G., Taentzer, G. (eds.): ICGT 2008. LNCS, vol. 5214. Springer, Heidelberg (2008)
5. Kastenberg, H., Kleppe, A., Rensink, A.: Defining object-oriented execution semantics using graph transformations. In: Gorrieri, R., Wehrheim, H. (eds.) FMOODS 2006. LNCS, vol. 4037, pp. 186–201. Springer, Heidelberg (2006)
6. König, B., Kozioura, V.: Counterexample-guided abstraction refinement for the analysis of graph transformation systems. In: Hermanns, H., Palsberg, J. (eds.) TACAS 2006. LNCS, vol. 3920, pp. 197–211. Springer, Heidelberg (2006)
7. König, B., Kozioura, V.: Augur – a new version of a tool for the analysis of graph transformation systems. ENTCS 211, 201–210 (2008)
8. Rensink, A.: Canonical graph shapes. In: Schmidt, D. (ed.) ESOP 2004. LNCS, vol. 2986, pp. 401–415. Springer, Heidelberg (2004)
9. Rensink, A., Zambon, E.: A type graph model for Java programs. In: Lee, D., Lopes, A., Poetzsch-Heffter, A. (eds.) FMOODS 2009. LNCS, vol. 5522, pp. 237–242. Springer, Heidelberg (2009)
10. Rensink, A.: The GROOVE simulator: A tool for state space generation. In: Pfaltz, J.L., Nagl, M., Böhlen, B. (eds.) AGTIVE 2003. LNCS, vol. 3062, pp. 479–485. Springer, Heidelberg (2004)
11. Rensink, A., Distefano, D.: Abstract graph transformation. Electr. Notes Theor. Comput. Sci. 157(1), 39–59 (2006)
12. Rieger, S., Noll, T.: Abstracting complex data structures by hyperedge replacement. In: Ehrig, et al. (eds.) [4], pp. 69–83
13. Rozenberg, G. (ed.): Handbook of Graph Grammars and Computing by Graph Transformations. Foundations, vol. 1. World Scientific, Singapore (1997)
14. Sagiv, S., Reps, T.W., Wilhelm, R.: Parametric shape analysis via 3-valued logic. ACM Trans. Program. Lang. Syst. 24(3), 217–298 (2002)
15. Smelik, R., Rensink, A., Kastenberg, H.: Specification and construction of control flow semantics. In: Visual Languages and Human-Centric Computing (VL/HCC), Brighton, UK, pp. 65–72. IEEE Computer Society Press, Los Alamitos (September 2006)

Author Index